A New Guide for
Better Technical Presentations

A New Guide for
Better Technical Presentations

*Applying Proven Techniques
with Modern Tools*

Edited by

Robert M. Woelfle
E-Systems, Inc.
Greenville Division

**IEEE
PRESS**

A Selected Reprint Volume
IEEE Professional Communication Society, *Sponsor*

The Institute of Electrical and Electronics Engineers, Inc., New York

This book may be purchased at a discount from the publisher
when ordered in bulk quantities. For more information contact:

IEEE PRESS Marketing
Attn: Special Sales
PO Box 1331
445 Hoes Lane
Piscataway, NJ 08855-1331
Fax: (908) 981-8062

Printed in the United States of America

10   9   8   7   6   5   4   3   2   1

ISBN 0-87942-283-1

IEEE Order Number: PP0277-4

Library of Congress Cataloging-in-Publication Data

A New guide for better technical presentations : applying proven
   techniques with modern tools / edited by Robert M. Woelfle.
          p.   cm.
      ''A selected reprint volume, IEEE Professional Communication
Society, sponsor.''
     Includes bibliographical references and indexes.
     ISBN 0-87942-283-1
     1. Communication of technical information.   I. Woelfle, Robert M.
   T10.5.N49   1992
   601.4—dc20   92-13001

Contents

Preface

THIS PRACTICAL GUIDE is essential reading for all engineers, scientists, and managers who want to maximize the effectiveness of their technical presentations. It is useful for presenters ranging from beginners to experienced speakers, and is logically organized to meet the needs of a typical user. The content has been fully updated to address the opportunities offered by modern computer-based techniques and tools while retaining and re-emphasizing the importance of basic presentation requirements.

Advancing technology in the areas of computer graphics, video, telecommunication, and electronic publishing has drastically changed the way presentations are prepared and given. In addition, changing audience interests, technological backgrounds, and sociocultural composition require presenters to adopt a more global perspective. These issues are addressed in sections devoted to planning and preparation, delivery techniques, visual aids, computer graphics, video, and multimedia. The individual articles were selected to help the reader increase presentation effectiveness, reduce preparation time and cost, enhance author and sponsor image, and select the best available tools. A glossary of special presentation-related terms is also included.

Introduction

FEW ENGINEERS, scientists, managers, and other professionals confronted with the requirement to present a technical briefing, explain an engineering proposal, describe a scientific breakthrough, or outline a management plan can draw upon formal training for the techniques needed to communicate effectively with an audience. Some individuals can prepare highly polished *written* material but cannot cope with the requirements of the real-time personal communication imposed by technical presentations. Other individuals can easily and effectively communicate with one person or a few people on a face-to-face basis but tend to freeze when required to speak in front of a large audience. The purpose of this guide is to make available the collected advice of experts, containing "how-to" information for engineers, scientists, or managers who want to maximize the effectiveness of their technical presentations.

Since the original edition of this book was published in 1975, the tools and techniques used to prepare and execute technical presentations have changed dramatically. The general availability of desktop computers and graphics software tools significantly increases the capabilities of individual authors to control the production of their own presentations. In addition, the simplification and broader availability of video production equipment increase the feasibility of using its dynamic capabilities for even routine technical presentations. And sophisticated multimedia formats, often considered to be useful only for sales and marketing, are now practical and desirable for many technical presentation applications. While the basic considerations for presentations—for example, planning, audience consideration, content integrity, delivery techniques, and visual legibility—remain essential factors in achieving success, their execution has been simplified and enhanced by evolving technology.

Technical presentations are a vital part of almost every phase of the governmental, industrial, and academic worlds. Highly sophisticated presentations are used by both large and small companies to sell their products or services to potential customers. Similar presentations are used to brief stockholders and potential investors on a company's financial status and planning. Presentations of various types and sizes are used by government agencies, private industry, and the academic community to outline and explain proposals for initiating new projects and to report progress and achievements in those projects as they proceed and after they have been completed. In addition, presentations are used for routine applications such as

1. defining the scope and requirements of a new engineering or manufacturing project;
2. describing a new management policy or organization;
3. presenting sales, marketing, and technical plans and achievements;
4. reporting the results of a laboratory experiment, design project, or analytical study;
5. describing the plans for a new building or special facility;
6. summarizing a paper at a technical conference or symposium; or
7. supporting various types of training and educational activities.

To be effective, a technical presentation must be tailored to the requirements and conditions of a particular application. This requires consideration of several variables, such as the

1. size and nature of the audience,
2. location and characteristics of the facilities to be used,
3. types and quality of the visual aids that can be prepared and used,
4. amount of time and money available to execute the presentation,
5. types and quantity of information that must be collected,
6. availability of existing materials and resources, and
7. length of time available for the preparation.

With so many variables to be considered, careful planning is an essential ingredient for all effective presentations.

The most important consideration in planning a technical presentation is the audience. Since the whole purpose of a presentation is to communicate a particular message, the presenter should know as much as possible about the particular audience and then tailor the presentation accordingly. For example, the background and interests of the audience are important considerations to avoid either "talking down to" or "over the heads of" the audience. Attitude is also an important factor to consider because the approach for a "friendly" audience would be significantly different from the approach for the same subject presented to an "unfriendly" or "skeptical" audience. A presentation to a "friendly" audience could be more informal and would require less supporting detail than addressing the same subject for a "skeptical" audience, which would probably demand more background and justification of the points presented. The size of the audience also has a significant impact on planning a presentation. For example, a presentation designed for a small group in an informal environment might be ineffective for larger audiences. The visual aids for a small group can be produced by most available techniques but legibility becomes a limiting consideration for large groups. Other factors that should be considered in analyzing the audience for a specific presentation include the anticipated physical and psychological environments, the length of the presentation and its relationship to other presentations, and the time of day scheduled for the presentation. Careful analysis of these and other similar audience considerations will allow a presenter to maximize the effectiveness of the presentation for his or her specific audience.

The way in which visual aids will be used is also important in planning and executing a presentation. The processes and equipment used to prepare and utilize visual aids have evolved rapidly over the past few years. An individual can choose from a wide variety of media, ranging from simple hand-lettered charts to elaborate multicolor slides, dynamic videos, or various combinations in multimedia formats. The primary visual aid forms currently in use are

1. various types of manually manipulated boards (e.g., chalk, marker, poster);
2. static projection using opaque materials, slides, or viewgraphs;
3. dynamic projection using movie film, video tape, or computer animation;
4. three-dimensional displays using mockups, models, demonstrations, or simulations; and
5. fully integrated combinations of various visual aid forms in a multimedia presentation.

The current trend in the evolution of these visual aids is toward providing better quality products faster, easier, and at lower cost. High-quality presentations still require access to professional illustration and photography capabilities and equipment to assure logical layout, legibility, and aesthetic balance. Many routine presentations, however, can now be prepared by an author using only desktop computer equipment and software that guides the user in producing acceptable output. Even projection/display equipment has been simplified so that it is more portable and easier to operate.

The most misunderstood aspect of technical presentations is their cost. To minimize cost, internal presentations are often prepared with little planning and review, using ''do-it-yourself'' visual aids. On the surface, this may appear to be an economical approach, but some of the ''hidden'' costs are often ignored or overlooked because of expediency or lack of sensitivity. For example, the lack of adequate planning and review usually results in a confusing and ineffective presentation that does more harm than good because the audience gets the wrong message. Preparation of the associated visual aids can also waste the valuable time of skilled technical personnel because they are not familiar with the best and cheapest techniques, and the results are often illegible and reflect an unprofessional image. In addition, internal presentations often end up as part of formal proposals, published reports, or technical papers, requiring the material to be redone by professionals at added expense.

Producing a professional quality technical presentation does not necessarily require more time and money than the ''do-it-yourself'' approach. For example, good viewgraphs and slides can be produced faster and at less cost than hand-lettered flip charts because computer-based equipment provides a variety of efficient tools for the preparation of original art and simplifies the incorporation of any required changes. In addition, the associated artwork can be stored and subsequently retrieved and used for a variety of other applications with minimal adaptation. Acquiring the capability to prepare visual aids, however, does require some investment for the processing equipment and/or the procurement of outside services, and such costs must be considered in choosing the best approach for a particular application. Each presentation requirement, therefore, must be assessed in light of the immediate and anticipated future applications, the nature of the information to be presented, and the available resources to select the most cost-effective approach.

The articles included in this guide are arranged in seven parts that address the

1. importance and fundamental applications of presentations;
2. planning and preparation required to develop an effective presentation;
3. visual aids used to support and enhance the effectiveness of presentations;
4. delivery techniques used to execute presentations;
5. computer graphics techniques, tools, and applications available to produce, and sometimes deliver, presentations;
6. video techniques, tools, and applications available to produce dynamic presentations; and
7. multimedia integration of various combinations of visual aids for special presentations.

The first four parts generally reflect the thought process an author would use in addressing a presentation requirement, while the last three parts describe special tools and techniques that can be used to enhance a presentation.

The individual articles were selected because they treat their respective subjects in a tutorial manner and describe techniques that are of practical value in solving the problems normally encountered in planning, preparing, and delivering a presentation. Some articles were retained from the original book published in 1975 and may reflect a dated writing style. Their content, however, is timeless in nature and merits repetition. In addition, articles treating the same or related subjects may appear to offer overlapping, inconsistent, or contradictory points of view but they should really be considered as alternatives when faced with varying requirements. Some articles are conceptual in nature to provide background and encourage creativity while others provide practical ''how-to'' instructions. Many of the articles also include checklists for selecting the best approach and equipment for specific applications. This guide, therefore, is an important reference tool for any engineer, scientist, or manager who values effective communications and is interested in his or her own professional advancement and growth.

Part 1
Importance and Fundamentals

PRESENTATIONS are the backbone of interdisciplinary communications among technical professionals. They can be an effective tool for selling an idea or product, outlining a plan or procedure, or explaining a problem and the recommended solution. To take full advantage of this tool, however, an engineer must be convinced of the importance of presentations, understand the human relationships involved, and appreciate the basic procedures that must be used. This part addresses the importance of presentations from several perspectives and outlines some of the fundamentals for preparing them.

The importance of presentations is outlined in the first chapter of Ed Hodnett's book, *Effective Presentations*. He indicates that effective presentations are the mark of a real professional and represent an avenue for advancement and professional growth. He also defines the various roles a person may be called upon to play in the development and execution of presentations. Although his book was published in 1967, the thoughts Hodnett expresses have not diminished with time and apply equally well to the collection of articles presented in this book.

Raymond Floyd's article, "Presentation fundamentals," provides a brief description of the key points that an engineer or scientist must consider when preparing for a technical presentation. Planning, organization, structure, and methodology are explored in a manner that allows an author to review his or her material and style to improve areas that may be weak. The information in this paper could also be used as a quick refresher to review the requirements of an effective presentation prior to starting the planning and preparation.

Planning, preparing, and executing an effective presentation involves consideration of many separate interrelated elements, such as purpose, audience, facilities, and available delivery time. Calvin Gould's article, "Anatomy of a presentation," identifies the individual elements that must be considered and shows their interrelationships in two flow charts. One chart lists the specific factors that should be considered and identifies related alternatives. The other chart analyzes the scope of each factor. He also identifies some of the alternatives available to presenters and suggests ways to evaluate them.

Joseph Cillo contends that a presentation is effective when the audience remembers what was said in the proper context and with the desired emphases. In his paper, "Some fundamentals for presenters," he indicates that an author needs to answer three fundamental questions in planning a paper: What is the objective? the audience? the environment? He also defines eight universal presentation "laws" that must be addressed and offers some suggestions for complying with them.

For the engineer or scientist in a hurry to acquire presentation skills, Ethel Curtis's "Three basic recipes for a speech" offers a brief but tested route for planning and executing technical presentations. It identifies the role of a presentation in a technical context, provides planning suggestions and three sets of organization guidelines, and compares the needs and rewards of various types of presentations. It also includes tips for rehearsing, use of humor, audience rapport, and delivery techniques.

In his article, "Making presentations that command attention," Ralph Kliem stresses that the key to giving effective technical presentations is developing the self-confidence that comes with being fully prepared. He indicates that the first step in any presentation is defining one's objectives, including the key points and desired results. He emphasizes the importance of analyzing the audience to determine their backgrounds, interests, and goals in attending the presentation. In addition, he presents suggestions for achieving maximum impact and controlling the results. He also includes a checklist to determine if you're really ready.

The Importance of Presentations

The word "presentation" is an example of how the English language flexes to the shape of new meanings. None of the dictionary definitions quite covers the specialized sense in which the word has come to be widely used in business, industry, government, and other professions.

A Presentation Is a Specialized Form of Communication

A presentation is first of all a talk that contains expository material, is often persuasive, and frequently involves graphics or other audio-visual aids. The use of audio-visual aids is usually the most distinguishing characteristic of a presentation, but the term is commonly used when no aids are present.

A presentation has the serious purpose of acquainting an audience with information about a specific matter, often with the intent of aiding or influencing decision-making. A talk to a service club about why school millage should be increased is a presentation; one about the outlook for a football team is not. A clergyman submitting his plans for a new Sunday school is making a presentation; delivering a sermon, he is not. The term is loosely used for many talks not considered to be presentations in this book—sermons, for instance. The distinctions are unimportant. The point is that the use of the term presentation for a specialized form of discourse has general sanction.

It is also common to hear the term presentation applied to proposals where sheets of paper are all that is tangible—for instance, when written matter supports a face-to-face request to a foundation for a grant of aid, a bid to a board of education to handle the cafeteria service, or a talk to a technical society about a new molding compound. By extension, such a written statement standing alone is commonly called a presentation.

Normally, then, the two distinguishing features of a presentation are the presence of graphics, written material, or other audio-visual aids and the pervasive effort to secure a predetermined, practical response from the audience.

You are free to dismiss the word as jargon, but salesmen, consultants, advertising men, engineers, architects, government officials, teachers, lawyers, doctors, clergymen, and other people in all walks of life make millions of presentations every day. To them the term is meaningful in designating the specialized nature of this kind of communication.

Reprinted from *Effective Presentations*, E. Hodnett, Ed., ch. 1, 1967, pp. 1–10. Copyright © 1967 by Parker Publishing Company, Inc., West Nyack, N.Y.

Good Presentations Are the Mark of a True Professional

As your professional career progresses, the more presentations (both oral and written) you will give, and the more the presentations of the people who work for you will influence your success.

You give oral presentations inside your organization, such as those of a market research director to his top management, a sales manager to his salesmen, or a surgeon to his internes. Much of your success professionally depends on the effectiveness of the presentations you make at group problem-solving sessions within your organization.

Outside oral presentations are given at professional society meetings, at customers' plants, and before civic bodies.

You give formal presentations in conference rooms, classrooms, and auditoriums, and you give informal ones over the telephone, in colleagues' offices, and across luncheon tables.

You give written presentations, as already noted, on the many occasions when written statements accompany oral presentations, as is habitual at professional society meetings, and by extension in all those memoranda, reports, letters, papers, and articles that meet the conditions of our definition of a presentation.

Your professional activities put a premium on your oral and written presentations. This is one of the bench marks of all true professions. If you are an accountant, an advertising supervisor, a physicist, an engineer, or a member of any other business or industrial group, the obligation to communicate in this special way is one chief characteristic that you have in common with members of the legal, medical, teaching, or other professions. In a true profession, as distinct from a trade, what you know and what you can do have to be implemented by your ability to communicate by means of effective oral and written presentations.

A professional, incidentally, is a person who knows exactly what he is doing and does it better than the journeyman performer. When someone says, "Gordie Howe is a real pro!" he does not mean that the great Red Wing plays hockey for money. True, the fellow who plays for pay or works at a craft for pay is more often a pro in this complimentary sense, than the fellow who does so for fun. The person who paints portraits for a living knows things about preparing the surface of his canvas, placing his subject, mixing his colors, choosing his brushes, and a dozen other crucial matters that the talented amateur rarely bothers to learn.

So with the person who makes professional presentations, though he may never be paid directly for them. He has studied the technique of giving presentations so assiduously that he knows *in advance* all the steps he has to take to ensure excellence. Even while he is in the middle of a presentation, he is aware of how well he is doing and what he needs to do to improve the response.

Successful Presentations Are an Avenue of Advancement

What we are calling a presentation often occurs at times of crucial importance to an enterprise. Therefore, if you make one excellent presentation—just one—it may lead to a significant departmental decision, increase your company's business, help you win a promotion, or elevate you to professional or public leadership. No one has ever been seriously handicapped professionally by lack of skill in after-dinner speaking or essay writing, but many careers have been compromised by poor presentations and many causes lost. The exceptional importance invested in presentations is the justification of this book.

Skill in presentations is just as practical a part of your job and your civic activities as any other. What you know and what you can do are not enough. A corporation president who goes before his board of directors to ask for authorization to build a $10,000,-000 plant cannot get by on his ability to run plants or sell goods. The $10,000,000 decision, with all of the consequences whichever way it goes, rests to a critical degree on how shrewdly he plans his presentation and how well he gives it.

The brilliant scientist, the winning attorney, the influential scholar, the popular public official are almost invariably masters of the presentation. Their success should prove how vital it is to your career that your presentations—in the broad sense of this book—be a totality of planning, practice, and habit.

A certain share of the presentations you give will be competitive. In most instances the competition is covert—you are simply on the program with several others. Yet as speaker succeeds speaker, the audience inevitably compares your presentation with the others. If you bumble, drone, or talk too long, you get low marks. If you are well organized, animated, and economical of time, you get good marks. Since the marks at most meetings run from B to D (with most around C), you have a splendid chance to cover yourself with glory by earning an A.

The impression you make on an audience is to a considerable extent passed on to the organization you represent. If you and your colleagues take the trouble to make your presentations all that they should be, after a while the word will get round: You can always count on those Jones & Smith fellows to give a first-rate presentation; they must be a grade-A firm.

Here is the case history of a presentation essentially as it happened. Petrox, a large petrochemical corporation, is reviewing its business at a luncheon it is giving at its New York headquarters for fifty members of the National Society of Security Analysts. The treasurer of Petrox asks John Ashwood to do a special presentation about the new biochemical division that Ashwood heads. He will have precisely 30 minutes, including a question period.

Ashwood calls in his controller, director of planning, and his director of public relations to help him plan his presentation. The audience is self-defining. These are the men and women who advise banks, mutual funds, insurance companies, and other large investors about the financial health and outlook of companies in which they have invested or may consider investing. In order to arrive at their judgments, analysts study individual company annual reports and other financial records, the economic environment of each company, and its competition. They are especially interested in the way a company is run and its potential for growth.

The situation analysis by Ashwood and his associates is a sobering one. The judgments made by these fifty security analysts can have a significant effect on how Petrox fares on the stock market and whether or not large investors hold, buy, or sell Petrox stock, and at what rate of interest Petrox can borrow millions of dollars for new enterprises. These actions affect the profitability of the company and the welfare of all employees. This background influences the planning of the presentation by Ashwood and his associates.

A secondary element in the situation is a personal one. The success of Ashwood's presentation will be measured not only by the reactions of the security analysts, but also by the treasurer's opinion of how well it enhances the standing of Petrox. Also at the meeting there will be other influential officers of Petrox and members of its board of directors. Ashwood's future is in the hands of these men.

Out of these considerations comes the chief message—the biochemical division of Petrox has had an excellent record of growth and profitability and has even brighter prospects for the future.

Ashwood spends hours writing notes, talking with associates, having slides made up, collecting sample products, and soliloquizing. Alas, the first rehearsal runs an hour and ten minutes. Ashwood burns up time on organization charts, slides of division plants, and, perhaps because of his training as a biochemist, explanations of the structure and characteristics of products.

But Ashwood is an honest, dogged worker. He has plenty of lead time, and he makes note of the suggestions from his staff. He holds three rehearsals with his staff, and he rehearses dozens of times by himself. When he gets up after the luncheon in New York, he is master of his material, and he knows it. He knows that the main elements are now rightly proportioned and exactly what emphasis to give the details.

He disposes of the basic facts about the division in three minutes with the aid of one slide. He emphasizes the high proportion of research and development. He discusses the five product lines by merely holding up a sample of each. Pie-chart slides indicate the large dollar share of the pharmaceutical market held by the biochemical division of Petrox. He reveals plans for entering new markets with new products. Then he gives the final ten minutes to an analysis of sales, reinvestment, and earnings with a projection of doubled growth within five years.

After the guests are gone, the treasurer says, "You did a great job, John." And he almost smiles. For weeks, stories come out in financial papers and business magazines, complimentary references to Petrox pop up, and appreciative letters from the analysts flow back.

Every Executive Is a Director of Presentations

The *you* in these pages is a person who makes presentations. In the broad sense of the word, you are an executive. But as your professional career advances, you will be increasingly responsible for presentations that others make. The degree of your involvement will vary. You may merely decide that a presentation should be made, or you may plan its general attack. You may work out all the details, or you may check the detailed plans of others. You may spend hours coaching someone who is giving a presentation, or you may be a final critic.

However it may be in a specific instance, you will fill all of these roles many times during your professional career. Even the youngest executive finds, with dismay, that his professional success is to a disconcerting degree in the hands of those persons who make presentations under his direction. Beyond your own self interest, you have a responsibility to help all professional persons working for you to become expert in making presentations; to do that, you must first master the subject yourself.

The title "director of presentations" does not exist, but every large organization should have someone with this responsibility. Few do. When the responsibility is recognized at all, it is usually spread among several persons, none of whom has the breadth of experience necessary to direct all phases of presentation-making. An organization considers itself progressive if it has an audio-visual aids department, but such aids are only one relatively mechanical aspect of the job of our hypothetical director of presentations.

In fact, every executive is a director of presentations. Sooner or later you will be one. First, you should have a broad general education, with more than the ordinary understanding of the processes of reasoning.

Second, you should be thoroughly acquainted with the goals, procedures, and policies of your organization. You should always be able to follow the reasoning in the light of the interests of the organization. You do not have to be familiar with all of the material covered in every presentation. However, it certainly helps if you are directing a financial talk to know what key terms such as standard costs, variables, and gross margins mean.

Third, you should be skilled in the organization and expression of thought.

Fourth, you should be knowledgeable in speech techniques.

Fifth, you should have a working acquaintance with graphics, audio systems, projection, and other presentation aids.

A person who acts in the capacity of a director of presentations is unlikely to be strong in all five of the areas noted. If, however, he starts with a broad general education and a conscious awareness of logical methods of reasoning, he can become adept in all the other areas. This process is not reversible.

To be an effective director of presentations, you should enjoy the confidence of the top management of your organization, or at least of that sector in which you function. A person guiding the presentations of a sales force or of a government agency cannot make critical judgments about the material being presented by a member of the organization, unless he can place it within the context of executive thinking.

You cannot direct presentations at the level of technique alone. Not only must you be able to analyze the strengths and weaknesses of a finished presentation; you must, if need be, plan the strategy, sketch the pattern of thought, develop the supporting evidence, and decide what aids should be used. You do not customarily do all this, but you should be able to do it. And you should have the tact and the acceptance from your colleagues to have your suggestions prevail most of the time.

Making a presentation is often a serious occasion—sometimes it is a traumatic one. As a director of presentations, you have to act somewhat in the role of a doctor. For your patient's sake, you must insist on early confrontation with the job in hand, revisions of first plans, rehearsals, changes, and more changes. Inexperienced persons, predictably, react negatively to this pressure. They think such perfectionism absurd; they may even grow hostile, thinking they are being pushed around.

If you are a good director, you will usually learn of such negative reactions when someone comes to thank you for help in what turned out to be a successful presentation. Happily, persons experienced in making presentations—the pros—are perfectionists themselves and welcome critical help without having their vanity hurt. They know that an excellent final presentation comes from gradual improvement of specific parts.

Skilful directors of presentations are as rare as Kirtland's warbler. But they will become more common, because they are badly needed. Meanwhile, as you take a hand in guiding the presentations of your associates, you have an obligation to become more professional in your guidance. You would not take on the job of directing an amateur production of "Gammer Gurton's Needle" without some knowledge of the theater and the business of play directing. You should take much more pains to be prepared to direct a presentation on which a great deal depends, than to direct a play on which nothing much depends.

In directing presentations, you should never forget that your main goal is to make each of the persons you work with self-sufficient. You are a teacher. Every preparatory session is a tutorial. Every presentation is a stepping-stone to better presentations. You are successful when your charges can do without you, or, at least, can do without you if they have to. You fulfill your role best when the persons you have helped become competent to direct others.

Mastery of Presentations Is Part of Your Professional Growth

This book of professional techniques deals with specifics, with what are the least difficult parts of making presentations. In the broader sense your effectiveness depends upon what kind of person you are, how much you know about your subject, and how well educated you are in general.

No matter how well delivered, the presentation of a consultant's recommendation based on unsound evidence or mistaken cause and effect, poor analogy, or any other logical fallacy is unsatisfactory. A technical report that has a serious mathematical error in it is unsatisfactory. A letter to a correspondent that overlooks one of his key questions is unsatisfactory. An otherwise good oral presentation marred by slovenly speech or appearance is unsatisfactory.

Effective Presentations is meant to be a practical guide to specific matters that habitually concern people in industry, business, public affairs, science, and the other professions. The suggestions are precise because the successful execution of any action, from a forward pass in football to sending a spacecraft to photograph a planet, depends on the exact management of many details.

Nothing suggested in this book is meant to be absolute. Some latitude is assumed—is indeed inherent in the philosophy of communication as adjustment to specific circumstances. Nevertheless, these procedures and techniques have been tested on the firing line of actual practice and following them should save time and improve professional performance. In the recommended readings and in your public library you will find more detailed statements about many of the matters discussed here.

If you have a special individual difficulty, do not coddle it. Do something about it. Study regularly by yourself. Take a night course. Join a community group such as the Toastmasters. Above all, ask for criticism and use it. The essence of professional growth is to take positive steps to develop yourself in all respects. You have ample opportunities to improve your presentations; make the most of them.

Your own personal growth is involved. You want to meet the

requirements of your job, and you want to meet the highest standards of your profession. But beyond that you want to be as effective a person as you believe you can be. Your own standards for yourself, the highest standards of your profession, your responsibility as an educated citizen, and your own self-directed efforts to keep growing professionally make it imperative for you to give serious attention to both oral and written presentations at all times.

Any well-run organization can set up a systematic procedure for handling presentations. First, all executives should be made aware of the importance of presentations and of their responsibility for all of those made by persons under their direction. All persons who normally make presentations should be given instruction in the various matters covered in this book. Competent help in planning presentations should be available. So should an adequate audio-visual aids department. No one should ever be allowed to make an official presentation without rehearsal and clearance. But in the end, the responsibility lies with you, the individual, to realize how important your presentations are to your organization and to your career and to lift them to the professional level.

Presentation Fundamentals

RAYMOND E. FLOYD, MEMBER, IEEE

Abstract—This paper provides a brief description of the key points an engineer or scientist must consider when preparing for a technical presentation. Planning, organization, construction, and presentation methodology are all explored in a manner to allow a speaker to review his or her presentation material and style and to improve any areas which appear to be weak.

ALMOST everyone can carry on a conversation, technical discussion, or other such activity in a one-on-one situation. If the audience is expanded to three to five people, but kept at an informal level, again almost everyone can continue to participate effectively. However, let's change that conversation or discussion to a technical presentation to be given for an audience of ten, 20, 50, or more. Suddenly the number of people willing, and able, to stand up and give that presentation has been significantly reduced. Is it because of a lack of expertise in the particular subject? Is it a conflict of job schedules and pressures? No! The simple answer is—fear of public speaking! Many studies have been made and each identifies public speaking as the number one fear today.

Perhaps the best approach to any type of public speaking is for the prospective speaker to remember the ease of conversation. Speaking to more than a few people can, and should, be oriented toward a conversational approach. The speaker can convey the feeling that each member of the audience is being informed on the subject matter. The key here is that the material and the speaker's style must match. Word usage, vocal control and variety, gestures, and visual aids must all blend into a harmonious approach to be effective.

Let's examine some of these areas and discuss some fundamental approaches to aid the new (or experienced) speaker in preparation and presentation.

WORDS

Regardless of the nature of the speaker's presentation, whether it is after-dinner entertainment or purely technical at a conference, word selection should be appropriate to the audience, the material, and the presenter. Frequently, speakers use unfamiliar or "big" words because they seem to lend stature to the presentation. In a short time, however, the audience detects the hesitation and stumbling over these words and at that point the speaker's credibility as "the expert" begins to fade. This does not mean that if an unusual

or long word is proper in the context of the presentation, the speaker should search for some simple one; use the most appropriate word for the situation. The speaker's word selection must be aimed at the average intellect of the audience and should take into consideration such other factors as time of day, number of preceding presentations, visual aids used, the room, and any other aspect that can adversely affect the comfort of the audience.

VOCAL FAULTS

A problem that typically arises is one the majority of speakers are not aware of. It centers on distracting vocal habits the speaker has developed without being aware of injecting them into the presentation. What are these vocal *faux pas*?

Think for a few minutes and perhaps they will come to mind. How many times have you listened to people punctuate their presentations with "uh," "OK," "you know," "right," or some such word or phrase? If you were to ask the speaker, or many of the audience, whether such utterances were used, the majority would say "no." They have learned to tune out such words (and perhaps even more). While it is difficult to prevent some slips like this, every speaker must make a conscious effort to minimize such utterances. It becomes fascinating after a while to see just how many *you know*'s a speaker will use, and, as a result, the audience's ability to concentrate on the information content of the material is lost.

One approach to overcoming such a problem is to have someone listen specifically for any such mannerism during practice sessions prior to the presentation. The speaker may be dismayed at the number of times this vocal trap is sprung, primarily during pauses between specific points or sentences. But he should not panic or give up. Simple concentration and keeping one's mouth closed during those in-between pauses will prevent such utterances. This takes an effort at first but becomes natural before long, and such vocal additives will disappear.

VOICE

Besides these vocal errors, a speaker must be concerned with voice control and variation. How many times have you listened to a speaker drone on and on in a monotone, slowly putting you to sleep? The human voice is a flexible instrument, one to be used with care and deference by a public speaker. The voice can provide an effective means of emphasis. By changing pitch, intensity, and pace, one can "show" points of emphasis to the audience. The length of a

Manuscript received Sep. 13, 1979; revised Nov. 29, 1979.
The author is an Advisory Engineer in the General Systems Div., IBM Corp., P.O. Box 1328, Boca Raton, FL 33432, (305) 994-4669.

Reprinted from *IEEE Trans. Professional Commun.*, vol. PC-23, no. 1, pp. 40–42, March 1980.

pause can be an effective indication of the importance of the topic just covered, or an alerting to the information that is to follow.

The use of pace and pause is difficult to master. The speaker must remember that the human mind can follow a presentation at a faster rate than the spoken word. As a result, if his pace is too slow or even too uniform, the listener's attention will drift. The speaker should vary the pace of the presentation, rushing through those parts that are easy to follow, slowing where needed to make a point or greater understanding is required. Pauses should range from one-half second to a maximum of three seconds, depending on the emphasis to be placed on the material being presented. Pauses are effective but as with any tool, the effectiveness is lost or dulled by continual use. Just as with any physical gesture, frequent long pauses will give the presentation a disjointed sound and will cause the audience to lose interest.

GESTURES

Gestures can also be effective tools in a presentation. There is controversy about the use of gestures in technical presentations although few people argue against the appropriateness of gestures in an entertaining speech.

The best advice is, if the gesture fits the presentation, then by all means use it! The key words here are *if it fits*. The gestures used must make sense to the audience and must intuitively fit the word picture the speaker is developing. As such, a gesture reinforces the point being made. An inappropriate gesture will have the opposite effect and generally will result in the loss of attention by some of the audience as they spend time trying to figure out what the speaker meant.

What kinds of gestures are we talking about? All kinds: a simple shrug of a shoulder, a shake of the head, a full sweep of an arm to encompass a broad category or group, the use of the fingers to enumerate some ideas or points made, or the slap of an open palm on the lectern to make a point of emphasis. (This latter technique should be used with caution, especially if the microphone is attached to the lectern.)

There are a few gestures that a speaker should avoid, gestures in the same category as those vocal *faux pas* mentioned earlier. In the same fashion, these gestures are unconscious actions, the speaker not being aware they are being used. For

Ann Jones

POINTS ONE
TWO AND THREE

Ann Jones

men, the most common one is to place a hand in the pants pocket and play with change or keys. In a large audience this may escape notice but in a smaller room it can be most annoying. The easiest way to prevent this type of annoyance is to empty one's pockets of any such temptation prior to speaking. Women have an equivalent problem if they have jewelry that jangles or flashes as they move about.

What other types of errors are there? A common mistake made by speakers is to rearrange objects on the lectern. Chalk, pencils, water glass, pointer, etc. are all there to be neatly aligned or placed in some other desired formation. Don't do it! The audience can sense the speaker's preoccupation and decide that he either doesn't care about them or isn't sure of his subject matter. In either case, he loses effectiveness.

Another common habit that speakers develop is playing with the pointer. It can be very distracting to the audience if the speaker twirls the pointer between points of emphasis or swishes it nervously. If the speaker is going to use a pointer, he must use it properly. Other habits that the speaker should be aware of include earlobe pulling, nose scratching, and eyeglass pushing, to name a few. All of these are unconscious signs of nervousness. The speaker should have someone look for these irritating mannerisms during practice sessions.

VISUAL AIDS

Many studies show that we remember things best when we receive multiple sensory inputs. In the case of a technical presentation, when we listen to the material *and* have visual aids to provide a second sensory input, we remember more, longer. With this in mind, what guides should be considered in preparing visual aids?

The first rule for visual aids is that the aid must be pertinent. If it fails to provide a reinforcement of the speaker's message to the audience, then it will detract from the effectiveness of the presentation. A second common problem with visual aids, especially overhead projector transparencies and slides, is that the speaker makes them too busy. He attempts to put the entire paper on one slide or foil—perhaps six graphs that are superimposed, or bar charts with margin notes, or any such cluttering effect. The key point for the speaker to remember when preparing visual aids is that they are meant to assist both the audience and the speaker. The visual aids are not meant to be a replacement for the speaker, nor a reiteration of his words.

The speaker should inspect the projector, screens, room layouts, sound system, etc. prior to the presentation. Such problems as obstructions, spare bulbs, focus, and screen placement should all be resolved. If he is going to handle the foils or slide projector controls, their operation must be clearly understood. If someone else is going to handle changes of the visual aids, the speaker must arrange a signal for such changes. This may not seem important in a small room where the speaker can simply say "Next slide please," but it is a problem in an auditorium where the projection booth is some distance away and sound-isolated.

Foils and slides are not the limits in visual aids, they are the more commonly used. The next most common visual aid is the flip chart. The same rules apply for the flip chart as for foils or slides. Keep them pertinent, clear, and simple. The shortcoming of the flip chart is that it rapidly loses its effectiveness as the size of the room or audience grows.

One key benefit of flip charts is that lightly penciled notes can be placed on them to key the speaker. Note *lightly penciled*—the notes should not be seen by the audience. In addition to flip charts, there are *real* visual aids. If the subject is about metal fatigue, for example, then some samples can serve as visual aids. If the speaker is discussing botany, then a variety of plants, flowers, etc. pertinent to the presentation would be appropriate. The keys to the successful use of visual aids are pertinence, clarity, and simplicity. The speaker who keeps these in mind will find that visual aids complement the presentation.

ORGANIZATION

Thus far, many words have been used to describe some aspects and cautions in the presentation. What about the structure of the presentation itself? The structure is quite simple, consisting of an opening, a body, and a close. The problem most speakers have is that they don't know what or how much goes into each part. Let's take a brief look at each section and review some guidelines for them.

Before the speaker can develop a structure for the talk, some idea of the time alloted on the program is necessary. In general, he should allow 10 to 15 percent of the time each for the opening and the closing parts of the talk. Thus, for a ten-minute talk, one to two minutes is appropriate for the opening remarks and the same for the close. The speaker then has between six and eight minutes to devote to the body of the presentation.

Opening

In the opening what does the speaker say? Quite simply, the opening must provide a "bridge" for the audience. That bridge can be background information, historical support, light comments, etc. The express purpose is to gain the audience's attention and to prepare them for the material that follows. If the speaker fails to do this, a large portion of the talk may be misunderstood as the audience ponders and wonders where the speaker is going.

Close

In the concluding remarks the speaker should provide a summary of the points he wants the audience to take away with them. He should not have to restate all the reasons why these points are important; that should have been provided in the body of the talk. The conclusion should seem to logically follow the body of the presentation and should finish at a recognizable point. It can be very disconcerting to the master of ceremonies and the audience for the speaker to finish and have no one aware, or sure, that he has. It should not be necessary for a speaker to begin concluding remarks with "In conclusion. . . ." The speaker should have built a bridge from the body of the talk to the concluding remarks, once again meant to prepare the audience.

Body

In the 70 to 80 percent of the time that remains, the speaker should use another "rule of thumb": A topic or statement cannot be adequately developed in less than one to two minutes. Allowing three minutes for the combination of opening and close, the speaker will find time to develop only three to six ideas in a ten-minute talk.

This limit trips most speakers and causes them to run overtime in their presentations. They simply try to cram too much into the time available to them. Don't do it! A program has been established and everyone has an allotted portion of that schedule. One speaker should not infringe on someone else's time. If another minute is needed, then that minute must be pared from the opening and closing remarks, chopped from the other main topics, or major surgery is necessary. It is the responsibility of the speaker to fit his presentation into the allotted time.

I mentioned that in the opening and in the body of the talk, the speaker should build bridges to the adjacent sections. The more complex the material, the more sound that bridge must be. Another word of caution, however. Once the bridge is in place, the speaker must not take a 90-degree turn in subject matter. If the speaker suddenly introduces new material the audience was not prepared for, he will lose the audience's attention for some time while they try to piece things together.

There is one final hint to aid in the quest for good reviews of all presentations: *practice, practice, practice!* The greater the number of times the speaker can give the pitch to a group who can provide a critical review, the greater the probability of success. During these practice reviews, the speaker must make the audience understand what he is looking for and that he is willing to accept criticism. If, for some reason, he does not have a group to review the presentation, there are alternatives. Through the use of a tape recorder the speaker can listen to the presentation, while a session before a mirror can provide feedback on the visual effect achieved. It is through these sessions that the speaker can perfect the presentation and be assured that the message received is the one meant to be given.

Anatomy of a Presentation

CALVIN R. GOULD

Abstract—A speaker may formally address an audience in the form of a speech, a presentation, or a lecture. The distinction between these forms is explained, and suitable applications for each are discussed. A great many of the requirements for effective communication by means of any of the three are common. A comprehensive breakdown of the factors involved and the choices available to the speaker is presented in diagrammatic form, supported by descriptive text.

INTRODUCTION

A DISTINCTION should be made between a speech and a presentation. For purposes of this paper an oral presentation is identified as information presented by one or more individuals requiring a combination of speeches and various aids such as movies, slides, and/or other support. To put it simply, a speech becomes a presentation if just one visual aid is required. While it may be more instinctive to identify a presentation with a seminar or a major proposal activity, the fact is, most meetings, large and small, are presentations needing a similar amount of attention for communications effectiveness. The distinction is important to make because a presentation requires so much more support than a speech. Perhaps a major reason for some poor presentations is that the presenter thought of his presentation as a speech and he planned it independent of any outside support; then when he realized he needed help it was too late for that help to be effective.

Another important reason for the distinction is the trauma caused by the term "speech." Many good presentations have been made by nontrained speakers. The value of a presentation is measured by its content more than by its showmanship, and its content is related to the knowledge and experience of the presenter on the subject he is presenting. If he looks upon his task as a conversation between him and his associates on the subject of his specialty, he will have a higher confidence level than if he thought of it as a speech before a large and strange audience.

ONE-TIME-ONLY STIGMA

Most communications problems occur in the low-budget one-time-only presentation. Because the presentation would normally be given one time, little professional attention is given to this critical area of information dissemination. Industry allocates high budgets in the promotion of its product to mass media for advertising and sales meetings. Also, it is not unusual for major corporations to spend $50 000 on their annual stockholders meeting. As a result, major commercial organizations do exist with professionals in the arts of advertising and theatrical productions to service industry in this specialized presentation field. Their highly paid staffs require large fees, making such services impractical for that ordinary but more frequent one-time presentation.

It is because of this that most presenters are left to their own devices. Still, the need for effective communications is equally as great, regardless of the size of the audience.

THE PRESENTATION

Good presentations are possible if individuals will make a sincere effort to communicate with their audiences. I believe people have a genuine desire to do their best and are constantly seeking ways to improve oral communications. Part of the solution is knowing what to do and what to look for.

The charts entitled "Anatomy of a Presentation" (Fig. 1) and "General Analysis of a Presentation" (Fig. 2) are meant to be complimentary road maps for any presentation. They were developed as an attempt to identify the ten primary considerations and to carry each point to its conclusion. Fig. 1 can be used as a checklist by the presenter to consider all aspects as they relate to his specific presentation.

As a general practice, about half of these points are overlooked until just before, during, or after the presentation, but too late to avoid the goofs which may be remembered long after the presentation has been forgotten.

An assumption is made with this chart that the objective of the presentation, its content and organization are functions the presenter is capable of developing. The ten primary points are usually given some attention, but most presenters would not have the experience to consider them in depth.

A further definition of each of the subjects on Fig. 1 will make it more useful.

Purpose

These categories cover most presentations peculiar to government–industry communications, but in a broad sense they can cover any presentation.

Briefings: These are mostly intracompany types of presentations, conducted as staff meetings, project reviews,

Manuscript received December 3, 1969.
The author is with the Martin Marietta Corporation, Orlando, Fla.

Reprinted from *IEEE Trans. Eng. Writing Speech,* vol. EWS-13, pp. 17–24, May 1970.

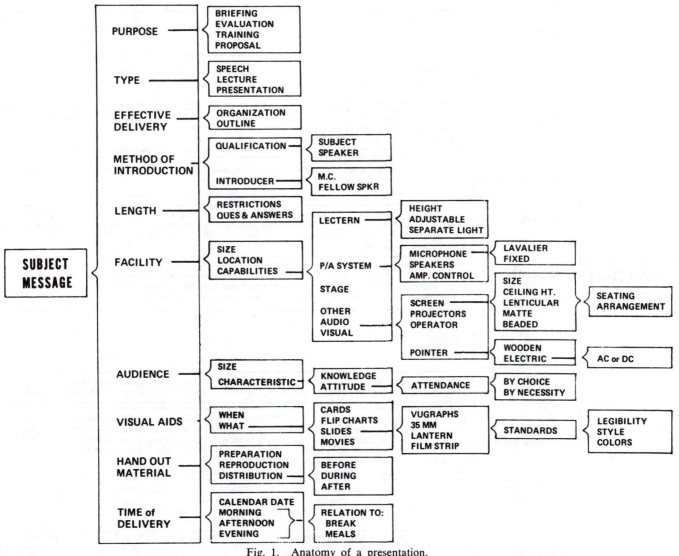

Fig. 1.   Anatomy of a presentation.

and board meetings to update subordinates or superiors on business activity, policy changes, etc. The atmosphere is informal, and brevity is essential. They usually need strong support by low-cost visual aids and projection hardware.

Evaluations: These meetings are held to appraise new concepts, develop comparative studies, or rehearse major presentations. They usually need significant support in visual aids, projection hardware, and handout material.

Training: These meetings include educational activities such as manufacturing training, company-peculiar activities, orientation, and outside technical meetings. In a strictly training environment, more complicated visual aids may be used in conjunction with handouts. For meetings outside the company, there usually is strong emphasis on company facilities.

Proposals: These appeals for customer and management support for new business involve a sales presentation with creative approaches by an individual or a team. They are characterized by directness and brevity but require significant audiovisual support.

Type

Establishing one of these three types directs the planning of other support.

Speeches: These usually are limited to 20 minutes and seldom use visual aids. The purpose often is entertainment, expression of opinion, or an attempt to sway the opinions of others. Attention should be paid to personal appeal, voice modulation, and succinctness. They may require a lectern and a public address system.

Lectures: These usually are concerned with education and training. The audience must be able to respond to the discipline established by the speaker, such as note-taking and reference to handouts issued before or during the lecture. More complex visual aids are called for. One must be careful not to confuse the lecture technique with that of the speech or presentation. The members of the audience must be prepared for this participation, or they will lose interest and everyone's time will be wasted.

Presentations: These are speeches supported by visual aids, which should be uncomplicated and useful to the audi-

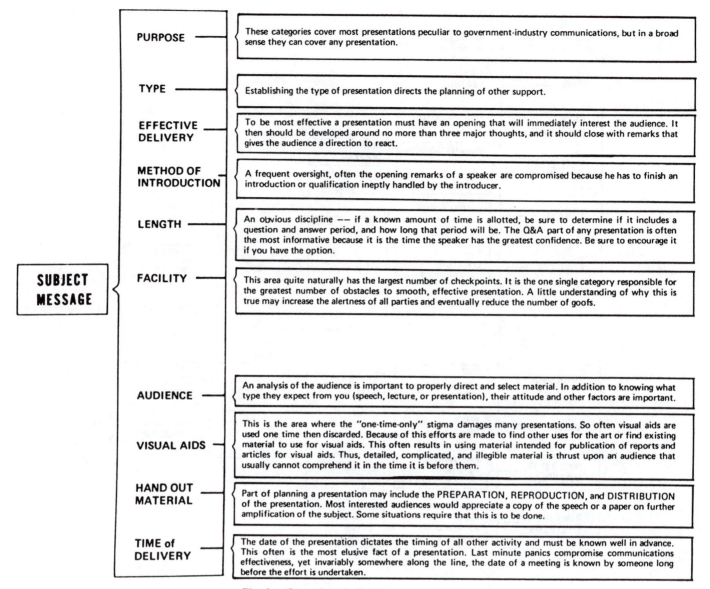

PURPOSE	These categories cover most presentations peculiar to government-industry communications, but in a broad sense they can cover any presentation.
TYPE	Establishing the type of presentation directs the planning of other support.
EFFECTIVE DELIVERY	To be most effective a presentation must have an opening that will immediately interest the audience. It then should be developed around no more than three major thoughts, and it should close with remarks that gives the audience a direction to react.
METHOD OF INTRODUCTION	A frequent oversight, often the opening remarks of a speaker are compromised because he has to finish an introduction or qualification ineptly handled by the introducer.
LENGTH	An obvious discipline —— if a known amount of time is allotted, be sure to determine if it includes a question and answer period, and how long that period will be. The Q&A part of any presentation is often the most informative because it is the time the speaker has the greatest confidence. Be sure to encourage it if you have the option.
FACILITY	This area quite naturally has the largest number of checkpoints. It is the one single category responsible for the greatest number of obstacles to smooth, effective presentation. A little understanding of why this is true may increase the alertness of all parties and eventually reduce the number of goofs.
AUDIENCE	An analysis of the audience is important to properly direct and select material. In addition to knowing what type they expect from you (speech, lecture, or presentation), their attitude and other factors are important.
VISUAL AIDS	This is the area where the "one-time-only" stigma damages many presentations. So often visual aids are used one time then discarded. Because of this efforts are made to find other uses for the art or find existing material to use for visual aids. This often results in using material intended for publication of reports and articles for visual aids. Thus, detailed, complicated, and illegible material is thrust upon an audience that usually cannot comprehend it in the time it is before them.
HAND OUT MATERIAL	Part of planning a presentation may include the PREPARATION, REPRODUCTION, and DISTRIBUTION of the presentation. Most interested audiences would appreciate a copy of the speech or a paper on further amplification of the subject. Some situations require that this is to be done.
TIME of DELIVERY	The date of the presentation dictates the timing of all other activity and must be known well in advance. This often is the most elusive fact of a presentation. Last minute panics compromise communications effectiveness, yet invariably somewhere along the line, the date of a meeting is known by someone long before the effort is undertaken.

SUBJECT MESSAGE

Fig. 2. General analysis of a presentation.

ence. Extensive support may be required from all other sources. Emphasis is on the conversational approach. A 20-minute limit is usually imposed on one speaker; however, two or three members of a team may speak in sequence on related topics.

Effective Delivery

To be most effective a presentation must have an opening that will immediately interest the audience. It then should be developed around no more than three major thoughts, and it should close with remarks that give the audience a direction in which to react.

Outline: Without a plan, the text of a presentation could be so jumbled that the audience would "tune out" the speaker before he made any significant points. To be sure that effective buildup of thought occurs in logical order, a basic outline must be developed.

Organization: Once the organization is decided on, it should be looked at objectively. The best way to do this is

to tape rehearsals for playback. This is also a good way to catch and correct grammatical errors, poor enunciation, slow or fast talking, and other speaking defects.

Method of Introduction

Often the opening remarks of a speaker are compromised because he has to finish an introduction or qualification ineptly handled by the introducer who has failed to relate his remarks satisfactorily to both speaker and subject.

Qualification: The person who is going to introduce the speaker should be given adequate information. The subject should be related to previous activities of the meeting, and the speaker should be introduced in a way that makes him obviously qualified on the subject.

Introducer: Establishing rapport is important; in team presentations be sure to establish ahead of time who will introduce the succeeding speaker: the master of ceremonies, fellow speaker, etc.

Facility

This area quite naturally has the largest number of checkpoints. It is the one single category responsible for the greatest number of obstacles to smooth, effective presentations. A little understanding of why this is true may increase the alertness of all parties and eventually reduce the number of goofs.

In "facility" the reference is to a remote facility not physically connected with the presenter's place of business. This can mean a hotel or a motel or meeting room in other establishments.

We can pretty well assume that there is at least one individual connected to that facility who knows the physical capabilities of the meeting rooms. This means he can inform a presenter of its size, comfort, electrical outlets, etc.; he can also usually explain the support capabilities in projection hardware, public address systems, etc. This person becomes the "expert" in the minds of those scheduling use of his facility.

Of those scheduling use of the facility, we can assume they know the subject of their presentation or the theme of the meeting, and thus, know what they want in terms of a general requirement, such as a projector to show slides or movies.

To the "expert" of the facility, the person requesting use of his facility is the "expert" on the presentation.

Now we have two individuals arranging a communication function who consider each other experts, and yet the result could be chaotic.

The fact is, neither one of the two parties generally arranging for meetings in a situation like this is an expert on communication. Seldom would either of them know the difference between a lantern slide projector and a 2- by 2-inch projector, a 16-mm projector and an 8-mm projector, a super 8 or standard 8, a lenticular screen or a matte-surfaced screen. Perhaps neither one would have the slightest idea what maximum legibility might be in particular circumstances.

As a result, the principals for whom the meeting is being held are the last ones to be considered. Who has really thought about the communication with relation to the audience? Who really cares if the screen is so located that half of the audience cannot see all of the image, and if what they can see is legible? Who really cares that there are mirrors and other hard surfaces to distort sound and cause distracting reflections? Who is concerned, ahead of time, if the meeting is held up to get a proper projector or replace a burned-out bulb, or if the PA system feeds back with certain speakers?

In the case of the facility expert, his experience is most likely related to the type of business wherein the room is located. If it is a hotel or motel he knows the capability of the food and liquor services, the number and size of the bedrooms, the proximity of ancillary services that can make the customers' visit pleasant, but none of these are directly related to the subject of communications.

In my experience I have seen few hotels that have a projection table. A typical 30-inch table is provided which is too low to project over the heads of the audience. The result is that the heads of some of the audience are silhouetted on the screen or the projectors are precariously balanced on makeshift stands. Many hotels today have a ceiling height of 10 feet, making it impossible, in some rooms, to have the screen high enough to permit the bottom one-fourth of the image to be seen by people in the back rows.

Most often the rooms are divided by partitions which do not sufficiently eliminate outside noises.

This problem is not limited to hotels and motels. The facility may be a conference room or an auditorium of a progressive business or organization that has the latest audiovisual hardware. Here again there can be problems, for this facility is usually designed to accommodate presentations peculiar to that organization. Presentations prepared outside of that organization could well include slides that would not fit the projectors, or be for front screen projection while the facility was designed for rear screen.

All involved personnel are usually searching for that nonexistent expert, the presenter expecting him to be the program chairman, the program chairman expecting him to be the facility coordinator, and the facility coordinator expecting him to be the presenter. In the long haul, it will behoove any person required to make a presentation to conduct his own facilities investigation and plan that presentation against its known capabilities. This thought includes checking far enough ahead of time to direct changes if necessary.

Size of Facility: The facility size determines the makeup of visual aids and the type of ancillary equipment used with them. The formula for image size dictates that the front row of seats be no closer to the screen than two times the image width; the back row no further than six times the image width. The area required should be calculated on the basis of six square feet per person, plus that needed for the stage and aisles. Projector lenses should be selected to permit projection from *behind* the audience.

Location of Facility: The availability and convenience of public transportation to the facility should be considered. In a facility with many meeting rooms the location and directions to the room selected should be available.

Facility Capabilities: In addition to establishing the basic comfort of the room, the furniture, the extent of darkness when room lights are out, and the noise level of equipment in the room and outside, the following hardware items take on importance of various degrees, according to the needs of the presentation.

Lectern: Few facilities have stand-up lecterns, and when they do, they seem to be designed for giants; a short speaker on stage could be hidden from the front rows. If the lectern height is not adjustable, have a riser (small platform) available. The lectern light should have a shield, properly adjusted to avoid light from the front blinding the audience. It should be connected to a power outlet that

Setups before a busy or bright background.

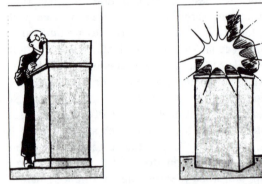

When lecterns are provided, they are usually for giants.

Unshielded lectern light.

Fig. 3. Lectern setups that alienate the audience.

is not switched off with the house lights or the power to other equipment. (See Fig. 3.)

Public Address System: The "ringing" or "screeching" often encountered in the use of public address systems is caused by the feedback from a loudspeaker to the microphone. Reducing the amplification or moving the microphone will eliminate this annoyance.

The available microphone may be fixed in position. This may be a problem if a speaker had planned to move about. The lavalier or neck-supported microphone permits the speaker to move about. However, when this is passed from speaker to speaker, the amplifier should be turned off to avoid "contact" noises.

The amplifier and its controls may be located in another room, making it difficult or impossible to vary the amplification for different speakers or for recorded material. An inconspicuous, local control should be arranged—if possible, at the back of the meeting room.

The Stage: An elevated stage, clear of obstructions, is to be preferred. Verify that the lectern, a table, or suspended lights do not block the screen for the projector. (See Fig. 4.)

The Screen: The screen should be located as close to the ceiling as possible. The bottom of the projected image should be at least four feet above the floor where ceiling height permits.

A beaded screen is adequate for motion pictures, color slides of normal contrast, and narrow meeting rooms. A

matte screen will do for close-up (overhead) projectors and extreme black-and-white contrasts. For wide meeting rooms a lenticular screen should be used.

The Seating Arrangement: In addition to the limitations on seating stated under Facility Size, the seating should be such that the audience does not face a large window, an entrance, or other distracting elements of the room. Screens and drapes may be used to mask such elements.

Projectors: Verify that the proper projector is available. Slide projectors may accommodate 35-mm (2- by 2-inch) slides or lantern (3¼- by 4-inch) slides. Many of these have remote control, but some require manual operation for each slide. Overhead (Vugraph) projectors also may be needed, or motion picture projectors for 16-mm film.

If possible, have a rehearsal with the projector operator. At least, give him the slides early with any special instructions or precautions.

Illuminated pointers will not work with rear-screen projection. If a battery-powered pointer is to be used, check the operation and have spare parts available; if ac power is to be used, verify that the power cord is long enough and that the power outlet selected remains "live" at all times.

Cards: Although cards (or posters) are suitable for table-top presentations and for small audiences, they are expensive to produce, awkward to display (if there are many), and difficult to handle. (Twenty 30- by 40-inch cards may weigh ten pounds.) The display easel must keep the card bottom at least 40 inches from the floor.

Audience

An analysis of the audience is important to properly direct and select material. In addition to knowing what type they expect from you (speech, lecture, or presentation), their attitude and other factors are important.

Size: This controls room requirements, which in turn control other factors of the presentation and facility.

Knowledge: Will they be able to understand technical depth and vernacular?

Attitude: This can significantly affect the manner in which the presentation is given. If the members of the audience are in attendance by necessity, the presentation has to change their feeling from "not wanting to hear and understand" to a feeling of "gratitude that they did attend." The members of the audience must be "rewarded" with information that will improve their thinking and affect them mentally, spiritually, or economically. If they are attending by choice, the speaker must return the compliment by giving them what was promised or what provoked their selection. He must not let them regret their decision to attend!

Visual Aids

This is the area where the "one-time-only" stigma damages many presentations. So often visual aids are used one time and then discarded. Because of this, efforts are made to find other uses for the artwork or to find exsiting mate-

Stage arrangements may block image to some people.

Few facilities have adequate projection tables.

Fig. 4.   Projection arrangements that alienate audiences.

rial to use for visual aids. This often results in using material intended for publication of reports and articles as visual aids. Thus, detailed, complicated, and illegible material is thrust upon an audience that usually cannot comprehend it in the time it is before them.

Visual aid art is a specialty in itself and the cost can be high. This area of the presentation should thus be well planned and the decision to use visual aids wisely made.

Transparencies can be made so cheaply and easily today that this technique, using a transparent overhead projector, has become very popular.

Unpracticed presenters often use many more visual aids than necessary. An economical approach is to rehearse with transparencies made cheaply and quickly from everything the speaker feels is necessary. This should be done regardless of the type of visual aid that will ultimately be prepared by an artist. The real need for the visual aid becomes clearer through rehearsals of this type. The end result will be the elimination of material that cost little to produce. Ignoring this approach may result in costly art production of visual aids that would not, or should not be used.

When: The presenter analyzes his needs only after the content has been developed. He uses them when words alone cannot develop audience understanding—when quick universal audience comprehension is required. He uses them for illustration purposes primarily—pictures as opposed to words, and charts as opposed to tables and figures.

What: Visual aids are for the audience and they must be legible and clear to all attendees. Selection of the type depends on the presenter and services available. Cards and flip charts are adequate for audiences within 25 feet of the speaker;[1] longer distances require projected slides.

Flip Charts: These can be used in the same way as cards. They are easier to handle and cheaper to produce. They are the only type of visual aid for which a vertical format is dictated, since they should be suspended from the short side. Suspension from the long side should not be attempted, because of the difficulty with "flipping." (See Fig. 5.)

Motion Pictures: Additional details are involved when movies are to be shown. The room must be darkened; a projectionist is required; proper timing is important. Projector noise may dictate reserving a projection booth, if one is available. Production of your own movie is specialized, time-consuming, and costly. If a good, short, existing film is applicable, its use is to be encouraged. If it

[1] 25 feet requires a minimum of ¾-inch letters and symbols.

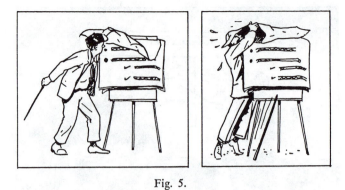

Fig. 5.

is a sound film, use a separate speaker placed behind the screen.

Slides: Vugraphs are the most economical visual aids to produce. Many office copiers can be used to make a transparency from any 8½- by 11-inch copy. A projectionist is preferable; if the speaker must operate the projector while he is speaking, he must not stand between the audience and the image. Vugraphs are exceptionally good for rehearsal of team presentations.

The most compact, universally used slides are 35-mm size. Remote slide changing, available with many projectors, permits the speaker more control. These require competent art and photographic capabilities, are relatively high in cost, and require more timing and planning. Changes (even elimination of material) are costly. Slides should be prepared only in the horizontal format.

When the projected image width is required to be 15 feet or more, lantern slides are preferred because of the brighter image available. These have all the disadvantages (e.g., cost, preparation skills) of 35-mm slides; in addition they are bulkier; the glass mounts are more easily broken; and automatic projectors are rarely available.

A 35-mm film strip makes possible instant image changes, but the cost of a strip is high, it is the least flexible visual aid, and it requires a special projector.

Standards: Art for visual aids is significantly different from any other commercial art. It is the one area when the artist is most effective if he concentrates on selling the speaker and his message. Time consuming artwork often becomes too detailed and conspicuous as an artistic accomplishment. What good have visual aids been that are remembered for their cleverness and artistic beauty if the message of the speaker is forgotten?

It is for this reason that basic standards should be established for visual aids that would eliminate illegible material and misuse of color, and encourage more simplicity and directness in visual aids.

Handout Material

Part of planning a presentation may include the preparation, reproduction, and distribution of handouts. Most interested audiences would appreciate a copy of the speech or a paper further amplifying the subject. Some situations require that this be done.

This may require significant support, especially if art, production, or photography is required, so it should be planned early and thoroughly.

The question of distribution relates usually to the type of function: speech, lecture, or presentation. Distributing such material *before* the talk may be necessary in the case of a lecture in a training session, for example, but it would be very distracting to circulate material before a speech or a presentation. Such nuisances would result as the sounds of the audience rustling paper, or their trying to read in a room darkened for projection purposes, or reading while the speaker is trying to hold their attention.

When it is necessary to distribute material, arrangements should be made ahead of time to acquire needed assistance and to allow for the time to do it.

Time of Delivery

The date of the presentation dictates the timing of all other activity and must be known well in advance. This often is the most elusive fact of a presentation. Last minute panics compromise communications effectiveness, yet invariably somewhere along the line, the date of a meeting is known by someone long before the effort is undertaken.

In addition, the time of day that the presentation is to be made is a factor that should be determined early. A speaker is able to make certain adjustments if he knows he has the handicap of a poor time on a program.

The worst times are right before lunch or at the end of the daytime activity. These segments are often compromised by poor timing and discipline of preceding events.

Another bad time is within the first hour after lunch when drowsiness creeps into the audience. Avoid, if possible, making a presentation that requires projection with the lights out for a meeting during this time.

The favorable times are the first session in the morning and just after coffee breaks in morning and afternoon. The audience is more alert for the first meeting and generally most comfortable after each break.

Some Fundamentals for Presenters

JOSEPH P. CILLO

A PRESENTATION can be a magic moment in which ideas are communicated and consensus is gained. When all goes well, it is a smooth interplay between presenter and audience, an enjoyable and uplifting experience for both.

The Macintosh, supplemented by emerging desktop presentation technology, offers an array of tools for producing this idyllic state. But many businesspeople still have trouble creating and giving effective presentations. Just having a Macintosh loaded with the latest software does not automatically lead to dazzling results. In our infatuation with technology, we sometimes skip over the basics of presenting.

What exactly is an effective presentation? Pragmatically, you've succeeded when your audience remembers what you said, in the proper context and with your desired emphases. To reach this goal, you should answer three fundamental questions before even beginning to create a presentation:

What is your objective? Is it to communicate status? To gain agreement with your point of view? To sell something? Before starting out on any trip, you need to know where you want to go.

Who is your audience? How knowledgeable are they on the subject of your talk? How many of them will there be? Are they fact- or emotion-oriented? Audience composition affects the content and structure of any talk.

What is the environment? Will you be in a small room or a large hall? Should you use slides, overheads, or the computer as a projector? The environment determines which medium and approach works best.

Only after these issues are clarified can you really get going. At this point, pay attention to several principles that, in practice, amount to universal presentation "laws:"

You're in show business. Like it or not, recognize that you have an obligation to be interesting, and being so can communicate your message compellingly. If you aren't, you may lose out to competition that is.

Keep it simple. A basic, fatal error in giving a presentation is losing your audience. Deliver complex ideas in a manner your listeners can comprehend.

Less is more. The apologia, "I would have written a shorter letter but did not have time," says it neatly. Craft your message to take less time or fewer words. In general, audiences have short attention spans.

Pictures beat words. Any time you can convert words into pictures, the end result will be more powerful. Sometimes you don't need words on a slide. For example, insert a crystal ball to communicate future uncertainty, or a bull's-eye for performance targets.

WYSIWYTH. What you see is what you take home. Prepare copies of the presentation, exactly as you give it, to hand out to the audience afterward, and tell them so. If they're taking notes, they're not listening closely.

When you've planned your approach and are ready to use the Macintosh, here are some techniques that work:

1. Use the Mac as a presentation factory. Make sure you have enough memory and disk space to do what you want to do. Assemble around the computer other people (if any) who'll create the presentation and work right on the Mac screen to produce outlines, storyboards, and other aids. Eliminate paper in your presentation. Keep copies of successful visuals and presentations on disk and use them as reference points or templates.

2. Allow one minute per visual. This rule of thumb works surprisingly well. If you have, say, one hour for a presentation, you actually present for only 25 out of 60 minutes (though this is plenty of time, given the typical attention span). Following the one-minute rule means creating 25 visuals to do the job.

3. Build complex visuals. In communicating complex concepts, use three or four visuals, each building on the previous one. For example, explain a management process by showing organizational interaction in progressively more detailed displays.

4. Highlight significant points. Some visuals, such as charts or graphs, must contain masses of information. Help your audience with emphasis or context by marking the most significant points.

5. Use a graphic on every page. Create a visual guide for your audience—an intuitive, emotional image of what you're communicating. Using a graphic as part of a title can be particularly striking. Let the Mac's graphics capability work for you. However, don't get carried away—one per page will usually do it.

6. Use other kinds of art. High-quality, useful graphics are available from some creators of clip art. If you have a scanner, try importing a photograph. An occasional large image with no accompanying text creates a "breathing space" and refocuses attention on you.

7. Number your pages. Remember to number each page for easy reference later when discussing your presentation with the audience.

8. Be careful with color. Color requires greater expertise. And following the WYSIWYTH principle, if you give a color presentation, you should hand out color copies. If you can't, give a black-and-white presentation. Also, remember that

Reprinted from *Macintosh Business Review,* pp. 33–34, Jan. 1989.

presentation handouts get copied; consider how colored pages will look when passed on in black and white.

Sooner or later, virtually everyone working in an organizational setting has to give some kind of presentation. Macintosh presentation software can help you organize ideas rapidly and pleasurably for professional results. As hardware technology, including scanners, film recorders, and projection devices, continues to evolve, lower prices will make using them more feasible. All that is required from you is an investment of time and thought to learn and apply the craft of presenting.

———————————

Three Basic Recipes for a Speech

ETHEL I. CURTIS, SENIOR MEMBER, IEEE

Abstract—For the engineer in a hurry to acquire speaking skill, this paper offers a tested route through preparation to platform performance. It identifies the role of the speech in the technical context, provides plan-ahead notes and a choice of three sets of simplified guidelines for speech organization, and compares the needs and rewards of the self-contained speech, the technical society talk, and the presentation. Finishing, polishing, and rehearsing are discussed, with emphasis on oral-versus-written words and the use of humor. The reader is directed to a friendly, supportive environment for speaking practice. Final delivery is treated, with tips on establishing audience rapport.

WHAT is so rare as an engineer who can explain his work to the world at large! As a matter of fact, how well can we talk to each other? Our work, at its best, does not speak for itself. Our voices must be raised over the hum of our hardware to amplify the message of our written words and carry the signal for action. Every day we are called on to speak.

A good speech takes practice. It takes a little confidence, a few platform skills, and a lot of labor. But wake up the back row and the hand that you get may sign your next contract or promotion. A good speech is well worth the effort.

Anyone who is willing to put forth that effort can master the art. In the past few years, I have seen many dozens of people (including myself) progress from a shaky start to polished platform performance. I remember an engineer newly arrived in the United States who mumbled his first speech so that not one word came through. He turned up a year later leading a technical session with poise and eloquence.

PLAN AHEAD

For the reader who asks "What am I going to do *today* if the call comes to speak?" the best advice is *start now* before the call comes.

• Plan your talk as you plan your technical project, using the same careful logic.

• Ask the relevant questions and as the answers come, fit them together in an orderly sequence.

• Underline the key points you want to make to your audience.

• Consider the questions they may ask and formulate the best responses you can.

• Review your diagrams, curves, tables, and computer printouts produced as working tools or report elements. How can you simplify them to serve as visual aids to a speech?

• Try out your rationale on an interested colleague and

benefit from his feedback. Discuss whenever you can. Make notes.

In due course, when you are asked to speak, you will be poised for quick response. Nothing breeds confidence like being prepared.

WHEN, WHERE, AND WHY YOU SPEAK

Potentially, an engineer may be called on for one of three types of talks:

1. Speech to a lay audience or a technical group.
2. Technical society talk summarizing a written paper.
3. Presentation in-house or to a customer.

These differ in their emphasis on data and interpretation (fact and emotion) and in the amount of support from hard copy displayed or left with the listener. Also, the concept of *trade-off* must be made clear to the lay audience. Where the demagogue paints the world in blacks and whites, the engineer must show the shades of gray.

Regardless of the occasion or type of talk, your first duty as a speaker is to orient your listeners. Before you dive into the nitty-gritty of your specialty, begin by answering these questions:

1. Why do we (should we) do this work (make this product)?
2. Where does it fit into the big picture?
3. How does it relate to the listener? Why should he prefer this to other alternatives?

In general, your purpose in speaking is threefold:

1. To interest—by means of colorful, meaningful description.
2. To inform—by hard logic and facts.
3. To stir to action—by inspiring faith—through the weight of your logic and facts, your reputation, or your persuasive delivery.

ORGANIZING YOUR SPEECH

For the engineer in a hurry, Table I presents a choice of simplified approaches to organizing a speech, i.e., the spoken words that carry your message. Part one, *Ingredients,* offers three versions of the traditional elements: title, introduction, body, and conclusion. Your choice among the three depends on whether your primary purpose is to interest, to inform, or to stir to action. But remember, in most cases all three purposes must ultimately be served by your speech.

Your title is important. It is the advance signal that stirs your audience before your first word is spoken. Compare, for example, your response to (a) "An Introduction to the Logic of Integrated Safety Systems for Electrical Power Plant

Manuscript received Aug. 13, 1979; revised Dec. 3, 1979.

The author is a licensed professional engineer and an Advanced Systems Engineering Specialist at Lockheed Missiles and Space Co., Sunnyvale, CA 94086.

Reprinted from *IEEE Trans. Professional Commun.*, vol. PC-23, no. 1, pp. 28–31, March 1980.

TABLE I
THREE BASIC RECIPES FOR A SPEECH

INGREDIENTS		
To Interest . . .	*To Inform . . .*	*To Stir to Action . . .*
A. Title—make it sing!	A. Title—descriptive	A. Title—promising
B. Introduction—wake 'em up!	B. Tell 'em what you're going to tell 'em	B. Here's the problem—have I got a solution!
C. Body—fill 'em in!	C. Tell 'em	C. How they both grew
D. Conclusion—memorable!	D. Tell 'em what you told 'em	D. Get busy on my remedy!
PREPARATION STEPS		
Directed Method	*Free-Association Method*	*Research Method*
1. Pick topic you like	1. Pick topic you know	1. Pick topic to learn
2. Think it over—what key points will you make?	2. Jot down main ideas of your plan-ahead notes	2. Consult reference list or library sources
3. Find a catchy title	3. Jell the speech purpose	3. Study best parts
4. Find a rousing opener	4. Label ideas as they fit choice B, C, or D above	4. Form your own views—set in B, C, D order
5. Find a good punchline	5. Set in logical order	5. Find gaps and fill them
6. Form speech—opener to key ideas to punchline	6. Plan title and opener	6. Plan title and opener
7. Redo—step up—polish	7. Write well-worded close	7. Write well-worded close
FINISH SPEECH		
Prepare detailed outline or write speech; rehearse with tape recorder		

Start-up and Control" and (b) "Explosion Can Be Hazardous to Your Health!"

A good speech is an art form, not nearly as simple as the ingredient list suggests. Depending on the complexity of your subject, the body alone may have to be as tightly plotted as a three-act play, with the introduction serving as a prologue and the conclusion forming the dramatic climax of the third act. For a very few of our speeches, this dramatic buildup comes naturally. For most, it requires work and rework in the course of preparation.

Begin by selecting the compatible set of *Preparation Steps* from Table I, depending on whether your subject is

One you know intimately and have digested thoroughly;
One you know well, but are still digesting; or
One you have yet to research (a real possibility in today's specialized, fast-moving technical world).

The *directed method* applies if you have your subject firmly in mind, with a quick fix on needed reference sources. This method starts with reflection: What are your main points? Your optimal story sequence for driving them home? Reflection brings inspiration: An intriguing speech title and opening lines . . . a punchline to underscore your conclusion and leave the desired impact.

Now you have the skeleton of your speech. All that remains is development of your thesis from opener through main points to punchline. When applicable, the directed method is the surest, fastest way to a good speech within the time limit (or before your audience escapes into reverie). There is always a time limit. A speech is necessarily a miniaturized form of communication.

The *free-association method* offers additional help in sorting out your thoughts on a familiar topic. It consists of collecting, recording, and arranging your main ideas from your project files, your plan-ahead notes, and your memory (plus needed references), as a prelude to the directed method. Thought

and inspiration are indispensable elements in your speech preparation.

The *research method* leans more heavily on the literature. It involves studying available references on a less familiar subject (preferably covering the spectrum of expert opinion), formulating your own viewpoint, recording and arranging your main ideas, then finding and filling gaps—before proceeding with the creative thinking of the directed method.

In its final form, your speech may be either written out in full or simply outlined in detail, for example on note cards. Speakers differ in their methods. Complete write-ups are highly useful for short speeches with tight time limits. Also, the written form facilitates final polishing. But that comes later.

THREE TYPES OF TALKS

So far, we have focused on the organization and content of your spoken words. How about backup to your speech in the form of visual aids and hard copy? What feedback can you expect from your audience?

The answers depend on which of three types of talks you are planning: a self-contained speech, a technical society talk, or a presentation.

The *self-contained speech*, generally broad in scope and variable in technical depth, has the least backup from visual aids and hard copy left with the listener, as well as the least benefit from audience feedback. The spoken words must carry the message. They must be clear, concise, colorful; rich in illustrative examples; free of unexpected ambiguities. Yet speeches have time limits, challenging the speaker to find a verbal shorthand for presenting his case honestly, fully, and fairly. Mastery of the self-contained speech carries over to lend quality to your other, aided talks.

The *technical society talk* is backed by a complete written paper. Generally specialized in scope but penetrating in technical depth, it uses slides or charts simplified from the paper.

However, the talk must be more than a mere reading of the paper—not only because of limited time but especially because of our limited rate of aural comprehension. (Most of us have faster reading ability.)

The speaker must use his best verbal skills—first to orient his listener, then to stress his main points, and finally, perhaps, to induce cooperative action. In return he may be rewarded by instant feedback of comments and questions as well as by technically profitable post-meeting response.

The *technical presentation,* either in-house or to a customer, generally has the highest immediate importance among speaking assignments. As such it merits full support from point-by-point charts and all hard copy and hardware required to demonstrate the worth of the project. Also, it merits all the skill the speaker can bring to provide convincing scope, depth, and emphasis.

In presenting, your charts are your cue cards. (Written notes may be minimal.) Two minutes per chart is a reasonably fast pace. If a single chart takes ten minutes for discussion, consider replacing it with several. If it takes an hour, you are giving a lecture. Finally, as presenter, you may look forward to the best feedback of all—the go-ahead or funding to keep your project alive.

One caution: If you have hard copy to hand out, plan to do so *after* your speech, not before. Otherwise you will lose your audience to your written material.

FINISHING, POLISHING, REHEARSING

Your preliminary draft is completed, your visual aids assembled. You are ready for final polishing and rehearsing—the secret weapons of the successful speaker.

Why polish? To make our words speakable. Consider the distinction between the written and the spoken word. Better still, let us begin by answering the question: Why speak?

With the written word, we can design great projects and record our plans in fine detail. But to get those projects rolling—to move men to action—we also need the spoken word. Both forms should be accurate. Both should be clear. Both present a point of view. And what is a point of view but an idea wrapped in an emotion? Here, the live performance—the human voice—excels in conveying emotion.

Try a sentence as simple as this: "The choice is obvious: On the one hand, we have nuclear, on the other, oil-fueled power." Say it several ways, varying your tone and stress, and you can represent both sides of the argument—and picture the hackles rising among your audience.

When you are prepared and self-confident, your voice will carry your viewpoint naturally, sincerely, and forcefully. Your fluency is best, by definition, if you have chosen words that flow easily from your tongue. Your final speech polishing should assure this as well as the judicious use of wit or humor.

On the choice of oral versus written words, Rodgers' and Hammerstein's ballads are a better source than Roget's *Thesaurus.* Generally speaking:

- Shorter is better.
- Phrases should be lyric-like, easy to say.
- Similes and metaphors help paint a picture.

- Redundancy is permitted. Two ways to tell it are better than one.
- Grammar should be impeccable, but
- Occasional vernacular provides welcome relief.
- Use of your listeners' "buzz words" helps create rapport.

Humor in a speech can be a breath of fresh air to reawaken audience interest. Often it wins you good will. But humor is a chancy thing. Take care when your barbs fall or you may offend the listener you would least like to lose.

If you are lucky enough to hear a comedian's one-liner that fits your case, record it instantly. For the most part, your best humor is situation comedy that grows out of your subject and highlights a major idea. A few pointers:

- Avoid the witty remark that puts down your audience or your competitor (or any social group).
- Avoid one that is a permanent put-down to you or your organization, but
- If you can proceed from your own comic dilemma to a solution that makes heroes of you or yours, go to it! This gives you license for almost unlimited bragging.

If you have second thoughts about your humor, try it out on your own small in-group until you feel comfortable with the telling.

Your first line of rehearsal, however, is at home before your mirror and tape recorder. Practice until your thoughts flow in natural sequence. Pick up small flaws and correct ambiguities. Become accustomed to using your voice as an instrument. Start with a conversational tone but allow your own interest and enthusiasm to come through and introduce emphasis and variation. (Objectivity does not equate to monotony, the one unforgivable sin in a speaker.) Learn to take a breath at the start of each major phrase. This air is your broadcasting medium. Give similar rein to your natural gestures. These, too, add interest and emphasis.

HELP FOR THE SPEAKER

When your call comes to speak, ask for detailed instructions on type and length of talk, available projectors, screens, and easels, as well as any special requirements on visual aids. Check out the room beforehand if you can. If you have charts, check for adequate lighting. Try out the microphone if one is available. Talk to the session organizer to determine exactly what subject coverage is required of you, who else will speak and on what, the technical level and special interests of your audience. Check with your company for format guides, release requirements, and specialized help in preparing your visual aids. Be sure that these reflect the same economy of words as your speech, in order to permit large type, easily read, and simple concepts, readily grasped.

Your best help in speechmaking will be found in actual podium practice within the fellowship of a friendly, supportive group. In a recent address [1], an officer of Manitoba Hydro of Canada told how his early years as an engineer were hampered by his tongue-tied syndrome when facing a group—even after he had collected "the world's largest library on how to speak." Introduction to a speaking club soon removed this barrier. He now handles public relations for his company within the highly communication-dependent power industry.

There are speaking clubs in your area. Consult your local paper or chamber of commerce and visit a few until you find one you like. Also, consult the bibliography [2] for references. A few technique-oriented practice speeches before a helpful, sympathetic audience will do wonders for your platform poise and delivery.

DELIVERY

This paper has stressed advance preparation as the secret of facing your audience with confidence. Preparation eliminates the so-called butterflies in your stomach, even those that seem to fly in formation. It brings out the ham in all of us—the positive pleasure in talking about a favorite subject to an intelligent, interested audience.

Good will is the best possible message-carrier. Consciously like your listeners, and they will like you. Speak directly to them and watch their reactions to make sure they are following your thesis. The better you know your material and the less you need notes, the better you can establish and hold rapport.

Above all, keep cool, warming up only as your talk gains momentum. Even if advance notice was lacking and you had only five minutes to snatch up your material and dash to the meeting room, keep calm. One hurried colleague of mine once arrived overburdened at the podium, knocked over the easel, and dropped his charts. But he turned to his audience with a friendly query: "Any questions so far?" The laugh that he raised kept his listeners solidly with him for the rest of the session.

Presently, you will have the floor. As you warm to your discourse, you will find your own formula for speaking success.

REFERENCES

[1] E. K. Stuhlmueller, "Address by International President," *Fall Conference, District Four Toastmasters*, Palo Alto, CA, November 1979.
[2] B. E. Fearing and T. M. Sawyer, "Speech for Technical Communicators: A Bibliography," *IEEE Trans. Prof. Commun.*, vol. PC-23, no. 1, pp. 53–60, Mar. 1980.

Making Presentations That Command Attention

The key to giving
an effective technical
presentation
is confidence.
You get it by being
fully prepared.

RALPH L. KLIEM
Boeing Computer Services
Seattle, WA

Ask any engineer or manager to specify the tasks that he dreads the most, and he will probably put technical presentations high on the list. The dislike is usually not due to an ignorance of the subject to be presented or a fear of speaking in front of a crowd of people. Instead, the big problem is uncertainty about how to prepare and present information in a way that will command attention.

Define objectives

Defining objectives is an obvious first step. But it is often not done at all or is done poorly. Part of the problem is that assignments are given without much explanation of what is specifically expected. For example, if your charge is to "give staff members a rundown on the *xyz* project," you had better find out what "rundown" means. Does it refer to a description of what the project involves, an update on project progress, or an explanation and evaluation of project targets?

Once you know exactly what your talk should deal with, the next step is to be sure all key points are cov-

Reprinted with permission from *Machine Design*, pp. 143–147, April 1987.

ered. Next, distinguish between essential and peripheral material, to give the presentation a sense of priorities.

Another problem with defining objectives is that presenters usually restrict their thinking to what should be covered. They do not consider how it should be covered. The "how" concerns the motivational aspects of the talk. For example, if you only want to inform people, all you have to present are certain facts and arguments in an interesting way. But to get people to act in a certain way on the basis of what you tell them, you must organize and convey information accordingly. Creating an impetus to action is now your principal objective; the information presented is merely a means toward that end.

Often, a presentation has more than one objective. Then, merely listing and dealing with each one is not an effective way to prepare for the talk. You need to know which objectives are the most important for proper emphasis. If there are no obvious priorities, subjects should be dealt with on the most logical or interesting basis.

Know your audience

Many technical people fail in this regard because they accumulate, organize, and convey information as though they are the audience. This usually means that almost everyone, except the speaker, is dissatisfied by the presentation.

The first issue to be faced when analyzing the audience is the level of expertise represented. If you are speaking to a group of technical specialists, you are free to use the concepts, terminology, and "buzz words" appropriate for the specialties involved. In fact, if you do not cast your talk in such terms, you will probably be looked upon as a novice or as someone who is gently ridiculing those who are in attendance. By using tech-talk, you cut through nonessentials and spend all of the time available communicating information and ideas that you know will be of interest to your listeners.

An audience of laymen, on the other hand, is a real challenge. Not only must you be able and willing to translate difficult technical con-

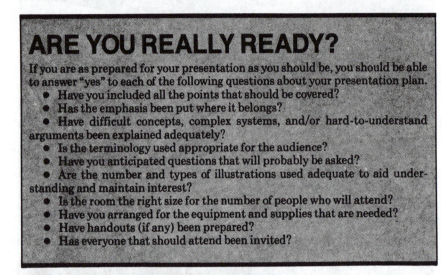

ARE YOU REALLY READY?

If you are as prepared for your presentation as you should be, you should be able to answer "yes" to each of the following questions about your presentation plan.

- Have you included all the points that should be covered?
- Has the emphasis been put where it belongs?
- Have difficult concepts, complex systems, and/or hard-to-understand arguments been explained adequately?
- Is the terminology used appropriate for the audience?
- Have you anticipated questions that will probably be asked?
- Are the number and types of illustrations used adequate to aid understanding and maintain interest?
- Is the room the right size for the number of people who will attend?
- Have you arranged for the equipment and supplies that are needed?
- Have handouts (if any) been prepared?
- Has everyone that should attend been invited?

If you don't speak in the language and to the needs of the audience, you will soon find that no one is listening.

DAVID RANKIN -87-

cepts into easy-to-understand language; you must also be careful not to give the impression that this is a chore. Many technically-knowledgeable speakers demean listeners by conveying the attitude that they are reluctantly trying to enlighten a group of peons. You can avoid such an attitude by realizing that you are just as ignorant of certain subjects as your lay-audience is of the subject you are addressing. Your purpose is to inform and, sometimes, to educate. You will succeed only if you communicate on the level of your listeners' understanding.

A mixed (tech/nontech) audience is the toughest of all. Assuming that both groups are equally important to your objectives in giving the presentation, you must present information and arguments in terms that everyone can understand. That means that the bulk of your presentation should be designed for the layman. However, there is nothing wrong with offering technical explanations or ramifications of what has already been conveyed in simple terms. Although such amplifications should be held to a minimum, they do let the technically knowledgeable people in the audience know that you are an expert in the matters involved.

Another way to satisfy the technical specialists in a mixed audience is to offer reports or other handouts for those who want technical details. For example, after describing the operation and benefits of a particular product design, you can comment that data sheets and mathematical analyses proving that the design is superior to alternatives are available on request.

The point is not to showcase your knowledge. Instead, you want the technically-knowledgeable people in the audience to understand that your rather simple presentation was designed that way on purpose and does not reflect a lack of expertise on your part. This can be emphasized at the beginning of the talk by telling your audience that "technical details" will be dispensed with in the interest of getting the basics across to everyone present.

If you are talking to decision-makers as opposed to lower-level personnel, your talk should deal with the concepts and terms that will make their evaluation easy.

The impact of the best talk will be greatly reduced if the size of the room is way out of proportion to the size of the audience.

This does not mean that you should skip or gloss over pertinent data, issues, and arguments. However, wherever appropriate, stress the potential decision-making consequences of the points you are making.

Plan for maximum impact

This stage of your preparation can make or break the presentation. The task here is to put your knowledge about objectives and participants into creating the mechanics and the setting for the presentation.

Outline: Many speakers believe that they have such command of their subject that they can "wing it" rather than plan what they will say. Then, when they actually give the talk, they lose their way. Their presentation becomes nothing more than a disconnected series of topics that confuses, rather than enlightens, the audience.

A good outline serves several useful purposes. First, it keeps you on the track you set out for yourself. Your talk will have a direction and a purpose. It also forces you to have a logical structure to the presentation. That means, be sure to have an introduction, a body, and a conclusion.

Many engineers ignore the need for structure. They start pumping out countless facts and figures without giving an overview of the situation, such as historical information or a statement and explanation of the problem. In failing to go from the general to the specific, they lose the audience. With an outline, however, you force yourself to inform and explain in a way that will result in maximum understanding and interest.

Be sure that you incorporate pertinent examples of the major points you intend to make into your outline. This practice is important in any presentation, but is a necessity in a technical talk. Without examples, hard-to-understand concepts and arguments will likely go over the heads of listeners. To be effective, the examples should be relevant to the jobs and knowledge of the people in the audience.

Many people speak directly from an outline. The advantage of this approach is that you will have a detailed roadmap of the talk in front of you at all times, assuring that you will never get lost. If you do not think that a complete outline is necessary, you can make notes from the outline and use them as your guide. But whatever route you take, remember, your purpose is not to "wow" the audience with your style. So you are better off making your job easier by having more rather than less information at hand.

Illustrations: Nothing can confuse an audience more than talking about technical issues without the aid of illustrations, especially when complex principles or systems are involved. For example, if you are talking about developing a large computing system consisting of many subsystems, illustrations should be used to show the subsystems and how they interact with one another. Letting the audience actually see what you are talking about is the best way to facilitate understanding.

To be effective, illustrations should be large, colorful, and numerous enough to capture and hold the audience's attention. Also, only pertinent information should be included. Members of the audience should not have to read too much to understand what is being shown. In addition, reading material should be presented large enough to be seen and understood easily by everyone in the room.

Facilities: The first thing to consider is where you will give your presentation. The right room is one that is not too big or too small for the number of people who will attend. Many technical presentations "bomb" because too many people are crowded into a small room, or too few people are sprinkled around a room that is too big. In either case, the audience is uncomfortable and their attention wanes. In a pinch, a room that is too big is better than one that is too small. The seats in a big room can be rearranged so that everyone is close enough to the action.

After the room is selected, make arrangements to have adequate equipment and supplies in the room. Common items include an overhead projector with spare bulbs, blackboards with extra markers, and an easel with adequate paper and markers. You may not need all of these supplies, but having them available when you need them is important. Many otherwise good presentations get off the track because the speaker has to search for equipment or supplies that he did not anticipate needing.

Stay in control

If your preparation has been thorough and rigorous, you are well on your way to a successful presentation. But the best preparation will go for naught if you do not control the actual proceedings.

Audience participation: Nothing is less stimulating than having to sit through a lengthy presentation. After about 45 minutes, people's attention starts to wander. If your presentation exceeds that timeframe, one way to maintain attention is to encourage audience participation. That might mean asking people questions or letting them ask you questions. By allowing questions, you maintain interest. Also, you gain feedback that will tell you whether you have kept or lost contact with the audience.

Many speakers do not solicit questions because they fear that they will be asked a question that they cannot answer. But this fear is minor compared to losing the attention of the audience. If you cannot answer a question, simply admit it and let that person know that you will get the answer for him by a certain time.

Mannerisms: Using distracting mannerisms is another way speakers lose control. The list of such mannerisms is endless, but com-

Poor speaking mannerisms — such as looking down when talking or fiddling with a pointer — will divert attention away from what is being said and toward what is being done.

MAKE INVITATIONS COUNT

Notifications to members of the audience are usually thought to be relatively unimportant and are often taken care of as casual afterthoughts. This is a mistake. You should allow enough time in your planning to think over who should attend and to prepare a formal notification of the presentation.

The best method for notifying attendees is a memo. At a minimum, the memo should specify the date, time, and place of the talk; identify the topic; list the attendees; and stipulate whether or not advance preparation is necessary.

The memo serves as a permanent reminder of the presentation and assures that attendees will be ready to participate. It also gives the people invited an opportunity to inform you if they cannot attend. On the basis of such feedback, you can either reschedule the presentation or invite substitutes for the people who cannot attend. ■

There is only one thing worse than giving a presentation without illustrations: using illustrations that are too small or too complicated to be seen or understood.

monly includes waltzing around the podium, using a pointer as a sword, toying with eyeglasses or other personal items, and talking to a blackboard. If you fall into such habits, the audience will watch what you are doing rather than listening to what you are saying.

Speakers succumb to distracting mannerisms because the habits make them feel more comfortable by using up their nervous energy. One way to control them is to practice the presentation in advance. Practice allows you to identify your faults before you give the talk. Another way is to direct your nervous energy into the presentation, itself, in the form of enthusiasm. Being animated in a constructive manner will leave little occasion or energy for distractive behavior.

Using poor eye contact will also hinder a presentation. Do not bury your head in your notes or stare at something at the back of the room. Force yourself to look at the audience, and focus on particular people now and then. This tells members of the audience that you are interested in them and helps to establish a rapport between you and your listeners.

Handouts: If practical, do not give the audience handouts before or during the talk. Such handouts direct the audience's attention away from what you have to say and toward what is in the documents. Unless the material is used during the presentation, avoid distributing the material until the end of the talk.

End game: Perhaps the most likely point in a presentation for a speaker to lose control is at the end. Most presentations have no real ending; the speaker simply stops talking. Executed correctly, the end game focuses attention on essentials and generates excitement about what you have said.

Summarize major points: A summary is very important because major points can easily get lost in a maze of technical details. By bringing the audience's attention back to essentials, you provide them with a lasting perspective of the major points. A good conclusion will help the listener to "trim off" the details.

Ask for questions or comments: This encourages feedback from the audience to determine whether or not you have made your points adequately and provides you with the opportunity to clarify any confusing aspects of your presentation. Also, by asking for questions, you are telling members of the audience that you are interested in what they have to say, and not just in hearing yourself talk.

Be available for discussion: One of the biggest mistakes speakers make is to disappear after they are through. Instead, you should linger after the formal part of the presentation to allow people to approach and talk with you. This practice proves your interest in the subject and the audience's understanding. It also provides you with the most candid feedback about how well you did. Lessons learned in this informal part of the program can be incorporated into your next presentation. ■

Part 2
Planning and Preparation

LOGICAL PLANNING and adequate preparation are absolute requirements for effective presentations, but they do not happen by accident. A wide range of planning and preparation techniques can be used, most of which are based on simple common sense and apply equally well to all forms of communication. The articles in this part describe planning and preparation techniques that are particularly applicable to technical presentations. Some of these articles address general techniques that are relatively timeless and apply to broad-based considerations, while others focus on specific requirements.

In his article, "Stand up and be heard," Maurice Broner describes the basic communication process of presentations and suggests techniques for planning, preparing, and rehearsing presentations and checking the adequacy of the physical arrangements. He also stresses that audiences remember both good and bad presentations, the good ones for their content and the bad ones for their execution. Although his article was originally published in 1964, the ideas presented are still appropriate for today's presentation requirements.

Because a presentation cannot be studied and understood via repetition, as can a written paper or report, it must be instantly intelligible to the audience. In his article, "Time to improve our oral presentations," Max Weber addresses the pros and cons of reading a paper versus delivering it from memory. He also presents seven steps for helping a speaker be more effective, including techniques for delivering a presentation and providing the related visual aids. In addition, he identifies some common faults in presentations and outlines ways to eliminate them.

The need to analyze the specific audience and situation for each particular presentation is stressed by W. A. Mambert in "Analyzing the specific audience-situation," Chapter 4 from his book entitled *Presenting Technical Ideas*: *A Guide to Audience Communication*. He identifies the factors that should be considered in analyzing an audience in terms of a typical member profile, general awareness, and the objective, attitude, and relationship of the audience to the presenter. He also suggests techniques for analyzing the situation in terms of the audience's objective, size, and authority; physical considerations such as the site, environment, and available facilities; and psychological considerations such as the time of day and duration of the presentation. In addition, he describes the best ways to use audience feedback advantageously.

In his paper, "Audience requirements for technical speakers," Robert Perry contends that a presenter has an obligation to provide the audience with a stimulating and provocative transfer of information. To do this, the presenter must consider the limited capacity of any audience to receive, process, and assimilate information; the need to screen the few ideas that can be effectively transmitted; and the requirement to overcome an audience's inherent disinterest. Perry also describes a method for using storyboards to address audience interests and objectives in planning a technical presentation.

Richard Lindeborg, in his paper "A quick and easy strategy for organizing a speech," explains how to use a word processor to simplify the preparation of technical presentation outlines. The outline form he suggests, coupled with the simplicity and speed of handling changes on a word processor, provides rapid visual feedback and guides an author in expressing ideas in an original, simple, and clear manner. Word processors also simplify the execution of repeated revisions and allow experimentation to achieve the best organization, visual display, and natural cadence.

In his paper, "A problem-solving approach to preparing professional presentations," Eric Skopec describes a four-step approach for preparing presentations. The first step involves analyzing the purpose of the presentation, the nature of the audience, and the demands of the situation. The second step consists of devising a strategy to integrate the problem definition with the author's knowledge of the subject. The third step involves executing the presentation plan using the conventional techniques of organization, explanation, and visual representation. The fourth step consists of comparing the resulting presentation relative to the original plan and modifying it to correct discrepancies, incorporate improvements, and reflect other feedback.

Based on the assumption that many technical presentations begin as a scientific or technical paper, Henry Ruark, in his paper "Planning and developing an audiovisual presentation," focuses on the process of paper-to-presentation conversion. He explains that this process basically involves selecting the main points, illustrating them with clarity and imagination, and complementing the visuals with narration rather than written text. He also emphasizes the need to rethink the sequencing of information to accommodate the linear nature of an oral presentation as well as the fact that an audience cannot easily refer back to earlier information for explanation and clarification.

In his paper, "Technical talks: They're not as hard as you think," David Adamy addresses the apprehensions of first-time presenters. He explains that the preparation for a technical presentation should begin with a decision about the audience reaction the author wants to achieve, followed by a determination of the subject scope to be covered. He stresses that only a few main points should be addressed, and they should be emphasized by appropriate repetition. He also includes a checklist and flow diagrams to define the steps for preparing a successful presentation.

Several considerations that engineers must address in preparing a technical presentation are described in "Crucial decisions for technical speakers" by Joseph Miller and William O'Hearn. These considerations include the purpose of the presentation, the audience's background, the scope of the subject, the organization and explanation of the content, and the method of delivery. To be effective, however, an engineer must *want* to share his or her ideas with the audience and present them so that the audience can grasp them in the way in which the presenter intended them to be understood.

In his paper, "Considerations for international presentations," Robert Woelfle focuses on the increasing demand for technical presentations in international environments. Because one of the principal criteria for an effective presentation is to address the background and interests of the audience, international presentations require special attention. Woelfle identifies and describes the special factors a presenter must consider in the areas of oral and written language, translation techniques, cultural values and attitudes, body language, and physical arrangements. He also explains that international presentations are further complicated by the fact that the nature and impact of these factors vary from country to country.

Stand Up and Be Heard

MAURICE A. BRONER, SENIOR MEMBER, IEEE

Summary—Presentations are effective tools to convey ideas. A presentation is either good or bad. A good one leaves a message. A bad one can be effctive also, since audiences remember that it was bad long after the message is forgotten. Planned presentations should employ both hearing and sight senses. Controlled tests indicate that people remember ideas longer under these conditions. Presentations should contain these elements: purpose; background; data generation, assimilation, and analysis; graphics; rehearsal; and delivery.

INTRODUCTION

Harrington [1] in his survey and analysis of speech courses taught in engineering schools, wrote, "The inability of the average engineering graduate to convey his ideas or plans to others is not only a handicap to himself, personally, (but) it is one reason that the profession is not recognized as one of the learned professions today. Many engineers devote all of their time to professional study and advancement, but fail to realize that their ideas are of little value until they can be transmitted to others in a lucid and forcefull manner."

Business is people, and people are controlled through effective communications. Management communicates policy and direction while the engineers and scientists must communicate the technical know-how. They transmit much of their know-how through papers, technical articles, drawings, and process reports, but more and more, engineers and scientists are finding it necessary to communicate with management, colleagues, and customers through the medium of the oral presentation. Frequently insufficient attention is paid to the gentle art of persuasion. Engineers and scientists feel that their work speaks for itself and that only the results must be chronicled. But, how many people read these chronicles? To propagate an idea, the scientist, the research engineer and especially the project leader must stand up and be heard. J. W. Wiggens, General Manager of Thiokol's plant at Huntsville, Ala., stated about his employees, "We want . . . technical people who are articulate in projecting clearly the capabilities we know they possess. They should be equally (as) skilled in making lucid presentations to individuals as to large audiences."

An oral presentation is only as good as its material, its preparation, and the techniques used to put it across. Many good ideas (maybe some of yours) have been lost because of poor presentation. The ABCs of an effective presentation are accuracy, brevity and clarity; but the job is more complex than this statement implies.

This paper outlines an engineer's approach to the preparation and delivery of an oral presentation. It discusses briefly the communications process, offers a route to follow from concept to delivery, and includes guides and checklists to assist during the preparation process.

THE COMMUNICATION PROCESS

Our approach to an effective presentation may be clearer by a broad examination of the human communication process.

Transmission

Human beings gain knowledge through exposure of their senses to stimulation. The more senses involved—the more knowledge gained. Although all five of our senses provide avenues of communication, the effective presentation generally relies on a combination of hearing and vision with an occasional assist from the other senses.

A properly planned presentation makes maximum use of both the receiver's sight and hearing sensors because words alone or pictures without words cannot adequately convey your story. Good visual aids can carry the burden of your presentation if they are properly organized, smoothly introduced, and effectively employed.

The U. S. Air Force has conducted a number of controlled tests on the ability of human beings to see, hear, absorb, and retain information communicated to them by words only, by pictures without words, and by a combination of words and pictures. A major result of these tests was a comparison of the amounts of information retained for three days after being presented by the three methods, *viz.* 1) listening to words alone, 2) viewing pictures alone, and 3) listening to words while viewing supporting pictures.

The curves plotted in Fig. 1 show that an audience will remember much more 72 hours after a presentation if both words and pictures are used, 6½ times more than when only words are used and 3¼ times more than when pictures without words are used.

Message Structure

The unit idea is the basis for oral communications and is the smallest part of a presentation that has real meaning by itself. In engineering terms, the unit idea is a functional subsystem of a larger system. It must be developed so that it can perform its specific function before it can be used in the system.

In a presentation words may be likened to components

Manuscript received April 15, 1964. This paper was presented at the International Conference, Aerospace Electro-Technology, Phoenix, Ariz., April 21, 1964.

The author is with Lockheed Missiles and Space Co., Sunnyvale, Calif.

Reprinted from *IEEE Trans. Eng. Writing Speech,* vol. EWS-7, pp. 25–30, Dec. 1964.

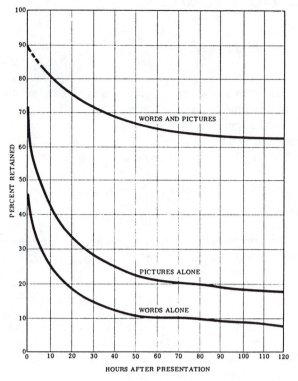

FIG. 1—Audience retention of information.

of a system. A number of components (words) are assembled into a functional subsystem (unit idea) and several subsystems are then assembled into a system (oral presentation with a purpose).

Words are spoken one at a time and they are received by a listener the same way, one at a time. Your listener stores the one-at-a-time words in his memory until a unit idea is comprehended. If you, as the sender, do not select and use your words properly, the unit ideas will fail to develop and you will be left with one or more subsystems not operating. With a nonoperating subsystem your entire system performance is degraded.

It is at the unit idea level that the visual aid is valuable to carry a great part of the burden of communication. As the words used to describe the unit idea must be simple, so must be the picture used to visualize it. Each picture should have but one message. The idea represented must be recognized instantly. If the audience must figure out what the picture means or why it is being used, the picture is a visual distraction rather than an aid. Both complex word descriptions and complex pictures confuse and bore. Either confusion or boredom will steal an audience. This means one or more subsystems has become nonoperational and the system (presentation) has broken down.

PREPARING THE PRESENTATION

In recent years presentations have become something of a habit within industry. Also the government frequently requires presentations as part of a contract. Like all habits, presentations may be good or bad. It is all too easy to go

through the motions of a presentation and fail to meet the objective. Although there has been communication, the speaker may have communicated ambiguity, boredom, and confusion. Only a portion of the message, a distorted message, or a completely different message may have been received.

The presentation can be a most effective tool for the engineer or the scientist if he will recognize its strengths and limitations and will learn to use it properly.

Once the decision to make an oral presentation has been made, there are four steps that must be taken: 1) planning, 2) writing, 3) rehearsing, and 4) readying.

Let us examine each of these steps as it affects our total effort to communicate.

PLANNING

The greatest single enemy of a good presentation is lack of adequate preparation. Planning is the cornerstone of preparation and falls logically into four tasks: 1) determine the purpose, 2) analyze the audience, 3) gather and analyze the data, and 4) outline the presentation.

Determine the Purpose

A precise statement of the purpose or objective should be developed early in the planning and should be used as a periodic check throughout the preparation to insure staying on the track. Ask yourself: 1) Why am I making this presentation? 2) Is my purpose to educate or convince? 3) What reaction do I want from my audience? Direct Action? Approval? Amusement? 4) How do I want the audience to feel when I am finished? Inspired? Aroused? Encouraged? Relaxed? Alarmed?

Analyze the Audience

Some knowledge of the background, interests and objectives of an audience is necessary in preparing an effective presentation. There might well be a discussion of "re-entry heating" with either the local Rotary Club or with professional thermodynamicists. The approach will be different in each case. Ask yourself: 1) What is the background of my audience? Technical? Management? A mixture? 2) How much does my audience already know about my subject? How much background must I establish in my presentation? Am I safe in using jargon of the trade? 3) What does my audience want to hear? Do they have known biases? Alignments? 4) Does my audience contain a key person? Who is he and how does his presence affect me?

A critical question to be considered during the preparation should be, "How long should I speak?" The answer lies in an analysis of the audience and the occasion. If a time limit has been provided, stay within it. Running overtime is most discourteous to an audience and to other speakers who might follow. Also, it indicates lack of planning. There are some general rules. Many clergymen agree that any part of a sermon that runs over twenty minutes is lost. No one likes to sit more than 45 to 60 minutes without a break, and preferably a change in subject and voice.

Remember, don't overrun your time. If for some reason, an interference cuts into your time, pace yourself so that you do finish on time.

Gather and Analyze the Data

At this point select the main point or points that will be made in the presentation. The main points should be limited to two or three with four as a maximum if they are closely related. It is better to fix two or three points firmly than to stretch coverage too far.

You should consider all resources of subject matter at your disposal. Search for material that will fit together in a logical manner to develop your main points. Don't hesitate to relate your ideas in terms of personal experience, providing it is pointed and relative. If some of your points must depend on opinion, search for authoritative books, papers and articles for quotes of recognized authorities to back up your opinion. (Unless, of course, you are the pre-eminent authority.)

After the initial data-collecting effort, arrange material in a logical sequence of main points, each with supporting subpoints and illustrations. Conduct additional research as necessary to fill in gaps.

Now it is time to re-examine the material in the light of purpose and audience analysis, and to discard any data that is not pertinent. Remember that you can lose an audience in too much detail.

Examine the material to be sure that it suits a specific audience. If the audience is mainly management, avoid precise technical details and stick to management problems. For instance, in selling management on a new project you might ask: What is the concept you are selling? How much will it cost? How could the project be paid for? How should the project be organized? What organizational adjustments are indicated in the company structure? Whom do you recommend to manage the program? What are your suggestions for manning the project? What return can the company expect on its investment?

If the audience is technical, concentrate on the technical concepts, and don't over-emphasize management and cost considerations.

Outline Presentation

During the data collection and analyses process, the material probably has suggested an outline for its presentation. Rough out a presentation in outline form moving logically from the known to the unknown, or from the general to the specific; use subpoints to support main points and arrange main points to lead to a conclusion (objective).

In thinking through the points and subpoints in an outline, consider which ones might lend themselves to the effective use of visual aids. Sketch simple pictorial ideas

that could support your words. Be sure each picture is simple and makes only a single point or subpoint. Avoid the use of words (other than the required titles and labels) on your visual aids wherever possible. *Reading words on the screen will distract from hearing and comprehending.*

WRITING AND ILLUSTRATING

In outlining the presentation build the framework which will provide size, form, and position. The writing can be considered the flesh, muscles, and skin which will give contour, function, and beauty to the finished product.

Written Script

In writing the script for a presentation there is a cardinal rule which frequently is overlooked or ignored. *It should be written in oral expression.* This is quite different from the written expression you generally find in a technical paper or article. You are writing words to be spoken rather than read. So write them as they would be spoken to an associate.

Speaking Notes

Ideally you should know your presentation so well that you require no notes to jog your memory. Often you will require notes to assist you in making your presentation. Some speakers use 3 × 5-inch or 5 × 8-inch cards. Others prefer an annotated script while still others use notes scratched on a piece of paper. You use the method most comfortable to you.

Visual Aids

The final design of visual aids must evolve during writing since many words will refer to graphics. Also, the final design may be influenced by the type of visual aid that will be used.

Determine the type of visual aid or combination of aids best suited to your presentation after considering the size of your audience, presentation room, and budget. The spectrum of aids usually available are: charts, slides, models, movies, equipment, apparatus, and samples.

Your budget may determine how elaborate the visual aids can be, but good visual aids need not be elaborate art work, and conversely, elaborate art work does not insure good visual aids.

Most presentations must "fly" with charts or slides. Well prepared illustrations can aid materially in the effectiveness of a presentation. Conversely, poorly prepared ones may be worse than none at all since they detract from the message. You should consider the size of the auditorium when selecting the kind of visual aid you will make. The rule is simple: be sure your aids are large enough for all the audience to see all the details. Charts may be fine for small groups. For larger groups it generally is advisable to use projectors (slide or overhead).

Color and background are very important. Colored slides are pleasing to the eye and tend to hold the attention of the audience. Avoid monotony by varying the color of the background. (Thirty minutes of black lines on stark white background will put an audience to sleep). If you cannot obtain colored slides, try using negatives which will give white lines on a black background for illustrations such as block diagrams, circuits, and organization charts. A mixture of backgrounds will improve a presentation. You can tone down light backgrounds with shading films (such as Zip-a-tone) for relief and emphasis.

Occasionally art work prepared for color slides also must be used in black and white photography for the printed page. Such a case can demand a compromise in color selection so that the art work may serve both purposes.

Certain colors which are pleasing to the eye and which convey information dramatically sometimes photograph badly. Reds, deep blues, dark greens, and browns cannot be separated from black lettering. Black and white film used in graphics work is sensitive to the blue end of the visible spectrum and quite insensitive to the red end. Therefore, light blues reproduce as white while reds reproduce as black.

Normally, a Graphic Arts group will be aware of the basic rules for good graphics, but here are a few DO's and DON'T's in preparing art work for slides, and these may help in understanding the artist's problem and serve as an aid in making slides in the absence of professional help.

1) Show as few curves as possible on a slide——certainly not more than two or three unless they represent a family of curves.

2) If possible, use colors when several curves must be shown; use different colors for each set of related curves.

3) Any lettering should be large enough so that every one in the audience can read it. Omit notes, explanations, and other material best supplied by the speaker. As a guide, five-inch high image area of original art requires letters ⅛-inch high; Fig. 2 shows the minimum lettering size as a function of height or width, whichever dimension governs the final art. Capital block letters shold be used unless standard symbols require other treatment. You should avoid fancy lettering, which is difficult to make and as difficult to read.

4) Make lines of sufficient weight that they will not be lost or blurred in size-reduction and reproduction. Use ink or chart tape for line work because pencil lines are difficult to reproduce. Fig. 3 shows the minimum line weight as a function of original size of art work.

Neatness and cleanliness are important because the photo techniques for slide reproduction record all pasteups, construction lines, and smudges.

REHEARSING

The greatest single enemy of a good presentation is lack of adequate preparation.

Practice

Rehearsal does not actually follow the writing step but

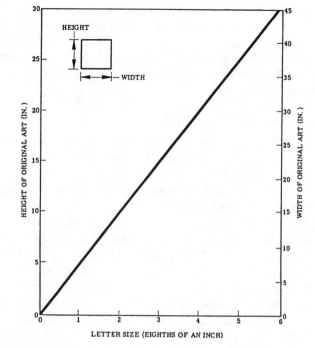

Fig. 2—Lettering size as a function of original art height and width.

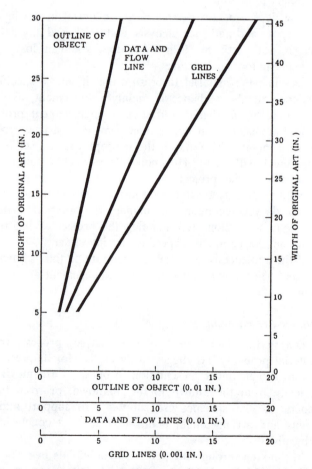

Fig. 3—Line weight as a function of original art height and width.

rather overlaps it considerably. Rehearsal should start during the writing phase to insure good oral expression and to find out if your visual aids do their job before they are put in final form.

Good delivery demands repeated rehearsal and only through practice can you develop in your delivery the subtle changes in expression and pace that can make your presentation dynamic.

The most profound work, the best conceived presentation, and the greatest idea that mankind ever had may not be received if the delivery is poor.

One way to kill a good presentation is to deliver it in a poor speaking voice. It is the speaker's responsibility to his audience to enunciate clearly, to talk in a pleasing voice, and to talk loudly enough so that a listener in the farthest corner of the room can hear without straining.

A device for rehearsing by yourself is to talk in front of a full-length mirror and record your voice on a tape recorder. The mirror will allow you to see yourself as an audience would see you, and the recording will reveal any obvious rough spots in your delivery. Rehearsing for the presentation is like learning a part in a play. Recognize your cues and know what follows.

Speaking Mechanism

Biologically, the speaking mechanism is more useful for eating than talking. The tongue and vocal cords protect the lungs from irritations. Man is born with the ability to eat and breathe, but he must learn to talk. This, perhaps, is the reason why some people speak badly. They have never learned to talk correctly. Where articulation is improper, breathing probably is incorrect or speech impediments exist that have not been overcome. Proper coaching and practice are necessary to overcome this.

Criticism

Entering the final phase of rehearsal, you need constructive criticism from someone else. You are too close to the effort to analyze your own work effectively. Ask your colleagues to listen to your talk. You are much better off obtaining criticism from your close associates than from your final audience where the stakes may be much higher. If time permits make several preliminary presentations, analyze the criticism, and incorporate any changes you feel are important.

Checklists

During rehearsal, critics can help if they have a checklist of things to note as you practice. The following checklist is divided into two parts. The first part applies to how the speaker handled the subject; the second part covers the delivery techniques.

Checklist On Subject Handling

1) Did the speaker arouse and sustain interest in the subject?

2) Did the presentation move along at a good pace without undue pauses or lags in thought?

3) Was there good distribution of time between the various points?

4) Was the transition from one point to the next smooth so that the presentation was coherent?

5) Was the subject completely covered and were the main points clear?

6) Did the speaker hold the attention of the audience?

7) Were the visual aids adequate, pertinent and clear, even from the back of the room?

Checklist on Delivery Techniques

1) Was the speaker poised, confident, and in control of the presentation?

2) Was the speaker enthusiastic about his subject?

3) Could the speaker be heard and understood clearly?

4) Did the speaker appear to be talking directly to the audience rather than to the screen, blackboard, or charts?

5) Did the speaker use trite expressions or over-use certain words or phrases?

6) Did the speaker have any distracting mannerisms?

7) Did the speaker present a neat appearance?

Presentations often prompt questions from the audience. During a presentation to a large group, courtesy generally dictates that questions be held until the end of the presentation. In planning, allow a reasonable time for the discussion of questions following the presentation. When speaking to small groups it is not uncommon for members of the audience to interrupt the speaker with a question. In either case be prepared to answer or otherwise handle any question that may be asked by an audience. Thinking out in advance questions that might be asked and preparing answers in your mind can save awkward moments.

Generally it is good practice to repeat any question asked before the answer. This lets the questioner know that you understand and also gives the audience another (and perhaps a better) chance to hear the question.

READYING

Readying means all of the things that must be procured, arranged, prepared, and considered between the dress rehearsal and the final presentation. Possibly there is nothing quite so frustrating as to find a vital arrangement missing at the last moment. There is only one sure way to avoid such a situation, consider and arrange for each and every detail yourself.

Advance Notice Checklist

1) Have I notified everyone concerned of the time, place, and nature of the presentation?

2) Have I explored the various ways of stimulating thought and creating advance interest in my presentation?

Presentation Analysis Checklist

1) Have I given sufficient emphasis to the key ideas on which I must get acceptance?

2) Have I fixed the objectives of the presentation in my own mind and identified the main points I want to stress?

3) Am I so familiar with my subject that nothing can sidetrack me from my main points?

4) Have I analyzed my audience so that I know what to expect? a) Does my presentation answer the principal questions I might expect about the subject discussed? b) Have I considered what other questions the audience might ask, and am I prepared to answer them?

Final Arrangements Checklist

1) Have I secured a suitable room? a) Are the ventilation, heat, air conditioning, and lighting facilities adequate? b) Are the chairs, ashtrays, tables, desks, etc. adequate? c) Is there a lighted lecturn?

2) Will there be refreshments? What? Where? How? When?

3) Are my visual aids ready? a) Is the easel for my charts in the presentation room? b) Do I have a pointer? Where is it? c) Is the slide or movie projector in the room and is it in working order?

4) Have I made all necessary arrangements with my assistants and does each of them know what he is to do? Is the projectionist familiar with the equipment?

5) Have I enough copies of all materials which I expect to distribute to the attendees? How and when will I distribute?

Self-Evaluation Checklist

A final checklist for self-evaluation of your presentation after the fact can help to prepare you for the next time you must give a presentation. Most elements of this self-analysis list can be useful following your last dress rehearsal.

1) Did I make all necessary preparations?

2) Did I start on time?

3) Did my presentation meet all of the objectives? a) Was my presentation progressive? Did I stay on the subject and were all points directed toward the conclusion I wanted to arrive at? b) Did I maintain audience attention and interest? c) Did I make full use of my visual aids? d) Did I make a satisfactory conclusion?

4) Did I rephrase all questions and did I answer them adequately?

5) Did I maintain control at all times or did I allow the audience to take the discussion away from me? a) Was I resourceful in handling group questions? Was I quick to grasp and develop comments and questions from the audience? b) Was I able to handle over-aggressive questioners skillfully and tactfully? c) Did I permit any of the questions to lead to an argument?

6) Did the audience leave with my message?

CONCLUSIONS

A good presentation is not an easy job and its preparation should never be taken lightly. However, a systematic approach combined with practice can produce creditable results.

Simple checklists can be very helpful in three areas: 1) getting comments on your rehearsal, 2) ensuring readiness for the presentation, 3) self-evaluation.

The actual few minutes of a formal presentation may reflect years of background, months of study and data accumulation, and days or even weeks of preparation. In these few minutes is focused your opportunity to affect management decisions and even the expenditure of large sums of money. It is a vital few minutes to you. Make the most if it.

REFERENCES

[1] C. B. Harrington, "A Survey and Analysis of the Speech Courses Taught in Colleges and Departments of Engineering in the United States of America," Ph.D. thesis, Southern Methodist University, Dallas, Tex., p. 7; 1950.
[2] H. Weiss and J. B. McGrath, Jr., "Technically Speaking," McGraw-Hill Book Co., Inc., New York, N.Y.; 1963.
[3] W. P. Sandford and W. H. Yeager, "Effective Business Speech," McGraw-Hill Book Co., Inc., New York, N.Y.; 1960.
[4] Visual Communications Div., Management Services Office. "Presentations." Wright-Patterson Air Force Base, Ohio; 1960.
[5] G. M. Lovey, "Briefing and Conference Techniques," McGraw-Hill Book Co., New York, N.Y.; 1959.
[6] U. S. Air Force, "Presentations," Washington, D.C., p. 4; 1956.
[7] Lockheed Missiles and Space Co., "Effective Chart and Slide Preparation, Sunnyvale, Calif.; revised September, 1963.
[8] Monroe, "Principles of Speech," Scott, Foresman and Co., Chicago, Ill.; 1958.

This paper was presented at the Phoenix Conference on 21 April 1964. It is being reprinted here by special arrangement with the editor of IEEE Transactions on Engineering Writing and Speech. (Revised paper received by PTGAS 5-4-64)

Time to Improve Our Oral Presentations

Max Weber

NO MATTER HOW WELL WRITTEN *a paper may be, the oral presentation of that paper may fall completely flat. The speaker often forgets that what he says must be instantly intelligible to his audience rather than ultimately understandable, the way it can be to a person who is reading what was written. This article discusses some of the common faults in the oral presentation of papers and suggests means for eliminating these faults.*

As remote as the possibility may seem to you, eventually there is a good chance that you will write a paper and it will be accepted for presentation at a meeting or conference. Perhaps you will have discovered something new, such as a new punctuation mark (like the interbang*) or another use of the comma or hyphen. Or perhaps your boss has told you that your company needs some inexpensive publicity, and the best way to get it would be for you to present a paper at the next STWP International Technical Communications Conference. Or perhaps you have tired of hanging around the office and want your company to send you to a conference, but your boss tells you the only way he can justify the expense of sending you there is for you to present a paper.

Hopefully, you will not be like many another author who feels that his work is finished when he has completed the writing of his paper. Often a good, or even outstanding, paper has been made to seem, at best, mediocre because of the author's inadequate preparation for the oral presentation. The author forgets that when he changes from a written to an oral medium of communication he must also change the way in which he presents the material to his audience. He forgets that he cannot deliver a speech in the same way in which he delivers a piece of merchandise. A speech is delivered by means of sound waves and light waves, and the methods used by the speaker determine what the listeners get out of the presentation, how well they get it, and how long they remember it.

As a speaker, you must remember that an oral presentation is not an essay on its hind legs. An oral presentation of a paper is different enough from the written presentation to require different techniques. Only when you have taken the trouble to make yourself aware of the different techniques involved will you be able to make a truly effective oral presentation.

MODE OF DELIVERY

The first (and probably most important) problem you must resolve concerning the oral presentation of your paper is the mode of delivery to use. There are three methods of making the oral presentation of a paper. You can read the paper, you can memorize it and recite it word-for-word to your audience, or you can present it extemporaneously (with or without notes).

Reading a Paper. Let us first consider the advantages of reading a paper. Every paper includes certain key points that you want to be sure to impress on your listener. Most written papers are carefully thought out, well organized, and nonrepetitious. Therefore by reading your paper, you are assured of making a well-organized presentation, and you know you have omitted nothing you meant to tell your listeners. In addition, you may become nervous and flustered in front of an audience, and looking out on a sea of eyes staring back at you tends to magnify your nervousness. By reading your paper, you can help to maintain your self-composure by not having to look at your listeners. Also, many of us do not know what to do with our hands while we are speaking. Reading your paper permits you to keep

Work performed under the auspices of the U. S. Atomic Energy Commission. Based on a presentation given by the author at the STWP 16th International Technical Communications Conference, Washington, May 1969.

*Conceived for the rhetorical question, the "interbang" is a combination of the question mark and exclamation point.

Reprinted with permission from *Technical Communication,* published by the Society for Technical Communication, vol. 17, pp. 6–11, First Quarter 1970.

your hands occupied shuffling the pages of your manuscript back and forth. Finally, an important advantage of reading your paper is that it gives you a good excuse for letting your paper run overtime. After all, how can anyone expect you to read a paper and watch the clock at the same time? (If you are questioning whether these last items mentioned are really advantages, then perhaps you are beginning to understand that it's pretty hard to find many real advantages in reading a paper.)

Let us now consider some of the disadvantages of reading a paper. To understand the main disadvantage we must recognize the basic differences between oral and written presentations. When someone reads the published version of your paper, he can set his own pace for absorbing the material. He can go back and reread what you wrote or he can sit back leisurely to digest one point before he goes on to the next one. When you make your oral presentation, however, everything you say must be instantly intelligible to your listeners. Since your listeners cannot reread what you say, you have the responsibility of repeating things you have said to be sure you have not lost your audience. This means not only avoiding the reading of your paper, but also maintaining good eye contact with your audience while you are talking. The

best way to tell how well you are understood is to watch the faces of your listeners. When you realize your listeners are not taking a point you have made, you can repeat it in another way. In a written account, it will be boring to be repetitious. In a spoken account, on the other hand, you are under a moral obligation to put your key ideas in several ways to be sure the audience has grasped the point. If all you are going to do is read, you might just as well print your paper, circulate it to your audience, and spare them the trouble of assembling in a meeting hall.

Suppose your paper includes tables, diagrams, and other illustrative material. How can you possibly talk directly to your audience about this material if you keep your eyes glued to your paper? You must be able to move around and point out the things you are illustrating, and you cannot do this while reading. When pointing to something, you must stand to the side that permits you to point while continuing to face your audience, and you cannot do this while you are reading. Perhaps I can best summarize the disadvantages of reading your paper by repeating what one wit is supposed to have said: "Reading a speech is like courting a girl through a picket fence. Everything you say can be heard, but there's not much contact."

Memorizing a Paper. Some people feel that by memorizing their presentations, they can be sure to include everything they want to say, while still maintaining good eye contact with the audience. Perhaps you have felt that you are safe with a memorized speech. Remember, though, that when you memorize a speech, instead of having to remember a simple thought pattern, you are now required to remember hundreds (or even thousands) of words. Even if you succeed in memorizing your speech, the chances are that, unless you are a professional actor, your memorized speech will sound memorized. Undoubtedly, many people do memorize their speeches as a way of not having to read them. The only people who deliver memorized speeches effectively are those who spend years practicing their delivery and are willing to expend unlimited labor to attain artistic perfection. Memorizing may be useful for special exercises in voice training, or for improving bodily action, but not for original speaking.

The main reason for not memorizing a speech is the fallibility of the memory. Practically everyone has had the experience of having had something perfectly memorized and then had his mind go blank part way through the delivery. This even happens to many professional actors, but a professional actor is trained to improvize to cover up a lapse of memory. Unfortunately, the average individual has not had this training. Is there anything worse than having a speaker's mind go blank in the middle of a memorized speech? It's bad enough on stage when an actor forgets his next line. If he cannot think of something else to say, a prompter is standing in the wings. But heaven help the speaker whose mind goes blank when he is delivering his memorized speech. There isn't a lectern in the world large enough to hide behind when this happens.

THE READER

Making an Extemporaneous Presentation. Let us now consider the extemporaneous presentation of your paper. As you may know, an extemporaneous speech is one in which the ideas are firmly fixed in mind, but the exact words are not memorized. It can be with or without notes; this usually depends on how well you know your subject, how well prepared you are to make your speech, or how forgetful you tend to be in front of an audience.

There are advantages and disadvantages to making an extemporaneous presentation. To me, the basic advantages of extempore speaking are that you maintain good eye contact with your audience (which you cannot do if you read your speech) and you are relieved of the burden of wondering what to do if you forget something (as can happen if you try to recite a memorized speech).

By maintaining eye contact with your audience, you are establishing good liaison with your listeners. When listeners know you are speaking to them, rather than to a manuscript, they are bound to pay attention to your talk and you are thus assured that they will get the most out of it. Perhaps even more important, if you watch your audience while you talk, you can tell how well they follow what you are saying and you can find out when you need to repeat or rephrase what you have said until it is understood.

The essential thing about a talk that justifies bringing you and your audience together (rather than just sending them a copy of your paper) is the emotional contact. If you have to find your words as you speak, you are forced to think along with your audience and this keeps you from going too fast. Even more important, you get the exhilarating feeling of rapport. As Sir Lawrence Bragg points out, "A lecturer who reads is earthbound to his script, but the lecturer who talks can enjoy a wonderful feeling of being airborne and in complete accord with his audience. It is the greatest reward of lecturing."[1]

One disadvantage of the extemporaneous speech is that, in your effort to clarify some points by being repetitious, you stand an excellent chance of consuming your allotted time before you have covered everything you meant to include. Although this appears to be a serious shortcoming, remember that your paper is being published in the proceedings of the conference. Actually, then, the purpose of making an oral presentation is not to tell your audience everything, but to get them interested enough in what you said to want to go read your paper.

Another disadvantage—and perhaps the reason the extemporaneous presentation is not employed more often—is the time required for its preparation. Good extemporaneous speaking means that you have thought deeply about your subject, you know the details intimately, and you have put them in order in your mind. You are speaking as someone who knows everything he must say and the proper place to say it. Certainly, this type of preparation cannot be done in the time it takes to prepare to read a paper. But I feel the improvement in the quality of the presentation is worth the extra time required in preparing for a good extemporaneous presentation. Perhaps we should say, as Dr. Bragg says one may be tempted to say, "The worst spoken lecture is better than the best read one."[2]

PREPARING FOR PRESENTATION

Once you are sold on the idea of speaking extemporaneously, you are faced with the problem of preparing for the presentation. If you spoke to a dozen different experts on speech making, you would get a dozen different formulas telling you how to prepare to make an effective oral presentation. Rather than set myself up as an expert among experts, I would simply like to quote from one of them. William Norwood Brigance[3] suggests that the following seven steps will help a speaker be more effective:

1. Choose a subject of interest to you, and one that can be made interesting to the audience.
2. Do not try to cover the whole subject, but select one specific part for your central idea.
3. Phrase the central idea into a purpose-sentence so you will know where you are going.
4. Make a list of two to four main points of your central idea.
5. Obtain specific and interesting supporting material for each of these main points.
6. Organize the speech into an outline.
7. Practice delivering the speech until you have it well in mind.

I would like to elaborate briefly on each of these steps, particularly as they relate to the oral presentation of a paper.

1. *Choose a subject of interest to you, and one that can be made interesting to your audience.* This should not be difficult. Certainly, you would only write a paper about something of interest to you.

Just as certainly, if the subject is not one that can be made interesting to your audience, it should not have been accepted for presentation. As some sort of guide, however, I suggest that you (1) pick a subject that interests you and about which you know a great deal and (2) pick a subject your audience knows something about, but wants to know more.

2. *Do not try to cover the whole subject, but select one specific part for your central idea.* Perhaps the most serious error you can make is to try to cover your entire paper in your oral presentation. No one can object to thorough coverage of your subject in the paper you write for publication, but for the oral presentation you must narrow the subject so you can cover it interestingly and specifically. Pick one part you can cover thoroughly within your allotted time, and make your presentation interesting enough that the people in the audience will be motivated to read your complete paper afterward.

3. *Phrase the central idea into a purpose-sentence so you will know where you are going.* Once you have decided that you will not read your paper, you must avoid the danger of swinging too far in the opposite direction and making an oral presentation that seems like an aimless conversation. To be effective, your presentation must move toward a goal. Therefore keep in mind where you are going, and put it down in writing as part of the preparation for your presentation.

4. *Make a list of two to four main points of your central idea.* Many speakers, in boiling down their papers for an oral presentation, make the mistake of skimming over all the material at hand and reducing each point by an equal amount. The result is almost invariably a hodge-podge of short, seemingly unconnected trivia. Carefully appraise the points you have covered in your written paper. Then pick only two to four main points, develop them thoroughly, and tie them together so your audience understands the relationship of these points to your central idea.

5. *Obtain specific and interesting supporting material for each of these main points.* Each main point in your presentation is basically the statement of an idea or principle. But your statement alone is worthless unless it is supported or elaborated. If you want your listeners to remember and believe what you say, and especially if you want them to take some kind of action as a result of what you say, you must plan to support your ideas by a succession of details, examples, comparisons, or illustrations.

THE THREE-STEP APPROACH
INTRODUCTION   DISCUSSION   SUMMARY

TELL THEM WHAT YOU'RE GOING TO TELL THEM   **TELL THEM**   **TELL THEM WHAT YOU TOLD THEM**

6. *Organize the speech into an outline.* How detailed your outline needs to be depends primarily on the amount of experience you have had as a speaker. Novices at public speaking have been known to prepare extremely detailed outlines of their speeches containing up to about 60 percent of the number of words to be included in the full speech. You will undoubtedly want to develop your own system for preparing the outline of your speech.

If the written version of your presentation was organized with an eye to its oral presentation, then the outline you used for the paper to be published may be appropriate for the oral presentation. (At least, let's hope you wrote your paper from an outline!) On the other hand, you may want to prepare a completely new outline. A good plan to follow is to use the three-step approach: (1) you tell them what you're going to tell them (the introduction), (2) you tell them (the discussion), and (3) you tell them what you told them (the summary).

7. *Practice delivering the speech until you have it well in mind.* Once you have organized your speech into an outline, you are ready to practice the delivery. Try reading your outline silently and then aloud, each time without backtracking. Then rehearse the actual delivery without using the outline and again without backtracking, but be sure to check your outline afterward to see what you may have omitted. How many times you need to rehearse depends largely on how experienced you are as a speaker and how nervous or forgetful you tend to become in front of an audience. Dr. Brigance[3] recommends rehearsing five to ten times. In planning your rehearsal, be sure to stand while rehearsing, pick a time and place where you will be free of interruptions, and try to rehearse in a room that is as close as possible to the size of the room in which you will be making your presentation. (This last requirement may be the most difficult one to fulfill. You may not be able to determine in advance the size of the room in which you will be present-

ing your paper. Even if you can determine the size of the room, you may not be able to find a room of that size which you can use for rehearsing.)

If possible, arrange for a "dry run" in front of some of your colleagues so they can provide feedback as to how you sound and look. Another device used by many successful speakers is to practice the speech with a tape recorder; this is especially helpful in eliminating tongue-twisting or easily slurred expressions that may otherwise be overlooked. Try practicing in front of a full-length mirror so you can see what you look like to others. The ideal arrangement would be for you to record your entire practice session on video tape; then you can provide yourself with a complete picture of how you look and sound.

MAKING THE PRESENTATION

If you have done a thorough job of preparing for your oral presentation, the actual presentation itself should proceed smoothly. To make your presentation completely effective, however, you need to be careful of some of the niceties involved in changing your presentation from a mediocre one to an outstanding one.

Perhaps the greatest single factor that sets the amateur speaker apart from the professional is the way each one starts to speak. Before you speak, pause and get ready. Make sure of your posture, take a deep breath or two, look at your audience, and smile (if you can). Wait a few moments until everyone is looking at you, and then address yourself to the chairman, other panel members, distinguished guests, and finally your audience.

Now you are ready to start your talk. To be effective, your opening remarks must be something that will get the attention and good will of your audience. You can attain this goal in several ways. You can say something that will establish common ground between you and your audience. Or you can pay the audience an honest compliment. Or you can make some reference to the occasion or the surroundings. Or you can refer to matters of special interest to your audience. Or you can start out with some pleasantry or humor (but be sure it is short, apt, and funny). Or you can make direct reference to the significance of your subject (only when the listeners' interest is really intense).

Should you decide to use notes, you want them to be as unobtrusive as possible. Make them no longer than necessary, and put them on 3-by-5-inch cards. When you have completed using one of your cards, instead of turning it over, slide it aside so your audience is not distracted by the movement of the card.

Try to look dignified, yet natural. If you feel nervous, you're in good company; many highly successful singers, actors, and speakers admit that they never really overcome their feelings of stage fright. The problem is not necessarily to overcome stage fright, but to reduce it, control it, and give better speeches because of it.

Try to speak as naturally and as conversationally as you can. Be careful about speaking in a monotone; learn to modulate your voice (listening to yourself on a tape recorder can be very helpful in cultivating voice modulation). Be sure to enunciate clearly, and avoid running words together. Speak slowly (beginning speakers usually go too fast); if you plan your speech properly, you should not have to rush through it.

Watch out for the distracting sound *uh*. When you pause between phrases or sentences, don't say anything; avoid the *uh* habit. Well-spaced pauses can be very

effective; they give you a chance to stop and think of what to say next, and they give your listeners a chance to catch up with you and consolidate their thoughts.

Learn to make effective use of gestures and other body actions. Every speaker gives two speeches at the same time—one with words, the other with action. A speaker who uses no action puts his audience into a state too near sleep for them to listen alertly. If you use controlled, communicative action, your listeners will find it easier to listen. On the other hand, don't use distracting motions. I'm sure you have seen speakers who button and unbutton their coats, put their hands in and out of their pockets, rock up and down on their toes, play with objects on the lectern or table, play with a piece of chalk, or use other uncontrolled motions that do nothing but twist the audience into fitful, distracting responses. Use actions, but not distracting ones.

Be sure to maintain good eye contact with your audience. Look slowly around the room, not neglecting anyone. If you use notes, look at them only briefly. If you use charts, slides, or other illustrative material, look at them only long enough to orient yourself—then face your audience and talk to them, not to your illustrative material.

The conclusion of your talk should have one purpose—to give your proposition a lasting effect so your listeners won't just walk out and forget it. To be effective, your conclusion must move swiftly; make it short . . . make it motivated. You might conclude by presenting a challenge for the audience to do something. Or you might conclude by using an apt quotation to reinforce your theme. Or, if your speech is complex or your purpose was to present information, you might conclude by drawing the important points together in a condensed, unified form. Or you might conclude on the high note of looking forward into the future. Or you might end by simply rounding out your thoughts with a few swift strokes. Whatever method you use in your conclusion, do it quickly and remember never to conclude by thanking your audience; they thank you (with their applause) for what they have gotten from you—you do not thank them.

USING VISUAL AIDS

Perhaps the most used, and abused, feature of many talks is the use of visual aids. The first decision you must make regarding visual aids is whether they will really be helpful with what you are trying to express or are just some fancy trimmings that will do nothing but impress the audience with the importance of your paper.

From some of the conferences I have attended, I get the impression that at least half of the visual aids used could just as well have been omitted without detracting one bit from the value of the talk. Before you decide to use any visual aids, be sure they are worth the time and expense of preparation. Slides mean extra trouble to you in arranging for them to be projected correctly, and inconvenience to your listeners of having to sit in a darkened room so they cannot take notes. Generally, slides or charts of graphs or numerical tables (so frequently used in technical presentations) should be shown sparingly, if at all, primarily because the audience will not have the time to absorb what you have shown. Include such graphs and tables in the written version of your paper being published in the pro-

ceedings of the conference, summarize your findings in your talk, and tell your listeners where they will find the detailed materials you are discussing, but do not show charts or slides that do nothing but get your audience lost in a maze of details.

Once you have decided which illustrative material is essential for your talk, there are some considerations you should keep in mind. Here are five points that will lead to better visual aids and your presentation of them:

1. Be sure that everything is large enough to be clearly seen by everyone in the room. If you are using charts or diagrams, be sure the lines are heavy and broad. (Very rarely is anything that is drawn on a blackboard clear enough to be seen by those in the back of the room. As a general rule, then, avoid using blackboards.) If you are using slides, be sure the projected image is large enough to be seen easily.

2. Avoid crowding too many details into one illustration. When preparing charts or slides, omit details and include only the bare essentials. Otherwise you lead to confusion, distract attention, and provoke curiosity. Your audience can usually read no more than 10 to 20 words without losing your thread of thought. When you are planning to explain several things or a series of steps in a process, do not put everything on one diagram. Instead, use several diagrams, each containing only one idea and each as simple as possible. And remember that graphs and maps are almost always better than tables of results.

3. Be sure to talk to the audience and not to your illustration. You may find your illustration to be a very welcome refuge from the eyes of your listeners. But if you fix your eyes on your illustration, you invariably wind up mumbling to it instead of concentrating on your audience. Look at your illustration only when pointing to something specific on it, and learn the art of keeping a pointer aimed at what you are discussing while you maintain eye contact with your audience.

4. Avoid standing between the audience and your illustration. Be sure to stand out of the way so your audience can see what you're talking about. If the audience is close to you, you may have to stand as much as 3 to 4 feet to one side. Naturally, this means using a pointer to be sure you stand far enough away to keep out of the audience's line of vision.

5. Do not let an unused chart distract your audience. Keep charts out of sight until needed, and remove them as soon as

you have finished referring to them. If you cannot keep your charts out of sight, be sure to cover them until you are ready to use them.

In addition to the above items, here are a few points to keep in mind concerning slides. First of all, be sure the screen is large enough to allow the projection of an image that can easily be seen. Don't expect your audience to have perfect eyesight, and be sure that what you are showing can be seen easily by everyone, even those without 20/20 vision.

Secondly, be sure the room is dark enough for all details to be seen easily. (Color slides require a much darker room and a brighter screen than black-and-white because the brilliancy of hue in color pictures depends on the darkness of the room.) Test the room ahead of time and check for special precautions such as the use of draperies, curtains, or shades, and be sure that corridor lights, exit lights, and lectern lights are shielded to avoid shining into the eyes of your audience.

Thirdly, be sure that no one in your audience is seated directly behind the projector. The escaping light from the projector makes it hard to see the image on the screen. If you cannot place the projector behind your audience, move them out of the V-zone of light behind the projector.

Finally, pay special attention to making yourself heard and understood. Remember that you are speaking in the dark

so the audience cannot read your lips or see your gestures; you must depend on your voice alone. (If you operate the projector yourself, you may not be speaking from the front of the room. This means you must use enough additional energy in speaking to build up room resonance so that the direction of voice is secondary to its easy audibility.) Keep in mind that noise of the projector can mask the intelligibility of your words, so you must speak loud enough to override the projector noise.

I have avoided giving specific rules about how much to include in your illustrations and how large to make your lettering. Details of this type should be left to your graphic-arts group; they are the experts in this area. The biggest mistake you can make in ordering charts or slides for a talk is not to discuss them in advance with your graphic-arts people. If you do not have a graphic-arts group and must arrange for the artwork yourself, be sure to include among your books at least one that spells out the requirements for good charts and slides. For example, Ulman and Gould's book *Technical Reporting*[4] has an excellent chapter on "Visual Presentation of Information." They point out that "the American Chemical Society says that each slide should be limited to the presentation of one idea; that a slide should contain not more than 20 words, and preferably not more than 15; and that a table should have no more than 25 or 30 data." They further tell you that "the Society of Automotive Engineers propounds the highly valuable rule that no letter or figure shall be less than 1/40 the vertical height of the copy." This means that if you draw your slides horizontally on 8½-by 11-inch paper and fill up most of the sheet, you should not draw any character with anything smaller than a 175-size (0.175-inch) Leroy or Wrico guide.

If you feel you still need specific instructions on the preparation of visual aids for your talk, you might order a legibility slide rule ($5 each) from Calvin R. Gould, Audio/Visual Coordinator, Martin Marietta Corporation, Orlando,

Florida. The device is described in the *Proceedings of the STWP Fifteenth International Technical Communications Conference.*[5]

CONCLUSION

A good oral presentation is not easy, and its preparation should never be taken lightly. However, a systematic approach, combined with lots of practice, can produce creditable results. The few minutes of your oral presentation may reflect years of background, months of study and data accumulation, and days or even weeks of preparation. In many instances, these few minutes can affect management decisions and the expenditure of large sums of money. These few minutes are vital to you; make the most of them.

ACKNOWLEDGMENT

The drawings in this article were prepared by Joseph J. Bauer of Argonne National Laboratory.

REFERENCES

1. Lawrence Bragg, "The Art of Talking about Science," *Science, 154:* 1613 (Dec. 30, 1966).

2. Lawrence Bragg, "More on the Art of Talking about Science," *Nuclear Applications, 4:* 282 (May 1968).

3. William Norwood Brigance, *Speech, Its Techniques and Disciplines in a Free Society,* Second Edition, New York: Appleton-Century-Crofts, Inc. (1961).

4. Joseph N. Ulman, Jr., and Jay R. Gould, *Technical Reporting,* Revised Edition, New York: Holt, Rinehart and Winston (1959).

5. Calvin R. Gould, "Legibility Slide Rule," *Proceedings of the STWP Fifteenth International Technical Communications Conference,* May 8-11, 1968, Los Angeles, Calif.

6. Marx Isaacs, "Effective Speech is Effective Communication," *Technical Communications,* Vol. 16, No. 2, Second Quarter 1969.

ANALYZING THE SPECIFIC
AUDIENCE-SITUATION

A successful idea presenter invariably exhibits that characteristic described in terms such as audience awareness, an understanding of human nature, empathy, or in some way denoting that he really understands how to reach his audience. Behind his success is the simple fact that he took the trouble to analyze thoroughly both the specific audience and the situation. To be effective, he must have a clear picture of the actual people to whom he addresses himself and the specific conditions under which he will do so. As has been seen, his analysis has deeper roots, beginning in a foundation awareness of the basic concepts of human behavior, drives, and needs—an insight into so-called human nature. He is not only empathic with his audience's basic human drives and needs, but he is also aware of and establishes a rapport with the specific interests and motives which bring them together as *his* audience. Even further, as a two-way communicator he receives on-the-spot "feedback" from both the individuals composing his audience and the environment of the presentation situation; and he begins by asking certain specific questions.

THE AUDIENCE

When the average technologist is assigned or invited to make a presentation, his first reaction is probably emotional; he feels the first nervous tingle of stage fright. His next thoughts will undoubtedly be concerned with *what* he is going to say. Presumably he has been told to whom he is going to speak. More accurately he has been told the *name* of the audience—not *who* the members really are. Few hosts take the trouble to furnish an audience analysis with their invitations. It is up to the presenter to find out by asking the right questions of the right people and through any other means that he can. But the responsibility is his, and the questions which he can ask are numerous. He might begin by asking what the typical audience member is like:

Who is he?
What does he seem to want and need?
What does he really want and need?
What are his "hot spots"?
What does he like?
What is his self-image?
What is his language?
What is his level of awareness of the subject to be presented?
What is his image of the presenter?
What breaks his preoccupation barrier?
Is he looking for wealth, romance, acceptance, survival?

As varied as the possible questions might be, he can give some order to his specific analysis.

Forming an Image of a Typical Audience Member

Age. Both the average and the span are significant since age affects ability to understand and comprehend. It affects vocabulary use. A younger audience generally is more susceptible to a dynamic, more emotional delivery, whereas older members, having been around, are more wily and objective toward enthusiasm and dynamism and are skeptically looking for the facts (they think).

Education. Formal education and practical experience come under this heading. A group of experienced foremen or technicians, for example, may be better educated than a group of young engineers fresh out of college. There is a difference between being college trained and being educated. The adept presenter therefore considers both schooling and experience in his analysis and in his presentation chooses vocabulary, draws examples, and uses thought patterns that fit his audience.

Sex. Men and women differ in their interests, motives, and responses. Some drives and reactions overlap; yet a good presenter consciously considers the sex of his audience—especially with respect to humor, examples, criticism, condescension, and vocabulary.

Occupation. This can affect the slant of a message. For example, a description of the features of a new product or process to technical marketing representatives might be outlined in terms of selling points or expected sales potential in order to maintain their interest. Yet the same features might be described to a customer as benefits or savings; to the maintenance force as major and minor maintenance factors; or to the organization's executives as accomplishments of the research force, budget factors, or personnel justifications. Occupation is also a good indicator of income level, cultural interests, and vocabulary recognition—in fact, the whole point of view from which the presenter proceeds.

Affiliations and memberships. These can indicate the interests, motives, prejudices, and preconceived ideas of audience members. They can provide valuable clues to comments or expressions of opinion which the presenter should either avoid or include to let the audience know he is aligned with them (if he can honestly do so). An excellent example of this might be found in a presentation demonstrating a certain phase of automation. The reader might ask himself: How would I present the automation of a certain machine operator's job to the following?

A group of the operators
A labor union group
A group of customers
A mixed group, including competitors
The foremen responsible for implementation
A group of newspaper reporters
Organizational management

What is Audience Level of Awareness?

The factors just discussed will give the presenter a good beginning toward finding out how much his audience knows about the subject he is to discuss. Should it be approached from an elementary point of view, or will the discussion of basic concepts bore most of the audience? Which technical terms will have to be given with built-in definitions? And the definitions should be built in; that is, they should never be given in condescending terms. Which information will be completely new to this audience? Which terms and concepts can be assumed as already understood? What if a computer-programmed "numerical controller" were added to the drilling process outlined in Chapter 3? And the audience consisted of computer programmers who had never seen a turret drill in operation; or they knew the meaning of point-to-point and continuous path drilling but had no concept of the fact that a computer can "communicate" with a turret drill?

The significant point here is that the presenter must ask himself this level of awareness question. It is perhaps more meaningful than the other audience questions, important as they are, because level of awareness has a direct bearing upon objective development—the central, pivotal, developmental factor of the whole presentation. As stated in Chapter 3, the presenter's main goal is always to advance his audience from one state to another. In advancing their awareness, he must come as close as he can to the beginning point. If he conducts a thorough analysis of his prospective audience he will have a fairly accurate awareness profile of them. One of the best methods is to plan and aim the presentation at the level of what he considers to be a typical audience member, based on this total analysis.

What is the Audience's Objective?

In addition to the basic personal objectives and drives of audiences in general, the presenter can assume that this specific audience is bound together by certain interests in common for this occasion. Have they come together to work, to buy, to learn, to play, to share knowledge, or to advance a cause? It is surprising how many technical presenters ignore the main aim of the gathering insofar as their individual presentations are concerned—a direct result of narrow preoccupation with subject matter over the weightier matter of audience awareness.

The technical presenter can well take a page from the book of the familiar door-to-door salesman, perhaps one of the best practical audience analysts around. Before he even attempts to get inside the door, let alone make his presentation, he "qualifies" his prospect. He asks a series of carefully phrased questions designed to find out whether or not the need and desire for his product are present, if the prospect can afford it, and if the spouse is at home to co-sign the contract. The technical presenter should qualify his "prospects" in much the same way by asking the basic question:

"Are the audience's and presenter's objectives sufficiently aligned and similar to justify the presentation?"

Audience Attitude Toward the Presenter

The presenter should consciously ask himself "How does this audience feel toward me, my subject, my objective if they know it, and the group I represent?" The attitude of a given audience will encompass a wide range of opinions and feelings on an individual level. The audience as a group will be either interested or uninterested in a subject. They will be either friendly or hostile, sympathetic or apathetic, believing or unbelieving. The analysis process plus the audience's responses during the presentation will contribute much to the presenter's awareness of how the group feels. It is his job to adapt his presentation to the predominant group attitude. Generally analysis will reveal an attitudinal beginning point parallel to the level of awareness beginning point just discussed. In fact, the two usually will be closely linked. The presenter should attempt to reduce to a word or two what he objectively believes his audience's basic attitude to be. Although there are many variations, the following list should serve to set him thinking in the right direction.

apathetic	favorable
hostile	unfavorable
interested	believing
undecided	unbelieving

curious	sympathetic
friendly	condescending
unfriendly	prejudiced

His evaluation should never be so conclusive, so hard and fast, that it cannot allow for error in judgment or a change based on further analysis. The main purpose of a list like this is to point out the importance of the presenter's *awareness* and sensitivity to the audience's fundamental attitude. Thus, one of the questions he must ask is, "What is this audience's beginning *attitude* toward me and my presentation?"

Relationship of Audience to Presenter

Audiences present varying degrees of formidability for a presenter. His peers, for example, affect his disposition one way, his subordinates another, his superiors still another. But whoever they are, when they become his audience, he in effect becomes their leader. He must burn his own emotional bridges behind him and take the stand necessary to accomplish his objective. His own ego can be a detriment if it prevents him from taking an unequivocal stand or causes him to take a condescending approach toward those "beneath" him. He must for the moment free himself from whatever existing relationship he has with this group. He becomes the presenter; they are his audience. The only rules that he follows are those of good presenting.

THE SITUATION

There is an obvious overlap in analyzing the presentation situation as opposed to the *people* in the audience. Nor are there any rigid rules of analysis. The important thing is that as many questions as possible be asked—and answered. Allowing for this overlapping, the following questions outline what a presenter should know about the presentation occasion.

Why is the Presentation Being Given?

Those who decided to include this presentation in a program or series of events presumably have an inherent purpose; more specifically, what do *they* expect of the presenter; what is their objective? A good presenter will ask, taking careful note of what they say. Many times he will find that they themselves are not sure. Nevertheless, before he accepts the assignment or invitation, the person who is going to do the presenting must have a clear understanding with his host of why this presentation is being made. And it *must be articulated*, written down, and approved if at all possible. There are far too many presentations given just because someone thought the subject "ought to be included in the program." In fact, there are far too many programs for the same reason. It goes right back to the matter as stated in the beginning of Chapter 3. If the reason for a technical presentation cannot be *stated* and written down in terms of a desired audience response, there can be no deliberate act.

If the host does not or cannot state the reason for the presentation, the presenter, knowing what he is about and that he is in business to accomplish specific objectives, should do so and "with all due respect" submit it to his host for formal concurrence.

How Large is the Audience?

Audience size dictates many things to the presenter. For example, will a microphone be needed? Does he know microphone technique? Audience size dictates the types of visual aids to be used, the number of handouts to prepare, whether or not a question period can be allowed, and similar matters. Knowing audience size also helps the presenter to set himself mentally for his delivery. Any experienced speaker knows that it takes a different, perhaps more style-conscious, disposition to face a large audience than the intimacy a smaller group permits.

Where Will the Presentation Be Given?

Many a speaker has spent anxious moments just before going on hunting for the address or room number, only to arrive breathless and with a little extra nervousness to add to his regular portion of prepresentation jitters. Experienced presenters never ignore such details; as experienced planners they assemble all pertinent logistic information well in advance. They also arrive early whenever possible—for a number of reasons besides sitting in one of the chairs. If the presentation is out of town, there are problems of transportation and housing, which should be seen to well in advance, with arrival times, connections, and reservations all fully verified. It is true that many of these details are automatically taken care of by committees and secretaries, but a professional *makes sure.*

How Long is the Presentation?

This is a seemingly inconsequential fact but it actually has a major influence on the presentation. Most presenters try to cover far too much material in the time allotted, mainly because their selection of topic is too broad or general. Can the desired objective be accomplished in the allotted time? How much detail must be included to adequately prove the main point of the presentation? The emphasis here is on *restricting* and *narrowing* the presentation objective because rarely can a presentation be too narrow in terms of subject matter. The average presentation will infrequently be more than an hour and a half long.

From the presenter's point of view this will seem extremely short. As he begins to gather information he will find that there almost always will be far more material than he can possibly cover. Therefore he must restrict and narrow his subject to a point where what he does cover will be treated in sufficient depth. This is also part of the reason for choosing a specific objective. The time limitation will bear directly on what that goal is to be, with a single main thought sequence pattern to guide both presenter and audience to its accomplishment. By fitting his objective to the time limit imposed upon him, the presenter thus draws that much closer to reducing the theory of making his presentation a deliberate act to the reality of actually doing so.

What is the Presentation a Part of?

A symposium? A workshop? A sales proposal? Whom does the presenter follow? Who comes after him? What is the main theme of the program? Answering such questions guarantees that the presentation will fit, that it will not represent an embarrassing departure from what other presenters have been doing and saying. If it is to complement, elaborate upon, or contrast with other presentations, it is necessary to know what they will cover. It is often wise to obtain a copy of the program in advance, perhaps to contact the other program participants or even to call a general meeting or rehearsal.

What Time Will the Presentation Be Given?

Both the time of day and place in the total program should be considered. The earlier in the program that he appears, the more latitude the presenter will have in use of information commonly known by or available to the other presenters. The earlier in the day and in the program, the more alert and interested will be the audience. Conversely, lateness in the proceedings will govern delivery technique to maintain attention. It will restrict the new information available, but it can also provide valuable grist for the presenter in terms of reference material to relate and integrate his presentation with the others. Some of his points may be preproved—or refuted before he gets a chance to make them. Consulting the participants who precede him can therefore be advantageous. If he cannot get their material beforehand, the next best thing to do is to listen to them as they speak, making appropriate notes or comments in his own script or outline.

What Facilities are Available?

It is most disconcerting to have notes and no lectern to set them on, slides and no projector, chalkboard notations and no board to write them on, a board but no chalk, easel charts and no easel. These things happen! Again, a true professional takes no chances. Even if he delegates, he checks. Early arrival will afford him the opportunity he needs to do so.

What is the Audience's Authority?

Trainees cannot make purchasing decisions. Normally, laboratory technicians do not decide upon budget matters. Executive committees do. It makes little sense to lead an audience to a decision which it has no power to make. Many a technical sales presentation has been wasted on minor company functionaries, when the decision maker was not even present in the audience. Thus, the experienced presenter fits what he is asking an audience to do to what they *can* do. He establishes an attainable objective—not only in terms of the conditions and limitations within which he must work, but in terms of the people with whom he is dealing. And once again, the matter of objective comes to the fore as the key element in making the presentation a deliberate, finite, accomplishable act.

ANALYZING AUDIENCE FEEDBACK

The audience furnishes a presenter valuable information during the presentation itself in a kind of two-way conversation. In order to take full advantage of this feedback, he must remain objective, emotionally relaxed, and fully aware— while he is concentrating on his presentation as well. The presenter who is emotionally unsure of himself cannot effectively do all of these things and still concentrate on his presentation, its objective, and his delivery technique. There is only one way to be sufficiently relaxed and confident in front of an audience. He must know that he "has the goods," that he fully understands the subject, and where he is going. Out of this basic confidence will come relaxation and naturalness, the freedom to observe the audience, and the ability to *control* the presentation situation. There is no substitute for confidence.

Remain Objective

Once a presenter permits an audience to affect him on a personal level, to anger or intimidate him, or to make him feel that he is being patronized, he loses the ability to analyze them objectively and to control them. If he understands the mechanics of his own attitudinal reactions, he will not permit this to happen. A look of disbelief or disapproval from a member in the front row will be duly registered but never taken to heart. If he asks a rhetorical question and gets an unexpected reaction, he will contain it and continue. Having fully analyzed this audience, his chances of running up against such a situation are reduced to a minimum.

Do Not Overrespond

Our emotions also tend to cause us to prejudge and emotionally evaluate an audience, often in direct contradiction to the facts. For example, in a recent presentation, as is my practice in establishing audience contact, I continually scanned the audience to detect the friendly faces, the hostile ones, the apathetic ones, etc. One man in particular attracted attention. Throughout the presentation he stared directly at me with a look of distinct hostility on his face to the degree that it actually became distracting. I felt certain that here was a skeptical individual who saw through my presentation devices and was in effect looking at my bare soul. Yet, at the end of the lecture this individual came forward with the statement, "I've always had a real problem in conveying my knowledge of my subject to an audience. It seems that no matter how well I know my subject, I just can't seem to establish a proper audience relationship. I was extremely interested in how you did it. Could you give me some additional advice on how to establish audience contact?" Needless to say, I was quite surprised. Actually this individual turned out to be the most interested member of the entire audience. Complete misinterpretation of audience feedback!

What specific communication mistake did this illustrate? In the areas of emotions and feelings, it showed that no matter how practiced one is in making presentations, one must continually strive to counteract the natural human tendency to overrespond emotionally. It also showed that it is never really possible to completely understand people. Yet this in no way cancels out the value of practical, clinical, objective-as-possible *audience awareness.*

Observe Audience Behavior

Shuffling feet, yawns, general restlessness, glances at watches—or rapt attentiveness—all are things which should be consciously noted by the person at the podium. No man has a right to bore people in today's busy world. Yet some presenters ramble on despite the fact that every audience indicator tells them that as far as the audience is concerned the presentation is over. It is far better to call an unscheduled break and regroup forces than it is to continue without audience contact.

Build In Response Points

These may be periodic questions to be mentally or orally answered by the audience. For example,

"How often do *you*, the reader, visualize a typical audience member as you prepare a presentation?"

A response point also may take the form of humor. It may be a pause. It may be, "Are there any questions so far?" Many experienced presenters have developed habitual interjections such as: "Okay?" or "Do you see?" These can easily become offensive to an audience. The main purpose of such response points is to keep the audience thinking with the presenter. They help to ensure attention, as well as carrying the audience along in what might be called a pattern

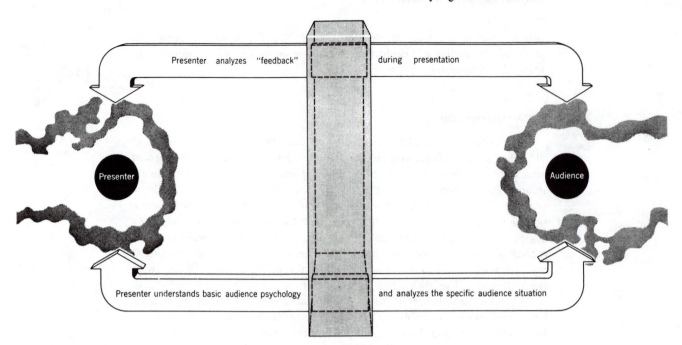

Presenter analyzes "feedback" during presentation

Presenter

Audience

Presenter understands basic audience psychology and analyzes the specific audience situation

Preoccupation barrier

Figure 4-1. As discussed in the text, the *professional* communicator engages in intentional, deliberate, clinical analysis on a level not normally found in everyday communicating situation. He tries to see what is really going on in the interchange between him and his audience. Although no one can do this completely, simply by functioning in this "aware" state the presenter greatly increases his chances of penetrating the preoccupation barrier and communicating with his audience.

of cumulative affirmation. Insofar as the presenter is concerned, questions have the two following important functions:

1. To get information.
2. To lead people in the direction he wishes them to go.

EVALUATION OF OBJECTIVE ACCOMPLISHMENT

This is the closing phase of the presenter's analysis and provides a natural transition into the next chapter, which discusses objective development. It consists of a simple question:

"Did the presentation accomplish its stated objective?"

SUGGESTIONS FOR FURTHER READING AND STUDY

Garn, Roy, *The Magic Power of Emotional Appeal*, Prentice-Hall, Englewood Cliffs, New Jersey, 1960. Read critically, can help the presenter condition himself to asking the right kinds of questions about his audience.

Hall, Edward T., *The Silent Language*, Doubleday, New York, 1959. Our environment, time, and space have "voices" for those who would hear. This book provides valuable insights for the audience analyst.

Hall, Edward T., *The Hidden Dimension*, Doubleday, New York, 1966. For those who would continue their study of the "voices."

Hollingsworth, H. L., *The Psychology of The Audience*, Macmillan, New York, 1935. An older work but still good background for analyzing an audience.

Lang, Kurt, and Gladys E. Lang, *Collective Dynamics*, Crowell, New York, 1961. A text for the serious student of group action.

Linton, Ralph, *The Study of Man*, Appleton-Century, New York, 1936. Today's psychologists harken to this as a foundation study.

Martin, James G., *The Tolerant Personality*, Wayne State University Press, Detroit, 1964. Centers upon a study of prejudice. Background for the serious reader.

Sanford, William P., and Willard H. Yeager, *Effective Business Speech*, McGraw-Hill, New York, 1960, pp. 95-108. A basic text.

Audience Requirements for Technical Speakers

Abstract—Suggests adoption of a practical methodology for translating technical papers to effective technical presentations. Definitions describe the applicable meeting environment. An integrated set of visuals and associated text provide an example of a presentation while developing the methodology. Basic audience requirements (or constraints) which the presentation designer must accommodate are shown to be the very limited capacity of any audience to process and assimilate information, the need for screening the few ideas that can be transmitted, and the necessity for surmounting audience "significance blindness" by incorporating the ideas as thesis titles on every visual.

INTRODUCTION

THE eighteen illustrations in this paper are presentation visuals, scaled as required by layout frame specifications. Used in a presentation, each visual would be read aloud as disclosed, and the purpose of the graphics explained. The significance of the thesis would then be addressed in words similar to those actually used. The 1,500 words of visual-related text require an average of about 40 seconds per visual to speak. Allowing time for the initial description of each visual, this paper provides an example of what could be a 20-minute presentation.

DEFINITIONS

Audience A collection of professionals capable of being motivated to acquire and apply information. The burden of motivation must fall upon the speaker, since the majority of a given audience can be shown to be browsing and sampling presentations. Statistically, two-thirds will know little more than the title and three-fourths will know nothing of the speaker and his work.

Reprinted with permission from a booklet originally designed and written for the use of the American Federation of Information Processing Societies, and subsequently revised and published by the author; copyright 1975 by Robert E. Perry.

The author is the Principal of Perry Communications, specializing in technical marketing communications, 852 W. Las Palmas Drive, Fullerton, CA 92635, (714) 879-6058.

Chairman A senior professional completely cognizant of the contributions and attainments of the members of the technical session or panel he has been designated to guide. The chairman's function is to inform an audience of the speaker's qualifications and the significance and relevance of the topics addressed. As a moderator he unobtrusively makes certain all discourse is succinct and meaningful.

Layout Frame A template or form so proportioned that the mandatory lettering size within the frame insures that the designer cannot exceed the audience constraint known as the "threshold of confusion." It further insures a legibility standard for an image area which provides the brilliance to meet the constraint that an audience should never be in a dimly lit or darkened room.

Panelist One of a small group of professionals especially qualified to discuss a selected topic within the group and with an audience. Because the goal is to stimulate an instructive dialog, each panelist must state a topic-related position in the form of a brief and provocative presentation.

Paper A comprehensive written treatment of new or tutorial information. A paper is intended for publication as a reference and is organized to be of use to a professional peer in advancing or applying an aspect of the art.

Presentation A time-constrained oral/visual development of selected information in which a speaker's spontaneous commentary is cued by specially structured visuals. The objectives are to interest a peer group in a paper or special field of interest, to update fact, and to stimulate discussion.

Speaker A professional especially qualified to make a significant contribution in his field of interest, and strongly motivated to design and deliver a presentation which will move an audience to discuss and apply his work.

Reprinted from *IEEE Trans. Professional Commun.*, vol. PC-21, no. 3, pp. 91–96, Sept. 1978.

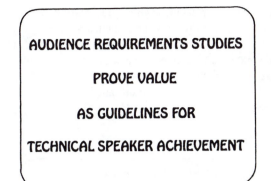

AUDIENCE REQUIREMENTS STUDIES

PROVE VALUE

AS GUIDELINES FOR

TECHNICAL SPEAKER ACHIEVEMENT

You probably will find it necessary to change your perspective of the audience/speaker relationship to achieve excellence in technical presentation. The more enlightened sponsors of professional meetings hold the heretical view that the audience does not exist for you. You are there for the audience, and its wishes and constraints have the force of a command upon you.

The above statement is the thesis of this discussion. It reflects a very simple concept: That is, as a speaker you have a substantial obligation to provide your audience a stimulating and provocative transfer of information. Because the pivotal element *is* the audience, extensive studies have been conducted concerning the constraints and requirements a speaker must meet if he is to be effective. The findings have proven to be of significant value when used as guidelines for presentation design and delivery. Real-world technical speaker achievement testifies to the worth of the approach.

A PRESENTATION IS FOR HUNDREDS...
BRIEFLY SAMPLING THE FARE

AN AUDIENCE POSES SEVERE REQUIREMENTS
FOR TECHNICAL SPEAKER ACCEPTANCE

The first command is that by overwhelming audience preference, you will be making a presentation, not . . . repeat *not* . . . reading a paper. The distinction is that a presentation is for the hundreds who are in your audience to sample the technical fare you (and your paper) have to offer. Your words are temporal and serial. These characteristics alone indicate that presentation fact density necessarily must be low.

The speaker's entire function must be to meet frequently inflexible audience requirements for acquisition of information from a presentation. The emphasis upon the audience is entirely valid. Unless audience acceptance is good, no goals can be met; neither the sponsoring society's, nor the profession's, nor yours.

A TECHNICAL PAPER IS FOR ONE...
WITH TIME FOR DIGESTION

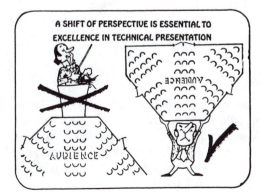

A SHIFT OF PERSPECTIVE IS ESSENTIAL TO
EXCELLENCE IN TECHNICAL PRESENTATION

In contrast to your presentation, your paper is intended for an audience of one with the time and inclination for digesting the data. Your printed words are spatial and referable. For that reason, fact density in a paper can be very high. Paper and presentation complement one another; the virtue of the presentation is that it permits response to audience reaction.

Because your audience can be shown to have a limited rate at which it can absorb information, and because it is there to sample the worth of your material, technical presentations have one primary objective. They are intended to accent and update the provocative features of your work, so that your colleagues will open a dialog with you and read your paper.

The first step in developing a presentation from your paper is to rework the title. A paper's title usually is so phrased as to intrigue only a reference librarian. Couched in the deadly passive voice, it is excellent for Key Words In Context indexing, but provides no hint of the goals, findings, or intrinsic interest of your topic.

On the other hand, a good presentation thesis title will work for you to provoke audience interest. Bear in mind that the first audience requirement is that each of its potential members would like to have some idea of just why he should join his fellows to create and *be* an audience. The term "thesis title" is important and will be explained in a moment.

The best way to determine just why an audience should listen to you is to prepare your final visual immediately after you have developed the visual that carries your presentation title. The last visual is where the proven claims concerning the high-audience-interest features of your work are digested. The "Conclusion," when prepared early, serves to give you a lead on just which ideas should be selected for discussion in the body of your presentation.

Remember to graphically identify your topic and your specific interests early in your presentation. The best time is immediately after your title has been shown. This is an important point, because while your presentation title may attract an audience, it is only your timely revelation of "what" and "why" that will induce them to remain.

When material is transferred directly from a paper to a visual it will fail to meet audience requirements for several reasons: It will be too complex; it will be illegible to most of the viewers; it will carry a label and not a title (or worse, neither). Showing a table of data, for example, and labeling it "Table" falls into the category of the "Horse" chart. It's much as though you had a picture of a horse, and helpfully labeled it "Horse." Such primer-type labels offer little audience aid for understanding even the best of presentation visuals.

Having a title, a description of a topic, and a conclusion, it remains only to structure pertinent ideas to connect presentation thesis and conclusion. The optimum rate of presentation for these ideas can be shown to be one per minute. If you target for the usual 20 minutes you will therefore select a maximum of some 17 salient ideas, and convert them to thesis titles for 17 visuals.

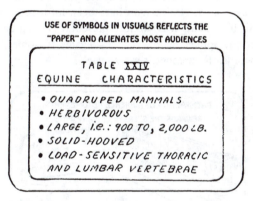

It must be emphasized that your audience requires a thesis title for your presentation and for each of your visuals. The human mechanism assimilates graphics and pictorials far, far faster than alphanumerics. Use of symbols and phrases in visuals reflects written paper usage and, because it is inappropriate to the presentation medium, alienates most audiences. For that reason, tables and illustrations drawn from your paper *must* be reworked for use in your presentation.

Generally, a thesis title takes the form of a simple declarative sentence. A picture of a swaybacked horse would be titled, not "Horse," but "A Swaybacked Horse Has Impaired Road Clearance." Unlike a label, a thesis title makes a statement, proved by the illustrative data. Audience and speaker both benefit.

Design of visuals may be approached from two directions: thesis to graphics or graphics to thesis. In the latter case, thesis invention can be a severe test. When the question is asked, "What pertinent thesis could these data prove?" the answer often is "None!"

By definition, a thesis is a statement which must be proved. You can, using your paper, marshall a formidable array of facts to argue your case. *Don't!* Select a maximum of six. The audience short memory saturates at six facts per visual. More will confuse them and waste your message. Worse, a trivial fact may replace an important one in the long memory.

You may, however, hold a given thesis title for several successive visuals, and develop it six facts at a time. In terms of graphics, six facts translate to no more than two or three curves on a graph (the designations of the coordinates rate as "facts") or six blocks in a diagram.

Many presentation designers find it difficult to meet the audience requirements for simplification. However, 200 years ago Vauvanargues said, "When thought is too weak to be simply expressed, it's clear proof that it should be rejected."

The audience requirement that presentation information be developed in limited volume, at a limited rate, is met almost automatically through use of the "layout frame." The size scale the frames require can only be met if the facts supporting a given thesis title are appropriately limited.

Use of the layout frame also assures legibility of your data for your back-row VIPs. Of course they're all VIPs, but in many presentations the illegibility of the visuals suggests that the speaker believes all the audience beyond the fourth row to be of no consequence.

Designing to the layout frame scale also results in a slide which can be projected to give a relatively small, and thus a very bright image. This meets another audience requirement. People have a pronounced distaste for being plunged into total darkness at irritating intervals. *Your* meeting room need only be dimmed.

The layout frame further insures audiences against poorly structured presentations. When the completed draft frames are pinned to a wall in sequence, or "storyboarded," it becomes very easy to check your continuity and coverage.

Perhaps the most important application of the storyboard is its use before a critical audience to determine the persuasive impact of your presentation. This exercise serves as a time check as well. Present the information as you normally would, using the draft frames to cue yourself as you would if you were using the completed visuals.

Your audience requires that you read or describe all the material on each visual as it is disclosed. You then expand upon key points. Unless you have been informed otherwise, you will be making a 20-minute presentation. At an average rate for the spoken word, that aggregates to some 2,000 to 2,400 words. Remember: time yourself!

Satisfaction of audience requirements, you will find, rewards the speaker in many ways. You'll not actually receive any trophies from the audience, but the value of the awards is nonetheless significant and substantive. Aside from the possible material rewards, there will certainly be one return of significant value. That will be the knowledge that you, as a professional, accepted a demanding and difficult assignment, that you prepared for it, and that you performed it well.

Of course, you still have a few steps between a successful storyboard dry run and post-conference relaxation. Those steps relate to production of slides (a matter of mechanics) and platform performance (a matter of rehearsal). The emphasis in this discussion has been upon design, because given a well-designed presentation, the medium and the message will win audience acceptance for all but the most singularly inept of speakers.

When the completed draft frames are pinned to a wall in sequence, or "storyboarded," it becomes very easy to check your continuity and coverage.

A Quick and Easy Strategy for Organizing a Speech

RICHARD A. LINDEBORG

Abstract—Organizing thoughts—arranging ideas in an effective or logical order and stressing them appropriately according to their importance—is an important part of writing an effective speech on a technical subject. However, it takes so much time and effort to organize material manually and revise it until it is clear that many writers give up before producing a clearly written speech. Writing in a modified outline form on the word processor helps solve this problem. The outline form, coupled with the simplicity and speed of making changes on the word processor, makes it easy to respond as you write to the visual feedback of your writing, almost forcing you to express ideas in a well organized, simple, and clear manner. For many writers, the speed and ease of the word processor is the only thing that makes repeated revision possible. On the word processor, you can experiment repeatedly with the visual display of the text and keep revising until the display—and the text—reflects the best organization of the material and the natural cadences of the spoken word. The result is a speech that guarantees that the listener will understand the message and stay awake through the speech.

THE development of a speech can be like the development of the land masses of the earth. Theory says the land masses came from a single, primeval continent and then drifted apart. Similarly, the idea behind a speech may be unified, with a fairly simple shape. As the speech develops, this grand, unified idea fragments, with the pieces slowly drifting apart. Finally, the speech resembles the geography of the earth: ideas are scattered like islands around the surface, but four-fifths of it is all wet.

The speech may include unbelievably massive Eurasian ideas that you could never hope to traverse in the lifetime of a single speech; huge ideas, resembling the American continents, held together by narrow Panamanian isthmuses of ideas; African ideas—full of potential, but politically unstable in places and largely undeveloped; Australian ideas— beautiful but embellished with flora and fauna found nowhere else; and Antarctic ideas—easy to define, but impossible to warm up to. You need an atlas just to find your way around a speech you thought was going to express an idea directly.

CURING BAD ORGANIZATION

If you have experience writing anything, let alone speeches on technical subjects, you already know that good writing is hard work. Good writing requires good organization and much rewriting. You need knowledge of how to organize the speech, and you need to have the time and energy to organize, reorganize, write, and rewrite. Textbooks, journals, and teachers provide the knowledge of rhetoric required to

Richard A. Lindeborg is with the Forest Service, U.S. Department of Agriculture.
IEEE Log Number 9037664.

produce a well-organized, clearly written speech. The word processor greatly reduces the time and energy required.

Simplifying and organizing used to take stacks of index cards, a good outline, and—if you could spare the time—many drafts. But now, with the aid of a word processor, you have the ability to change the outline as you go along, making repeated revisions to the text and the way it is displayed on the page. Revisions that were once too much trouble to contemplate can be done quickly. This helps you organize the material and keep it simple, resulting in a speech that is easier for the speaker to deliver with conviction and easier for the audience to understand.

THE QUICK AND EASY WAY

One writing technique that works well on a word processor is typing the speech in a modified outline form, breaking down every thought into manageable phrases, and determining where in the outline the idea belongs. This controls the complexity of every phrase and its position in the speech. Fig. 1 shows a passage from a speech on research and development (R&D) developed using this technique.

Place the most general or most important phrase or sentence first within each related group of phrases or sentences. This method of organizing will produce a speech in what is called the deductive style of writing, where the general idea is stated first, followed by more specific ideas. This style of writing is most common in journalism, where it is important to get the main ideas to the reader or listener quickly. The deductive style works well in speeches on technical subjects because it allows the speaker to make his or her point quickly. Hearing the main point up front allows the listener to place the supporting evidence in perspective and evaluate the supporting ideas as the speech progresses.

As you write in outline form, place each phrase on its own line or two of type, separated from other phrases and indented to show its relationship to the phrases preceding it and following it. Despite breaking sentences into phrase lines and indenting the phrases different amounts, you can still punctuate most of the material as you would in traditional manuscript typing. Standard punctuation makes the outline easier to follow. It also helps produce a traditional written text of the speech if you want to distribute the speech to the press, the audience, or others.

Experimenting with ways of arranging material in phrase lines and in outline form would be cumbersome without a word processor, but it could be done. The amazing thing is not that a speech can be done in outline form, but that this

Reprinted from *IEEE Trans. Professional Commun.*, vol. 33, no. 3, pp. 133–136, Sept. 1990.

The United States is not among the leaders in the percentage of Gross National Product devoted to R & D.

Last month, Business Week listed the figures as 1.8 percent of GNP for the United States,

versus 2.6 percent of GNP for West Germany and 2.8 percent of GNP for Japan

That is for nonmilitary R & D.

These two strongly competitive nations are outspending us by 50 percent in R & D.

Research is too important for us to allow it to suffer this neglect.

Fig. 1. Sample page of the outline form used for speeches. Three levels of indentation are used. The outline shown here is printed on a laser printer, using desktop publishing software. The left margin is wide to allow space for revisions and the speaker's delivery notes.

particular method of outlining allows the physical act of writing (the act of keyboarding) to dictate the writing style of the speech. On the screen, ideas that are not expressed in short, simple clauses and phrases grow to multi-line entries that are obviously not in character with the other material on the display screen. Splitting complex ideas into simpler, shorter parts is easy and quick on the word processor. Other ideas may nest deeper and deeper in the outline, indented farther and farther to show their subordination to main ideas. On the screen, this symptom is an obvious signal that something is terribly wrong with the organization of that part of the speech. The word processor makes it easy to add, delete, and rearrange points until the problem is corrected. Properly organized material fits the outline. Nothing else

does. Without the word processor, you might not notice these problems. Even if you did notice them, there is a good chance you would shrug your shoulders and move on, rather than take the time to split up ideas onto separate index cards or rewrite a handwritten outline.

After you have written the speech in this outline style, you can print it out using a large, easy-to-read typeface and clear formatting. Fig. 1 is an example of output printed on a laser printer using desktop publishing software. This sample is printed in 12-point type (boldface except for minor points) on the upper one-half to two-thirds of the page. This point size is large enough for most speakers to read without stooping or squinting at the text. A 3-inch margin on the left keeps the number of words on a line within reason and allows room for

R & D in the world

R & D in the United States

R & D in this industry

R & D in this company

Progress in conducting an R & D project

Steps needed to gather this data

This week's work assignments

Fig. 2. Seven levels of ideas for a speech on R&D. The final speech should probably use only three levels of ideas. Which three levels to use depends on who the speaker is and who will be in the audience. The points generated for possible inclusion in the speech should be sorted. Those that seem to fit within the three levels selected for the speech should be arranged in outline form. The other ideas should be set aside in case the structure of the speech changes.

changes or for the speaker to add delivery notes. Restricting the text to the upper half of the paper allows the speaker to read the text without lowering his or her head. This is good for two reasons: it helps the speaker maintain eye contact with the audience; and it allows unrestricted flow of air past the vocal chords. A speaker with a lowered head will mumble into the lectern.

GETTING STARTED

No one should give a speech without having an objective. Before replying to an invitation, develop a list of reasons for accepting or declining the invitation and a list of points that might be covered in a speech to the particular audience involved. If these lists reveal a good reason for speaking, accept the invitation. Otherwise, refuse it.

Once the engagement has been accepted, the speaker, the speech writer, and one or two technical experts can meet to discuss ideas for the speech. The important thing is to have several minds involved in generating ideas. If you are writing the speech for yourself, you may also be the technical expert. In this situation, you should meet with a couple of your peers. The participants should start this meeting by discussing the characteristics of the audience, their interests, motivations, preconceptions, and biases relative to the topic. They should discuss the occasion, the strengths and weaknesses of the speaker, and strategies to tie the listeners' purposes and priorities to those of the speaker. They should continue developing objectives for the speech, start on a list of major topics to cover, think of specific points to make, and start a list of topics that need to be researched in order to write the speech. Properly arranged relative to your strategies, all of these points become the outline for the speech.

ARRANGING IDEAS

Whether you jot down ideas at home or meet with a group to toss out ideas, you may get confused about the relationship between the general ideas. Some ideas are obviously global in

scope; others seem like trivial details; most are somewhere in between. Depending on the stature of the speaker and the nature of the audience, some global ideas may be too general to belong in the speech. Similarly, many of the details may be too specific to be of interest in the speech.

Fig. 2 shows several levels of points for the material considered for inclusion in the speech on research and development. The president of a trade association, speaking at a national convention, might use world research and development as the main theme for a speech, U.S. research and development in several industries as the main points, and U.S. research and development by several companies as specific examples. The head of research for a single company, speaking to the board of directors, might use U.S. research and development in that one industry as the main theme, research and development by several companies in that industry as the main points, and individual research projects in those companies as specific examples. A leader of a research team within a company, speaking to other team leaders, would speak in even narrower terms.

In general, one of the first things you as a speech writer need to do with a list of ideas is to come to grips with how to organize the individual ideas into tiers of points, each tier subordinate to the one above. You should try to include no more than three levels of ideas in a single speech. Your first step is to determine which levels are proper for the speech you are writing. Temporarily move items that seem too global or too trivial to the end of your list, and move the remaining ideas around until they seem to form a logical structure for the speech. This kind of shuffling can be done with index cards, but it is easier and more efficient on a word processor.

Use the main topic in the title of the speech, the introduction, and the conclusion. Use each of the main points as the heading for a section of the speech, and present each as the first group of thoughts in its section. Finally, use the specific examples within each section, following each with as much explanation as is appropriate for the audience and the occasion. This strategy puts the overall speech in the deductive style—starting with the most general ideas and moving to the most specific. This style is well suited for talks on technical subjects.

As you try to arrange ideas this way, it may become obvious that you have some related details with no main point to put them under or a main point with no details under it. These holes in your outline will enable you to come up with the missing items or, in some cases, show you that your ideas are leading you to a totally different structure for the speech than you originally envisioned. One explanation for this kind of "drift" is that you are straying from your topic. In this case, you can file the irrelevant portions of the outline for use in some future speech on that topic. Another explanation is that an unexpected, but entirely acceptable, approach to the topic has forced itself on you. The new approach may require you to reassess the value of all your ideas. You will be glad that you saved all the ideas, rather than throwing the ones that at first seemed to be outside the scope of the speech. This kind of shift in emphasis is fairly common in the early stages of organizing a speech.

FILLING IN THE BLANKS

Once you seem to have the overall structure of the outline in place, you can work on adding supporting points, working in illustrations, and writing an introduction and conclusion. At the same time, you can use the outline as a template for the writing style of the speech. You have to put any new piece of material someplace in the framework that already exists. Like a paleontologist assembling a dinosaur, you have a sack of bones and a mental picture of what the beast should look like. If a bone doesn't fit, perhaps it belongs to another dinosaur.

As the outline fills in, it becomes easy to see what ideas are properly placed, what ideas are in the wrong place, and what ideas have no place. Indent subordinate ideas under the main ideas to making relationships clear. Use short phrases on separate lines to keep the ideas clear and make them easy for the speaker to deliver.

Ideally, you should be able to begin each idea on its own page. If you use 12-point type with an extra half line of space between lines and one and a half spaces between phrases, you can fit a bit more than 20 lines of writing on a full page, or 13–15 lines on two-thirds of a page. A page usually provides enough room for a phrase or sentence summarizing the idea that will follow on that page, plus room for phrases or sentences supporting that topic. When you organize the page this way, the entire page is in the deductive style, moving from the most general phrase to the most specific.

As you add a point to the speech, you can actually see, as you type the words into the word processor, whether or not they are phrased properly. Bureaucratic expressions, overly technical language, sentences that are too long and complex—all become painfully obvious as soon as you type them. They just get too long. As the machine wraps a sentence to the third line, you know something is wrong. The structure is getting too complex.

When you are faced with an idea that has become a long, complicated monstrosity, first try to break it into parts. You can break it into equal parts, indented the same amount but separated by a blank line, or you can break it into unequal parts, some indented more than others. Most of these long sentences are fairly easy to break down; identifying them is hard when you type in paragraph form, but easy when you type in outline form. Other long sentences seem to defy cleaving into parts. Generally, these are the bureaucratic boulders that have no place in the pea gravel of your speech. You can find ways to crush the essential boulders so they can be used, but you should load up most of these bureaucratic boulders and ship them off to other speech writers, who might still have use for them.

REVIEW AND REVISION

After you have produced the basic speech, let the speaker review it and suggest changes. If you are the speaker, then you will have to review it yourself or let a peer review it. You then work the changes into a revised version of the speech. This process of review and revision can be repeated as needed. You should give the technical people who have been doing the research for the speech an opportunity to review it, as well as anyone who needs to review policy issues. The word processor makes it easy to work new material into the speech between reviews.

If you are writing for someone else, do not be disappointed if the speaker makes changes. The final speech should be the speaker's creation; the speech writer and the technical staff should support, not override, the speaker. This relationship between the speaker and those producing the speech is similar to the relationship between a research scientist and the scientific writers and editors who transform research findings into readable reports. The speech writer, like the scientific writer or editor, helps put the ideas in a simple, well organized form; but the ideas—the essential part of the speech or article—belong to the speaker, just as the ideas in an article belong to the scientist.

SUMMARY

If you go astray in crafting a speech, you can wind up like Mary Shelley's famous medical scientist. You finish grafting together all of the parts that are handed to you and wind up with something that is full of ideas but virtually uncommunicative. You are stuck with a monster whose heart is in the right place, but whose behavior is socially unacceptable. And as Dr. Frankenstein found out, it takes more than good intentions to make the public accept your point of view.

Outlining and revision are the only ways to avoid the problems of poorly organized speeches—but methods of organizing and writing take so long that many people quit before the job is one. Working in a modified outline form on a word processor makes the task fast enough and easy enough that you can actually take the time to do it.

When you take the time to organize and rewrite, you can produce speeches that use direct language; present major ideas clearly and concisely; follow a clear outline for combining major ideas with background, evidence, or example; and avoid jargon, technical language, and bureaucratic double-talk.

The word processor gives you a quick and easy way to organize your thoughts and revise your speeches. Organizing and rewriting are essential. The speed and ease of writing on the word processor make it practical to keep revising until the structure is logical and the writing clear enough to keep "continental drift" from fragmenting the grand, unified idea behind your speech.

A Problem-Solving Approach to Preparing Professional Presentations

ERIC W. SKOPEC

Abstract—This paper describes a systematic approach to preparing presentations that uses a four-step problem-solving method familiar to most professionals. Step 1, understanding the problem, involves analyzing the speaker's purpose, the character of the audience, and the demands of the situation. Step 2, devising a plan, consists of selecting a strategy to integrate information from the first step with the speaker's knowledge of the subject. Step 3, carrying out the plan, uses conventional techniques of organization, explanation, and visual representation to compose a draft of the presentation. Step 4, examining the solution, compares the draft with the plan and modifies the presentation in response to feedback.

ROGER WILCOX of the General Motors Institute tells the story of a young engineer making his first oral report [1, p. 7]. As the story unfolds, readers witness the engineer expending enormous effort completing the assigned project and preparing a written report. However, he fails to prepare an oral presentation and the results are disastrous. The moral of the story is that preparation is the key to making successful oral presentations.

The story is a good one, but I would add that not all forms of preparation are effective. The last-minute frenzy that often precedes an oral presentation may leave the speaker so shaken that the results are almost as bad as those in Wilcox's example. In fact, if I were rewriting Wilcox's text, I would balance his example with one from my consulting practice. I spent the better part of one morning working with an executive preparing a presentation in support of his operating budget for the coming year.

Our ostensible purpose was to polish his delivery but in three hours we were interrupted eight times: three times by subordinates submitting additional view graphs, twice by his superior asking if he was finished yet, twice by subordinates reporting that they couldn't find materials he had asked for, and once by his secretary commenting that their copy machine was out of order so distribution copies would be delayed until another machine could be located. At the close of our session, I was nearly run over by another subordinate rushing in to report that he had "found another book." My example is as extreme as Wilcox's—and all this activity took place less than 24 hours before the presentation was to be delivered.

Neither the young engineer from the first story nor the engineering manager from the second is a typical professional. However, I know that many people use hasty, careless, or unsystematic means to prepare oral presentations, procedures that would be unacceptable in any other aspect of their work. There are many reasons for this condition and I don't believe that the speakers are wholly responsible. The important point is that such haphazard procedures don't allow speakers to make the best possible use of their unique abilities and knowledge.

In contrast, a systematic procedure would improve the use of composition time by presenting tasks in an ordered sequence. Unfortunately, the readily available literature says very little about the speech composition process. There are bits and pieces scattered throughout various professional literature but there are few plausible syntheses to guide professionals in effective use of composition time.

It is easy to learn a procedure that uses steps with which you are already somewhat familiar. Most of the professionals with whom I work are extremely skilled at solving problems and I assume that you are too. The problem-solving approach capitalizes on your abilities by characterizing preparation of oral presentations as problems to be solved through the use of appropriate materials. As I was developing this approach, one of my clients—a senior project engineer with 35 years of experience—introduced me to a fascinating volume that many of you probably know already: G. Polya's *How To Solve It*. Polya reduces problem-solving activities to the following four phases [2, pp. 5–6]:

> First, [he says] we have to *understand* the problem; we have to see clearly what is required. Second, we have to see how the various items are connected, how the unknown is linked to the data, in order to obtain the idea of the solution, to make a *plan*. Third, we *carry out* our plan. Fourth, we *look back* at the completed solution, we review and discuss it.

The purpose of my presentation is to introduce you to a problem-solving approach to preparing professional presentations. I describe the activities that accompany each of the four phases defined by Polya.

UNDERSTANDING THE PROBLEM

Understanding the problem is the first step. Presentation problems are defined by three considerations: your purpose in speaking, the character of your audience, and the demands of the situation.

Reprinted with minor changes from the *Conference Record* of the IEEE Professional Communication Society Conference held in Boston, MA, October 13–15, 1982; Cat. 82CH1830-9, pp. 68–74, IEEE Service Center, 445 Hoes Lane, Piscataway, NJ 08854; copyright 1982 by the Institute of Electrical and Electronics Engineers, New York.

The author is a consultant and associate professor, Speech Communication Dept., Syracuse Univ., Syracuse, NY 13210, (315) 423-2308.

Reprinted from *IEEE Trans. Professional Commun.*, vol. PC-26, no. 1, pp. 30–35, March 1983.

Purpose

You should always begin by identifying your purpose because it is the dominant concern. Without a specific purpose, there is no reason to make a presentation. Whenever you are unsure of your purpose, ask yourself what you want the audience to do as a result of hearing your presentation. Although the question invites many specific answers [3, 4], most fall into one of four categories.

The first category consists of fulfilling an assignment given by someone else. Your purpose is completing the task and you should be satisfied when the audience recognizes that you have done a good job in completing the assignment as it was given to you. This is an interesting situation because displaying yourself is as important as presenting the subject. You may have little to gain but much to lose because a well-done presentation may produce few rewards whereas even minor flaws are noticed.

The second category involves explaining a concept and you should be satisfied only when your audience comes to understand a novel concept, idea, or process. Purposes of this kind generally call for tutorial presentations or lectures.

The third purpose is generating interest in a project or report. To appreciate the significance of this special function, remember that information overload is an increasing problem in most organizations and that many executives read less than half of the reports routinely routed to them. This is more than an individual problem and we know that much time and effort are wasted when executives are not aware of activities in all sections of their operation.

Finally, your purpose may be generating support for a proposal. Proposals are central to many professional activities and you should be satisfied only when the audience acts favorably on your request.

These purposes range from displaying your own abilities by fulfilling an assignment to generating support for a proposal to which you are personally committed. Things are seldom as simple as this characterization suggests and most presentations involve multiple purposes. However, selecting a dominant purpose facilitates preparation and makes it possible to marshal your resources for the best possible chance of achieving favorable results.

Audience

The second task in defining the problem is determining the audience to which you will speak. The term "audience" is generally taken to mean everyone present but there are frequent occasions in which a more specific focus is instructive. Of the people present, who is in a position to affect the success or failure of your presentation? Occasionally, everyone present is important, but more often than not a relatively small group of decision makers is the critical portion of the audience [4–6].

Once you have identified the audience of concern, you should consider their expectations about the subject. Audience expectations and frame of reference are frequently used to refer to the specific objectives of an audience in listening to a presentation but I use them in a more general sense: Expectations refers to the whole body of experience and knowledge that governs an audience's ability to process information. Attention to such expectations is often identified as a means of persuading. Of course, that is one of the more important functions but it overlooks a fundamental consideration. Only by knowing the background an audience brings to a presentation can a speaker anticipate what portions of a message will be received, how they will be ordered in the minds of the audience, and how they will be interpreted [7, 8, 9].

Studies of perception tell us a great deal about audience behavior:

- Receivers consciously process fewer than one cue in ten thousand and large portions of a message are inferred from what the receiver thinks will be discussed [10, p. 57].

- The average audience member remembers less than half of the material presented [11, 12].

- Only one recipient in four can summarize a message with reasonable accuracy [13].

- Even simple directions will be ignored if they are not identified as vital elements [14].

- Even long-standing relationships include substantial areas of misunderstanding because participants have not established common expectations [15].

- Recall of even simple geometric shapes is influenced by the names used to describe them [12].

Situation

The demands of the situation are the final factor in defining the presentation problem. They usually limit what you can hope to accomplish. For example, my time on this program is limited to ten minutes and I know that I can make few changes in your established behavior. However, I do believe I can generate sufficient interest in my presentation to encourage you to read the full paper.

DEVISING A PLAN

The second step in preparing a presentation is devising a plan. This is an area in which exceedingly bright people often excel because they are able to look beyond the immediate details to visualize a completed product. Many work in an intuitive fashion that the rest of us may never be able to comprehend, but it appears that their thought processes involve identifying models or analogues that can be adapted to the needs of a particular situation. We can use a related procedure by examining several models and choosing the one that most nearly fits our needs. Although there are numerous gradients and alternatives, most presentations can use one of four strategies: ritual, tutorial, report, and argumentation.

Ritual

The strategy I call ritual originates from members of every organization having developed standard ways of doing things. These approaches constitute the culture of the organization [16, 17] and members describe these approaches whenever they tell an outsider ''this is how it is done here.'' Recent research has demonstrated that culture plays a vital role in maintaining organizations and that any individual's contributions to an organization are often judged by the extent to which he enacts appropriate rituals.

Although the term ''ritual'' has a slightly pejorative connotation, there are times when ritual presentations are most appropriate. These cases arise when you are required to make presentations on behalf of your supervisor, greetings and tours for vistors, and orientations for new employees. Other occasions for ritual presentations include cases in which audience expectations create a predetermined pattern from which you dare not depart. For example, financial and project status reports often follow such carefully prescribed lines that any effort to do something different would be wholly inappropriate.

Ritual patterns depend on the specific organizations that sanction them and it is virtually impossible to describe them in a generic way. However, the stakes may be high enough to warrant an effort to identify the ritual patterns of your own organizations. You have several sources of information. Examine models or records of past presentations, suggestions and directives from your superior, and corporate folklore about how such things are done.

Tutorial

Tutorials are used whenever your purpose calls for explaining novel concepts to your audience. The essential element of a tutorial is a sequential presentation beginning with familiar ideas and building to the novel or unique ideas in easy steps. The basic principle is to use the known to explain the unknown and there are several common sequences. The order in which you learned or developed the concept is often acceptable but other sequences may be more instructive for your audience.

Sequences include whole-to-part, general-to-specific, and popularized-to-specialized [18, 19]. Your selection should be based on your knowledge of the subject and the audience. The sequence and pace at which you introduce new material determine the success or failure of a tutorial presentation.

A flow diagram showing the ideas and the relationships between them is a tremendous aid in planning a tutorial. Begin by specifying the concept to be explained and work backwards to concepts with which the audience is familiar [18].

Report

Reports are commonly used when an oral presentation accompanies submittal of a written record. Examples include progress reports, project summaries, and convention presentations. In all of these cases the overriding concern is the relationship between the oral report and the written record [4]. Of course, the reporter can simply begin reading at page one and go on until someone stops him, but doing so makes poor use of the speaking opportunity and may alienate an audience that can read the report on its own. Moreover, reading the written record fails to capitalize on the unique functions of oral and written reports.

Written reports are more or less permanent records of your work and their purpose is to document your results. In contrast, oral reports are transient descriptions of your findings and their function is to generate audience interest in your work. To achieve this end, your oral report should focus on the most distinctive, novel, or significant feature of your work. Anyone who wants the full story can go to the written report for details and that is exactly what you hope will happen.

Notice that this strategy differs from the common practice of making the oral report an abstract of the written report. An abstract is akin to reading the big heads in an outline and saves time by omitting details. In contrast, an oral report includes detailed information of particular interest to the audience and saves time by omitting features that are commonplace but which must be documented in the written report.

Argumentation

I call the final strategy argumentation—not because the speaker picks a fight with the audience but because a carefully stated and documented argument is often the best way to get audience support for a proposal. This strategy calls for a clear statement of what you want an audience to do and reasons for doing it. The one difficulty is that reasons you find attractive may not convince an organizational decision maker. Very simply, decision makers see things in different frames of reference from those of us ''in the trenches.'' The solution to this difficulty is to describe your proposal in terms that interest the decision maker.

Finding the point at which a description is adequate may be difficult but two guides are appropriate. First, an audience doesn't need to see how something works to appreciate its significance. A friend with 30 years' experience as a public relations specialist summarizes this principle by saying that ''You can tell someone what time it is without telling him how your watch works.''

Second, we know that decision makers are interested in anything that has a direct bearing on the ability of their organization to meet its goals [4, 20, 21]. The central concerns of any decision maker include dominant goals, maintenance goals, and delegated goals. Dominant goals reflect the essential nature of the organization and often center on profit or return on investment; maintenance goals involve conditions necessary for an organization to survive—and personnel and facilities are central elements; delegated goals are specific functions assigned to a unit within an organization and they vary with the type of organization and the place of a unit within the organization.

An argumentative presentation capitalizes on these insights by stating proposals in terms that decision makers understand and by demonstrating that acceptance of the proposal will enhance the organization's ability to meet its dominant, maintenance, and delegated goals.

CARRYING OUT THE PLAN

The third step in preparing a presentation is carrying out the plan. Here we are on familiar ground because the relevant techniques are well known and because this is the point at which most people begin. Since this is the point at which most people begin, you may ask what we have accomplished so far. The answer is quite a bit. If you've been working through a presentation with me, you've decided what you want to accomplish by speaking, you've identified features of the audience and situation that may affect your presentation, and you've selected a strategy to use in making the presentation. With this planning completed, composition is a relatively simple task and you can concentrate on what you know best: the particular subject about which you will speak.

The only potential difficulty is that you may lose sight of your plan but I think you can direct your composition in a manner to minimize that danger. At this stage in the process, an outline is your best tool because it helps you see what materials you need and how they will fit together. This doesn't mean that you must work from an outline in presenting the speech, but it does mean that you can govern composition more effectively by setting up the major components of your presentation in a skeletal form.

As you know, a well-structured presentation has an introduction, a body, and a conclusion. Each of these units has a distinct function, and awareness of these functions helps you decide how to allocate the materials you plan to present. The introduction should prepare the audience for the substance of your presentation. This is an appropriate place for materials that indicate the importance of your subject and for historical or other background material that will help the audience appreciate the significance of information to come. In brief, most of the things that you think should be mentioned but that are not directly relevant to the purpose you selected can be included in the introduction. And the final part of the introduction is a preview of the topics you will discuss in the body of your presentation.

The body of a presentation is a sequential development of the major topics called for by your plan. You can help the audience recall your material by focusing on three or four major statements around which other information is organized. These statements should be the essential elements of your presentation—the things you want the audience to remember even if they forget everything else.

The conclusion summarizes the major ideas and closes by emphasizing whatever material is most likely to move the audience in the direction called for by your purpose. Notice that with the conclusion you have stated your major ideas for a third time. The value of the maxim, "tell 'em what you're going to tell 'em, tell 'em, and tell 'em what you've told 'em," is supported by countless studies and a good deal of common sense.

After you've created the skeleton of your presentation, it is time to add meat to the bones. The flesh of a presentation occupies much time in basic speaking courses but a review of your [engineering communication] literature tells me that you already know the techniques pretty well [22, 23]. You know how to use definitions, examples, comparisons, and contrasts to make your ideas clear; you know how to use statistics, quotations, and empirical studies to generate support for your proposals; and you know how to use visual representations to make your ideas memorable and to clarify complex relationships.

Knowledge of your subject furnishes far more material than you could hope to present in most situations, and knowledge of your audience helps you choose items that will be effective. You can even cover your bets by composing several outlines reflecting different assumptions about the audience. No matter how well you think you know the audience, there is always room for surprises and it is wise to be prepared. There isn't room here to describe the technique, but you might want to familiarize yourself with the branching outlines described by Olbricht [18].

The next task in carrying out your plan is preparing the text from which you will speak. Your options include key-word outlines, full-sentence outlines, and complete manuscripts. Your literature includes several lively debates about the merits of reading and other forms of presentations and I will add my own prejudices to the fray. It seems to me that both reading and extemporaneous modes of presentation have unique advantages and skilled speakers should be flexible enough to choose a form appropriate to the situation. Extemporaneous modes emphasize spontaneity and give you maximum opportunity to respond to the concerns and interests of the audience. Manuscripts emphasize thorough preparation and provide assurance of considerable accuracy of statement. This is a trade-off that you always face and your own skills and preferences are essential in choosing the form you use.

EXAMINING THE SOLUTION

The final step in the process is checking your results. This is really the time for fine tuning your presentation and should involve more than practicing delivery. Practicing delivery is important, and almost everyone could do more, but several things have higher priority.

Many people get so involved in composition that they lose sight of their purpose, the audience, the situation, and their strategy. Sit back with your completed draft and compare it with the decisions you made in defining the problem. Is the presentation consistent with your purpose? Does it speak to the audience you've identified? Is it consistent with the situation? Will it fit in the time allotted?

Next, examine the fit between your presentation and the strategy you've selected. Does the presentation address all of the important issues? Does it use an appropriate sequence? Are the major ideas adequately explained? Do the materials support the essential concepts? Are sufficient and appropriate visuals planned? Are the phrasing and "stance" consistent with your strategy?

This is also the time to call in outside advice. The outsider can be a subordinate (but be sure to ask one with the confidence to disagree with you), a friend or professional associate, your boss, or even an independent consultant. Several of my clients believe the advice of an outside consultant is essential at this point and I do my best to confirm their belief. However, most people can check their own drafts if they take the time to work systematically.

At the risk of sounding pedantic, I have seldom seen a presentation that could not be improved by more thorough checking. All good things must come to an end, however, and checking must give way to practice. The amount of practice required depends on the complexity of the subject, the demands of the situation, and your ability as a speaker. A few items can be altered after practice begins but you should be free to concentrate on delivery. Nothing generates anxiety more than a cluster of last minute revisions. I believe you can make better presentations by using a well-practiced delivery than with a manuscript revised and edited right down to the final moment.

CONCLUSION

In this essay, I've described a systematic approach to preparing professional presentations. The approach capitalizes on your problem-solving abilities and follows the phases described by Polya, including definition of the problem, selecting a strategy, carrying out the plan, and checking the results. The greatest advantage of this procedure is that it makes it possible to produce respectable presentations even in adverse conditions. I'm sure you all know the feeling when inspiration isn't there and when time is limited. In such a situation, the problem-solving approach gives you a handle on preparing the presentation by making it easy to find a starting point.

Most people recognize the value of a systematic procedure but I occasionally hear people argue that the nature of their work makes it difficult or impossible to prepare presentations in such a systematic manner. They point out that reports may be assigned at the last minute, that all results may not be available until shortly before delivery, or that some other feature of their situation won't permit the use of a highly structured approach. I understand these concerns but I disagree that a problem-solving approach is inappropriate. In fact, the more adverse the situation, the more important it is to use a systematic procedure to govern our activities. The greater the pressure, the more important a systematic procedure is because it reduces the probability of error that often results from haste and uncertainty.

I'M SORRY, HE CAN'T BE DISTURBED. HE'S PREPARING A REPORT FOR THE BOARD OF DIRECTORS.

REFERENCES

1. Wilcox, Roger P. *Oral Reporting in Business and Industry.* Englewood Cliffs, NJ: Prentice-Hall; 1967.
2. Polya, G. *How to Solve It* (2nd ed.). Princeton, NJ: Princeton Univ. Press; 1971.
3. Humes, James C. *Roles Speakers Play.* New York: Harper and Row; 1976.
4. Skopec, Eric W. *Business and Professional Speaking.* Englewood Cliffs, NJ: Prentice-Hall; 1983.
5. Rogers, Everett M.; Agarwala-Rogers, Rekha. *Communication in Organizations.* New York: Free Press, 1976.
6. Schoenberg, Robert J. *The Art of Being a Boss.* Philadelphia: J. B. Lippincott; 1978.
7. Berger, Peter L.; Luckman, Thomas. *The Social Construction of Reality.* Garden City, NY: Doubleday, 1966.
8. Bettinghaus, Erwin P. *Persuasive Communication* (3rd ed.). New York: Holt, Rinehart, and Winston; 1980.
9. Clevenger, Theodore, Jr. *Audience Analysis.* Indianapolis, IN: Bobbs-Merrill; 1966.
10. Haney, William V. *Communication and Interpersonal Relations* (4th ed.). Homewood, IL: R. D. Irwin; 1979.
11. Trenaman, Joseph. M. "The Length of a Talk" (unpublished paper). London: British Broadcasting Corp. Research Unit; 1951.
12. Weaver, Carl H. *Human Listening.* Indianapolis, IN: Bobbs-Merrill; 1972.

13. Maloney, Martin. "Semantics: The Foundation of All Business Communication." *Advanced Management*. 1954; 19:26–29.
14. Burns, T. "The Direction of Activity and Communication in a Departmental Executive Group." *Human Relations*. 1954; 7:73–97.
15. Boyd, B. B.; Jensen, J. M. "Perceptions of First-Line Supervisor's Authority: A Study in Superior-Subordinate Communication." *Academy of Management Journal*. 1972; 15:331–342.
16. Deal, Terrence E.; Kennedy, Allan A. *Corporate Cultures*. Reading, MA: Addison-Wesley; 1982.
17. Pacanowsky, Michael E.; Putnam, Linda L., Editors. "Interpretive Approaches to the Study of Organizational Communication." *Western Journal of Speech Communication*. Spring 1982; 46:114–207.
18. Olbricht, Thomas H. *Informative Speaking*. Glenview, IL: Scott, Foresman; 1968.
19. Tucker, Charles O. "An Application of Programmed Learning to Informative Speech." *Speech Monographs*. 1964; 31:142–152.
20. Etzioni, Amitai. *Comparative Analysis of Complex Organizations*. Glencoe, IL: Free Press; 1975.
21. Perrow, Charles. "The Analysis of Goals in Complex Organizations." *American Sociological Review*. 1961; 26:854–866.
22. Mambert, W. A. *Presenting Technical Ideas*. New York: John Wiley and Sons; 1968.
23. Woelfle, Robert M., Ed. *Guide for Better Technical Presentations*. New York: IEEE Press; 1975.

_____

Planning and Developing an Audiovisual Presentation

HENRY C. RUARK

Abstract—Many audiovisual presentations begin with a technical paper, but neither of the following extremes yields a successful talk: showing the original illustrations as slides or foils while reading the paper, or illustrating every point in the paper while narrating additional detail. The successful paper-to-talk conversion requires selecting only the main points; illustrating them with imagination and clarity; rethinking the order of presentation to accommodate the linear nature of an oral presentation—the audience cannot refer to an earlier portion, underline phrases, annotate the margins, etc.; complementing the visual aids with narration rather than written text; and developing supplementary materials for clarification, questions, and challenges.

START with this inexorable, inescapable fact: Any audiovisual (AV) experience, especially for specialized and well-grounded audience members, is simply not the same as the experience your colleagues obtain by reading your professional paper.

An AV presentation is a specialized form of communication. For some professional purposes, it can be far more effective than print alone or language-read—when it is well-prepared and skillfully presented. AV presentation can carry essential detail, supplemented by complementing narration, in a context and with continuity different from any professional paper, through visualization and selective linearity, and thus more interesting.

PAPER VS. PRESENTATION

The written professional paper, even when read aloud, is an essential communication instrument for very specific and extremely detailed and precise purposes—for which it has been painstakingly developed over many years and in many disciplines. Strong professional papers are tightly structured. They are most often written not only for understanding and information transfer, but also in such a way that information cannot be misconstrued or misunderstood.

In some disciplines, primarily the physical and biological sciences, writers work under conditions of agreed and precise language (one word—one meaning) simply not possible in other disciplines such as the social sciences and economics.

Indeed, there are professional fields noted for the involute and prolix language of their papers, leading to the accusation that not even those within these disciplines can read some of the papers.

Neither physical nor social science papers translate well into AV-media form without insightful adaptation even though they may work admirably for professional study, where concentrated attention can be devoted to every nuance of language and where the reader can refer to other sources, search

Reprinted with permission from *Functional Photography*, vol. 15, no. 2, p. 26, Mar. 1980. Copyright by PTN Publishing.

The author is a Contributing Editor of *Functional Photography* and head of Learning Media Associates, 200 E. Chestnut St., Suite 1511, Chicago, IL 60611, (312) 787-6225.

out supporting information, and even re-read as often as required.

It is lack of this insightful adaptation which leads to presentations that reduce a strong subject attraction to such painful levels that the audience is driven away, frustrated, rather than being effectively informed and motivated towards still further professional interest and attention.

You simply don't have that concentrated attention and continued communication effort in most professional program opportunities; there's not time enough nor strength enough, even given the most willing group. Reading and study don't occur in the context of the professional meeting, at that level.

Another important factor is the difference in communication purposes between concentrated individual learning attention and group overview of important information. What you do in a meeting is not what you accomplish with your paper.

When you convert your professional paper (precisely written as above, naturally) to provide you with AV ammunition for considerably different communication purposes, under much looser and less concentrating conditions controlling your audience members, here's what you *don't* do:

Don't undertake to read your entire original text over a few randomly chosen slides, crowded with illustrations intended for the printed page. That way lies boredom, inattention, and the curses of your colleagues (even if not openly expressed).

Don't construct complete and overwhelming visualization, further extended by even more detailed narration covering every sentence, paragraph, and point in your paper. That way lies even greater audience boredom—and you may perhaps even achieve outright rebellion in the seminar room or conference theater.

Your colleagues *can* read. To communicate your information in such painfully detailed and precise mode, you'd better simply allow them to do so; that's the function of the professional paper, in print.

PAPER INTO PRESENTATION

Here's what you *can* do, to build an effective AV program, briefing your precisely expressed and completely detailed paper right down to footnotes and documentation:

1. Consider carefully this question: "Why am I doing this AV briefing?"—and give yourself an honest answer.

Is it to present significant new findings? Is it to persuade your peers towards a particular course of action? Is it to instruct them in a new technique of operation?

Your purposes for the AV program are not necessarily the same as for your paper; for some of those purposes, you may proceed much differently than for others. You'd better know precisely where you want to take your audience, before getting out there in front to lead them.

Reprinted from *IEEE Trans. Professional Commun.*, vol. PC-23, no. 4, pp. 179–181, Dec. 1980.

2. Now reverse your point of view: This time, ask yourself "Who, really, is my audience—and why are they going to spend this valuable time with me?"

If you are going to bring them something, and expect them to receive it well, you'd better think through what they can get out of your program besides your commanding presence and scintillating personality.

The most successful AV programs are always those planned and prepared for a specific audience, custom-built to suit their background knowledge and levels of understanding, their professional attitudes and needs, and even their personal preferences.

Since these are your peers and colleagues, you are in an excellent position to isolate, identify, define and then develop those particular points.

3. Go one step further: Anticipate the reactions you should expect from your group to this report, at this time, presented in this way.

Then you can be professionally prepared not only for the inevitable questions but also for any difficulties in understanding; for expanded and extended detail if needed; for possible objections; and even for clarification of any foundation knowledge you must assume with any group.

4. Find out the physical setting for your AV program presentation; that will guide and shape your choice of AV media and visual format and organization within those media.

Is it a major presentation before a large audience? Is it a working session with a smaller, less formal group? Or is it a friendly seminar or information exchange among well-informed colleagues?

Every one of those situations demands a differing AV approach, probably different AV equipment, and most certainly a different style of presentation and depth of preparation on your part.

5. Consider the necessities imposed by your paper's professional content. Some material may require motion for proper demonstration. You may have high levels of complex technical data requiring charts, graphs, tables, diagrams or special illustrations for effective transmission and correlation.

Motion means film or video, requiring introduction and adaptation of sequences even if they are readily available already from your project. Effective AV graphics require careful planning and meticulous production and suggest either 35-mm slides or the overhead projector, or perhaps a combination to provide the best of both formats.

Without the proper visual presentation for each of the major content elements, you may suffer a "fate worse than death"— you can end up sounding ridiculous. It is impossible to tell your critical audience of peers and colleagues "Now, think hard how this would work ..."; they expect *you* to show them and explain precisely and concisely why and how it does work.

You can avoid that fate very simply by careful choice of visual media and equally careful planning and production of every visual involved. With no pun intended, your professional image is right out there in plain view, and it won't help your professional credibility or the circulation and use of your paper if you fail to be both illuminating and clear.

What works in print via pages for reading simply won't cut it for an AV briefing because of fundamentally different

viewing parameters, differing legibility requirements, limitations on space and composition within the AV media formats, and the impact of contextual factors already mentioned.

You must plan and produce your AV elements for the AV viewing-listening situation to get your presentation on screen properly and thus achieve your professional purposes.

If this AV-adaptation process sounds like more work, that's because it is ... and well worth your extra effort since you'll end up communicating in extremely effective fashion about your professional paper, thus spreading the impact of your major project so reported.

If your project is worth reporting in a professional paper, then it is worth the additional effort to produce an effective AV program about it, too.

You've already accomplished the most difficult part when you wrote the paper. Everything in the AV process described is a tried-and-tested adaptation of your paper's content into another medium, more powerful and flexible than print alone for many purposes.

PRESENTATION

Now here's how you can effectively extract the most important content of your paper for your AV program:

- **First read the paper!**

That may sound both unscientific and revolutionary since you wrote it. But this time read it with your AV program in mind, and remember that your AV purposes now will most likely be quite different from those that controlled development of the paper itself.

Select from each major section *only* those main points that must be made; then plan to make each point with visual impact and essential supporting narration, to achieve your new purposes.

List each such main point and outline just the most cogent key words of required narration. You can put each visual and its keyword-required language on a card or sheet, separately. Use simple sketching for each such visual; outline graphics, paste up photos, use stick figures. All you want is a new playing card for each visual element.

Don't look now, but you are "storyboarding"—analyzing the content of your paper for its absolutely essential visual elements and including only those equally essential language components (words) which must accompany each essential visual.

- **Second, shuffle and redeal those storyboard playing cards.**

Try new ways to reorganize and redevelop your report. All it costs you is some shuffling time and this is where the selective linearity and the flexibility and power of the AV mode can be brought most effectively into play.

You are thus free to rebuild, reconnect, redevelop with those essential visual elements and their supporting narration. You can provide for essential linear development while selecting only those elements truly visually necessary—an entirely different "game" than you have to play within a professional paper and its discipline-demanded requirements.

You may find that you can summarize visually with key points, right at the beginning; then reiterate by visual repetition (different images) and careful, more detailed, but still visually rapid development.

You will almost surely find that you can move much more rapidly from main point to main point in the AV mode, since the enforced linearity of written language no longer traps you, as it inevitably does on the page written-to-be-read.

Storyboarding allows you to escape that linear-development trap which is a rigidifying requirement for the traditional discipline-endorsed professional paper.

It is a highly flexible tool which allows you to "play around" with the major required elements of your report until you find the most striking and effective way(s) to fit them into an AV presentation. Never underestimate the value of this playing around—it is the heart of creativity.

- **Third, make some key decisions.**

You have now thought about the AV-oriented capabilities of the major visualized elements in your report. You've also explored the multitudinous ways in which they can be combined, through storyboarding, to achieve effective AV presentation.

Now choose, in light of all that's gone before in this AV-adaptation process, the specific and particular ways and organization in which you want to present your materials to your colleagues.

Organize your materials (via storyboard cards) into that approach and continuity which achieve your choices. Here's where "pace and flow" are best considered: You need to pace your development (movement through content) rapidly enough to command the attention of your audience without running away from them and without such painfully slow and detailed progression that they doze right off on you.

You also need to make sure there is good flow from visual to visual, and within the narration. "Flow" is the transition from sequence to sequence and within the sequences themselves. Careful narrative writing can provide graceful and effective flow and cover otherwise awkward jumps from visual to visual—but you should also provide additional images if required to visually ease the cut from one concept to another.

- **Fourth, now that you have a real blueprint for AV production, go ahead and produce.**

The blueprint is the hardest part of any AV project to achieve; the actual production steps themselves, in any AV medium, are matters of technique and careful composition much more easily managed once you've decided what you must have to do the visualizing job right.

What happens next depends upon your access to AV production techniques and skills. If you can, enlist the aid of professional photographers, artists, and writers to aid you in the actual AV production work.

PRODUCTION

Some persons with excellent working relationships with such staff people may prefer to bring them into the AV-adaptation process at the storyboarding stage; others find that they must judge content and value situations themselves.

Your AV professionals can be extremely helpful and can work very well from your storyboard materials; follow their suggestions on any point of AV technique but hold out for your own best-considered approach and coverage of content.

Many times you'll need to be your own AV producer, photographer, and even graphic artist. Don't panic—everyone working in AV production had to start sometime, somewhere. No AV technique is any more difficult than those you've already conquered if you are writing professional papers within any discipline.

You can produce an entirely competent and effective AV presentation using very simple amateur-level techniques: 35-mm copy photography, handmade overhead projector visuals, simple finger-skill art, and carefully recorded audiotape sound.

You can find plenty of guidance in trade and professional journals and in any number of excellent references and "how to" books.

If you have access to project-record photography, either still or motion-picture, or to videotaped materials, or perhaps even to high-speed-analysis visual media, you can select and incorporate such footage or images in a variety of ways to strengthen your AV presentation.

Whatever you finally do, don't forget to provide additional visualized materials to accommodate questions, challenges, additional details, and clarification of basic information.

Plan these as separate AV modules, each complete on its own and available outside your main sequence. Set up your slide module in its own tray; arrange your overhead visuals in modular fashion, by topic, for quick selection of the visual image you may need.

For most "here's my answer" materials, you come off stronger and more effectively if you simply "talk over" the visual rather than run taped narration with it. But for some more detailed modules, such as review of basic information, it is nice to be able to say "I just happen to have here a short section on that . . ." and flip on tape-narrated visuals. You guarantee correct and comprehensive review-display that way.

Keep all modules short, succinct, bare-bones simple; plug 'em in rapidly when needed and then ad lib any comments that tie the materials tightly into your basic presentation or the group situation.

CONCLUSION

Audiovisual presentation is not nearly so tough nor trying as this may have read; that's another restriction of material via printed page. The best way to find out how to do AV adaptation of professional-paper content is to do it.

You'll find out that you may make much more impact in one or several such presentations than you ever did with the original article.

Don't be surprised—that's show biz; it really works and pays off. Just look around you at the current social scene for all the proof you'll ever need.

Technical Talks: They're Not as Hard as You Think

DAVID L. ADAMY

Abstract—Preparation for a technical briefing should begin with a decision about the reaction desired from the audience, followed by a determination of the scope—breadth and depth—of the subject which can be covered. Only a few main points should be included and these should be worked into the introduction and the conclusion as well as the body of the talk.

SOONER or later, every engineer faces the task of making a technical presentation. You just can't earn your salary as an engineer without making others aware of your work—and the written word doesn't always do the job. Sometimes people have to hear directly from your lips. This particularly applies when you must sell your ideas to your customers or managers, or calm their fears about some aspect of your work.

Even veteran engineers approach technical briefings—like any other creative engineering work—with great respect. But let's face it; the new engineer who prepares for that first briefing often approaches it with feeling closer to stark terror. Only the experience of making (and, incidentally, living through) a few presentations will reduce the terror to respect. There are, however, a few basic principles (Table I) that can make that first briefing, and every one following it, successful and satisfying experiences.

WHAT DO YOU WANT THE BRIEFING TO ACCOMPLISH?

One of the unfortunate facts of life is that most technical briefings are not presented well. Frankly, many are boring. We engineers tend to emphasize technical accuracy and completeness of our subject matter and give little or no thought to what we want to achieve with the briefing and how to best present our subject matter to the audience.

Before you even begin to prepare a briefing, sit back, put your feet up, and ask yourself "What do I really want to accomplish?" You are not ready to decide what to tell the audience until you decide what you want the audience to *do* as a result of your presentation. Do you want to tell them to buy something? Do you want them to be convinced that your design is sound, or that your project will be completed on time and within budget? Or do you just want them to understand some technical concept so that someone else can convince them to take some further action? Note that all of these questions relate to the reaction of the audience to your talk, rather than to the talk itself.

Once you have set your goals for the talk, consider specifically what you want the audience to remember. Think

Reprinted from *MicroWaves*, vol. 18, no. 4, p. 101, Apr. 1979, with permission of the author. Figures and table added later.

The author is owner of Adamy Engineering, 556 Weddell, Suite 9, Sunnyvale, CA 94089, (408) 747-0474. This paper was written while Mr. Adamy was with Antekna Corp.

TABLE I
STEPS TO A SUCCESSFUL BRIEFING

1. **GOALS:** What do you want to accomplish?
2. **KEY POINTS:** What do you want the audience to remember?
3. **SCOPE:** How much of the subject can you adequately cover?
4. **OUTLINE:** Introduction, body, and conclusion
5. **VISUAL AIDS:** Keep them legible and simple
6. **REHEARSE:** Polish the briefing and gain confidence
7. **RELAX:** Your audience wants you to do well

back to the last technical briefing you heard and you'll probably recall only three or four things that were said. You may have been given a handout with much more information, but you might as well have received the handout in the mail. The only direct effect of the briefing itself is that set of three or four things you remember—which may or may not have been the most important things the speaker had to say. By deciding ahead of time what three or four things you want to make sure are remembered, you can build your talk around them and use them to help achieve the goals that you have set for audience action.

HOW MUCH OF THE SUBJECT CAN YOU ADEQUATELY COVER?

Once your objectives are set, you're finally ready to start thinking about the subject matter. But before you get deeply into the details, give some thought to the scope of the material you want to cover. This will be limited first by the goals you want to accomplish, second by the technical level of your audience, and finally by the amount of time you have available.

A certain technical depth will be required for your audience to be able to take the actions you want them to take; any further depth is just a waste of everyone's time. It is sad but true that although we engineers love to relate the minute details of our work, few people really want to hear about them. Too much technical detail—which often obscures the important points that need to be made—is the most common error made by engineers giving technical briefings. Remember to carefully evaluate just what level of coverage is required to accomplish your goals.

It should go without saying that you must tailor your briefing to the technical level of your audience. Many of us get so involved in our work, however, that we forget there are people who do not have our level of expertise in our particular specialties. If most of the members of your audience (or particularly those you must convince) are managers, they're probably not used to dealing in deep technical terms. So give them a break, be careful to define your terms, explain basic

Reprinted from *IEEE Trans. Professional Commun.*, vol. PC-23, no. 1, pp. 22–25, March 1980.

concepts, and use "buzz words" only when required. This does not mean that you are talking down to your audience in any way; you are just presenting the information to them in a form and at the level that is most useful to them.

Finally, find out how much time has been allocated for your briefing and be sure that you finish within that time. One of the most annoying things that a speaker can do to an audience is to run overtime. The acronym BABE might help keep this in focus: the Brain can Absorb only what the Backside can Endure. But equally as annoying as running overtime is stopping abruptly at the end of your time limit without tying up the loose ends of your presentation. Be sure that you limit either the breadth or the depth (or both) of your subject matter so that it can be adequately covered in your available time.

TELL 'EM, TELL 'EM, THEN TELL 'EM AGAIN!

Every briefing, no matter how short, should have three distinct parts: an introduction, a body, and a conclusion. The old saying goes, "Tell 'em what you're gonna tell 'em, tell 'em, then tell 'em what you told 'em"—which is excellent advice, even though a little oversimplified. Actually, each of these sections of the talk should also accomplish certain unique functions.

The introduction grabs your audience's attention, introduces the subject matter, and sets any basic premises required for the rest of the talk. It is often desirable to open your talk with a grabber, that is, some sort of dramatic statement that tells your audience why the subject matter of your briefing is relevant to them.

Depending on the situation and your relationship with the audience, you might want to relax the group with a joke or some other type of humor, but only if it relates directly to the subject matter you will be covering and if you are comfortable with it. If your audience doesn't know you or what your credentials are relative to your subject, you should cover this in the introduction. In setting the basic premises for the talk, let your audience know what limitations you are placing on the subject matter and if there is any other background information they need to appreciate or relate to the rest of the talk.

Most importantly, think back to those three or four points you want the audience to remember from the talk, and work them into your introduction.

Many experienced speakers advise that you should actually write out what you plan to say in your introduction so you can carefully plan those first impressions and get your talk off to a strong, snappy start.

In the body of your talk develop those three or four main ideas and expand and impress them upon the audience through examples and backup information. Except in very special circumstances, it is best to organize the material in the body of your talk from the general to the specific, so that you first give the big picture and then break the information down into more limited pieces.

Since the nature of the material in the body is highly dependent upon the specific subject matter you are presenting, little more can be said about it here. Just keep in mind that you should carefully think out the way you want your audience to

receive your material and how it will help support the goals of the talk.

In your conclusion summarize the talk ("tell 'em what you told 'em) and make a pitch to your audience to take the action required to accomplish your goal for the talk. Be sure to repeat those three or four things you want the audience to remember and to finish with some sort of strong statement to cement the talk into their minds. As with the introduction, many speakers find it useful to write out the conclusion.

A general rule of thumb is that the time allotted for your briefing should be divided with about ten percent devoted to the introduction, ten percent to the conclusion, and the balance to the body. However, in a short talk (let's say five to ten minutes) it might be more appropriate to devote 20 percent of your time to the introduction and 20 percent to the conclusion. Another important point about time allocation is that the average adult interest span is supposed to be only about seven minutes (and you'll probably find it to be shorter than that right after lunch). Therefore, be aware that some sort of a change of pace, vivid illustration, or other type of attention refresher about every five minutes will keep your audience more interested in what you have to say.

VISUAL AIDS: KEEP THEM LEGIBLE AND SIMPLE

It is virtually impossible to give a meaningful technical presentation without visual aids. The best type depends upon the audience, the subject matter, the nature of the briefing room, and, of course, on what is available to you. However, whether you're using a blackboard, an overhead projector, 35-mm slides, three-dimensional models, or anything else, two principles must be observed if your visuals are to be effective. First, they must be legible. Second, they must be simple enough so that the audience can grasp their content in the short time you will be talking about them.

To be legible your visuals should use print that is large enough and colors that have enough contrast to be easily read or recognized by everyone in your audience. If the people in the back row can't read the board or the screen, part of your audience will be asleep before you finish.

By limiting the amount of material on any one visual, you allow the audience to quickly read it and still have time to hear what you have to say about it. This applies just as much to material written on a blackboard as it does to previously prepared slides; the slides can (and must) be completely laid out while you are preparing your talk.

A fairly good rule is to have one slide per minute of your talk. Many speakers use up to two per minute. In practical terms, this means that each slide should contain about three, but certainly no more than six, pieces of information. So if you are presenting written information, a single slide should have three to six words or phrases. A block diagram should have no more than six blocks. A functional diagram should have no more than six functional elements. A photograph should be composed so that your eyes focus quickly on the important element you want the audience to see. To explain a complex concept, like the system shown in Fig. 1, a set of nested visuals as shown in Fig. 2 is much more effective than one big, busy diagram. First use a simple overall-concept slide,

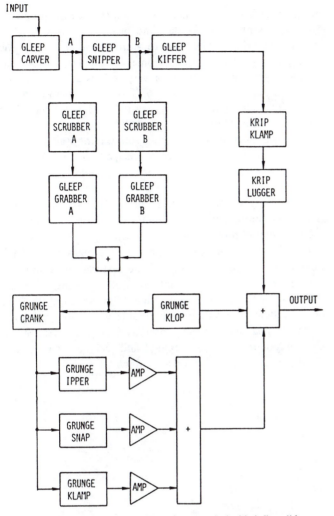

Fig. 1. A typical, much too busy, technical briefing slide.

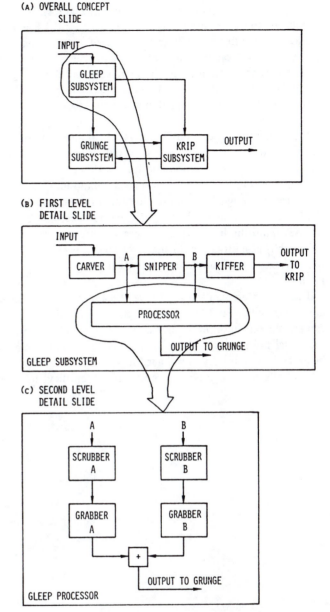

Fig. 2. Nested visuals to explain the system of Fig. 1.

then follow it with several slides showing more detail on each part of the system. After each set of detail slides you can return to the concept slide to keep the audience from getting lost.

When you have finished selecting your slides, carefully arrange and number them so that you can present them to your audience without any fumbling or shuffling and so that when you drop them on the floor two minutes before you are to go on, you'll be able to put them back together.

Keep in mind while you are preparing your visual aids that you will be *presenting* them to your audience, not reading them. The material that goes on the slide should reinforce and clarify what you plan to be saying to the audience while they are reading or observing each slide.

Rehearse to Polish Style and Gain Confidence

The actual presentation of your briefing is, of course, the payoff for all of your preparation and it deserves careful consideration. Naturally, the quality of your delivery technique will improve with experience, but right from the start, proper speaking notes, rehearsal, physical arrangement of the room, and your appearance can pay dividends.

It is a good idea to have some speaking notes available to

you, particularly in your first few presentations. Those notes, ideally no more than two pages, should include the important points in your introduction, the main points of the body of your speech, and the important points of your conclusion. It's also a good idea to have a list of your visual aids (by number and subject) in your notes. A good set of slides can lead you through your talk well enough that you don't need actual notes on the lectern but it's still good to have the list of slides in case you need to find a particular slide to answer a question.

It is hard to overemphasize the importance of rehearsing your talk, both to improve its quality and to give you confidence. Start by flipping through your visuals all by yourself to be sure the thoughts flow logically. Next, stand up and deliver your talk to your office or bedroom wall a couple of times to give your butterflies time to start flying in formation and to get your thoughts connected with some words. Finally, if you are going to give a very important or formal briefing, rehearse your briefing for a small group of people who will give you some constructive feedback and will allow you to judge audience reaction to the things you say and the way you say them. The earlier you do your researching, the more chance you'll have to change things in your talk if you decide it could be improved.

A few minutes spent on the physical arrangement of the room in which you are to give your talk can be very worthwhile. If you are not responsible for the room (for example, when presenting a paper at a technical symposium), at least arrive early enough to look over the room before you go on. Find the buttons on the lectern. Figure out where the screen will be. Find the light switch if necessary. Get a good handle on the physical situation so that you don't get caught stumbling around when you should be concentrating on giving your pitch.

If the room preparation is your responsibility, be sure there are enough chairs and that they are arranged so that everyone can see the screen or blackboard on which you will be presenting visual information. Be sure there is chalk on the blackboard and a pointer if you will need it.

People listen with their eyes even more than they do with their ears, so consider your physical appearance when you are giving a presentation. Sandals and beads may be fine when you are presenting a few ideas to your coworkers, but if you are trying to convince some customers or the top management of your company to take some very serious action based on the things you are going to say, or if you want them to be impressed by your technical or managerial ability, you would do well to dress more formally.

The single most important thing you can know about public speaking is that your audience wants you to do a good job. Even on those rare occasions when there is a heckler in the crowd, the audience itself is not hostile; in fact, those listening are probably just very glad it's you up there talking and not them. You know your subject matter, you are well prepared, and the briefing room is all set up ready to go. So just look at your audience as the friendly people they really are and tell them what you have to say. If hand gestures, voice modulation, and other technical aspects of public speaking come naturally to you, by all means use them. But don't feel that you must use them unless they are natural to you in that situation; otherwise you may appear stilted and clumsy. Above all, relax; if you're properly prepared and feel your audience is friendly, you are bound to do a good job.

Crucial Decisions for Technical Speakers

JOSEPH B. MILLER and WILLIAM F. O'HEARN

Abstract—This paper presents several concepts which technical speakers must consider in the preparation of a speech. Most speakers believe that preparing for a speech requires only that their ideas be placed in some orderly pattern. But the concept of sharing ideas requires technical speakers to make two crucial decisions before preparing their pattern of organization. First, they must decide just what purpose they are trying to accomplish by speaking. Second, they must consider how knowledgeable their audience is. After these decisions have been made the subject of the speech will have to be explained and limited considering both audience and time. A conversational quality of delivery should be used.

SPEAKERS who must present either technical or nontechnical information often focus their entire preparation for speaking on constructing their message. Usually they concentrate on the four important aspects of speaking: (1) limiting the subject to make it manageable, (2) organizing the subject into its component parts, (3) providing a comprehensive and clear explanation of each part, and (4) delivering ideas effectively, in a manner comfortable for the speaker.

Presenting information, in other words, is generally thought to require a speaker to focus attention on the preparation of the content of the message. Most speakers believe that preparing for a speech is simply a matter of putting one's ideas into an orderly and complete pattern and giving those ideas to an audience in some effective way. In this way speakers are thought to be able to discharge their duty to "present information."

But that attitude overlooks the single concept which is most important to public speaking. This concept is that the objective of a speaker is to share his or her ideas. "To share," in this context, means that the ideas of the speaker are understood by an audience *in the way in which a speaker intended them to be understood.*

It follows then that the speaker must consider two other facets of the speaking situation *before* a speech is constructed: the purpose for giving this speech and the extent of knowledge of the audience being addressed.

DECISIONS

The Purpose for Speaking

"Purpose" here is intended to mean that the speaker must decide, before composing the message, how the audience should respond to his ideas. The speaker will desire either of two responses:

1. The response "I understand" or "I understand more fully." When presenting either new information or a new analysis of information, the speaker is presenting an informative speech. If the speaker's purpose is to inform, heavy emphasis must be placed in preparation on clarity of organization and on explanation.

2. The response "I agree" or " I'll do it." When the speaker presents information so the audience will choose among alternatives, or agree with a particular point of view on a controversial matter, or do something, the speaker is presenting a persuasive speech. If the speaker's purpose is to persuade, information must be prepared to support the point of view the speaker is advancing.

Before proceeding to construct the speech the speaker must decide which of these two responses really defines his purpose. Once that determination has been made, he has a frame of reference to use in developing the speech. The statement of specific purpose on which the speaker decides should then explain what the subject is and what the speaker's purpose is in speaking.

The Audience's Background

The way the speech is put together depends on the knowledge of the subject that the audience has at the time the speech is presented. Before beginning to construct the speech the speaker should judge how well informed the audience is. The speaker should also know what point of view the audience holds regarding the information to be presented.

There are a variety of methods which the speaker can use to gain an understanding of the audience. For example, when making a presentation before a group of one's business associates, the speaker can simply rely on past experience with the group to recognize the extent of their knowledge. Sometimes the speaker may not be familiar with the audience. The most common and the easiest method of gaining an understanding of that audience is to talk to a representative sample of the people who will be in the audience, asking specific questions that will reveal the extent of their information and the orientation of their opinions. For example, discussions with those who arranged the meeting, or with the person who will act as chairman for the meeting, or with a person who has attended previous meetings of the group will help to give the insight needed on those occasions.

These conversations can lead to conclusions regarding the terminology that can be used, the amount of explanation that will be necessary, and the type of examples to be used. It may

Manuscript received Sep. 13, 1979; revised Nov. 28, 1979.
Dr. Miller is Associate Professor and Chairman, Department of Communications, and Dr. O'Hearn is Professor of Physics and Assistant Dean of the College of Arts and Sciences at John Carroll University, University Heights, Cleveland, OH 44118, (216) 491-4911.

Reprinted from *IEEE Trans. Professional Commun.*, vol. PC-23, no. 1, pp. 14–17, March 1980.

THE WIZARD OF ID reprinted by permission of Johnny Hart and Creators Syndicate, Inc.

be necessary to reconstruct a speech, varying the approach used, each time the presentation is given.

If the speaker is trying to influence the audience by seeking to change its opinion or to cause it to act in a manner different from its previous behavior, then it will be necessary to discover the goals and needs the audience has, in addition to its background knowledge and opinions. By knowing these, it is possible to organize a speech to demonstrate how those goals and needs can be better satisfied by adopting the change the speaker advocates.

PREPARATION

Decisions about the purpose for speaking and the knowledge and beliefs of an audience set the stage for the four elements to be considered in constructing a speech. The concept of sharing ideas demands that the purpose of the speech and the audience to which the speech will be delivered be kept clearly in mind as construction proceeds.

Limiting the Subject

The expectation, interest, and knowledge of the audience concerning the subject determine how it is to be narrowed to fit into the speaking situation. As an example, an engineer discussing the design of a new motor at a professional meeting should be able to discuss the technical problems and overall performance of the motor thoroughly in a given period of time.

On the other hand, if he were making a presentation about the same motor to a group of upper level managers he could not go into technological detail in depth. Most of the managers would probably lack the interest, knowledge, and understanding of the technical processess. In this case the speaker would present an overview of the design criteria and the performance characteristics of the finished product.

In these examples, the specific purpose would be limited by the audience. For the professional meeting the subject might be limited to "Design problems encountered and the methods of solution used to obtain optimum performance for the motor." For the management meeting it might be limited to "Specifications and how the performance characteristics of the motor meet them." These examples show that the same subject must be approached in different ways because of differences in audiences.

As a practical matter the speaking situation also carries a time limit. In a professional meeting, a management conference, or any type of speaking situation the speaker must narrow his subject to make it possible to discuss the subject fully within the time limits. For example, in the management conference mentioned previously the speaker may be able to explain only how the torque output meets the demands expected of the motor. Other meetings may be necessary to discuss other specifications such as the energy efficiency of the unit.

Thus the speaker faces limitations in constructing the speech: the background of the audience and the time available. Both of these factors must be weighed and balanced.

Organizing

Because audiences, by their previous experiences, are conditioned to expect movement of a discussion to an objective (the specific purpose), it is useful to divide the specific purpose into progressive parts. Each part must then be discussed, individually. There is no surefire way in which the specific purpose can be partitioned; the manner of division depends on the subject, the speaker, the audience, and the occasion.

Nonetheless, there are some partitions that a speaker should examine. For example, some subjects lend themselves to a chronological division. Others may be separated by a systems approach in which each part of a process is examined on a step-by-step basis.

But probably the most common method of division is a topical pattern, in which no single part is more important than another. In this case the role of the speaker is to decide on the sequence of presentation of the parts. The criterion for determining which of the several parts is discussed first is tied to the analysis of the audience. The speaker must decide which sequence will sustain and build the interest and understanding of the audience in his subject. The result of that decision is, as a general rule, the way in which the parts should be ordered.

One exception should be noted: In instances in which information is being presented to persuade an audience it is also necessary to consider the opinion of the audience toward the subject at the time the presentation is made [1]. If the audience is known to be in favor of the point of view being discussed, the speaker can attempt to intensify that point of view. If the audience is known to be opposed to the point of view being advanced, the function of the speaker is to gain a common ground or agreement on objectives so that later efforts to persuade can build on these objectives. With either type of audience, presenting both sides of a controversy usually helps develop the speaker's credibility and thus helps persuasion.

One widely used pattern of organization is the "motivated sequence" [2] which divides speeches into five parts: attention, need, satisfaction, visualization, and action. Each of these steps requires that the specific purpose be carefully analyzed and divided, considering the information to be used and the beliefs and attitudes of the audience toward the subject.

Explaining

Analysis of the audience must also be applied to this facet of speech construction. Clearly, an audience that is familiar with the subject does not have to have terminology explained but it may require that examples be used to clarify or to reinforce the statements made. The speaker must usually use several types or forms of explanatory material, recognizing that the explanation is for the benefit of the audience, not for the benefit of the speaker. When an engineer explains torque output to a group of upper level managers he will probably have to explain his terms carefully. To do so he may cite characteristics of other motors produced by his company or make comparisons of torque output in similar motors produced by competitors. Tables of data and graphs of performance could be used to show differences or to point up similarities.

Of course, it is possible to include too many details in a speech. That possibility is minimized by a careful analysis of the audience before constructing the speech. If advance planning does not result in an adequate explanation, changes can be made even during the presentation. If the nonverbal reaction of the audience suggests that too many unneeded pieces of explanatory information are being included, the speaker must be facile enough to eliminate some details as he progresses. The opposite is also true: More can be added if the audience reaction suggests that more is necessary.

Explanatory information serves another purpose in a speech: It helps sustain interest. Introducing a statistic or a visual aid in the form of a model or a flow chart [3] is a way of directing the attention of the audience toward a particular idea or toward the speaker. Explanations not only clarify, they also reinforce and sustain interest.

Delivery

Delivering a speech requires a speaker to demonstrate and use the same qualities that are characteristic of good conversation. Eye contact, called directness, is essential in both conversation and public speaking. So are changes in the voice, speaking rate, and volume if ideas and examples are to be emphasized. Use of gestures, even changes of facial expression, is a sign to an audience of the extent of the interest and enthusiasm of the speaker about the subject. He must signify to the audience through his delivery that his interest in the subject is strong. Only when the audience senses that interest, and concludes from the delivery that there is reason for *them* to be interested, can the delivery be considered effective. Just as a conversation requires directness, emphasis, and animation, so too an effective speaker should strive to demonstrate those same qualities to an audience.

The principle to be accommodated through an effective delivery is one of change. Experienced speakers recognize that the attention span of an audience is limited and selective. Basic principles of psychology suggest that attention to a single stimulus lasts for only a very short time, perhaps 20 to 40 seconds. At the end of that time the speaker (stimulus) must vary his delivery so that the attention of the audience is renewed and does not wander to a distracting stimulus. In other words, the speaker must change one of the factors of delivery.

To accomplish this change the speaker must contact a different part of the audience, or increase or slow down his rate of delivery, or adjust his volume, or change his facial expression, or use a gesture. All of these factors renew or sustain interest. Of course, it is also possible to build interest by changing the *type* of material being presented. Thus, an additional reason for using explanatory material is to promote interest and to help make the delivery more effective.

Another aspect of delivery, one which facilitates change, is the *method* of presentation. Some speakers prefer to use an outline as a guide to their speaking, others prefer to use a manuscript. The outline provides maximum opportunity for eye contact and for changes in rate and volume; it clearly lacks the specificity of the manuscript. On the other hand, the manuscript has the specificity but it reduces the likelihood of directness and emphasis. When a speaker has tried both methods, recognizing that each has advantages and disadvantages, he is in a position to select whichever method is more effective. Presenting technical data may require that a manuscript or a series of notes be used to supplement an outline to reduce the risk that the speaker may present data inaccurately. But as a general principle the speaker should use whatever method is most effective, whatever method enables the audience to understand ideas in the way the speaker wants them to be understood.

REFERENCES AND NOTES

[1] A concise explanation of the principles of adapting to audience opinion can be found in R. Ross, *Speech Communication*, 4th ed., (Englewood Cliffs, NJ: Prentice-Hall, 1977), p. 287.

[2] D. Ehninger, A. H. Monroe, and B. Gronbeck, *Principles and Types of Speech Communication* (Glenview, IL: Scott, Foresman and Co., 1978), pp. 142–161; contains the most recent discussion of the "motivated sequence."

[3] A recent issue of this TRANSACTIONS, vol. PC-21, no. 3, Sep. 1978, contains several articles dealing with the construction of visual aids.

Considerations for International Presentations

ROBERT M. WOELFLE

ENGINEERS and scientists are increasingly required to give presentations in international environments, for example, at conferences attended by representatives from other countries, at trade shows in countries other than their own, at business meetings of multinational companies, or other similar situations. Because one of the principal requirements for an effective presentation is to address the background and interests of the audience, such situations demand special attention. These situations are characterized by a requirement for the presenter to consider audience- and location-related factors such as:

1. oral and written language,
2. cultural values and attitudes,
3. body language, and
4. physical arrangements.

The process is further complicated by the fact that the nature and impact of these factors vary from country to country. The purpose of this paper is to identify the key considerations for international presentations and describe methods for addressing them.

ORAL AND WRITTEN LANGUAGE

Language is an important factor in assuring effectiveness of any presentation, even those that do not involve an international audience. Many presentations by English-speaking authors to English-speaking audiences, however, fail to communicate because of confusing content organization, poor word selection, improper grammar, and/or the use of unfamiliar jargon or slang. When the audience's knowledge of the presenter's language is weak or nonexistent, the process of communication is even more complicated.

Language considerations can impact both the written and oral versions of a presentation. If a presentation must be published in a conference proceedings or other similar document, the translation should be performed by a professional who is familiar with both languages involved and the general subject. The use of a knowledgeable professional is important because the published version becomes a permanent record, and any errors have a long and highly visible life span. The need for professional assistance is intensified by the difficulty of the translation, the complexity of the subject matter, and the sensitivity of the international situation. Other factors that may impact the complexity of a published translation are the alphabet(s) involved, the need for special characters or symbols, and the mechanics of typography. Some suggestions for simplifying translations are presented in Table 1.

TABLE 1
SUGGESTIONS FOR SIMPLIFYING TRANSLATIONS

- Avoid abbreviations, contractions, and acronyms.
- Avoid idiomatic expressions, slang, and jargon.
- Use a writing style that is relatively formal.
- Use grammatical constructions that are simple and direct.
- Eliminate all unnecessary words, sentences, and paragraphs.
- Limit the scope of vocabulary.
- Avoid examples that are uniquely American—keep them simple and direct.
- Use graphics to replace text whenever practical.
- Place callouts on line art so they can be overlaid by callouts in the translated language.
- Do not include people in photos or drawings unless required. If used, they should reflect the culture of the audience.

The process of translating a publication can be eased significantly and the effectiveness of the communication improved by emphasizing the use of graphics. If carefully planned, this approach can also improve the associated oral presentation. Pictures generally have a universal meaning with little or no translation required, and many internationally recognized graphic standards can be tailored to specific applications. In addition, graphics can communicate complex subject matter more efficiently and effectively than a corresponding verbal description. Care must be exercised, however, to develop graphics that will communicate effectively without offending the audience.

When text is translated, it can expand by as much as 20 percent, and it can affect both printed and oral materials. Sentences become longer, table and figure captions require more space, and completed documents become thicker. The increased size of translated text may cause problems unless expandability is designed into the original. Planning for this expansion is especially critical for space-limited items such as slides, video tapes, and other graphics.

When a presentation is to be translated orally, the translator should be provided with an outline of the presentation, a reasonably complete copy of the text, and a copy of the visual aids keyed to the text. This approach will give the translator an opportunity to practice any special vocabulary for the presentation and research any unfamiliar words. If possible, the visual aids should be presented in the audience's language. Translating the visual aids ahead of time will increase their effectiveness, reduce the workload of the translator, and favorably impress the audience. If they are not translated, however, the presenter should pause after each visual aid is displayed to allow the translator to read it to the audience. In addition, slang, jargon, and regional speech mannerisms should be avoided. Puns and other types of language-oriented humor should also be avoided because they are often impossible to translate with the humor intact.

The oral translation of a paper may be accomplished using either simultaneous or delayed techniques. For a delayed translation, someone who speaks both languages will listen to a short passage of the presentation and repeat it to the audience while the presenter waits. This approach requires patience and cooperation of the presenter, translator, and audience, and significantly increases the time required for the presentation. The presenter needs a good set of notes clearly annotated to show when pauses are required. The presenter must pause often enough for the translator to accurately remember what was said but not so often to be distracting for the audience. The best approach is to complete each thought in a single sentence and then pause. Pausing in the middle of a sentence can cause difficulty for the translator because the sentence structures for the two languages involved may vary significantly.

Even though the translator delivers the text, the presenter should speak directly to the audience. In most situations, individuals in the audience will have enough knowledge to understand all or part of what the presenter is saying. More important, the audience will receive the nonverbal part of the message (gestures, voice inflection, and facial expression) directly from the presenter without translation. During the translation, the presenter should listen intently to the translator and pretend to know what is being said even when unfamiliar with the language.

A smoother and more comfortable situation for both the audience and the speaker is provided by simultaneous translation. This approach requires the translator to listen to the presentation and speak to the audience in their own language with a delay of only one or two seconds. In this situation, the translator sits in a soundproof booth and listens to the presentation through earphones. He or she then speaks into a microphone, translating what is being said while the audience listens to the translator through earphones. The translator mimics the presenter's voice inflections, adds explanatory comments if needed, and reads the visual aids if they have not already been translated. When a member of the audience has a question, it is spoken into a floor microphone and the presenter listens to a reverse translation through earphones. After a little practice, a presenter should be able to ignore the direct translation and speak as though the audience were listening directly.

Cultural Values and Attitudes

Culture is defined as the way of life of any group of people. Because different groups have different norms, people from one culture often react defensively to the behavior of people from other cultures. Since people are generally ethnocentric, they feel their way is right—that those not of their culture are misguided, ignorant, and/or corrupt. Because people's enculturation is so strong, many cannot imagine doing things differently from the way they learned. As a result, a presen-

ter must be sensitive to the culture(s) of an audience to avoid confusing and/or offending them. Some important cultural considerations are addressed in the following paragraphs.

Attitudes and Values

Attitudes toward time, wealth, material gain, work achievement, and other comparable life-style elements vary from culture to culture. For example, most cultures outside the United States take more time to achieve similar tasks, spend more time talking about personal and other nonbusiness matters during business activities, and worry less about schedules and punctuality than do most Americans. Americans also tend to be more self-oriented and independent, while people in other cultures are more group- and company-oriented. In many countries, people believe that money and property are signs of evil and corruption, while the cultures of capitalistic countries are based on the profit motive. In Germany, humor is used only in informal circumstances, and is never used to lighten a speech. The Dutch prefer a person to be direct rather than deviously polite. The French like to ignore official regulations, while the Italians take pride in breaking them. A clear understanding of an audience's attitude and values, therefore, is necessary to assure a presentation reflects a positive perspective.

Social Organization

The importance of role relationships, such as family ties, personal friendships, and business relationships, also vary from culture to culture. In addition, a person's ancestry, birthplace, caste, race, age, sex, and education can affect one's status in certain cultures. In many Asian and Latin countries, for example, the extended family is the norm. In the Far East, strong interpersonal relationships must be established between managers and employees before a smooth working environment can be achieved. In the Middle East, male businessmen have difficulty dealing with female representatives from other countries. In many countries, only the rich are fortunate enough to receive a formal education. A presenter, therefore, must consider his or her perceived role relative to the audience to establish an effective relationship and avoid potential embarrassment.

The proper exchange of business cards is very important in many countries. In Japan, for example, the correct approach is to carefully look at a card after accepting it, observe the title and organization, acknowledge with a nod that the information has been digested, and perhaps make a relevant comment or ask a polite question. When presenting a card in most Asian countries, it is important to use both hands and position the card so that the recipient can read it. Other suggestions for using business cards are presented in Table 2.

Religion

Religion provides people with a basic philosophy about their existence and strongly influences individual behavior in

TABLE 2
BUSINESS CARD IMPORTANCE

The ultimate passport in the international arena is the business card—proof that you really do exist. Even a casual exchange of names between tourist and native usually calls for it—any business contact demands it. When your name is in writing, it is easier to comprehend and retain. Rank and profession are also taken seriously in many foreign countries. In Asia, for example, it is not so much who you are as where you are in the management structure of any particular meeting or transaction. Suggestions:

- The card must include your name, your company's name, and your position plus any titles, such as vice president, manager, or associate director. Do not use abbreviations or acronyms.
- If you are going where English is not widely spoken, the reverse side of the card should be printed in the local language.
- In most of Southeast Asia, Africa, and the Middle East (except Israel), never present the card with your left hand.
- In Japan, present it with both hands, and make sure the type is facing the recipient and is right-side-up.

most cultures. Since some religions reject technology, science, and/or common business or social practices, sensitive situations can easily occur in work and professional environments. Even colors, numbers, and symbols can have special significance in certain cultures (Table 3). A presenter, therefore, must be aware of important religious or social taboos to avoid offending his or her audience.

Technology

Depending on the status of technological evolution of a culture, the people in that culture will approach situations differently. Since technology can influence language skills and thought patterns, people from a culture in which systematic, analytical thinking is discouraged may have difficulty coping with a high-technology environment. In addition, some cultures consider technology to be unnatural, immoral, and/or socially unacceptable. As a result, presentations structured to emphasize technical subjects may not be well received in such cultures.

Politics

The political environment of a culture also influences the attitudes and values of the individuals within that culture. People who come from cultures that promote democracy are generally more independent and expect to participate in decision making, while those who come from a culture with more restrictive political systems are more group-oriented and resist participating in decision-making processes. The latter also tend to be more suspicious and less open-minded with regard to new concepts and opportunities.

Laws

The laws of a country tend to reflect the attitudes and values of that culture. For example, a person who lives in a country that has strict laws relative to the foods, beverages, clothing, or social activities allowed may be offended by the permissiveness of more open cultures. People from countries that have laws restricting profit, discouraging competition, and/or allowing bribery could perceive business transactions differently than individuals accustomed to open capitalistic practices.

BODY LANGUAGE

Actions often speak louder than words, and sometimes they say the wrong things. Eye contact, hand gestures, touching, bowing, and other actions, if inappropriate for an audience's culture, can result in disaster. For example, the Japanese have an aversion to casual body contact. While most Japanese who come to the West make the concession of shaking hands, at home they prefer the traditional bow from the waist. In most Latin countries, hugging is as commonplace as a handshake, and the French sometimes add a kiss on the cheek.

In Japan, eye contact is a measure of the respect one person has for another. What a Westerner considers an honest look in the eye, an Asian takes as a lack of respect and a personal affront. Even when shaking hands or bowing—especially when conversing—only an occasional glance into the other person's face is considered polite.

Attitudes toward punctuality also vary greatly from one culture to another. Germans, Japanese, and Romanians are very punctual, while many Latin countries have a more relaxed attitude toward time. In the Middle East, punctuality is very important for a visitor but the host may not be on time. In addition, prior appointments are essential in the Middle East, but visitors may find other people present and several meetings occurring simultaneously.

The interpretation of body language varies widely from one culture to another. Something with one meaning in one culture may mean the opposite somewhere else. Several examples of various gestures and the different ways they are interpreted are listed in Table 4.

PHYSICAL ARRANGEMENTS

All presentations require careful planning to assure effective handling of the physical arrangements, such as the room configuration, audio/visual equipment availability and operation, and supporting services. For international presentations, however, detailed planning of physical arrangements is especially important because nothing can be taken for granted. For example, the electrical power in most other countries differs from U.S. standards in frequency and/or voltage. As a result, equipment carried by a presenter from his or her home country may not operate properly at the presentation

TABLE 3
CULTURAL ATTITUDES TOWARD NUMBERS, COLORS, SHAPES, AND SYMBOLS

AFRICA

- **Chad**—Number: Odd numbers have negative overtones, with 13 especially unlucky. Even numbers are considered positive. Color: Black and red are unlucky colors; white, pink, and yellow are positive colors. Symbols: Figures of pigs are considered unclean.
- **Congo**—Symbols: Animal images and/or names connote bad luck and should be avoided.
- **Ethiopia**—Color: Black has a negative connotation. Bold, bright colors are preferred to soft tones. Symbols: Religious symbols should be avoided.
- **Ghana**—Number: 7, 11, 13, and 17 are considered unlucky. Color: Black is considered unlucky. Bright colors are preferred.
- **Kenya**—Number 7 or any number ending in 7 is considered bad luck.
- **Liberia**—Color: Black has a negative connotation while bright, bold colors are preferred.
- **Libya**—Color: Green is considered a positive color. Symbols: Images of pigs and the female anatomy should be avoided.
- **Madagascar**—Color: Black has a negative overtone, while bright colors are preferred. Symbols: The owl should be avoided as it is used as a sorcery emblem.
- **Morocco**—Number: 13 is negative while 3, 5, 7, and 40 have positive connotations. Color: White is a negative color, while bold colors and green, red, and black are positive. Symbols: A six-pointed star should be avoided.

ASIA

- **Afghanistan**—Number: Negative numbers are 13 and 39. Color: Positive colors are red and green. Symbols: Pictures of pigs and dogs should be avoided.
- **Hong Kong, Korea, Taiwan**—Number: Negative numbers include 4, 13, 38, 49, and generally odd numbers. 606, 914, and even numbers are positive. Color: Negative colors are white, black, and gray. Red, yellow, and bold colors are preferred. Shapes: Circles and squares are positive, while triangles are negative.
- **India**—Number: Numbers ending in zero are negative. Multiples of 3, 7, and 9 are positive. Color: Negative colors are green, yellow, red, orange, and bold colors.
- **Japan**—Number: 1, 3, 5, and 8 are considered positive numbers and 4 and 9 are negative numbers. Color: Black, dark gray, and black and white in combination should be avoided. Muted shades are preferred over bold colors, with red and white and gold and silver good color choices. Religious symbols and shapes should be avoided. Symbols: Patterns of pine, bamboo, and plum are considered desirable.
- **Malaysia**—Number: Negative numbers are 0, 4, and 13. Color: Black is a negative color if used alone. Red, orange, and bold colors are considered positive.
- **Pakistan**—Number: Negative numbers are 13 and 420. Color: Black is a negative color. Positive colors are green, silver, and gold. Bold colors are preferred to muted colors. Symbols: Writing or pictures considered obscene or religiously offensive are banned, i.e., the Star of David, pictures of dogs or pigs.
- **Singapore**—Number: 4, 7, 8, 13, 37, and 69 should be avoided. Color: Black is a negative color. Shapes: The shape or profile of Buddha is objectionable when used commercially. Symbols: Refrain from using religious words and symbols.

EUROPE

- **Czechoslovakia**—Number: 7 is a positive number. Color: Black has a negative connotation, while red, white, and blue have positive overtones.
- **Denmark**—Number: 13 is a negative number: Color: Red, white, and blue are positive colors. Symbols: Hearts, particularly at Christmas, are favored.
- **Germany**—Number: 13 is a negative number. Color: Red and combinations of red, black, and white or brown should be avoided. Symbols: Emblems resembling the swastika are legally banned. Religious symbols should also be avoided.
- **Greece**—Number: 13 is a negative number, and 3 and 6 are positive numbers. Color: Black is a negative color, and bold yellows, greens, and blue and white are positive colors. Symbols: Western Europe and U.S. identifications are favored over Mideastern or Moslem symbols.
- **Italy**—Number: 17 is a negative number. Color: Purple is a negative color. Bold colors preferred for foods and soft tones for clothing. Symbols: Female and religious figures are considered in poor taste.

LATIN AMERICA

- **Argentina**—Number: Odd numbers have negative overtones, with 13 especially unlucky. Even numbers are considered positive. Color: Black and red are unlucky colors; white, pink, and yellow are positive. Symbols: Figures of pigs are considered unclean.
- **Colombia**—Number: 3, 5, and 7 are positive, while 13 is negative. Color: Bright and bold colors such as red, blue, and yellow are favored over soft tones.
- **Guatemala**—Number: 13 and 14 are considered negative numbers.
- **Nicaragua**—Number: 7 is considered a lucky number, while 13 is unlucky. Color: The flag colors—blue-white-blue horizontal stripes—should be avoided. Shapes: The triangular shape should be avoided because it is closely identified with the national symbol. Symbols: The religious implications of the cross should be respected.
- **Peru**—Color: Bright colors, such as red, fuchsia, and yellow, are preferred.
- **Uruguay**—Color: Color combinations of two local football teams should be avoided, i.e., yellow and black, and red, white, and blue.
- **Venezuela**—Number: The numbers 13 and 14 are considered unlucky. Color: Venezuelan flag colors in the same sequence, i.e., yellow-blue-red, are prohibited from use. Symbols: Religious motifs should be avoided.

MIDDLE EAST

- **Saudi Arabia, United Arab Emirates, Iraq, Kuwait, Bahrain, Iran, Qatar, Yemen, and Oman**—Color: Pinks, violet, and yellow are negative colors, while brown, black (especially offset by white), greens, dark blues, and reds and white are positive colors. Bold colors are favored over soft. Shapes: Round or square shapes are preferred. Symbols: Religious signs, the six-pointed star, a raised thumb, and Koranic sayings should be avoided.
- **Egypt**—Number: 3, 5, 7, and 9 are positive numbers, while 13 is negative. Color: Red and green on white or black backgrounds, orange, light blue, and turquoise are desirable colors. Contrasts in strong shades are preferred. Dark colors, especially violet, are disliked. Shapes: Pyramid-shaped objects and intricate designs utilizing Islamic and Persian art are preferred.

TABLE 4
INTERNATIONAL GESTURE DICTIONARY

- **Eyebrow Raise:** In Tonga, this gesture means "yes." In Peru, it means "money" or "pay me."
- **Blink:** In Taiwan, blinking the eyes at someone is considered impolite.
- **Wink:** Winking at women, even to express friendship, is considered improper in Australia.
- **Eyelid Pull:** In Europe and some Latin American countries, this gesture means "Be alert."
- **Ear Flick:** In Italy, this gesture signifies that a nearby gentleman is effeminate.
- **Ear Grasp:** Grasping one's ears is a sign of repentance or sincerity in India. A similar gesture in Brazil—holding the lobe of one's ear between thumb and forefinger—signifies appreciation.
- **Nose Tap:** In Britain, this gesture means secrecy or confidentiality. In Italy, it is a friendly warning.
- **Nose Thumb:** One of Europe's most widely known gestures, this gesture signifies mockery.
- **Nose Wiggle:** In Puerto Rico, this gesture means "What's going on?"
- **Cheek Screw:** This gesture is primarily an Italian gesture of praise.
- **Cheek Stroke:** In Greece, Italy, and Spain, this gesture means "attractive." In Yugoslavia, "success."
- **Fingertips Kiss:** Common throughout Europe, particularly in Latin countries (and in Latin America), this gesture connotes "aah, beautiful!"
- **Chin Flick:** In Italy, this gesture means "Not interested" or "Buzz off." In Brazil and Paraguay, it means "I don't know."
- **Head Nod:** In Bulgaria and Greece, this gesture signifies "no." In other countries, it means "yes."
- **Head Tilt:** In Paraguay, tilting the head backward means "I forgot."
- **V Sign:** In most of Europe, this gesture means victory when you keep your palm facing away from you. The same gesture with the palm facing in means "Shove it."
- **Beckon:** Using a finger to call someone is insulting to most Middle and Far Easterners. In most of these countries, Portugal, Spain, and Latin America, someone is beckoned with the palm down, fingers or whole hand waving.
- **Fingers Circle:** This gesture is widely accepted as the American "okay" sign, except in Brazil, Greece, and Russia, where it's considered obscene. In Japan, it signifies "money."
- **Fingers Cross:** In Europe, crossed fingers commonly mean "protection" or "good luck."
- **Fingers Snap:** In France and Belgium, snapping the fingers of both hands has a vulgar meaning.
- **One Finger Point:** In most Middle and Far Eastern countries, pointing with the index finger is considered impolite.
- **Thumbs Up:** In Australia, this is a rude gesture; in most other places, it means "okay."
- **Hand Sweep:** In Latin America and the Netherlands, a sweeping or grabbing motion made toward your body, as though you were sweeping chips off a table, means that someone is stealing or "getting away with something."
- **Waving:** In Greece and Nigeria, this gesture is a serious insult, and the closer the hand is to the other person's face, the more threatening it is considered. In Europe, waving "goodbye" is done by raising the palm outward and wagging the fingers in unison. Waving the whole hand back and forth can signify "no," while in Peru, that gesture means, "Come here."
- **Arms Fold:** In Finland, folded arms are a sign of arrogance. In Fiji, it shows disrespect.

site. Even television/video standards vary from country to country. Most equipment-related problems can be avoided by using the equipment and operating personnel already available in the audience's country. A presenter, however, must verify that the necessary equipment is really available when required, and that any accessories brought from home, for example, slide trays, video tapes, computer discs, or electronic printers, will be compatible.

In many countries, the room arrangement may have special significance. Audience seating may need to reflect business or social status, and certain configurations may need to be avoided because of their cultural implications. Colors, shapes, and symbols used in supporting visual materials, posters, signs, handouts, furniture, and decorations, as well as any refreshments served, must also be carefully selected consistent with the audience's cultural preferences. In addition, special arrangements may be necessary to provide facilities for translators and their equipment.

RELEASE APPROVAL

When a presentation is to be given in the speaker's own country, certain approvals are usually required to release the associated information to a specific audience, especially when it involves matters of military, political, economic, and/or business proprietary significance. If the same or similar information is to be presented in an international environment, however, the approval process is significantly more complicated, and approvals must be secured on a case-to-case basis for each country in which a presentation is to be given. In the United States, approvals may need to be obtained from the Departments of Defense, State, and/or Commerce or other government agencies, as well as the presenter's company, university, or sponsoring agency. In all cases, the presenter is responsible for obtaining the required approvals before giving the presentation and/or distributing related documents.

Part 3
Visual Aids

VIRTUALLY ALL PRESENTATIONS involve the use of some form of visual aid, such as posters, slides, overheads, and video. The techniques and equipment available to prepare and use visual aids have evolved and expanded dramatically over the past 15 years due to advances in technology and the increased demands of new applications. As a result, engineers preparing to give presentations are confronted by a diverse array of visual aid options. The various visual aid techniques that are available and the considerations associated with using them for specific applications are addressed in this part.

In his paper, "A survey of visual aids," Robert Woelfle identifies and describes the primary types of visual aids in current use and presents the advantages and disadvantages of each. The specific visual aids addressed include boards, flip charts, posters, slides, overheads, computer displays, video, and multimedia. He also includes a matrix that provides a guide for selecting visual aids based on application-oriented factors.

Illegible and incomprehensible visual aids often result from a failure to consider the sizes of lettering and symbols in the projected image when the original artwork is being prepared. In his article, "Visual aids—How to make them positively legible," Calvin Gould identifies typical legibility problems encountered in visual aids and describes mathematical techniques that can be used to avoid them. He also defines the legibility standards normally used by industry and lists numerous references that provide additional background.

In his paper, "Visual aids—How to keep participants' attention," Robert Pike offers suggestions for using both nonprojected and projected visuals. He presents considerations for planning the room arrangement, including size, layout, lighting, power, and acoustics. He also provides guidelines for producing and using transparencies and slides.

Slide design should normally be done by a professional designer. Engineers, however, do not always have access to, and/or budgets for, professional designers. In his paper, "Designing slides," Jack Reich describes how modern desktop computer graphic systems and specialized presentation software can be used by engineers to prepare professional-level slides for many "standard" applications. He also presents several suggestions for preparing and evaluating slide content as well as related design, typography, production, and use.

While Gerald McVey addresses the use of dynamic media in his paper, "Legibility in film-based and television display systems," the information he presents is also applicable to slides and overheads. He defines the significance of the legibility of alphanumeric and graphic information in terms of symbol size, width-to-height ratio, stroke width, spacing, brightness contrast, contrast direction, and color. He also addresses the interrelationships of these factors with display system characteristics, environmental and human factors, and research study results, and offers some guidelines for visual design and presentation.

The human visual system's capacity and capability to process color can be applied as design criteria for preparing visual aids used in technical presentations. In his paper, "Using color effectively: Designing to human specifications," Gerald Murch reviews key elements in the visual domain of color, encompassing the visual, perceptual, and cognitive modes, and he develops a series of recommendations for effective color usage based on these elements. He also includes charts listing the best and worst color combinations.

The widespread availability of color in both traditional and electronic production techniques offers many opportunities for enhancing technical presentations. In his paper, "How to use color functionally: 53 tips," Jan White emphasizes that color is a visual language that should be used with purpose. It must be used to enlighten and add value, not just to dazzle and catch the eye. To realize the maximum benefits from the use of color, he presents a list of 53 points to consider when producing a presentation. For example, he recommends identifying a recurring theme with a color to help establish the relationship flow among visuals.

In his paper, "Looking good in public," Roger Parker describes how an engineer can use desktop presentation software to enhance the persuasive power of a technical presentation. He addresses the application of such software to the production of transparencies, slides, on-screen presentations, and videotape, and provides tips for maximizing the effectiveness of each application. He also explains ancillary capabilities, such as planning and recordkeeping tools, speakers' notes and prompts, and hardcopy production aids.

In his paper, "Producing foreign language presentations," Dean Fitch explains that technical presentations involving foreign languages and cultures require special planning and delivery. He stresses the importance of identifying the language and cultural needs of an audience, and indicates that solid language skills provide an edge in executing an effective presentation. He includes a list of "do's" and "don'ts," and addresses production considerations, tools, and alternatives.

A Survey of Visual Aids

ROBERT M. WOELFLE, SENIOR MEMBER, IEEE

INTRODUCTION

THE different types of visual aids that can be used to enhance the effectiveness of technical presentations include:

1. boards,
2. flip charts,
3. posters,
4. slides,
5. overheads,
6. computer displays,
7. video, and
8. multimedia.

The first three types of visual aids, that is, boards, flip charts, and posters, are directly displayed and viewed by the audience. Although these types of visual aids are often considered obsolete, they are still valuable alternatives for some applications. Since the basic artwork for these visual aids is also the end product seen by the audience, considerable care must be exercised in its preparation and handling.

The next three types of visual aids, that is, slides, overheads, and computer displays, use indirect techniques. With these techniques, the basic artwork is produced by a photographic or electronic process, and the end product is seen by the audience as a projected image. These three types of visual aids, therefore, allow a high degree of flexibility in preparing and handling the original art.

The last two types of visual aids, that is, video and multimedia, require special skills, equipment, and processes, and they are relatively expensive to produce. Their effectiveness in communicating complex messages and/or realistic situations, however, often offsets the related production costs. In addition, audiences have come to expect the dynamics and sophistication provided by these media.

Each type of visual aid involves different techniques for preparation and utilization, and each offers specific advantages and disadvantages. The best type of visual aid for a particular presentation depends on tradeoffs of several factors, including the

1. experience and capabilities of the presenter,
2. type of material to be presented,
3. amount of time allocated for the presentation,
4. size and nature of the audience,
5. size and layout of the meeting room,
6. availability of presentation equipment, and
7. availability of visual aid preparation resources.

Only the presenter can select the best type of visual aid for his or her application. To provide some guidance for making this selection, however, the following paragraphs describe the characteristics of the most common types of visual aids, and define the advantages and disadvantages of each.

BOARDS

The simplest and most direct type of visual aid is the chalkboard or markerboard. It is inherently limited to simple symbols, key phrases, and the most elementary types of illustrations that can be developed as the presentation evolves. In addition, the resulting visual aid is temporary, and a permanent record is not usually produced.

A pad and easel can be used in much the same way as a chalkboard or markerboard, that is, simple phrases or drawings can be developed as the presentation progresses. A pad and easel, however, offer several advantages over the basic chalkboard. For example, light preprinted lines on the paper can be used to provide guides for developing the desired visual aid. In addition, completed visuals are not erased, but simply turned over the back of the easel so they can be recalled if desired, and any complex visuals can be prepared in advance.

Electronic copyboards are similar to a marker board, except that they can automatically produce copies of the image produced. These boards usually consist of two to five plastic panels, and information is written on one panel at a time. As the information on each panel is completed, the resulting image is captured by a digital sensor, and the resulting electronic signal is used to produce a selected number of copies. This process is repeated as necessary for each visual produced during a presentation.

Other types of boards that can be used as visual aids include flannel or flock board, magnetic boards, and various types of special boards using proprietary materials such as "Hook and Pile" and "Grab-Fab." Key words and phrases, simple drawings, and individual parts of more complex drawings are preprinted on separate strips of art board with cloth or metal glued on the back to provide the desired adhesive properties. As a presentation progresses, each strip is positioned by the presenter in a predetermined sequence and location until the desired visual is completed. These strips are then removed and replaced by a new set for the next visual aid.

Summaries of the advantages and disadvantages of board presentations follow.

Advantages

- Because the presenter can add, delete, or change words and drawings as he proceeds, boards offer flexibility and two-way real time communication with an audience.
- A board-type presentation can be prepared with little or no advance notice since pictorials develop as the presentation progresses and often result from audience feedback.

Disadvantages

- A permanent record is often precluded. Usually the initial visual presentation must be erased or covered in order to make room for subsequent visuals.
- Portability is limited. Advance arrangements are required to ensure a board is available where the presentation is to take place.
- The quality of the resulting visuals is poor. Details cannot be shown well with the available tools and time, and it is hard to draw precisely on vertical surface.
- Execution of the visual process tends to be distracting and limits the presenter's ability to devote adequate attention to the audience.
- Coverage is limited—subjects can be visually presented only in brief text and diagrams.
- The time a presenter spends preparing a visual during a presentation is time that should be devoted to oral communication.

FLIP CHARTS

Flip charts are usually prepared on a heavy paper or other flexible material, and are mounted so they can be turned like the pages of a book. They are used in a manner similar to the pad and easel, but they are prepared before the presentation rather than during it. As a result, they are neater and easier to read, and complex drawings and tables can be accommodated without special treatment. Since flip charts are relatively easy to prepare, they are still a popular form of visual aid for informal presentations.

Flip charts are prepared in various sizes, ranging from 11″ × 17″ to 30″ × 40″, depending on the nature of the material to be presented, the anticipated size of the audience, and/or the anticipated viewing distance. The smaller charts are often mounted in portable ring-type freestanding binders for tabletop presentations.

Summaries of the advantages and disadvantages of flip charts follow.

Advantages

- Quickest to prepare as a "rush job."
- Most economical; inexpensive base material available anywhere.
- Easily transported; charts are rolled, folding easel fits into small package.

- Changes easily made by splicing new information into existing charts.
- Simplest to set up and use.
- Can be reproduced to provide handouts.
- No special lighting or equipment is needed.
- Highly reliable—not subject to equipment failure.

Disadvantages

- Flip charts get dirty and ragged with repeated handling.
- Photographs and complex artwork cannot be easily incorporated.
- Adhesive-backed lettering, tapes, and color tints tend to fall off.
- The number of charts is limited to approximately 30 or 40; more will not turn over conveniently.
- The sequence cannot easily be altered, and additions are difficult to incorporate.
- The audience is limited to the number of people who can easily read the conventional maximum 30″ × 40″ chart. Larger sizes are unwieldy and impractical.
- Standard flip-chart packages are too long to fit conventional files.
- The simplicity of flip charts often leads presenters to select them as "easy way out" when other methods of visual presentation would be more effective.

POSTERS

Posters or placards are also a relatively simple direct type of visual aid. They are prepared on heavy-weight illustration board, and usually range in size from 18″ × 24″ to 30″ × 40″. Here again, the size of the poster depends on the nature of the material to be presented and the size of the anticipated audience.

The techniques used to prepare posters are similar to those used for flip charts. Due to their rigidity, however, posters allow more latitude with regard to artistic treatment. For example, photographs, prints, symbols, and even lightweight three-dimensional objects can be attached to a poster. In addition, phototype or other forms of machine-prepared type can be used for text. Posters can also be embellished with self-adhesive color tapes and preprinted symbols and lettering.

Since posters are inherently more durable than flip charts and allow the use of more sophisticated artistic treatments, they are usually employed for repetitive-type presentations. They are also an excellent form of visual aid for regularly scheduled meetings because they facilitate the use of periodic changes to show growth or progress. Since they are not bound as flip charts are, they also allow the presenter more sequential flexibility. Their use, however, is usually restricted to internal presentations because they are relatively heavy and cannot be rolled, thereby limiting their portability. In addition, they are labor-intensive to prepare, and major changes are difficult to incorporate.

Summaries of the advantages and disadvantages of poster presentations follow.

Advantages

- Posters are usually produced by illustrators, resulting in a higher quality image.
- Much wider range of illustration techniques is possible on rigid artboard than on thin flip-chart materials.
- Photos, complex art, and special stick-on materials can be safely used.
- Posters can be rearranged, old posters removed, and additions substituted with ease.
- Ordinary easels or any surface against which posters can be propped can be used to set up this type of presentation.
- Illustration board is tough, durable, and can accommodate minor changes.

Disadvantages

- Posters are bulkier and less portable than other presentation media.
- Repeated handling tends to degrade their quality.
- Major art changes often require complete new poster (flip charts can be spliced).
- Posters require more handling during a presentation than other media, which may be distracting to audiences.
- Storage can be a problem—posters are large and can warp unless stored flat.

SLIDES

Slides are prepared using a combination of manual illustration, computer graphics, and photographic techniques. The basic artwork can be prepared at any convenient size with the proper aspect ratio. As a result, typeset copy, photostats, computer printouts, and various types of artist's aids can be used to produce the required artwork rapidly and at minimum cost. In addition, color and special optical effects can be easily incorporated. This process also minimizes the effort and time required to incorporate changes. When the required artwork is completed, it is photographically copied to produce a film transparency, which, in turn, is mounted in a frame to produce a completed slide.

Slides can be produced in a variety of sizes and shapes, but the most common form consists of a 35-mm film frame inserted in a $2'' \times 2''$ mount. Other sizes include 31 mm $\times$ 33 mm (super slide), 18 mm $\times$ 24 mm, and $2\frac{1}{4}'' \times 2\frac{3}{4}''$ (lantern slide). In addition, a special Polaroid transparency film can be used to produce slides $2\frac{1}{8}''$ square. In general, the larger slide formats provide better resolution for detailed information. Regardless of the format size, slides can be effectively used for audiences ranging from a few people around a table to thousands in a large auditorium.

Using slides for presentations requires the availability of special equipment for both production and projection. As a minimum, slide production requires at least a reasonably good copy camera, lighting equipment, a film processor, and a light table. The production of slides using computer graphic techniques requires a desktop computer with adequate storage capacity, presentation software, and a film recorder. A magazine-loaded projector with remote control capability and a fairly large screen are required to display slides during a presentation. Without ready access to such equipment, using slides in presentations may involve outside service organizations, which can be expensive and inconvenient.

For individuals and organizations that routinely prepare and give presentations, the advantages offered by slides more than offset the cost of the related equipment. For example, production of the basic art is comparable to other visual aids, but changes can be handled faster and at less cost. In addition, the basic art can be easily stored and retrieved for use in other presentations. The inherent flexibility offered by slides also allows a presentation to be tailored to a specific need at minimum cost. The sequence and mix of slides can be easily varied from one presentation to another to suit different audiences and applications. New slides can also be added to a presentation right up to the last minute without affecting the other slides. In addition, the same set of slides can be used for both large and small audiences simply by changing the projection setup.

Summaries of the major advantages and disadvantages of slide presentations follow.

Advantages

- Virtually any subject matter or art treatment can be accommodated.
- Although ideal for large groups, slides can be presented to any size audience.
- Full color can be incorporated with minimum extra cost.
- Slides can be added, deleted, and interchanged to accommodate the needs of individual presentations.
- Using a remote control, a presenter can change visuals with minimum audience distraction.
- Slides are small, easily carried, mailed, and handled.
- Since 35-mm projectors are widely used, suitable equipment can be borrowed or rented easily in most places.
- Any number of duplicate slides can be provided from the original at low cost, enabling multiple presentations to be held without degrading the original art.

Disadvantages

- A preliminary dry run is needed to ensure that slides are sequenced correctly and the projector, screen, and other equipment are properly arranged.
- The presentation room must be darkened, reducing the interchange between the presenter and audience.
- The failure of projection equipment can delay or disrupt a presentation.

- The production of slides requires special equipment or access to a service bureau.
- The time required for slide film processing precludes the incorporation of last-minute changes.

OVERHEADS

The overhead projector, unlike other projection systems, produces an image indirectly by means of a mirror and a condensing lens. A powerful light source in the base of the projector is directed up through a horizontal glass table holding the selected overhead transparency, and the resulting image is reflected by the mirror and focused through the lens to produce the image on a screen. This approach allows the presenter to face the audience and simultaneously operate the projector. It also produces an image with sufficient intensity to preclude the need for darkening the room. Since the transparencies used for overhead projection are up to $10'' \times 10''$ in size and are readily accessible during a presentation, the presenter can add animation by using such devices as movable cutout arrows, multiple overlays, three-dimensional objects, and other similar devices. The presenter can also add information directly on the transparency by using a crayon or grease pencil.

The artwork for overheads, or viewgraphs, can be prepared in either of two ways, depending on the processing to be used. If the transparency is to be produced using photographic techniques, conventional opaque artwork similar to that prepared for slides can be used. If a diazo process is to be used, the original artwork must be translucent. The latter process is considerably cheaper than the photographic process, and a wide range of colored foils is available to produce the transparencies. In addition, multicolors and various special effects can be achieved through the use of overlays, special markers, or color film.

Summaries of the advantages and disadvantages of overheads follow.

Advantages

- Since speaker faces audience while visuals are presented, information can be readily exchanged with audience.
- Projection takes place in room with normal lighting, encouraging audience participation.
- Overheads are handled by speaker, who can write, draw, or point out items on projected image.
- Step-by-step processes and cumulative buildup situations are easily presented through multiple overlays attached to the overhead base.
- Economical materials and reproduction processes can be used.
- Animation can be added through motion of transparent elements, pointers, and polarization techniques.
- Overheads can be quickly produced. Information can be typed or drawn using readily available tools; only a few minutes are needed to run off transparencies ready to project.

- Overheads are easy to read, handle, place on the projector, and focus.
- Varied production techniques can utilize hand-rendering on film, duplicate printed or typed material, and photocopies. Enlarged halftone positives, graphic overlays, and color transparencies can also be used.

Disadvantages

- Older overhead projectors require some room darkening.
- With exception of one or two new lightweight portable projectors, most overhead projectors are bulky, heavy, and must be stored close to presentation site.
- Overhead projectors typically produce distortions of the projected image unless the screen is carefully set up exactly perpendicular to projected beam.
- A presenter needs more practice in handling overheads than other visuals because manipulating, pointing, and marking are part of the presentation, and remote controls are not available.
- The overhead handling process can be distracting to the audience.
- Full-color continuous-tone photographs are expensive for general use because of the large size of film needed.
- Since overhead projectors are not as widely available as 35-mm projectors, special arrangements may be needed to secure the required equipment.

COMPUTER DISPLAYS

The most recently developed visual aid involves the use of desk-top computers to directly drive either an LCD projection panel or video-type projector. Both approaches eliminate the need for intermediate transparencies (and their processing), and allow near-real-time control of the visual-aid content by the presenter. While the use of computer displays requires access to some expensive equipment, it provides a presentation with maximum flexibility and negligible degradation of image quality.

An LCD projection panel is an electronic device that is placed atop an overhead projector in place of a transparency and is connected to a computer. Computer-generated text and graphics are displayed on the normally clear panel using liquid crystal technology to produce either a monochrome or color image. This image is then projected onto a screen in a manner similar to a transparency. The scope of the images that can be projected is limited only by the capabilities of the computer and related software. These devices produce a screen image that is comparable to overhead transparencies in quality, clarity, and color, with the added benefit of being able to execute dynamic effects such as fades, wipes, and limited animation.

As an alternative, video-type projectors can be connected to a desk-top computer through a special interface. Such a projector uses video technology to produce a screen image comparable to a slide. Like the LCD panel approach, the scope of the images that can be projected is limited only by the capabilities of the computer and associated software. In most cases, the presenter can remotely control the pace of the presentation similar to a slide projector. Some software programs also offer random access so that a presenter can skip to a particular image or use branching to address one of several alternate images, depending on audience reaction. Some programs also allow the presenter to interactively change the image content and/or the use of various dynamic effects.

Summaries of the advantages and disadvantages of computer displays follow.

Advantages

- Computer displays eliminate the need for intermediate transparencies and the associated photographic processing.
- Complete presentations can be stored on a single floppy disk instead of individual slides or overheads.
- Images can be easily changed, allowing presenter interaction during the course of the presentation.
- Special dynamic effects not available with slides or overheads can be included in a presentation.
- Visual aid sequence and timing can be automatically controlled, or a presenter can randomly access images as needed during a presentation.
- Prerecorded video can be included as desired in a presentation.

Disadvantages

- Relatively expensive and bulky computer and projection equipment is required.
- Equipment setup and operation are more complex than for slides or overheads, and its operation is less reliable.
- The image quality is not quite as good as with a slide projector, and it can be degraded more easily by improper equipment setup and/or adjustment.
- Special software is required.
- Complete utilization of computer display capabilities requires a presenter to have a higher level of skills, including computer literacy, artistic sense, and audience rapport.

VIDEO

As the cost and complexity of video equipment is reduced, it becomes a more viable alternative for providing visual aids in technical presentations. Thanks to television, audiences have come to expect a fairly high level of quality and sophistication in presentations, and video technology offers the tools to meet their expectations. Many companies routinely produce videos for customer relations, training, marketing, employee relations, and other similar applications. The resources used for these applications, for example, cameras, recorders, editors, special effects generators, and studios, can be readily adapted to produce video for technical presentations.

Video presentations can be produced using information from either conventional camcorders or computers, or a combination of both. Camcorders are generally used to capture real-life scenes comparable to photographs, while computers are used to produce text, line graphics, charts, and artist concepts, including special effects such as animation, fades, dissolves, zoom, and rotation. Materials originally produced for other presentations, such as slides, overheads, and posters, can also be incorporated. Synchronized audio can also be easily recorded directly on a video tape.

One of the principal advantages of video is the dynamic opportunities it offers, that is, people, equipment, and processes can be realistically displayed. Action can be shown in accelerated, normal, or slow motion. Inanimate objects can be animated; and various types of simulations can be produced. The ability to incorporate audio eliminates or reduces the need for a presenter, and ensures the reliable repetition of the message. The zoom capabilities of video can also be used to focus on key points, a capability not readily available with other media.

While video can be used for virtually all types of presentations, its use for large audiences is hampered by equipment limitations. A normal video monitor screen is too small to be viewed by more than a small group. This limitation can be overcome by using multiple monitors or a video projector. Large-screen projection systems can display video images to audiences up to several thousand people, but the resulting image quality is substantially less than the image produced by a slide projector. These systems are also complex and expensive, limiting their use to fixed installations. Smaller video projectors, comparable to slide projectors in size, are also available, but their image quality is also less than other visual aids.

Summaries of the advantages and disadvantages of video follow.

Advantages

- Dynamic situations can be shown in a realistic and interesting manner.
- Activities can be recorded in real time from a remote location.
- The material being recorded can be immediately viewed (instant replay).
- The program content can be accurately repeated from one presentation to another.
- Quick-reaction turnaround and almost unlimited flexibility in program content structuring can be provided.
- Program can be executed without professional assistance, freeing presenters for other tasks.
- Material from other presentation media, such as slides, photos, and films, can be readily accommodated.
- Tapes can be copied quickly and inexpensively, and they can be stored and retrieved easily, with a relatively long shelf life.
- Video provides a highly effective communication medium with which virtually everyone is familiar.

TABLE 1
VISUAL AID SELECTION MATRIX

Visual Aid Type	Quality Level	Audience Size	Production Flexibility	Presentation Flexibility	Equipment Requirements	Relative Cost
Boards	Low	Small	Low	High	Minimal	Low
Flip charts	Low	Small	Medium	Low	Minimal	Low
Posters	Medium	Small	Low	Medium	Minimal	Moderate
Slides	High	Medium to large	Medium	Low	Medium	Moderate
Overheads	Medium	Medium	Medium	High	Medium	Low
Computer displays	Medium	Medium to large	High	High	High	Expensive
Video	High	Small to large	Medium	Low	High	Expensive
Multimedia	High	Large	Medium	Low	Complex	Very expensive

Disadvantages

- The required playback and projection equipment tends to be cumbersome to transport, and a quality image is difficult to provide for large audiences.
- The required production and projection equipment is relatively expensive.
- Personnel with special skills and training in the video production field are required.
- Changes are difficult to incorporate at the last minute or during a presentation.

MULTIMEDIA

Multimedia has traditionally implied the integrated combination of more than one visual-aid medium, for example, slides, video, or films, in a single presentation. With the advent of sophisticated computer technology, however, the definition of multimedia has been expanded to encompass associated capabilities such as laser and compact disks, touchscreen, still video, LCD and video projector screens, computer-generated graphics, animation, interactive video disks, synthesized sound, and video walls.

The basic concept of multimedia is to select and integrate the best features of each visual-aid medium for a specific presentation. This allows a presenter to capitalize on advantages and avoid the disadvantages of each, to meet the specific requirements of the applicable program content and duration, audience, available equipment and facilities, location, and support needs.

Summaries of advantages and disadvantages of multimedia follow.

Advantages

- The individual visual aids in a presentation can be selected to optimize the effectiveness of the presenter's message.
- Individual media can be selected to accommodate varying situational requirements.
- An audience's interest and expectations can be better accommodated.

Disadvantages

- The equipment and processes involved are relatively complex and expensive.
- Presentation planning, production, and control requirements are more complicated.
- The presenter must be more skillful and devote more attention to rehearsing the program.

SUMMARY

The selection of a particular type of visual aid or combination of media depends on several factors such as the desired quality level, audience size, production flexibility, presentation flexibility, equipment requirements, and cost. A matrix of these factors is presented in Table 1.

Many other factors, such as presenter experience and capabilities, presentation objectives and audience interests, presentation location and facilities, time available for preparation and execution, content review and security requirements, and record retention requirements, also impact the selection of visual aids. Each of these could be the subject of a separate paper, and many are addressed elsewhere in this book. No one type of visual aid is best for all applications, and the visual aids for each presentation should be carefully planned to optimize the audience's ability to receive and understand the associated information.

Visual Aids--
How to Make Them Positively Legible

CALVIN R. GOULD

Abstract—Illegible and uncomprehensible visual aids often result from confusion about sizes of lettering and symbols between the original artwork and its projected image. This paper discusses legibility standards used by industry and basic mathematical exercises to assure legibility of copy on visual aids used as original art or as reproduced slides.

S INCE visual aids are meant to help people in an audience understand information presented to them by a speaker, why is the audience so frequently bombarded by material that has the adverse effect of confusing or disturbing them?

Why do presenters introduce a slide with words like "As you can see from this chart . . . " when they know nobody beyond the front row can make it out? Better yet, why do they start with "I know you can't read this but . . . ?"

The answer may be that we are print-oriented in our thinking while we address a picture-oriented audience. In this new age of audio-visual communication we feel compelled to use visual aids when we speak, even though they may be a verbatim copy of the speech.

We know better than to use illegible or incomprehensible visual aids. We must learn to think more graphically, yet we know cost and time prohibits converting all words to pictures. We can improve our presentations tremendously if we can be virtually certain the audience can read and comprehend the visuals we put before them. All we need is a positive way to determine good from bad . . . answers to simple questions like – "What is a good visual aid?", and "How can I be sure of legibility for a given size audience?".

THE PROBLEM WITH VISUAL AIDS IS . . .

Though it may be difficult to find concurrence among opinions on what makes a good visual aid, we unanimously recognize practices that cause ineffective and illegible material.

A poor visual aid generally has one or more of the following faults:

1. It is too crowded
2. Color has been grossly misused
3. Line detail is too light
4. Letters and symbols are hard to read.

Manuscript received January 3, 1973.
The author is with Martin Marietta Aerospace, Orlando, Florida.

Too Crowded

These faults are often attributable to the user who expects too much from each visual. Copy or detail too voluminous for one visual must be more carefully thought out, edited down to its simplest form, then broken into several visuals if it is still crowded. Many experienced producers recommend that only one point be made with each visual.

Color Misused

Problems with color start when we think of using it for color alone. Color should be thought of as a function of the visual – cool backgrounds for cool subjects, and halting color for alarm. Multiple colors should have sufficient contrast when used together. Dark colors used in curves often blend together so that they all look like black lines from a distance. Pastel lettering on a white background will seem to disappear.

Line Detail Too Thin

Line detail that seems heavy enough to the person preparing a visual may look very thin from a few feet distant. How many times have we blinked our eyes to try to make out words being scribbled on a chalkboard by a speaker?

Illegible Copy

Generally, lettering and symbols are hard to make out because they are too small for the circumstances, a situation often caused by the first fault mentioned – too much information crowded into the visual.

A Base for Legibility

We can mathematically establish a legibility limitation for lettering sizes for given distances, but there are at least six other precautions to take before legibility is assured by size (considering basic titling, numbering, callouts and copy).

1. Use Block Lettering – A simple block letter, without serifs, should be used for copy on visual aids. The most acceptable design in type font is called Bold Gothic. Do not slant (italicize), use all capital letters.
2. Use a Proper Density – To avoid making a letter too thick or too thin use a standard line width 1/7 of its height.

Reprinted from *IEEE Trans. Professional Commun.*, vol. PC-16, no. 2, pp. 35–38, June 1973.

3. Space Properly – The space between words should be at least the width of a "W" letter, the space between lines no less than 3/4 the height of the letters. Use letter spacing for copy as opposed to an even space between each letter.
4. Stay with Horizontal Lettering – Vertical alignment, as often seen on the sides of Curve Charts, is hard to read, as is stacked lettering (putting letters below each other).
5. Don't mix colors in words.
6. Don't place copy on dark background nor use light colors for copy on light background.

As long as these admonitions are not seriously violated, it is possible to develop a standard approach to lettering that can virtually assure legibility.

MATHEMATICAL DEVELOPMENT OF LEGIBILITY

Legibility of letters and symbols will be assured if their height is 1/32 inch for every foot of distance from the viewed letter. A 1 inch letter is legible from 32 feet, a 1/2 inch letter from 16 feet, etc. (See Figure 1).

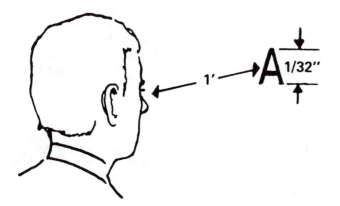

Fig. 1. For every foot (distance viewed) the letter height should be 1/32"

If the viewing distance is known it can be divided by 32 to determine the size of material to be used. Thus, for a viewing distance of 40 feet, the letter size required is 40/32 inch or 1 1/4 inch.

Optometrists use a letter that subtends a 5 minute arc, viewed from 20 feet, to measure eyes against the 20-20 visibility standard (See Figure 2).

Fig. 2. Geometry of a letter subtending an arc. Optometrists use for "x" a 5 minute arc to measure eyesight. This paper recommends enlarging to a standard of a 9 minute arc.

They use a chart with different sizes of black lettering on a white background. This is called a Snellen chart (See Figure 3). The conditions for viewing a Snellen chart in an optometrist's office are more ideal than in a typical

presentation, however. Considering different light values, backgrounds of different contrast, and eyesight variations of a mixed audience in a presentation, it is recommended that a larger size be used. The Eastman Kodak Company, in their pamphlet "Legibility Standards for Projected Material" states, "To ensure accurate recognition of symbols they should subtend at least 9 minutes of arc from the position of the farthest viewer." The 1/32 rule is based on a letter that subtends a 9 minute arc.

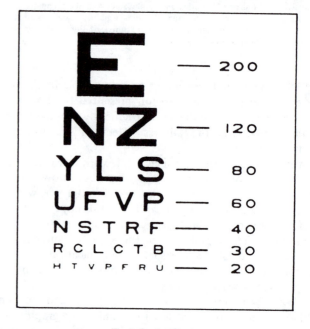

Fig. 3. Snellen Chart

LEGIBILITY FOR SLIDE ART

Establishing legibility for slide art requires that the enlargement factor be known, i.e., how much larger will the image be over the size of the original art? If the art work is 10 inches wide and the projected image is 5 feet (60 inches), the enlargement factor is 6 (See Figure 4).

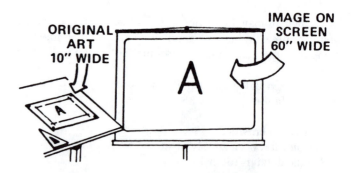

Fig. 4. Example illustrated here shows that the image-original art ratio is 6:1, thus the letter "A" on art work is 1/6 the "A" in the image.

The next step is to determine the size letter required for the greatest distance from which the 5 foot image is to be viewed. Let's assume the distance is 30 feet. Legibility for 30 feet requires a letter 15/16 inch (30/32). With the image being 6 times the size of the original art, the 15/16 inch letter in that image would have to be 1/6 of 15/16 inch; therefore, 5/32 inch in the art original.

Working in the other direction, lettering on original art merely has to be multiplied by an enlargement factor to establish the projected size from which the legibility can then be computed.

A dimension of 1/8 inch is often the minimum size of lettering on slide art work. This is equivalent to capital letters of 12 point Gothic and it is also the size of most standard typewriters. Let's consider 1/8 inch letters made on art 10 inches wide, the image to be 60 inches wide. One-eighth inch multiplied by an enlargement factor of 6 is 3/4 inch. The legibility of a 3/4 inch letter is 24 feet.

Generally, the only thing certain to an artist is the size of the letters on the original art. This is dictated by the input provided. But, an art department can absolve itself of the blame of producing illegible material if it adopts a policy of informing the user of the legibility of the slides. If the input forces 1/8 inch lettering in art 10 inches wide, a statement like "these slides are legible from a maximum distance of 24 feet when projected to a width of 5 feet" could be included as a note, or as an additional slide in the series.

A UNIVERSAL STANDARD

The facility used for a presentation dictates sizes to be used for legibility, but it may be hazardous to prepare a presentation to "fit" a specific facility.

Direct visuals, namely flip charts, cards and displays have a restriction based on the smallest lettering within them. Projectuals, on the other hand, can be enlarged to fit the circumstances.

Screen widths are generally determined by room length. A 6W/2W standard has been established as an image size most acceptable to quick scanning by an audience. This rule is generally followed by architects and audio-visual specialists when constructing conference rooms.* Under this standard, the last row of viewers should be no further than six image widths from the screen, and the front row of viewers should be no closer than two image widths (See Figure 5). Therefore, if producers follow the 6W standard in art work for slides and prepare all work in a horizontal format, legibility can be reasonably assured, regardless of the size of the facility.

CONSTRUCTING LETTERING
FOR THE 6W STANDARD

Recall an example referred to earlier which considered original art dimensions 10 inches wide, an image 5 feet wide, and a viewing distance of 30 feet. This example followed the 6W rule (6 × image width or 5 feet = 30 feet). The combination dictated a minimum letter size of 5/32 inch. If the original art was for 35 mm slides, the height of the original art would be 6 3/4 inches. (Aspect ratio of 35 mm slides is .676:1.) A 5/32 inch letter on that

size art is 1/43 of its height. With this in mind, legible art for 35 mm slides, against the 6W screen standard, can be prepared any size as long as the minimum height for the lettering is held to a size 1/43 of its height (based on horizontal format).

Because of the different aspect ratio of art for lantern slides (sometimes called "3 1/4 x 4") and filmstrips this letter changes to 1/46 of the height and with vugraph art the figure is 1/50 of its height.

IMAGE WIDTH vs SCREEN SIZE

We are inclined to think of image size as screen size. If all the computation to establish legibility is based on screen sizes and the full width of the screen is not used in projection, the minimum material will still be below the legibility standard. In practice, the image seldom fills the screen. Many screens manufactured today are square, but the ceilings of most rooms are below 10 feet. The bottom of an image should be 4 feet from the floor.

Mounted screens are usually 1 foot or so below the ceiling. Portable screens seldom stretch 10 feet high. Thus, less than 60% of the height of the 70 inch square screen is safe for projection (See Figure 6). When vertical slides are mixed with horizontals, the image must be reduced about 30% to project into a safe area. This is why many audiovisual people insist that all slides be photographed in the horizontal format (See Figure 7).

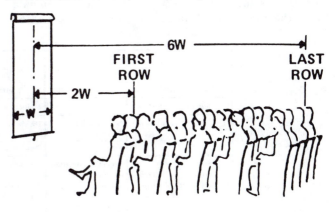

Fig. 5. Illustration of the 2W/6W formula used in projection setup.

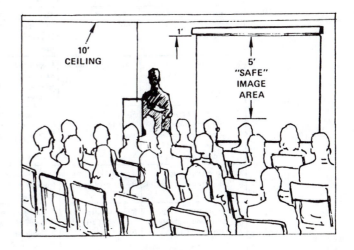

Fig. 6. Typical screen installation shows how lower portion of screen will be obscured by heads of audience.

*The typical portable screens in hotels are 70 inches wide, and most conference rooms in hotels are 30 to 40 feet long, which permits application of the 6W rule (70 inches x 6 = 35 feet).

WHEN VERTICAL AND HORIZONTAL ART FOR SLIDES IS MIXED, ALL ART SHOULD BE PHOTOGRAPHED TO USE SLIDES HORIZONTALLY

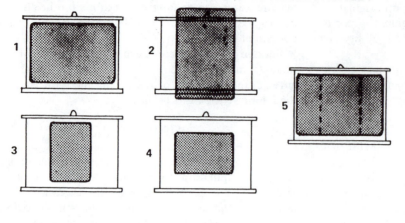

1. INDICATES IMAGE AREA WITH PROJECTOR POSITIONED FOR MAXIMUM SIZE.

2. VERTICAL IMAGE WITH PROJECTOR IN SAME POSITION.

3. VERTICAL IMAGE WITH PROJECTOR REPOSITIONED —

4. NOTE HORIZONTAL SLIDES ARE REDUCED BY APPROXIMATELY 30%.

5. BY FOLLOWING ABOVE POLICY MAXIMUM SIZES ARE ACHIEVED WITH MIXED SLIDES. COMPARE DOTTED LINE (VERTICAL IMAGE) WITH FIGURE 3.

Fig. 7. Illustration of logic of horizontal format.

SUMMARY

If all basic design criteria are properly employed, i.e., use block lettering of reasonable density, sufficient color contrast, etc., then legibility is a question of relating letter height to distance viewed.

As long as the letter used subtends at least a 9 minute arc to the most distant viewer, the relationship will be sufficient to assure legibility to all members of the viewing audience.

The size for such a letter can be postulated from any one of several different facts usually known about a presentation. However, if nothing is known except the size of lettering in the original art, the limit of that art can be positively established.

The chart that follows can help the reader review methods to compute a letter size subtending the 9 minute arc based on specific conditions.

KNOWN CIRCUMSTANCE	TO DETERMINE	SOLUTION PROCESS (FOR A LETTER SUBTENDING A 9 MINUTE ARC)	EXAMPLE	
			ASSUME	DEVELOPMENT
I Size of viewed letter	Legibility limitation	Multiply letter height by 32	Letter is 3/4"	3/4" x 32 = 24'
II Distance to last viewer	Minimum letter height	Divide by 32	Distance is 30'	30' ÷ 32 = 15/16"
	Size of image	Divide by 6 (for 6W formula)	Distance is 30'	30' ÷ 6 = 5'
III Size of letter in original art and probable width of image	Legibility limitation when projected as a slide	1. Divide image width by width of original art 2. Multiply answer No. 1 by height of letters in art 3. Multiply answer No. 2 by 32	Letter is 1/8" Image width is 60" Art work is 10"	1. 60" ÷ 10" = 6 2. 1/8" x 6 = 3/4" 3. 3/4" x 32 = 24'
IV Letter size dictated by input or typefaces available	Safest method to assure legibility	Make height of art 43 times height of letter and in proportion to slide format (horizontal layout)	Letter is 1/8" for 35 mm slide	43 x 1/8" = 5 3/8" Art for 35 mm slide should be no larger than 5 3/8" x 8"
V Existing slides and width of images	Legibility limitation	Project slide considered to have smallest letters. Measure height of that letter in image. Multiply by 32.	Smallest projected letter is 3/4"	3/4" x 32 = 24'
VI Letter too small when test projected on screen size planned for the presentation (knowing room size)	Required adjustment to presentation condition	1. Compute letter size required (see II above) 2. Enlarge image size by % factor required to achieve legibility*	Projected letter is 3/4" but should be 1"	1" − 3/4" = 1/4" 1/4" ÷ 3/4" = 1/3 Image must be 33% larger
VII To enlarge image	Acceptable legibility for existing slides	Move projector further away from screen or use shorter focal length lens		
VIII Letter size for legibility and probable width of image	Minimum size letter on original art (slides)	1. Divide width of art into width of image 2. Divide answer No. 1 into letter size	Screen is 60" wide Letter for known viewing distance is 15/16" Art is 10" wide	1. 60 ÷ 10" = 6 2. 15/16 ÷ 6 = 5/32"

* If this solution results in an image too large for the room, the alternatives are rearrange seating to accommodate slide limitation, get a larger room, or re-do art to accommodate facility.

VISUAL AIDS

How to Keep Participants' Attention

ROBERT W. PIKE

Research has proven that it's possible to learn much more in a given period of time when visual aids are properly used. Studies at the University of Wisconsin have shown an improvement of up to 200 percent when vocabulary was taught using visual aids. Studies at Harvard and Columbia show a 14 to 38 percent improvement in retention though the use of audiovisuals. And studies done at the University of Pennsylvania's Wharton School and at the University of Minnesota demonstrate clearly that the time required to present a concept was reduced up to 40 percent and the prospect of a favorable decision was greatly improved when visuals were used to augment a verbal presentation.

In his book *Presentations Plus*, David A. Peoples, a consulting instructor for IBM, says that people gain 75 percent of what they know visually, 13 percent through hearing, and a total of 12 percent through smell, touch, and taste. A picture, he says, is three times more effective than words alone, and words and pictures together are six times more effective than words alone.

NONPROJECTED VISUALS

Flip Charts

Flip charts are widely used—and misused—in the meeting/training process. They can be used effectively to create real-time visuals (that is, they're drawn as the presentation takes place), or they can be developed in advance.

Basically, flip charts are suitable only if your group is relatively small, 15 to 20 people at the most.

Here are six simple tips that can increase the effectiveness of a flip chart.

1. If you have basic points to make, pencil them in lightly on your various sheets of chart paper before the presentation. Then you can use your markers to write in the words so everybody can see them. In a sense, your chart paper becomes a very large set of notes or an outline.

2. Prepare some of your charts in advance, and cover the basic points with cut strips of paper. Then, as you get to each point in your presentation, you can simply remove the paper strip you've taped over the particular point on the chart.

3. Use a variety of colors beyond the basic black, red, and blue that we so often see. Consider not using these common colors at all. Try bright colors that people aren't used to seeing. For example, I use colors such as red-violet and blue-green—colors that are easily seen and just a bit unusual.

4. Leave the bottom third of your flip-chart sheets blank. This will allow people in the back to see your entire page and, as you post flip charts on the wall, it leaves space to go back and add more information.

5. Brighten the chart's visual appeal. You can do this by underlining or boxing key words. Use color, graphic designs, and geometric shapes to add visual interest to your chart.

6. Use flip-chart pages to record information. For example, during a brainstorming session, write key words that reflect each contributor's ideas. Better still, have two or three participants record on flip charts ideas as they come up. This allows you to maintain control of the group, clarify ideas, etc., while someone else does the writing. Give participants who will be recording material two different colored markers, and ask them to alternate colors as the ideas are presented.

PROJECTED VISUALS

Whenever you're going to use projected visuals, you must carefully plan, in advance, where to locate your projector

Illustration 1

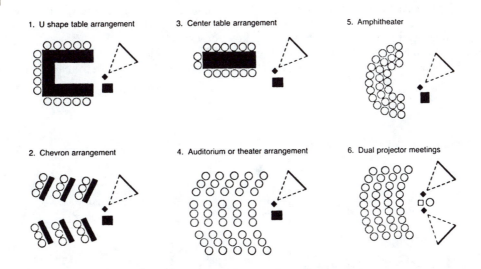

1. U shape table arrangement
3. Center table arrangement
5. Amphitheater
2. Chevron arrangement
4. Auditorium or theater arrangement
6. Dual projector meetings

(whether 16mm, video, overhead, opaque, or slide projector). Though these 20 guidelines may be most appropriate for the overhead projector, nearly all apply to all projected visuals.

1. Each person should be able to see every visual you use easily. The bottom of the screen should be at least 42 inches off the floor (see Illustration 2). This height is higher than the tops of the heads of 95 percent of your seated audience.

2. For best visibility, place the screen in a corner and angle it toward the center of the room. This is particularly true for overhead and opaque projectors, but it applies to other visual projectors as well. You don't want a large, blank screen to be the center of attention when you're not using a projected visual. If you choose to write on certain visuals, this placement allows you to do so while facing your audience and not blocking their view.

Illustration 1 shows some suggested room arrangements that can facilitate visibility. You'll notice the chevron is similar to classroom style, but it permits participants to see one another. The auditorium and dual projector arrangements feature seating in an arc for the same purpose.

Illustration 2

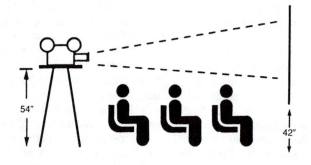

3. Your projector should not obstruct your audience's view of the screen. If you're using a 16mm or 35mm projector, place it on a stand so the audience has an unobstructed view of the entire visual. Special lenses are available that permit you to place the projector all the way at the back of the room on a high stand (normally 54 inches high) and project over the heads of your seated participants. Illustration 2 depicts this arrangement.

If you're using an overhead projector, which is placed in front of your participants, make sure it does not block your participants' view. Place an overhead projector on a surface

Illustration 3

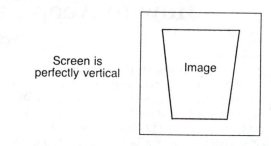

Screen is perfectly vertical

Image

large enough to hold at least two stacks of visuals and other equipment and props you may use in your presentation. This arrangement allows you to move prepared visuals from stack A, those you haven't used yet, to stack B, those already used.

4. Avoid image distortion or "keystoning"—as shown in Illustration 3—by having the projector beam meet the center of the screen at a 90 degree angle. Keystoning occurs when the projector is placed low, in order to keep it out of the audience's line of vision. It is caused by the difference between the distance that bottom and top rays of light have to travel between the focus of the screen. The farther the ray of light has to travel, the wider the image. You can avoid this effect by tilting the screen forward at the top or backward at the bottom so that the distances traveled by the various rays of light are about the same at all points on the screen (see Illustration 4).

5. Whenever possible, use a matte surface screen for greatest image visibility and seating breadth. Beaded (or lenticular) screens reflect more light to viewers in the center of the room, but the image dulls and drops off when viewed from the sides of the room. A beaded screen is fine if you have a long, narrow room, but a matte screen will serve you best in most situations.

6. To select the right screen size and seating arrangements, remember the two-by-six rule:
- The distance from the screen to the first row of seats should equal twice the width of the screen (2w).
- The distance from the screen to the last row of seats should equal six times the width of the screen (6w).
- No row of seats should be wider than its distance to the screen.

7. The projected image should fill the screen completely. If your screen is adjustable from top to bottom, raise the bottom

Illustration 4

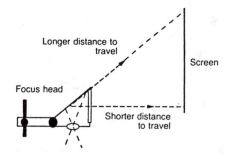

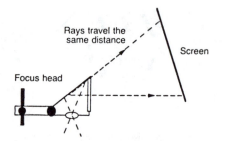

of the screen in order to frame the image you're projecting.

8. Always carry extra accessories. An extension cord, three-pronged plug adapter, masking tape, marking pens (for an overhead projector), blank sheets of write-on film, and spare projector bulbs. The availability of these "spares" may mean the difference between showing the visual portion of your presentation and not being able to.

9. Know the room you're going to use. Know the location of light switches, electrical outlets, and heating and air-conditioning controls, and know how to use them. Which light switch do you use in order to darken the room partially? Can you turn off the lights in the area immediately in front of your screen?

10. Know the location of the telephones and the name and number of the individual you can contact in case of an equipment emergency.

11. Set up and check out your equipment in advance. Check the screen for keystoning. Test and focus the projector. If you're using sound, check the levels. If you're using an overhead projector, clean the glass stage of the projector with a soft cloth and water or lens cleaner. Tape any cords to the floor. Know how to change your projector lamp if necessary.

12. Check your visuals.

13. Make sure the air vent on your projector is clear. If it's blocked, the lamp will overheat and burn out sooner.

14. Check your projector and screen placement by drawing sight lines at the top and bottom of the screen for both the audience (being careful to allow the sight line to extend over the heads of people seated in the front rows) and for the projector itself.

15. If you're going to use a back-of-the-room projector, you may want someone else to operate it for you. If it's a slide projector, avoid having to say "Next slide, please" by giving the projectionist a copy of your presentation. Underline the words you'll use as cues for changing the slides, and rehearse these with your projectionist. Alternatively, you might give your projectionist a visual cue, such as touching the knot of your tie or your necklace. Make sure you and the projectionist know what to do if a slide jams, the film breaks, or something goes wrong with the videotape.

16. Clean the lens of the projector.

17. For larger rooms, consider a high-intensity projector that will project lots of light over a large distance to a large screen.

18. Short focal-length lenses can be used to produce a large image on a screen from a short projection distance. This is useful for viewing large, impressive images in a shallow, wide room.

19. Long focal-length lenses can be used to produce small images or in places where long projection distances are required for relatively small images.

20. To facilitate viewing, seat people in a fan-shaped area of about 70 degrees with the center perpendicular to the screen.

ROOM ARRANGEMENTS

Picking a Room

Whenever possible, hold your meeting in a room appropriate to the size of your group. If, however, you have a choice between a room that's a little too small and one that's a little

too large, select the latter. There are ways to make a too-large room look smaller and cozier, but there's nothing you can do about overcrowding. The ideal room ratio is 1 foot of width to 1.2 feet of length. For most meetings, 12 to 20 square feet should be allowed per person.

Seating

Be sure the chairs are comfortable; uncomfortable chairs can decrease your audience's attention span. Whenever possible, get chairs with armrests. Do you have enough chairs and tables for the number of participants you're expecting? Are the exits clear once the room setup is complete?

Screen Viewing

Avoid or minimize any obstructing posts or columns that interfere with visibility either of the speaker or of the group members. Everybody should be able to see clearly when seated.

INSIDER'S TIP: Avoid placing an aisle in the center of your audience. The center offers the best seating for viewing, so place the aisle to either side.

Lighting

Make sure the room is well lit but not so bright that lights distract your listeners with glare. The room should be dark enough for projection yet light enough for note taking. Check to see if the lights can be dimmed or switched independently. If not, can the light bulbs immediately in front of your screen be removed in order to darken the area around the screen? Use incandescent lights with dimming systems if possible. Check and label all light switches, and assign a person to turn lights on and off on cue. Light sources that create flicker can cause distraction and discomfort.

Power

Are electrical outlets adequate and conveniently located? How much extension cord will you need to bring power to your equipment? You should know if the current is AC or DC. Will it run your equipment? If you haven't brought an adapter with you, you may have problems with two- or three- pronged plugs that don't match up to the outlets. How much electricity will your meeting require? Overloading the line running to your meeting room will bring your class to a halt.

Check to see if outlets are switched and fused separately from the room lights and if spare fuses and standby circuit breakers are ready. An electrical outlet should be located near the head table to permit plugging in any equipment used by the speaker.

Acoustics

Bouncing sound waves get on the nerves of speakers and listeners, so check acoustics by clapping your hands together slowly. A brittle, ringing echo indicates poor acoustics. If you have this problem, you can reduce the bouncing sound waves by draping walls with fabric, placing carpeting on floors, or, if possible, finishing floors and ceilings with acoustic tiles.

Be sure that sound carries to all parts of the room. "Dead spots" can occur and can distract people from your message by making it difficult for them to hear. Make sure, also, that there is no interference from noisy mechanical equipment (e.g., projectors) placed too close to participants. Check to be

Illustration 5

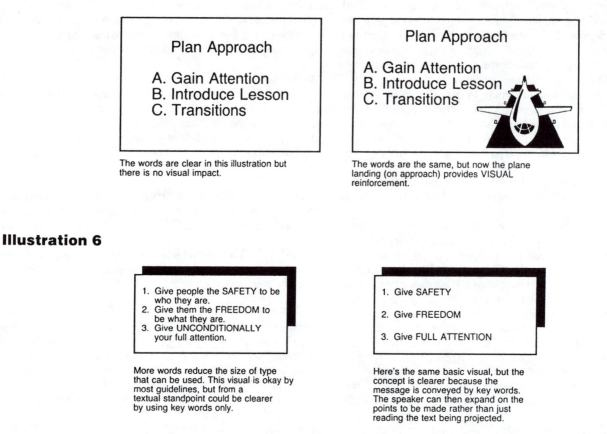

Plan Approach

A. Gain Attention
B. Introduce Lesson
C. Transitions

The words are clear in this illustration but there is no visual impact.

Plan Approach

A. Gain Attention
B. Introduce Lesson
C. Transitions

The words are the same, but now the plane landing (on approach) provides VISUAL reinforcement.

Illustration 6

1. Give people the SAFETY to be who they are.
2. Give them the FREEDOM to be what they are.
3. Give UNCONDITIONALLY your full attention.

More words reduce the size of type that can be used. This visual is okay by most guidelines, but from a textual standpoint could be clearer by using key words only.

1. Give SAFETY

2. Give FREEDOM

3. Give FULL ATTENTION

Here's the same basic visual, but the concept is clearer because the message is conveyed by key words. The speaker can then expand on the points to be made rather than just reading the text being projected.

sure that no loud sounds from outside the room will distract your class.

PREPARING TRANSPARENCIES

1. Limit your work area on the original to a maximum of 8 by 10 inches. This will ensure that, once your visual is prepared, you'll have room to mount and project it without having the framing block part of the visual.

2. Limit each visual to one idea. If the topic you're covering is more complex, you might want to use overlays. You can superimpose additional transparencies on top of a base transparency to build a concept or progressively present a complex issue.

3. Keep your visuals as simple as possible. Excessive wording or too-elaborate diagrams on a single visual not only compete with you, they become more and more difficult for the audience to read. Remember the six-by-six rule: No more than six lines per transparency and no more than six words on a line.

4. Be imaginative. Use illustrations, cartoons, graphs, maps, and charts whenever possible, instead of relying exclusively upon words or numbers, as shown in Illustration 5.

5. Use appropriate type sizes.

■ Use at least 18-point type (¼-inch high letters) or larger.

■ Use bold, simple typefaces.

■ Avoid ornate styles for maximum readability.

■ Vary the type size in order to illustrate the relative importance of information.

■ Never use a typewriter to create your masters.

■ Use the same type style for each series of transparencies.

■ Use upper and lowercase letters. Lowercase are generally more legible than all uppercase. Occasionally use only uppercase for contrast or for headings.

■ Vary the length of words and provide ample spacing between the letters, between the words, and between the lines of type. A good guideline: between each line of type, leave a space equal to the height of an uppercase letter of the size and style you are using.

■ Use tinted film to reduce lamp wear and colored markings to add realism and emphasis. Tinted films also can provide a means for color coding your transparencies. You might use the same tint for a series of transparencies on one topic, then change the background tint for another topic.

7. Use the space on your transparencies to make ad lib markings during your presentation.

8. Position your material on the upper part of your transparency. When you project it, your audience can view it readily since it will be on the top part of the screen.

9. Avoid using both horizontal and vertical forms. Many experts suggest using horizontal visuals exclusively for maximum visibility.

10. Choose your words carefully. When creating visuals, try to think in "bullets." Use active words and short phrases, as shown in Illustration 6.

11. Use—but don't overuse—color. A maximum of two to

three colors per transparency should be sufficient. More than that will make it difficult for the eye to focus on the important parts of the visual.

12. Avoid vertical lettering. A quick look at Illustration 7 shows the importance of this suggestion. Vertical lettering may look fancy, but it's very difficult to read.

Illustration 7

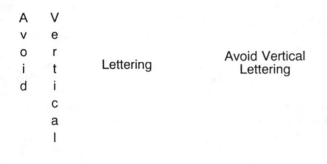

13. Use a maximum of two type styles on any single visual. Any more than two causes the type styles to compete with the information being presented (see Illustration 8).

Illustration 8

Don't use more than two type styles.

Don't use more than two type styles.

Don't use more than two type styles.

Don't use more than two type styles.

Don't use more than two type styles.

14. If you're using non-sequential items in a visual, don't number the items. Use checkmarks, bullets, boxes, arrows, etc.

15. Clip-art books can be a great source of copyright-free illustrations, giving you inexpensive, ready-made professional art.

USING TRANSPARENCIES DURING YOUR PRESENTATION

Very simple techniques can significantly affect your presentation of visuals. For instance, by turning the lamp on and off at appropriate times, you can control your audience's attention. Keep the lamp off when changing visuals; you won't have to look at lamp glare, and your audience won't be distracted if positioning is clumsy.

If the objective is to block the light but not turn if off, there are other alternatives. One is to place a sheet of paper or a file folder trimmed to size underneath your visual. Slide the sheet down to reveal information, and, when you're finished, slide it back into place on the projector stage to block the light while you're positioning the next slide.

Another alternative is to tape a 6-by-6-inch piece of

cardboard to the projector head so you can flip it out of the way when you want to project an image and flip it down when you want to block the light. If you add detail to a transparency during the presentation, use nonpermanent markers or write on a clear sheet of film placed over the prepared visual, so you can use the original again.

If an assistant is changing the visuals, determine in advance a system of signals so this individual knows exactly when to change the transparencies without your audience having to hear you call for each new one. Once again, it is important that the visuals be clean, in the correct order, right side up, and facing the correct direction before your class starts. Keep blank sheets of film beside the projects so you can ad lib on your visuals.

If a transparency is to be used repeatedly, make copies for projection and store the master. When your copies begin to wear, you can use the master to make additional copies.

Be careful not to obstruct the projector light beam if you move around the room during a presentation.

PRODUCING AND USING SLIDES

Many of the basic guidelines for production and effective use of overhead transparencies, including design and layout, can be applied directly to the use of 35mm slides. If you're going to use slides instead of overhead transparencies, consider the following 18 basic guidelines.

1. Edit slides beforehand to make sure you have the right ones, that they are in sequence, and that none are upside down or the wrong way around.

2. Decide if you want horizontal, vertical, or mixed slides. This affects the screen size and projected image size. In most cases, horizontal format slides (slides that are wider than they are high) should be used, because most screens are wider than they are high. Vertical slides or a mix of vertical and horizontal slides are better used on a square screen.

3. Design slides for the back row.

INSIDER'S TIP: If you can read your slide when you hold it in your hand at arm's length, your participants will be able to read it when it's projected.

4. For average back-row viewing distances, large letter height is advised. A maximum of five to six words per line and a maximum of five lines of copy per slide are good general rules of thumb. To determine maximum viewing distance, use eight times the height of the projected image.

5. For maximum effect, use 10 or fewer words per slide.

6. To make sure your lettering is of readable height, have letters measure ¼ inch on the finished slide.

7. Use strong, bold, sans serif typefaces for reading ease. Don't use all uppercase letters. Provide ample spacing between words and letters.

8. Choose simple words. Use active words and short sentences, and write in conversational style (write for listening). Words should reinforce visuals.

9. Make sure slides are clean. There are several commercial cleaning products available.

10. Make slides crisp and dynamic.

11. Rehearse your presentation.

12. Use illustrations, cartoons, and drawings whenever possible.

13. Make good use of color contrast. Use a dark blue, black,

or brown background. Use white, yellow, or red for the letters and pictures.

14. Use a remote-control projector, or have an assistant change slides so you can face the audience as much as possible.

15. Mark the forward button on the remote switch with white tape. This prevents going backwards and makes it easier to locate in the darkened room.

16. Start and end with a black slide so you won't shock the audience with blinding lights.

17. Depending on the type of projector used, mount slides in plastic or in metal and glass holders. Cardboard-mounted slides tend to bend and may jam the slide tray.

18. Limit the number of slides to the time available. Figure on 15 to 20 seconds per slide.

SUMMARY

Remember, visuals are instruments used in presenting a message, but they are not the message itself. Select your visual materials with care, aiming for simplicity and choosing the easiest and most effective transmission system. Familiarize yourself with the equipment you will use. A few minutes spent with the instruction manuals can save you embarrassment and ensure a smooth-running presentation.

Rehearse your presentation in advance. Review your visuals with your narration. Try to anticipate questions that might arise. You'll be rewarded for the time and effort you invest with greater self-confidence and poise. And your audience will experience a more effective presentation.

You may never have given a presentation before, but, with the help of good visuals and with conscientious rehearsal, you can begin to rank as a skilled presenter. Obviously, properly selected and presented visual aids can add significantly to the power and impact of your presentation. Remembering the six key Ps—Proper Preparation and Practice Prevent Poor Performance—will help you achieve the results you want.

Designing Slides

JACK REICH

Abstract—Slide design is a professional activity and should be done by a professional designer. The suggested ideas and visual material given to the designer are crucial to the design of an effective slide presentation. This description of what the designer does is meant to be helpful to the engineer in structuring and writing a presentation and in preparing visual material for the designer. Preparation and evaluation of slide material, design, typography, production, and slide use are discussed.

THIS paper discusses the design and production of slides. It does not tell you how to design slides. My remarks are largely concerned with what the professional designer thinks about and then does when designing slides. This should give the engineer enough information about slide design to help him or her structure and write a slide presentation and prepare the visual material for the designer in a way that will result in slides, principally 35-mm, that are effective and well-designed.

Preparing and Evaluating Visual Material

In preparing a slide presentation you may have clear and precise ideas about what the visual material should be, or you may have none at all and simply present the script to the designer for visualization. Either approach has its advantages and is workable, although the first is the one most often used. When a presentation deals with a technical subject there is usually plenty of material that could be shown, and it then is only a question of deciding what will be used. The designer can often be invaluable in making an objective appraisal of what is necessary and how it can be most effectively used.

Slides should express, support, or supplement the ideas the engineer wishes to communicate; and effective, well-designed slides present those ideas in as clear and direct a way as possible, free from visual complexity and cliché.

Careful organization and a logical structuring of the slides by the engineer can be of great help to the designer, although this should not be so rigidly done that the designer cannot contribute fresh ideas or propose a different organization.

The designer begins with an evaluation of each proposed slide, often in conjunction with the engineer. Some questions that are discussed at these sessions (and that should be addressed by the engineer when material is being prepared for the designer) follow.

· Is there a need for the slide at all, or is it there only because you think there should be "something visual"? Each slide should be a significant addition to the presentation, make a complicated or complex idea clear, amplify an idea, or add interest to an idea.

· Can the audience understand and absorb the information on the slide in the time they have to see it? Are there too many slides saying too little or too few saying too much? Too many slides or too fast a pace can be tiring. Holding a slide on the screen or having none there at all can be a welcome relief to the audience's mind and eyes.

· Are the suggested visual ideas just that—suggestions? The designer should be free to develop fresh and imaginative solutions rather than try to revive clichés such as bull's eyes, targets, visual puns, or stick figures, or be asked to use cartoons or amateurish drawings.

· Is it realistic to expect that an effective slide can be made from the material suggested? Often a statement or idea cannot be visualized and requires only a word or phrase on the slide to reinforce the text.

· Is material chosen and organized so that the design can be consistent and unified? Sudden shifts in direction or the interjection of unrelated styles—an illustration in the middle of an all-photographic presentation—are disruptive.

· Are there copyright problems? Copyrighted material cannot be reproduced without permission. Photographs, diagrams, and text taken from magazines and books are almost always copyrighted.

· Is the material in final and complete form? It is extremely difficult to design slides that have continuity and unity when the supplied material is incomplete or uncertain.

· Is the meaning of diagrams clear? The designer should not be expected to have an engineer's understanding of technical illustrations, special languages and symbols, or shorthand conventions.

· Is copy typewritten? Typed material is easier to read and presents fewer problems of deciphering and comprehension.

· Are grammar, spelling, and tabular material consistent and have they been checked for accuracy? Some technical material has rules that abridge ordinary syntax and spelling; the engineer should be aware of the exact meaning of the words and have a complete understanding of the visual material.

· When writing the script, were the slides always seen as structurally necessary to the presentation? When reading the script, do the suggested visuals intelligently support or amplify the words? Presenting two messages—for example, one in the text and another on the slides—usually means that the audience will either listen *or* look, missing one of the messages, or try to listen *and* look, and miss half of each.

This evaluation of *each* slide, as well as of all the slides *with* the script, can significantly reduce the time needed to prepare the slides, and can dramatically lower the cost of their produc-

Manuscript received March 31, 1978; revised April 28, 1978.

The author is Art Director for the Promotional Programs Dept., Data Processing Div., IBM Corp., 1133 Westchester Ave., White Plains, NY 10604, (914) 696-2515.

Reprinted from *IEEE Trans. Professional Commun.*, vol. PC-21, no. 3, pp. 108–110, Sept. 1978.

tion. Later changes in direction or corrections are costly and time-consuming and can almost always be avoided.

DESIGNING THE SLIDES

Design begins only after the designer and the engineer agree that each has a thorough understanding of the whole presentation—words and slides.

The purpose of slides is to give information in an orderly way without visual distractions and the necessity of constant reorientation by the viewer to a new or different slide format. It is important to establish a visual format with the first slides so that the audience can concentrate on content.

The designer uses a "grid" to develop a format for the whole presentation and to organize the elements of each slide. A grid is an organizational tool that places each element—text, diagrams, photographs, artwork—on the slide in a logical, orderly way. This grid is determined by the structure of the material to be shown on the slides, and the design of any one slide is determined by the structure of the most complicated slide. The grid evolves (in a rather complex way) from the basic type size used, the size and shape of the slide, and the character of the material to be presented.

The first step in developing a grid is a careful analysis of the material for all the slides to determine the number of common elements: headings, text, charts, tabular material, photographs. The fewer the number of these elements the clearer and more consistent the presentation can be, and the easier it is to design.

Such an analysis often results in a different or simpler organization: One slide becomes clearer when an important element is removed and made into a separate slide; two slides with redundant elements can be combined into one by eliminating the redundancy.

When this analysis is complete, a grid is drawn that allows all similar elements on all slides to be the same size and in the same position. This gives consistency to the design in the same way that grammar and syntax give consistency to a written text. This consistency is the key to good design.

TYPOGRAPHY

Because type is crucial to the grid and thus to the design, type selection is of utmost importance.

Photocomposition is generally preferred over hot metal such as linotype and monotype. It is less costly, more flexible, and its precision and clarity (it is a film process) allow the extreme enlargement necessary for slides with little or no loss of sharpness. Other methods can be and are used, such as minicomputer composers, sophisticated memory and reproduction typewriters, the ordinary typewriter, and even handwriting. Results range from near-professional to marginal quality, but all can be used successfully by a careful designer who knows their limitations.

Because of peculiarities of light when projected through small or narrow openings, sans-serif type should be used except when all the type is very large. Since mixing serif (or Roman) and sans-serif type is usually poor typographic practice, sans-serif type should be used throughout. Helvetica, Helvetica Medium, Univers 55, and Univers 65 are excellent type styles.

Once a type face has been selected it should be used exclusively. Most type faces are available in a range of weights and italic that can fulfill all typographic requirements. Fancy type faces should be avoided; their overdesigned character is inappropriate for the informational purpose of most slides.

Photocomposition machines have the ability to decrease or increase the space between the letters of a word. But in all cases "normal" letterspacing should be used. Negative—or tight—letterspacing, unless it is used with extreme circumspection, can result in type that is very difficult to read. On slides, reversed letters (white or colored letters on a black background) that are very close together allow the light projected through them to "spill over" from one to the next, causing a vibration-like fuzziness that further reduces legibility.

Flush-left type is simpler and easier to read and should be consistently used instead of centered lines of type. Indents can be used to indicate paragraphs, although a one-line space between paragraphs is usually a clearer indicator.

Type set entirely in capital letters is harder to read than type set in upper and lower case letters. Except for acronyms, copy should not be set in all caps. Their use for emphasis is rarely warranted; italic type should be used for this purpose, if it is needed at all. Except for proper names, only the first letter of the first word in headings and text should be capitalized.

Two sizes of type are usually all that is necessary for any slide show. A text size, a heavier (bold) version in the same size for headings, and a smaller size in the text weight for charts work well. If possible, before slide copy is written, the designer should determine the maximum number of lines that can be accommodated on one slide, as well as the maximum line length in number of characters (letters plus punctuation plus spaces between words).

Usually at this point, if the slides are many or complicated, a storyboard (a rough but meaningful and accurate sketch) is made of the whole presentation. Now is an opportunity to get a general idea of what the finished slides will look like. It is often evident that the presentation is going to be too long and weighty (or too short and trivial) or overly complicated and detailed. Now is the time to make changes; to make them at any later time is usually time-consuming and costly.

PRODUCING THE SLIDES

Slides can be physically produced in a number of ways, from source material ranging from computer-generated multicolored images to typewritten pages photographed by a 35-mm camera. The method described here produces professional slides of excellent quality at reasonable cost; it is a gelled Kodalith (trademark of Eastman Kodak Co.) method.

First, mechanicals are prepared: Type is set, diagrams are drawn, reproduction-quality proofs are "pulled," and all the elements of each slide are assembled on "boards" (heavy white cardboard) in their proper relationship just as they are to appear in the finished slide. The mechanicals are made at a convenient size for the work to be done—usually three or four times the final 35-mm slide size. This might be 10 cm X 15 cm since the height-to-width ratio of 35-mm slides is 2 to 3.

The finished mechanicals are usually reviewed by the engineer for accuracy—but, nominally, not for changes. Changes at this time indicate that there was poor conceptual organization of the material at the beginning and can create significant design problems as well as loss of time and money.

Next, the mechanicals are photographed (usually 1:1) on Kodalith film, a special film that produces a sharp, clear, opaque black negative. The negatives are then "gelled." Theatrical gels (colored transparent films) are placed over the elements on the slides that the designer wishes to be in color. The backgrounds of all the slides will be black and the type or artwork will be in whatever color is chosen.

The final production step is the photographing of the gelled negatives. This produces a negative strip of film from which the final slides or multiple copies of the slides are made.

Slides mounted in cardboard are usually adequate, although for repeated showings and for protection from fingerprints and damage, glass mounts may be worthwhile.

Using Slides

All slides should be either horizontal or vertical. A vertical format is rarely used, however, because most projection screens are square and the lower portion of the screen can be obscured by heads in the audience. A vertical image uses the full depth of the screen while a horizontal image of the same size needs only the upper portion. It is also easier to organize design in a horizontal format.

There are certain factors affecting slide legibility that should be kept in mind during their design and production: the size of the room in which the slides will be shown, the size of the audience, the size of the screen, and the ambient light. All affect legibility, sometimes in unfortunate ways.

Type that is readable at 7.5 meters may be difficult to read at 15 m and impossible at 30 m. Red type that is perfectly legible at 15 m in a completely dark room may be impossible to read at half that distance in a room with some ambient light. Red, blue, magenta, and green are particularly troublesome colors under certain conditions; the number of layers of gels, and the particular brand of gel, can also affect legibility. There is even a color, Surprise Blue, that projects as pink. Obviously, color selection can be a tricky business and is better done by someone with experience in slide making.

As a general rule, white backgrounds are not recommended because they tend to produce glare in a darkened room. White type against a black background also has this effect.

Finally, it is always better to use the original slides rather than duplicates (especially in cases of photographs of places or people) since each successive copy becomes darker and muddier and less true to the original colors.

Legibility in Film-based and Television Display Systems

Gerald F. McVey

THE LEGIBILITY OF ALPHANUMERIC AND GRAPHIC MATERIALS *employed in film-based and television display systems is affected by many factors, including such classical factors as symbol size, width-to-height ratio, stroke width, spacing, brightness contrast, contrast direction, and color. This article discusses these factors and their interrelationships with display-system characteristics and environmental and human factors, reports the findings of relevant research studies, and offers guidelines for the specialist in visual design and presentation.*

The legibility of alphanumeric and graphic materials displayed via a film-based display system (i.e., slide, overhead, or motion picture projection) or via a television system is the result of many factors: such classical factors as the size, form, and dimensions of the characters; stroke width; spacing; display brightness; contrast direction; and color,[1] and other factors such as characteristics of the display system, the viewing environment, and the visual capability of the audience.

There is a wealth of legibility research,[2] but most studies were funded under defense contracts and deal primarily with military problems such as target detection, pattern recognition, and map reading. Few are concerned with the problems that are relevant to the classroom, training center, and presentation spaces in general. These problems include symbol and word recognition and the accurate identification and interpolation of projected charts and graphs. Substandard projection continues to plague school classrooms and training rooms in business and industry.[3]

This paper presents a number of practical recommendations to assist the visual design and display specialist in improving visual presentations. These recommendations have been developed through analysis and synthesis of existing research and studies conducted by the author and his graduate students over the past decade.

LEGIBILITY DEFINED

Frequently, some confusion arises regarding the meanings of legibility and readability. These terms are interrelated, yet they are different and pose different requirements for the visual design and presentation specialist. Legibility is a measure of the ability to detect and discriminate between individual characters. It is a function primarily of visual acuity—the ability to see small detail—and is an essential prerequisite for reading. Readability is a component of comprehension and requires that the reader synthesize individual characters into meaningful units. It is very dependent on the pattern of eye movements elicited by different type styles, word spacing, and text layouts.[4]

The suggestions offered in this article are intended for projected visual displays and not printed textbooks and manuals. While both products require certain amounts of legibility and readability, the emphasis given to each of these factors is a function of their specific objectives. For example, designers of textbooks or manuals are guided by design principles different from those that guide designers of transparencies intended to supplement speeches given in an auditorium. In the latter case, the visual designers might be willing to sacrifice some reading speed if it would result in improved recognition accuracy for those seated in the less desirable viewing locations of the auditorium. Conversely, some type faces and design layouts recommended for film projection would result in lower reading speed and comprehension if used for printed text. For example, text set in IBM Composer Univers will produce very good white-on-black Kodalith slides but does not work very well as print in a book. It is legible, but its thin stroke width and static shape and spacing make it slow and difficult to read.[5]

Legibility owes its primary nature to a number of interrelationships among so-called classical factors. The chief among these, which we will concern ourselves with, are symbol size; symbol form, shape, and dimensions; stroke width; spacing; display brightness; brightness contrast; contrast direction; color; and resolution.

SYMBOL SIZE

Symbol size is the most serious constraint in display design.[6,7] The size of all elements in the visual field can be thought of in terms of the angle that they measure (subtend) at the eye. This is an important concept, for the area that a visual target or object covers on the eye's retina is a major factor in determining whether it will be detected and recognized.[8] It is also a convenient measurement, for it takes into account an object's inherent size and distance from the viewer. For example, a symbol having a height of 1 inch at the display surface will measure (subtend) a visual angle of 30' (minutes of arc) at the eye of a viewer 10 feet away. However, to a viewer 16 feet distant, this same symbol subtends 18' of arc, and to a viewer 20 feet away, it subtends only 15' of arc (Figure 1). While the human eye can discrimi-

Reprinted with permission from *Technical Communication,* published by the Society for Technical Communication, vol. 32, pp. 21–28, Fourth Quarter 1985.

nate a visual angle subtending as little as 10 seconds of arc under the best viewing conditions, a display should be scaled so that details subtend at least 1 minute of arc and so that a symbol's overall height subtends an angle considerably larger than that required for minimum recognition and readability.[9] Given a well-designed graphic displayed on a high-resolution display system to college-age students with normal or corrected-to-normal vision and an optimum viewing environment, one can expect 100 percent recognition accuracy with a symbol subtending 10' of arc or larger. Reduce this symbol size to 7', and recognition accuracy will be reduced to 95 percent under the same conditions.[10] Should these same condtions exist, but a standard broadcast or high-quality closed-circuit television system be used, then symbols of increased size (minimum of 15') should be employed to achieve the same results.

Table 1 was prepared to help determine the appropriate minimum symbol height (at the display surface) for different display conditions. The user must (1) select the viewing condition that most closely matches that of his or her audience, (2) obtain the symbol height for the viewing condition chosen (this symbol height is for a viewer 1 foot from the display), (3) determine the maximum viewing distance the audience will be from the display (the seat farthest from the screen), and (4) multiply the symbol height by the maximum viewing distance in the presentation room to ascertain the required symbol height at the display for that situation.

To determine the symbol height needed on original art work in order to produce the required symbol height at the display surface, the reader must determine the ratio of the width or height of the display surface (projection screen, TV screen, etc.) to the width or height of the original art work that is to be photocopied and multiply this by the symbol height at the display (calculation just described).

For example, suppose one wishes to know the minimum height of symbols (measured at the display surface) that is satisfactory for viewing closed-circuit television on a standard monitor by persons seated 30 feet away. From Table 1, class D is applicable, and a minimum symbol height (at the television monitor) of 0.052 inch is required for viewing by someone 1 foot distant. However, in our scenario, some viewers are as much as 30 feet away from the screen. Hence, 30 times 0.052 inch yields 1.560 inches as the required symbol height at the monitor.

Next, consider what size the symbols would have to be on the original art work (9 inches by 12 inches, for example) to produce this 1.56-inch height on the monitor. A standard 21-inch monitor has a width of 19 inches. The poster board, as noted earlier, has a width of 12 inches. Thus, the ratio of the poster-board width to the monitor width is 12/19, or 0.63. This ratio, 0.63, times 1.56 inches equals 0.9891 inch. In other words, the presenter will need a symbol approximately 1 inch high on the original art work to produce the required symbol height of 1.56 inches on the television screen. This symbol will subtend a visual angle of 15' of arc to a viewer seated 30 feet away and thus will meet the minimum requirement set forth in Table 1.

Table 1 applies to conventionally designed viewing environments, such as classrooms, lecture halls, and auditoria, where there is a modest amount of geometric distortion of the display at the side-most viewing locations. In unusually wide viewing environments, either symbol size should be increased, or viewing distances at side-most locations should be reduced in order to maintain legibility.[11]

GUIDELINE: Determine the minimum acceptable size of alphanumeric characters on the basis of the type of display to be employed, the anticipated viewing condition, and the expected distance of the viewer located farthest from the display.

FORM, SHAPE, AND DIMENSIONS

The factors in a symbol's design that affect its ability to be recognized and read include character style, simplicity, shading, amount of space surrounding the outline, aspect ratio, and stroke width.[12] Certain alphanumerics resemble each other so much that they are frequently confused, for example, G, C, and O; 3, R, B, and 8; and 2 and Z. To avoid confusion, particularly in a television display, a number of researchers have recommended the use of the Lincoln/Mitre symbols (Figure 2).[13] Other letter styles that have been recommended for television use are Leroy 3245.500 C template and 3233-6 lettering pen, and Artype No. 1158 Futura Medium.[14] Typefaces recommended for film projection include Univers, Helvetica Medium, Tempo Bold, and Futura Bold.[15]

It is also recommended that segmented (bar-matrix) numerals similar to digital readout symbols not be used in low-resolution systems, such as standard television displays. Similarly, although quite legible as type face in a book, serif symbols and letters having a variable stroke width are not recommended for television or film-projection display. The reason is that the thinner stroke elements that make up part of a symbol's design frequently are lost either in making the visual or in displaying it.

An important element of a symbol's form is its aspect ratio, that is, its relative dimensions of width and

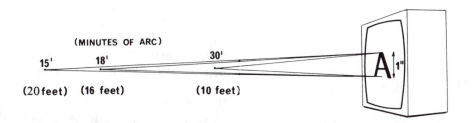

Figure 1. Symbol size in terms of visual angle subtended at different viewing distances.

Table 1 — Appropriate Minimum Symbol Size for Different Display Conditions

Class	Viewing Condition	Minimum Acceptable Symbol Size	
		Arc Subtended (minutes)	Height (in.) at Screen (Viewing Distance = 1 ft)
A	Art work displayed under ideal environmental conditions, standard 2 x 2 slides via high-resolution projection system in ideal viewing room. 95 percent recognition accuracy with college-age students with normal or corrected-to-normal vision.	8	0.028
B	Slide projection on good-quality lenticular, matte, or high-resolution rear-projection screen delivering adequate image brightness with acceptable image/non-image brightness ratio at screen. High-resolution television monitor. High-quality overhead projection on matte white screen.	10	0.035
C	Standard broadcast color television on high-quality monitor. 16-mm motion picture projection on lenticular or matte white screen. 2 x 2 slide projection on micro-beaded screen. Economy-type overhead projector on matte white screen.	12	0.042
D	High-quality closed-circuit television on standard classroom TV monitor/receiver. High-quality, high-resolution video display terminals, microfiche and film viewers.	15	0.052
E	1/2-inch video cassette or reel-to-reel recording displayed on typical classroom monitor/receiver. Classroom-type opaque projector. Standard video display terminals.	20	0.070
F	Standard TV projection (Schmidt Optical type). Art work or bulletin boards under poor viewing conditions.	30	0.105

NOTE: Adjust class selection downward one class for each 10 years the audience's age is over 40.
Adjust class selection downward one class if alphanumerics are to be displayed in color.
Adjust class selection downward accordingly for unusually wide viewing environments.

height. Recommended ratios are 2:3 or 5:7 for individual or narrow-sector viewing and 6:7 for wide-angle viewing. Condensed type faces should be avoided when the visual design is intended for extreme wide-angle view-

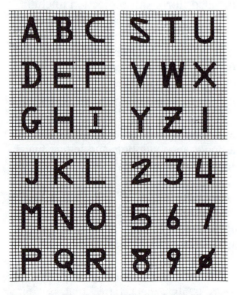

Figure 2. The Lincoln/Mitre (Mod X) font used for television display.

ing. They will appear to ''bunch up'' when viewed by those seated off to the side of the viewing sector. In such situations, a symbol slightly wider than it is tall may prove desirable.

GUIDELINE: Select a type face on the basis of the recognizablity of its characters, the stability of its form when photocopied or otherwise transferred to film, and the appropriateness of its aspect ratio for the width of the intended viewing environment. Recommended aspect ratios are 2:3 or 5:7 for usual viewing conditions and 6:7 for wide viewing sectors.

SPACING

Whenever there is a need for displaying more than one symbol at a time, and whenever the informational capacity of a display system becomes a consideration, spacing becomes a factor of importance.[1] Lines set too closely

together can be difficult to follow and can reduce readability. On the other hand, excessive leading is wasteful and interferes with normal eye-scanning movements. With too much space between the lines, the eye does not return smoothly to the start of the next line. It is generally recommended that spacing between lines be 100 to 150 percent of the symbol height.

More critical than the spacing between rows is the horizontal spacing between symbols.[16] Horizontal spacing should be at least 25 percent of symbol width for individual viewing and at least 50 percent of symbol width for group viewing.

The following are some practical and useful suggestions from Kemp and Dayton[17] regarding spacing:

GUIDELINES: 1. Separate lines within a caption so that adequate white space is left for ease of reading—about 1½ times the height of the lowercase letter m, measured from an m on one line to an m (or comparable letter) on the next line. 2. Space letters optically. Equal measured distances between all letters do not look equal. Make spaces look equal, regardless of measurement.
3. Allow 1½ letter widths for the space between words and 3 widths between sentences.

These guidelines are appropriate for individual viewing or viewers in standard viewing environments. For extra-wide-angle group viewing, the spacing recommended in suggestions 2 and 3 should be increased from 30 to 50 percent depending on the width of the viewing environment.

It should also be noted that spacing will be compromised if graphic material is not formatted to conform to the aspect ratio of the intended display medium. As a guide, Kemp and Dayton[17] recommend the following layout ''masks'' for conventional media: slides, 6 x 9 inches; filmstrips, 6 x 8 inches; overhead transparencies, 7½ x 9½ inches; and television, 6½ x 8½ inches. Obviously, if slides or overhead transparencies are intended for dual use, including television display, as they frequently are in teleconferencing sessions, then the designer should employ the television format in his production.

For those limited to using a typewriter to prepare their camera-ready art, it should be noted that pica type confined to a 3-inch x 4-inch area will produce alphanumerics of reasonable legibility for slides intended for either projection or television display.

UPPERCASE AND LOWERCASE

Uppercase letters are more legible than lowercase letters. Therefore, uppercase is recommended when individual alphanumerics, short statements, captions, and labels are to be displayed. Lowercase characters, however, have been shown to be easier to read, perhaps because they offer the viewer more identifying cues above and below the body of the letter—for example, compare "apple" and "APPLE." Thus, when proper names or extensive text is to be displayed, uppercase and lowercase letters should be used, as shown in Figure 3. Note that the height of the body of the lowercase letters is 3/4 the height of the capital letters, and lowercase letters with ascenders span the full height of the capital letters. Kodak[18] recommends that when lowercase letters are used along with uppercase letters, the minimum symbol size chosen should be based on the smallest lowercase letters.

GUIDELINE: Use all capital letters for labels, captions, and short statements. Use initial capital letters along with lowercase letters when displaying proper names and lengthy text. Select a font in which the height

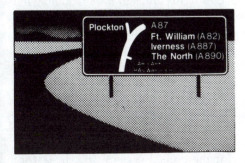

Figure 3. The use of uppercase and lowercase alphanumerics on highway information signs.

of the body of the lowercase letters is equal to 3/4 of the height of the uppercase letters.

STROKE WIDTH

Another important design element affecting the legibility of symbols is stroke width. Although the visual angle subtended by a symbol might be quite adequate and its form and aspect ratio appropriate, if the symbol's stroke width is either too broad or too narrow, legibility will be reduced.

Figure 4 is a graph adapted from Ollerenshaw,[8] in which letter weight is expressed as stroke width in a logarithmic scale. As shown by the sample letter "A," increases in stroke width improve visiblity and readability up to a point, after which there is a gradual reduction in visibility and a sharp decrease in readability.

Stroke-width recommendations depend upon illumination, contrast, and direction of contrast. Dark letters on a white background require a heavier stroke width than do white or luminous letters on a dark background. The reason lies in the visual process known as "irradiation," whereby white letters tend to enlarge and spread into their dark background, and black letters tend to thin out and disappear into their white surround. The brighter the illumination and the greater the contrast, the more disparity there will be between the recommended stroke width of white and black letters.[19]

Given normal viewing conditions, black letters on a light surround should have a stroke width of 1/5 of the symbol height. White letters on a dark background should have a stroke width (depending upon illumination) of 1/8 to 1/10 of the symbol height. For luminous letters, a stroke width of 1/15 of the symbol height will generally prove satisfactory.

The design of most back-lighted visual displays exemplifies ignorance of the irradiation principle. Invariably these displays are quite legible in the reflected light of the day, but at night their transilluminated symbols appear to enlarge to a point where legibility is practically lost. A more thorough and

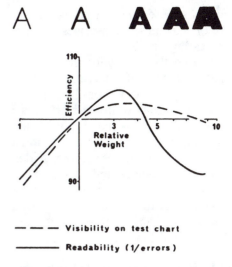

Figure 4. Relationship of stroke width to reading factors.

scientific description of this visual phenomenon may be found in Rubin and Walls.[20]

GUIDELINE: Consider the appropriateness of the stroke width for the anticipated contrast ratio, the direction of contrast, and the intended viewing situation.

BRIGHTNESS CONTRAST AND COLOR

When people speak of brightness, they are referring to either apparent (subjective) brightness or measured (photometric) brightness. A more appropriate term for measured brightness is luminance. Visual display specialists concern themselves principally with luminance, which, unlike apparent brightness, can be measured with an appropriate meter. A luminance level of 1 foot Lambert (FL) is the brightness emanating from a material that either reflects or transmits 1 foot candle (FC) of illumination. In this section the terms "luminance" and (measured) "brightness" are used interchangeably in keeping with other published material on the subject.[21]

Brightness contrast is the measure of the relative difference between the luminance of a symbol and its background. A contrast ratio between 3:1 and 10:1 is generally recommended for television displays. This represents between five and ten steps on a standard ten-step gray scale. The brightness contrast of projected visuals usually exceeds these ratios by at least a factor of 3 and in some instances by a factor as high as 10.

Frequently these ratios are excessive and negatively affect visual comfort. Research by McKittrick[22] and Des Rosiers[23] using white-on-black slides displayed on front- and rear-projection systems indicates that low contrast ratios of 2:1 to 4:1 are recommended for high image and background luminances, while higher contrast ratios of 5:1 to 10:1 are recommended for displays having lower image and background luminances.

GUIDELINE: Establish a brightness contrast ratio (the ratio of image luminance to background luminance) between 2:1 to 10:1 for high-contrast visual materials such as white-on-black slides. The contrast ratio should promote good legibility but not at the expense of visual comfort.

The direction of a symbol's brightness contrast is an important consideration when one is preparing materials for televised or film-based projected display. Light letters against a dark background (L/D) have greater visibility than do dark letters against a light background (D/L).[19] This difference is emphasized when ambient illumination is low. Also, Kodak[24] recently indicated that L/D slides are more comfortable to view and are less tiring to the eyes than D/L slides. This view, however, is not universally shared.[25]

Dark letters on light backgrounds have been shown to be more readable, especially under medium or high ambient illumination. Tinker and Patterson[26] indicate that this is due to the larger number of eye fixations required to read L/D visuals. However, recent experimental research by Hamilton[10] indicates that there is only a slight difference favoring the D/L slides, and that the selection of direction of contrast might best be made for reasons other than legibility. After reviewing the literature on the subject, I suggest the following guideline:

GUIDELINE: Prepare labels, captions, and titles as light letters against a dark background; prepare long statements and text as dark letters against a light background.

When colored materials are displayed with a black-and-white system, contrast is most important. Colored symbols should be at least five gray steps away from their background color in order to be legible.

The use of colored "gels" (unexposed but fully developed diazo film) on the backs of black and white transparencies is one popular method of cutting down on excessive contrast and "dressing up" the visual. Visuals produced in this manner have had varying degrees of success. Morton[27] found that white on black and black on white projectuals were more legible and preferred over (1) black symbols on a yellow background, (2) yellow symbols on a black background, and (3) white symbols on a blue diazo background. One of the problems inherent in this method of production is that the additive color film may reduce the brightness contrast of the visual to a point where much of its legibility is lost. Since the efficacy of this method is highly dependent upon the actual brightness of the display and the extraneous room light falling on the display, colored gels should not be used when either the image brightness of the display (open gate, i.e., with no visual in place) is lower than 5 FL, or when ambient illumination falling on the display creates a nonimage brightness greater than 1/10 of the image brightness.

However, if high-contrast slides or overhead projectuals have to be used in a totally darkened room (not generally recommended), then the use of an amber or yellow gel may be helpful. I have noticed that when using white-on-black Kodalith transparencies in a darkened room, viewers had difficulty keeping the projected text in focus. However, these same viewers had significantly less trouble when the same visuals were colored amber or yellow. It is not clear why this should occur; perhaps these monochromatic colors require less accommodative adjustment from the visual system because of their central position in the visual spectrum. This theory, however, remains conjecture until supported by appropriate research findings.

Some of the same viewers noted, with varying degrees of emphasis, that the use of a yellow gel seemed to make the black components of black on white slides appear "blacker," whereas the absence of the gel made the black areas appear "less black" or "more gray." Why this should occur, and in varying degrees for different people, may be explained by the individuality of contrast sensitivity, particularly in the presence of glare. The white light in the visual display probably scatters in the optical media of the eye and is superimposed onto the retinal image. A form of "veiling" glare is thereby produced that reduces the image contrast—or as expressed in this case, the apparent saturation of the black areas. Visual designers should keep in mind that sensitivity to contrast and glare is an individual thing, becoming quite pronounced in the older viewer because of the increased light-scattering caused by the increased opacity of parts of the eye that naturally accompanies aging.[4]

One of the practical problems associated with the use of high-contrast white-on-black slides is the Kodalith film on which they are customarily produced. Owing to the density of its emulsion, this film will absorb an excessive amount of heat from the optical system of a projector. This will cause the film to buckle far beyond the curvature anticipated with other slide film, and consequently even "curved field" projection lenses, which are designed to accommodate normal slide "popping," are unable to keep the full slide aperture in focus. One way to combat this problem is by mounting the Kodalith slides in glass and using a "flat field" projection lens. A better approach for both projection and comfortable viewing would be to use the Kodalith white-on-black slides as intermediates, adding color where desired, and duplicating them on Ektachrome film. The resultant density of the black in this film usually presents no problem for the optical system of a projector, and the range of contrast is comfortable to view regardless of the ambient illumination in the viewing environment.

The problem with Kodalith achieves major proportions when white-on-black transparencies made on this film are used on overhead projectors for a prolonged period of time. In this situation, the heat built up in the projector will cause the projector's plastic Fresnel lens to overheat and take on a distorted shape, and the optical properties will essentially be ruined.

Although colored displays are at a disadvantage in regard to minimum acceptable symbol size, they hold a number of distinct advantages over black-and-white displays, i.e., color coding of visual information and spatial separation of overlapping visual messages.[28] When colors are used in the design of alphanumerics, it should be remembered that the lens of the human eye is not color corrected and has to change its shape to bring different colors into focus. There is an accommodative adjustment of approximately 2 diopters when the eye shifts its focus from saturated blue to saturated red and vice versa. Recent research has shown that high-purity blue backgrounds require viewers to exert more accommodation (visual focus) to bring images into focus than the same images displayed on other colored backgrounds.[29]

Simultaneously employing saturated colors that are far apart relative to their electromagnetic wavelengths will require major visual adjustments and increase visual fatigue. This is not to say that the use of such complementary colors does not have its place. Complementary colors used in titles or attention-getting visuals can be very effective if not allowed to remain on the display screen too long. Analogous colors (i.e., colors that are close together relative to their electromagnetic wavelengths), while less arousing, are more comfortable to view for extended periods of time. It is a customary practice to employ colors in this manner. One visual design specialist suggests using complementary colors for displays with high visual impact and low information, and analogous colors as well as black and white for displays of low visual impact and high information.

Other facts about the visibility of col-

ors from Diffrient et al.[30] and Post [31]: (1) yellow is the most luminous and visible for the greatest distance; (2) orange and red-orange hold the most attention; (3) blue is likely to be hazy and indistinct and difficult to see when used to display small images; (4) the visibility of colors is greatly dependent upon the color of the display background; and (5) red and green are difficult to perceive when they are displayed outside a viewer's major field of vision (60°).

GUIDELINE: Use color in displays only when it helps the viewer process information with greater speed and accuracy or where its esthetics are required. Avoid the simultaneous use of highly saturated and spectrally extreme colors. Avoid pure blue for thin lines, text, small images, and visual display backgrounds. Avoid using red and green in the periphery of a large display.

Another factor for the visual designer to consider is that 6 to 10 percent of all males and 0.5 percent of females possess a significantly reduced ability to distinguish among certain colors. Commonly confused color combinations are yellow-reds with yellow-greens, or reddish-blues with greenish-blues.[32] Combining red with cyan and blue with yellow will help correct this problem.[33] Where there is some doubt as to the color-perceptual abilities of an audience, it is recommended that the following surface colors (pigments) be used: red, orange, yellow, blue, purple, gray, white, black, and buff.[34]

ILLUMINATION

The amount of illumination in the viewing sector and the area surrounding the television monitor will affect viewing accuracy and comfort. The background area surrounding the televised image should be between 1/3 and 1/10 as bright as the television image itself. Thus, given an average television image luminance of 20 FL, the area surrounding the television screen should be between 2 and 6 FL. In regard to overhead, 16-mm film, and slide projection, higher ratios between the image brightness and

nonimage brightness from the screen are required. These ratios range from 5:1 for simple high-contrast line material to 100:1 for high-quality color motion pictures.[35] Given the 18-FL brightness recommended by the Society of Motion Picture and Television Engineers for film-based displays and room lights on dimmers, this wide range of brightness-contrast ratios is attainable.

Determine the most appropriate ambient light level for a presentation in advance by having a trial run with the visuals. With a sample visual projected on the screen, raise or lower the room lights to the point that is visually comfortable without excessively degrading the legibility of the display. When this level is determined, mark the light dimmer position with drafting tape so that this level may be replicated at the time of the actual presentation.

Light fixtures (luminaires) should be placed high and out of the viewer's direct line of sight. Luminaires falling within the viewer's visual pathway to the display are a source of glare, and any form of glare is inimical to the seeing process. It is known that even a small glare source (5 FL) located 5° from the viewer's line of sight will reduce his visibility of the display by more than 50 percent.[36] In general, light sources should be kept away from the immediate location of the display; otherwise, their illumination will "wash out" the brightness and contrast of the images and thereby reduce legibility.

Incandescent luminaires should be equipped with microgroove baffles, and fluorescent luminaires should be equipped with either parahex diffusers or parabolic louver reflectors. These devices will help keep annoying glare out of the viewer's line of sight and from spilling over onto the critical display area. Wiring these fixtures to dimmers or to high-low-off switches will provide the user with the degree of control required for most visual presentations.

GUIDELINE: Design display environments so that it is possible to control illumination levels and directionality in a manner that creates

good contrast ratios between the display and the area immediately surrounding the display. Locate and equip light fixtures to distribute an appropriate level of glare-free illumination on the audience (for note taking) without spilling over onto the projection screen or the television monitors.

RESOLUTION

The resolving power of a medium refers to its ability to reproduce details found in the original scene. The finer the detail that this medium is capable of reproducing, the greater is its resolution. Film-based projection system specialists generally measure resolution in paired lines per millimeter; computer graphic specialists use pixels; and television researchers traditionally have measured resolution in terms of the number of active, horizontal lines per symbol height.[37] Display systems engineers are quick to note that resolution is only one of the optical properties contributing to the quality of a display and that there are better ways of measuring this attribute. These other approaches are too lengthy and complex to explain in this paper but are well described in the literature.[38, 39, 40]

An excellent study that reports on the optical properties, including resolution, of projection screens is that of Klaiber.[41] Although not recent, this study is still one of the best sources of information for those in the process of selecting either front or rear projection screens for their presentation facilities.

There does not seem to be any substantial difference between alphanumerics displayed on a high-quality matte white front-projection screen and those displayed on a high-quality rear-projection screen.[10] However, Menell[42] claims a superior legibility for rear projection because visual acuity is enhanced when projected images are viewed in a lighted room—a condition that is more feasible with rear-projection than with front-projection display systems.

With some of the more esoteric screens, differences in legibility emerge. Some of the "high-gain" rear-projection screens have a center-to-corner brightness ratio of more than 5:1, which creates problems for full-field legibility. Some of the lenticular front-projection screens used in movie houses can display text and images nearly as well as matte white screens if they are kept taut and free of dust, and if viewers are not allowed to sit too close to them. Standard beaded screens, on the other hand, very rarely produce acceptable image resolution for the display of text. When choosing a projection screen, the display specialist should always try different screen samples in the actual display environ-

ment before deciding on a purchase. Additional information relative to film-based projection equipment performance characteristics is reported in Bretz and McVey.[43]

Research indicates that legibility is so directly dependent upon resolution that there can be, within limits, an effective trade-off between minimum acceptable symbol size and display-system resolution. Television display is a good example of this. Figure 5, adapted from Erickson,[44] shows the trade-off between symbol size and degree of resolution as measured in horizontal lines per symbol height. As the horizontal scan lines per symbol are reduced, the symbol size (measured in minutes of arc) must be correspondingly increased in order to maintain the established level of 95 percent recognition accuracy.

GUIDELINE: Increase symbol size correspondingly to compensate for each degradation of the system, from the ideal, either in terms of design or display. Ω

ACKNOWLEDGMENT

The author wishes to express his appreciation to **Allen Burgess,** the Director of the Instructional Media Center at Boston University, for his helpful comments and practical suggestions and to **Thomas E. Pinelli** and **Virginia M. Cordle** for their advice, time, and effort in the technical preparation of this article.

REFERENCES

1. D.A. Shurtleff, *Design Problems in Visual Displays, Part I; Classical Factors in the Legibility of Numerals and Capital Letters*, ESD-TR-66-62, Project 7030 (Bedford, MA: MITRE Corp., June 1966).
2. G.F. McVey and C.J. Wiegeshaus, *Studies in Legibility* (Rochester, NY) The Center for Visual Literacy, University of Rochester, 1973.
3. Paul H. Preo and J.R. Sullivan, "A Survey of Classroom Projection Conditions," *Journal of the SMPTE* 82, no. 1 (Jan. 1973), pp. 24-28.
4. A. Cakir, D.J. Hart, and T.F.M. Stewart, *Visual Display Terminals* (New York: John Wiley and Sons, 1980).
5. D.A. Dondis, *A Primer of Visual Literacy* (Cambridge, MA: MIT Press, 1973).
6. Sidney L. Smith, "Letter Size and Legibility," *Human Factors* 21, no. 6 (Dec. 1979), pp. 661-670.
7. P.W. Cobb and F.K. Moss, "Four Fundamental Factors in Vision," *Trans. Illumination Engineering Society* 23 (1928), pp. 496-506.
8. R. Ollerenshaw, "Design for Projection: A

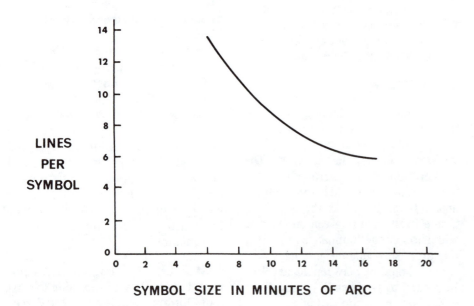

Figure 5. Trade-off between resolution and symbol size for 95 percent recognition accuracy.

Study of Legibility,''*Photographic Journal* 102 (Feb. 1962), pp. 41-52.

9. Lorrin A. Riggs, ''Visual Acuity,'' Ch. 11, in Clarence H. Graham (ed.), *Vision and Visual Perception* (New York: Wiley, 1965), pp. 321-349.

10. Mark A. Hamilton, *The Accuracy Recognition of Positive and Negative Symbols of Seven, Five, and Three Minutes of Arc on Front and Rear Projection Screens Under Self-Selected Illumination*, doctoral dissertation, School of Education, Boston University, 1983.

11. G.F. McVey, ''Television: Some Viewer-Display Considerations,'' *AV Communication Review* 18, no. 3 (Fall 1970), pp. 277-290.

12. D.Y. Cornog and F.C. Rose, *Legibility of Alphanumeric Characters and Other Symbols: II. A Reference Handbook* (National Bureau of Standards, Miscellaneous Publication 262-2, Feb. 10, 1967).

13. G.C. Kinney and D.J. Showman, *Studies in Display Symbol Legibility; Part XI. The Relative Legibility of Selected Alphanumerics in Two Fonts*, ESD-TR-66-116 (AD 639 750) (Bedford, MA: Decision Sciences Laboratory, Hanscom Field, Aug. 1966).

14. R.C. Matthias, The Production and Use of Television Aids for Medical Education, *J. Biological Photography Assoc.* 29 (May 1961), pp. 65-77.

15. A. Wright, *Designing for Visual Aids* (New York: Van Nostrand Reinhold Co., 1970).

16. M.N. Crook, J.A. Hanson, and A. Weisz, *Legibility of Type as a Function of Stroke Width and Letter Spacing Under Low Illumination*, WADC Tech. Report: 53-440, March 1954.

17. J. Kemp and D. Dayton, *Planning and Producing Instructional Media*, 5th Ed. (San Francisco: Harper and Row, 1985), pp. 122-123.

18. Kodak, *Legibility—Artwork to Screen*, S-24 (Rochester, NY: Eastman Kodak Co., 1980), p 5.

19. C. Berger, ''Stroke-Width, Form and Horizontal Spacing of Numerals as Determinants of the Threshold of Recognition,'' *J. Applied Psychology* 28 (1944), pp. 208-231.

20. M.L. Rubin and G.L. Walls, *Fundamentals of Visual Science* (Springfield, IL: Thomas Press, 1965), p. 168.

21. W.M.C. Lam, *Perception and Lighting as Form-givers for Architecture* (New York: McGraw-Hill Book Co., 1977).

22. J.L. McKittrick, *Viewer-Preferred Contrast Ratios for Projected Negative Lettering*, unpublished doctoral dissertation, School of Education, Indiana University, Bloomington, 1976, pp. 87-90.

23. E. Des Rosiers, *Brightness Contrasts and Illumination Levels for Close Sustained Visually Centered Tasks*, unpublished doctoral dissertation, School of Education, Boston University, 1975, pp. 118-126.

24. Eastman Kodak, *Reverse-Text Slides*, Publication No. S-26 (Rochester, NY: Eastman Kodak, Motion Picture and Audio Visual Market Division, 1982).

25. E.G. Dimond, ''On Cake-Colored Slides,'' *The New England Journal of Medicine* 308, no. 1 (Jan. 6, 1983), p. i.

26. M.A. Tinker and D.G. Patterson, ''Studies of Typographical Factors Influencing Speed of Reading: VII. Variations in Color of Print and Background,'' *Journal of Applied Psychology* 15 (Oct. 1931), pp. 471-479.

27. R. Morton, ''The Lantern Slide: Legibility and Production,'' *Photographic Journal* 108 (April 1968), pp. 89-97.

28. G.F. McVey, ''Putting Color Into Your Visual Presentation,'' *Photomethods* 19, no. 11 (Nov. 1976), pp. 57-59.

29. D.T. Donohoo, ''Accommodation During Color Contrast,'' *SID Digest* (May 1985), pp. 200-202.

30. Diffrient et al., *Humanscale 789* (Cambridge, MA: MIT Press, 1981).

31. D.L. Post, ''Effects on Color or CRT Symbol Legibility,'' *SID Digest* (May 1985), pp. 196-198.

32. L.M. Hurvich, *Color Vision*, (Sunderland, MA: Sinauer Associates, Inc., 1981).

33. W.J. Smith and J.E. Farrell, ''The Ergonomics of Enhancing User Performance with Color Displays,'' *Seminar Lecture Notes, Vol. II* (Plaza del Rey, CA: Society for Information Display, 1985), pp. 5.1-3—5.1-16.

34. C.A. Baker and W.F. Grether, *Visual Presentation of Information*, PB 111547 (Washington, DC: U.S. Dept. of Commerce, Office of Technical Services, Aug. 1954).

35. J.E. Kaufman, ed., *IES Lighting Handbook: Application Volume*, Illuminating Engineering Society of North America (Baltimore: Waverly Press, 1981), pp. 11-34.

36. Wesley E. Woodson, *Human Factors Design Handbook: Information and Guidelines for the Design of System, Facilities, Equipment, and Products for Human Factors* (New York: McGraw, 1981), p. 831.

37. Marsetta et al., *Studies in Display Symbol Legibility, Part XIV: The Legibility of Military Map Symbols on Television*, The MITRE Corp., AD 641 658 (Bedford, MA: The Clearinghouse, Sept. 1966).

38. H. Snyder, ''Image Quality and Observer Performance,'' Ch. 2 in *Perception of Displayed Information*, L. Biberman (ed.) (Plenum Press, 1973).

39. C.R. Carlson and R. Cohen, ''A Model for Predicting the Just Noticeable Difference in Image Structure,'' *SID Digest* (1978).

40. G. Verona et al., ''Direct Measure of Cathode Tube (CRT) Image Quality,'' *SPIE* 196, *Measurements of Optical Radiation* (1979), pp. 106-113.

41. R.J. Klaiber, *Physical and Optical Properties of Projection Screens*, AD 647 132 (Orlando, FL: Naval Training Devices Center, 1966).

42. J. Menell, ''Training Facilities and Equipment,'' Ch. 7 in R.L. Craig (ed.), *Training and Development Handbook* (New York: McGraw-Hill Book Co., 1976), pp. 1-26.

43. R. Bretz and G.F. McVey, '' AV Equipment—Beyond Basics,'' *Educational and Instructional Broadcast* (Sept. 1970), pp. 24-27.

44. R.A. Erickson, ''Visual Search Performance in a Moving Structure Field,'' *Journal of the Optical Society of America* 54, no. 3 (March 1964), pp. 399-405.

GERALD F. McVEY *is a professor in the media and technology program of the School of Education at Boston University.*

Using Color Effectively: Designing To Human Specifications

Gerald M. Murch

THE HUMAN VISUAL SYSTEM'S CAPACITY *and capability to process color can be applied as design criteria for color displays. This paper reviews key elements in the visual domain of color, encompassing the visual, perceptual, and cognitive modes, and develops a series of recommendations for effective color usage based on these elements.*

Color can be a powerful communication tool. Used properly, it can enhance the effectiveness of information. Improperly used, however, color can also seriously impair communication. Thus it is important to establish and follow some basic guidelines for effective color usage. Unfortunately, no one set of guidelines can cover all applications.

Up to now, color has been used almost exclusively in a qualitative rather than a quantitative fashion, i.e., showing that one item is "different from" another rather than displaying relationships of degree. A typical example would be color-coding each layer of a multilayer circuit board. Color serves to differentiate the layers but says nothing about their relationships. A simple quantitative extension of the multilayer circuit board might involve placing the layers in spectral order, with the first layer red, the second orange, and so on, following the popular mnemonic ROY G. BIV (red, orange, yellow, green, blue, indigo, violet). The demands for the proper use of color increase when color is used quantitatively to show progressing change.

Basically, effective color usage depends upon matching the physiological, perceptual, and cognitive aspects of the human visual system. This paper reviews some well documented aspects of these visual-system capacities and develops some basic principles that should allow us to improve graphics systems by using color properly.

PHYSIOLOGY OF COLOR

In understanding how we see color, it is important to realize that color is not a physical entity. Color is a sensation, like taste or smell, that is tied to the properties of our nervous system.

Figure 1 shows the light wavelengths to which the human eye is sensitive, along with the corresponding color sensed. The color sensation results from the interaction of light with a color-sensitive nervous system. Most importantly, individuals can have vastly different color-discrimination capabilities because of differences in the eye's lens, its retina, and other parts of the visual system.

The Lens

The lens of the human eye is not color corrected. This causes chromostereopsis, an effect that causes two pure colors at the same distance to appear to be at different distances. For most people, reds appear closer and blues more distant. In fact, short wavelengths—pure blue—always focus in front of the retina and thus appear defocused. This phenomenon is most noticeable at night when deep-blue signs seem fuzzy and out of focus, while other colors appear sharp.[1]

Lens transmissivity also has an effect. The lens absorbs almost twice as much energy in the blue region as in the yellow or red region. Also, a pigment in the retina's center transmits yellow while absorbing blue. The net result is a relative insensitivity to shorter wavelengths (cyan to deep blue) and enhanced sensitivity to longer wavelengths (yellows and oranges).[2]

As we grow older, lens yellowing increases, which makes us increasingly insensitive to blues. Similarly, aging reduces the transmittance of the eye's fluids, which makes colors appear less vivid and bright. Actually, age aside, there is normally a great deal of variation, with some people's eyes being very transparent and others' naturally yellowed. This variation alone contributes to differences in color sensitivities among individuals.[2]

The Retina

The human retina consists of a dense population of light-sensitive rods and cones. Rods are primarily responsible for night vision, while cones provide the initial element in color sensation.

Photopigments in the cones translate wavelength to color sensation. The range of sensation is determined by three photopigments—blue (445 nanometers, or nm), green (535 nm), and red (575 nm). Here, "red" is really a misnomer, because maximum sensitivity at 575 nm actually invokes the sensation of yellow.

Both photopigment and cone distribution vary over the retinal surface. Red pigment is found in 64% of the cones, green in 32%, and blue in about 2%. Additionally, the center of the retina, which provides detailed vision, is densely packed with cones but has no rods. Moving outward, the number of rods increases to eventual predominance; as a result, shapes appear unclear and colorless at the extreme periphery of vision.[3]

Reprinted with permission from *Technical Communication,* published by the Society for Technical Communication, vol. 32, pp. 14–20, Fourth Quarter 1985.

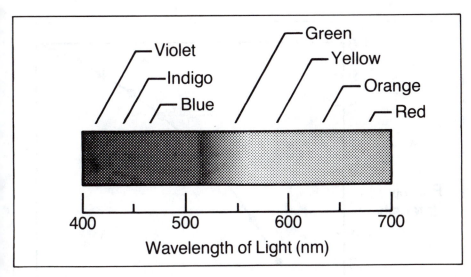

Figure 1. The color spectrum of human vision

Because of the cone and photopigment distributions, we can detect yellows and blues farther into our peripheral vision than reds and greens. Also, the center of the retina, while capable of high acuity, is nearly devoid of cones having blue photopigment. This results in a "blue blindness" that causes small blue objects to disappear when they are fixated upon.[4]

For the eye to detect any shape of a specific color, an edge must be created by focusing the image onto the mosaic of retinal receptors. An edge is the basic element in perceiving form. It can be created by adjacent areas differing in brightness, color, or both. Edges guide the eye's accommodation mechanism, which brings images into focus on the retina. Recent research has shown, however, that edges formed by color difference alone with no brightness difference, such as a red circle centered on a large green square of equal brightness, are poor guides to accurate focusing. Such contours remain fuzzy.[5] For sharply focused images, it is necessary to combine both color and brightness differences.

Also, for photopigments to respond, a minimum level of light is required. Additionally, the response level depends on wavelength, with the visual system being most sensitive to the center of the spectrum and decreasingly sensitive at the spectral extremes. This means a blue or red must be of much greater intensity than a minimum-level green or yellow in order to be perceived. Similarly, equal-energy reds might not appear equally intense.[3]

After the Retina

The optic-nerve bundle leads from the photoreceptors at the back of the retina. Along the optic-nerve path, at the lateral geniculate body, the photoreceptor outputs recombine.[2] Figure 2 diagrams how this recombination takes place.

Notice in Figure 2 that the original retinal channels—red, green, and blue—form three new "opponent channels." One channel signals the red-to-green ratio, another the yellow-to-blue, and the third indicates brightness. Again, we find a bias against the blue photopigments, since the perception of brightness, and hence of edges and shapes, is signaled by the red and green photopigments. The exclusion of blue in brightness perception means that colors differing only in the amount of blue will not produce sharp edges.[5]

Neural organization into opponent channels has several other effects, too. The retinal color zones, which link opponent red with green and opponent yellow with blue, provide an example. Opponent-color linking precludes visually experiencing combinations of opposing colors—we cannot experience reddish green or yellowish blue.[2]

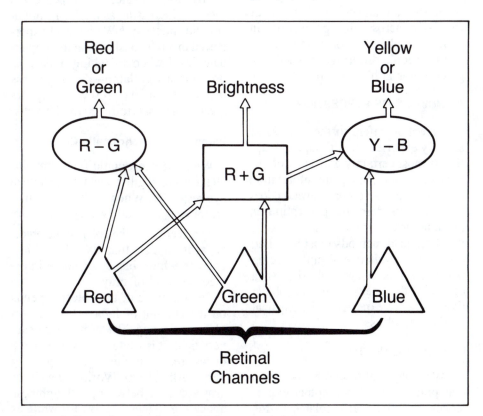

Figure 2. Processing of color input into opponent channels

Colorblindness

The term "colorblind" is an unfortunate summarization of the color deficiencies besetting only about 9% of the population, and only a tiny percentage of those deficiencies result in true blindness to color.[4]

Not all the causes of color-deficient vision are known; however, some are related to the cones and their photopigments. A rare form occurs when blue photopigment is missing. The best-known condition, however, is red-green deficiency, caused by the lack of either red or green photopigments. Lack of either photopigment causes the same color-discrimination problem; however, for people lacking red photopigment, long-wave stimuli appear much darker.

A more common case is that of photopigment-response functions deviating significantly from normal. In one form, the red photopigment peak lies very close to that of the green; whereas in another, green is shifted towards red. The net result is reduced ability to distinguish small color differences, particularly those of low brightness. Less extreme cases of response deviation, which occur regularly across the population in general, explain the common situation of two people differing on whether a given color is blue or green.[4]

PRINCIPLES OF PERCEPTION

Perception refers to the process of sensory experience. Although perception is most certainly a product of our nervous system, adequate information about the "higher order" function does not exist to describe perception in physiological terms. As a result, psychological methods must be relied on, the most valuable discipline being psychophysics. Psychophysics is a discipline that seeks to describe objectively how we experience the physical world around us.

Perception Is Nonlinear

Psychophysical research has shown that practically all perceptual experiences are nonlinearly related to the physical event. Figure 3 illustrates this

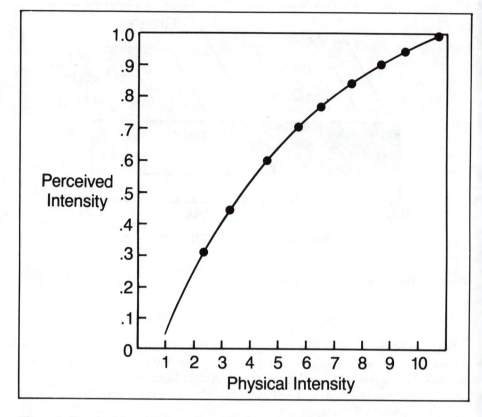

Figure 3. Perceived intensity as a function of physical intensity

by graphing perceived intensity versus physical intensity. The relationship is nearly logarithmic, with perceived intensity increasing as the logarithm of stimulus intensity.[6] We have all experienced this relationship on a more mundane level when switching intensities on a three-way lamp—the brightness increase from 50 to 100 watts appears greater than that from 100 to 150 watts.

Perception of Achromatic Color

White or achromatic light contains all the wavelengths to which the human eye responds. When such light strikes an object and all wavelengths are reflected equally, the color of the object is achromatic; that is, the object appears white, black, or some intermediate level of gray.

The lightness of the object depends on the amount of light reflected. An object reflecting 80% or more appears very light; it is white. Reflection of 3% or less results in the object appearing very dark (black). Various levels of gray appear in between, with lightness appearing to increase toward white as a logarithm of reflectance.[6]

Consider, for example, black, white, and gray automobiles. Each car reflects different amounts of light and, therefore, takes on a specific achromatic color. If the total amount of light illuminating the cars is increased, the lightness stays the same but the brightness increases. The white car stays white but becomes much brighter—perhaps even dazzling. Thus, lightness is a property of an object itself, but brightness depends upon the amount of light illuminating the object.

The most common example of all of this is black print on white paper. Changing illumination has little perceptual effect on the relative lightness of the paper or print, since the ratio of reflectance, or contrast, remains unchanged.

Perception of Chromatic Color

Objects that reflect or emit unequal distributions of wavelengths are said to be chromatic, i.e., to have a color. The color we sense derives from the physical attributes of the dominant wavelengths, the intensity of the wavelengths, and the number and proportion of reflected waves. Color identification also depends upon a multitude of learning variables, such as previous experiences with the object and association of specific sensations with color names. The sensation is also affected by the context in which the color occurs and the characteristics of the surrounding area or the colors of other objects.[7]

The study of the physical attributes of chromatic objects is always compounded by the experiences of the observer with colors in general. Because color perception is subjective, many of its aspects can be described only in the psychological dimensions of hue, lightness, saturation, and brightness.[8]

Hue is the sensation reported by observers exposed to wavelengths between approximately 380 and 700 nm. For the range between 450 and 480 nm, the predominant sensation reported is blue. Green is reported across a broader range of 500 to 550 nm, and yellow across a narrow band around 570 to 580 nm. Above 610 nm, most persons report the sensation of red. The best or purest colors—defined as those containing no trace of a second color—would indicate pure blue at about 470 nm, pure green at 505, and pure yellow at 575 nm.[4]

Hue, then, is the basic component of color. It is the primary determinant for the specific color sensation. Although hue is closely related to certain wavelengths, remember that hue is a psychological variable and wavelength a physical one. Although people with normal color vision will name a sector of the visual spectrum as red, disagreement will occur about the reddest red or where red becomes orange. Such disagreement reflects varying experiences with color as well as the intrinsic differences in the color mechanisms of each person's visual system.

Saturation is most closely related to the number of wavelengths contributing to a color sensation. As the band of wavelengths narrows, the resulting color sensation becomes more saturated—the wider the band, the less saturated the color.

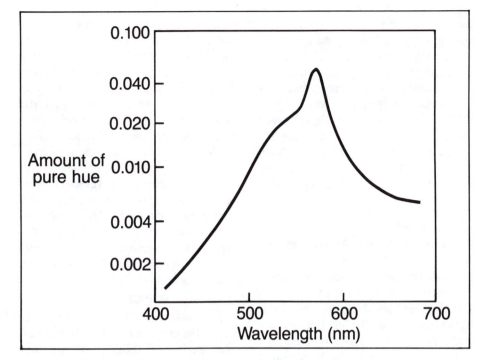

Figure 4. Perceived saturation at different wavelengths

Conceptually, a scale of saturation can be envisioned as extending from a pure hue, such as red, through less distinct variants of the hue, such as shades of pink, to a neutral gray in which no trace of the original hue is noticed. A measure of saturation discrimination can be obtained by starting with a neutral color and determining the amount of pure hue that must be added in order for the hue to become detectable. Figure 4 shows the results of such a study.[9]

In Figure 4, the abscissa indicates wavelength. The ordinate shows the proportion of pure hue that must be added to neutral color for the hue to become discernible. From this graph, it becomes obvious that substantial differences exist in our ability to detect color presence at different wavelengths. The largest amount of color required for detection is at the 570-nm stimulus. Such a yellow appears initially to be less saturated than any other pure hue and desaturates quickly as the wavelength distribution is broadened or as neutral colors are mixed with it.

Lightness, as mentioned previously, refers to the gamut of achromatic colors ranging from white through gray to black. By definition, achromatic colors are completely desaturated, since no trace of hue is present.

Just as with achromatic colors, the lightness of a mixed color also depends on the reflectance of the surface under consideration—the higher the reflectance, the lighter the color. As might be anticipated, monochromatic colors do not all appear equal in lightness. Some hues appear lighter than others even though their reflectances are the same. If, for example, observers are shown a series of monochromatic lights of equal brightness and are asked to rate them for lightness, a relationship similar to that shown in Figure 4 results. A monochromatic color 570 nm appears much lighter than all other wavelengths, and the lightness decreases rapidly as the extremes of the visual spectrum are approached.[9]

Brightness is another aspect of color perception. Increasing the illumination of both achromatic and chromatic colors produces a qualitative change in appearance that ranges from dark to bright. However, separation of brightness from lightness is often difficult, since brighter colors invariably appear lighter as well.[6]

Consider broadband light shining on a series of equal-reflectance surfaces, each reflecting a very narrow band of wavelengths. At low illumination, the surfaces are first perceived as gray. Increasing the intensity allows dark and desaturated hues to become discernible, with mid-spectrum wavelengths (555 nm) becoming visible at lower intensity levels. As the intensity increases, a broader range of hues appears, with extremely long and short wavelengths visisble only at high intensities.[3]

Colors in Context

Colors are also subject to contextual effects, in which adjacent colors influence one another. For example, a color on a dark background appears lighter and brighter than the same color on a light background. If a test field is neutral (gray) or dark and displayed on a colored background, the background induces color into the test field.[7] Red, for example, induces green into a neutral gray.

The size of a colored area also influences its perceptual properties. In general, small areas become desaturated and can show a shift in hue.[7] This creates problems when text is color coded, especially with blues and yellows, because they are susceptible to small-area color loss. Also, small areas of color can mix; red and green in smaller areas are eventually integrated by the visual system into yellow.

Individual Characteristics

Thus far, perception has been described in general terms as it applies to the typical human visual system. Yet all of us have our own perceptual idiosyncrasies that affect how we use color. For example, some people prefer highly saturated colors and others prefer muted tones.

It is important, too, to remember that color perception changes over time. We adapt to color with prolonged viewing. This results in an apparent softening of colors. As a result, there is a tendency to use highly saturated colors to offset adaptation. The unadapted viewer, however, sees the colors as highly saturated. Additionally, some research indicates that pure colors are visually fatiguing.[10]

Although we are still far from developing an aesthetics of color displays, some information has been compiled on color combinations that go well together and those that do not. Tables 1 and 2 present data from a study in which people were asked to pick the best and worst appearing colors on different backgrounds. Choices were made for both thin lines and for larger filled panels. The tables list those combinations preferred or rejected by at least 25% of the 16 subjects participating in the study. Obviously this is a small study sample, but it does represent the start of an understanding of the complex issue of color-display aesthetics.[11]

COGNITIVE PRINCIPLES

The least understood area of effective color usage is how to capitalize on our modes of thinking about, and associating with, color. This area of study falls into the domain of cognitive ergonomics.

Despite the infancy of this area of human-factors study, some initial observations prove useful in effective color usage. An example involves the functional use of color stereotypes: red for warning, green for go, and yellow for attention. Since we all have experience with these meanings, maintaining the relationship maps nicely into our expectations.

In color-coding graphed measurement data, variation of hue can quickly communicate important information. Portions of the data within a certain tolerance, or range limit, can be coded green; portions approaching a limit can be yellow; and excesses can be codes red. This procedure fits into the normal cognitive expectations.

For multiple graphs, where color is simply used for quick distinction between data, contrast is a big consideration. As a result, it is tempting to use red for one line and green for another. While this makes the data readily distinguishable, which is the goal, it can also bias an observer toward making some quality judgments about the data—the red data are bad or dangerous, the green are okay. Such biasing of the observer might not be what you intended. Similarly, the perceived magnitude of different colors

varies. A red square will be perceived as being larger than a green one of identical size.[12]

GUIDELINES FOR EFFECTIVE COLOR USAGE

On the basis of the preceding discussion, some general guidelines for color usage can be stated. They are listed here according to the area of their derivation—physiological, perceptual, or cognitive.

Physiological Guidelines

Avoid the simultaneous display of highly saturated, spectrally extreme colors. Reds, oranges, yellows, and greens can be viewed together without refocusing, but cyan and blues cannot be easily viewed simultaneously with red. To avoid frequent refocusing and visual fatigue, extreme color pairs such as red and blue or yellow and purple should be avoided. However, desaturating spectrally extreme colors will reduce the need for refocusing.

Avoid pure blue for text, thin lines, and small shapes. Our visual system is just not set up for detailed, sharp, short-wavelength stimuli. However, blue does make a good background color and is perceived clearly out into the periphery of our visual field.

Avoid adjacent colors differing only in the amount of blue. Edges that differ only in the amount of blue will appear indistinct.

Older viewers need higher brightness levels to distinguish colors.

Colors change appearance as ambient light level changes. Displays change color under different kinds of ambient light—fluorescent, incandescent, or daylight. Appearance also changes as the light level is increased or decreased. On the one hand, a change occurs because of increased or decreased contrast, and on the other hand, because of the shift in the sensitivity of the eye.

The magnitude of a detectable change in color varies across the spectrum. Small changes in extreme reds and purples are more difficult to detect than small changes in other colors such

Table 1 — Best Color Combinations (N = 16)

Background	Thin Lines and Text	Thick Lines and Panels
White	Blue (94%) Black (63%) Red (25%)	Black (69%) Blue (63%) Red (31%)
Black	White (75%) Yellow (63%)	Yellow (69%) White (50%) Green (25%)
Red	Yellow (75%) White (56%) Black (44%)	Black (50%) Yellow (44%) White (44%) Cyan (31%)
Green	Black (100%) Blue (56%) Red (25%)	Black (69%) Red (63%) Blue (31%)
Blue	White (81%) Yellow (50%) Cyan (25%)	Yellow (38%) Magenta (31%) Black (31%) Cyan (31%) White (25%)
Cyan	Blue (69%) Black (56%) Red (37%)	Red (56%) Blue (50%) Black (44%) Magenta (25%)
Magenta	Black (63%) White (56%) Blue (44%)	Blue (50%) Black (44%) Yellow (25%)
Yellow	Red (63%) Blue (63%) Black (56%)	Red (75%) Blue (63%) Black (50%)

Overall Frequency of Selection

Black	25%	Black	23%
White	20%	Blue	19%
Blue	20%	Red	17%
Yellow	13%	Yellow	13%
Red	11%	White	10%
Cyan	5%	Magenta	7%
Magenta	4%	Cyan	6.4%
Green	1%	Green	4%

Table 2 — Worst Color Combinations (N = 16)

Background	Thin Lines and Text	Thick Lines and Panels
White	Yellow (100%) Cyan (94%)	Yellow (94%) Cyan (75%)
Black	Blue (87%) Red (37%) Magenta (25%)	Blue (81%) Magenta (31%)
Red	Magenta (81%) Blue (44%) Green and Cyan (25%)	Magenta (69%) Blue (50%) Green (37%) Cyan (25%)
Green	Cyan (81%) Magenta (50%) Yellow (37%)	Cyan (81%) Magenta and Yellow (44%)
Blue	Green (62%) Red and Black (37%)	Green (44%) Red and Black (31%)
Cyan	Green (81%) Yellow (75%) White (31%)	Yellow (69%) Green (62%) White (56%)
Magenta	Green (75%) Red (56%) Cyan (44%)	Cyan (81%) Green (69%) Red (44%)
Yellow	White and Cyan (81%)	White (81%) Cyan (56%) Green (25%)

Overall Frequency of Selection

Cyan	24%	Cyan	23%
Green	18%	Yellow	17%
Yellow	16%	Green	16%
Magenta	11%	Magenta	12%
Red	10%	White	12%
White	8%	Blue	9%
Blue	8%	Red	7%
Black	3%	Black	2%

as yellow and blue-green. Also, our visual system does not readily perceive changes in green.

Difficulty in focusing results from edges created by color alone. Our visual system depends on a brightness difference at an edge to effect clear focusing. Multicolored images, then, should be differentiated on the basis of brightness as well as of color.

Avoid red and green in the periphery of large-scale displays. Because of the insensitivity of the retinal periphery to red and green, these colors in saturated form should be avoided, especially for small symbols and shapes. Yellow and blue are good peripheral colors.

Opponent colors go well together. Red and green or yellow and blue are good combinations for simple displays. The opposite combinations—red with yellow or green with blue—produce poorer images.

For color-deficient observers, avoid single-color distinctions. Colors which differ only in the amount of red or green added or subtracted from the two other primaries may prove difficult to distinguish for certain classes of color-deficient observers.

Perceptual Guidelines

Not all colors are equally discernible. Perceptually, we need a large change in wavelength to perceive a color difference in some portions of the spectrum and a small one in other portions.

Luminance does not equal brightness. Two equal-luminance but different hue colors will probably appear to have different brightnesses. The deviations are most extreme for colors towards the ends of the spectrum (red, magenta, blue).

Different hues have inherently different saturation levels. Yellow in particular always appears to be less saturated than other hues.

Lightness and brightness are distinguishable on a printed copy, but not on a color display. The nature of a color display does not allow lightness and brightness to be varied independently.

Not all colors are equally readable or legible. Extreme care should be exercised with text color relative to background colors. Besides a loss in hue with reduced size, inadequate contrast frequently results when the background and text colors are similar.

Hues change with intensity and background color. When grouping elements on the basis of color, be sure that backgrounds or nearby colors do not change the hue of an element in the group. Limiting the number of colors and making sure they are widely separated in the spectrum will reduce confusion.

Avoid the need for color discrimination in small areas. Hue information is lost for small areas. In general, two adjacent lines of a single-pixel width will merge to produce a mixture of the two. Also, the human visual system produces sharper images with achromatic colors. Thus for fine detail, it is best to use black, white, and gray while reserving chromatic colors for larger panels or for attracting attention.

Cognitive Guidelines

Do not overuse color. Perhaps the best rule is to use color sparingly. The benefits of color as an attention getter, information grouper, and value assigner are lost if too many colors are used. Cognitive scientists have shown that the human mind experiences great difficulty in maintaining more than five to seven elements simultaneously, so it is best to limit displays to about six clearly discriminable colors.

Group related elements by using a common background color. Cognitive science has advanced the notion of set and preattentive processing. In this context, you can prepare or set the user for related events by using a common color code. A successive set of images can be shown to be related by using the same background color.

Similar colors connote similar meanings. Elements related in some way can convey that message through the degree in similarity of hue. The color range from blue to green is experienced as more similar than the gamut from red to green. Along these same lines, saturation level can also be used to connote the strength of relationships.

Use brightness and saturation to draw attention. The brightest and most highly saturated area of a color display immediately draws the viewer's attention.

Link the degree of color change to event magnitude. As an alternative to bar charts or tic marks on amplitude scales, one can portray magnitude changes with progressive steps of changing color. A desaturated cyan can be increased in saturation as the graphed elements increase in value. Progressively switching from one hue to another can be used to indicate passing critical levels.

Order colors by their spectral position. To increase the number of colors on a display requires imposing a meaningful order on the colors. The most obvious order is that provided by the spectrum with the mnemonic ROY G. BIV (red, orange, yellow, green, blue, indigo, violet).

Warm and cold colors should indicate action levels. Traditionally, the warm (long-wavelength) colors are used to signify action or the requirement of a response. Cool colors, on the other hand, indicate status or background information. Most people also experience warm colors advancing toward them—hence forcing attention—and cool colors receding or drawing away. Ω

REFERENCES

1. G.M. Murch, "Visual Accommodation and Convergence to Multi-chromatic Display Terminals," *Proceedings of the Society for Information Display* 24 (1983), pp. 67-72.
2. L. Hurvich, *Color Vision* (Sunderland, Mass.: Sinauer, 1981).
3. G.M. Murch, *Visual and Auditory Perception* (Indianapolis: Bobbs-Merrill, 1973).
4. R. Boynton, *Human Color Vision* (New York: Holt, Rinehart and Winston, 1980).
5. F.S. Fromme, "Improving Color CAD Systems for Users: Some Suggestions from Human Factors Studies," *IEEE Design and Test* (Feb. 1984), pp. 18-27.
6. L.M. Hurvich and D.S. Jameson, *The Perception of Brightness and Darkness* (Boston: Allyn and Bacon, 1966).
7. J. Walraven, *Chromatic Induction* (Utrecht, Holland: Elinkwisk, 1981).
8. G.M. Murch, "Perceptual Considerations of Color," *Computer Graphics World* 7 (1983), pp. 32-40.
9. G.S. Wasserman, *Color Vision* (New York: Wiley, 1978).
10. L.D. Silverstein, "Human Factors for Color Displays," *Society for Information Display Seminar* (San Diego, California, 1982).
11. G.M. Murch, "Physiological Principles for the Effective Use of Color," *IEEE Computer Graphics and Applications* (Nov. 1984), pp. 49-55.
12. W.S. Cleveland and R. McGill, "A Color-Caused Optical Illusion on a Statistical Graph," *The American Statistician* 37 (May 1983), pp. 101-105.

GERALD M. MURCH *directs the activities of the Human Factors Research Group within the Applied Research Labs of Tektronix, Inc.*

How to Use Color Functionally: 53 EP&P Tips

By Jan V. White,
Electronic Art & Graphics Editor

The widespread availability of color is the latest advance in the development of traditional, as well as electronic, publishing. Most users are behaving like children with a new box of crayons.

Color must not be used merely to dazzle and catch the eye. Too often it is used for decoration, and such dressing up of the page only camouflages the underlying message. Instead, color must be used to enlighten and add value to the product. It is a material to be applied deliberately, not subjectively. It is a visual language that should be used with purpose. It should focus attention, explain relationships, analyze data. It should give visual order to information chaos. It can accomplish this by:
- Sharpening the delivery of a message by coding its elements.
- Ranking value by the sequence in which elements are noticed.
- Increasing the speed of comprehension by relating segments to one another.
- Establishing identity and character through consistency.
- Creating continuity through color-keyed associations.
- Enlivening the atmosphere of the product.

Here is a list of 53 points to bear in mind when handling color in print and presentations so your design investment pays the highest dividends:

1. Color is a valuable tool. Don't use it merely because it is available. Don't just decorate with it. Use it to sharpen the message: explain, highlight, emphasize, lead the eye.

2. Assign color deliberately. Decide what you want the viewer to understand first, then use color to make it obvious.

3. Use color to accentuate the positive: emphasize the benefits so the viewers' self-interest will attract them into reading the rest.

4. Use color to highlight instructions. Separating them from description makes manuals easier to use.

5. Categorize information so recurrent elements look different from the rest of the text. Run introductions and summaries, revisions, etc., in color.

6. Avoid running rules, bullets, bars and all other doodads in color unless there is a very good reason to do so. They are dull and hackneyed and add "colorfulness" for the wrong reason: its own sake.

7. Link separate elements with the same color. Shared color fuses them into a relationship.

8. Use the same color scheme for all signals and signposts, logos and instructions. The viewer should be able to recognize them quickly.

9. Reveal the structure of a publication by making major items such as chapter openers recognizable with color. Plan the product to allow for patterning.

10. Consistent color establishes personality. It provides

Reprinted with permission from *Electronic Publishing & Printing,* pp. 34–40, Jan./Feb. 1990.

Electronic Art World

continuity. It establishes familiarity.

11. Use color standardization as an element of any corporate identity program.

12. Choose a color that will help the viewer interpret the message. Beware of picking a color just because you "like it." The color must correlate to the message. For instance, a banana's yellowness is one of the characteristics of bananahood. Flecks of brown tell us the banana is ripening. An all-brown banana is rotten. A green banana is unripe. A blue banana is frozen. A purple banana was drawn by a child. A striped banana is deliberately shocking. A multicolored banana is a joke. A red banana is a plantain.

13. Keep color choices simple. The more colors used, the gaudier the result. Do you really want a fruit salad?

14. Four distinct colors are a practical maximum for the average viewer to remember. Two colors plus black are generally remembered best.

15. Use color coding consistently throughout a document or group of documents. Changing color without a reason merely confuses. Do not use a color to denote commands in one place and menu choices in another.

16. Choose colors that can be duplicated accurately and dependably in media besides print (overheads, slides, billboards, packaging, cloth wrapping, signs on trucks . . . wherever color recognition is a factor). The effect of color

depends on background, lighting and technologies. Check the likely results. Keep it simple.

17. Reinforce color coding with shape. (For instance, make the red line on a chart wider than the other lines). Redundancy is valuable because it attracts attention more powerfully, is more memorable, makes black-and-white copying more legible, facilitates translation from one medium to another, and helps people with impaired color vision.

18. The larger the tinted

> *Use color coding consistently throughout a document or group of documents. Changing color without a reason merely confuses. Do not use a color to denote commands in one place and menu choices in another.*

area, the paler the color should be. Large areas should be conservatively colored to keep the color from overwhelming the message. It is the message that matters, not the color.

19. The smaller the color area, the stronger the color can be. Accents should be bright in order to be noticed. But they must be small or they stop being mere accents.

20. Beware of the comparative brightnesses of colors. Warm colors appear closer, cool colors farther away. Dark colors appear heavier than light ones. Thus, an area looks

smaller in a dark color than the same-sized area in a light color.

21. All colors vary in appearance relative to their surroundings. The same color looks darker on a light background, lighter on a dark background. The same color looks warmer on a cool background than on a warm background. Don't forget the texture and the color of the surface on which the colors are printed. Uncoated paper makes colors look darker than coated paper does. Look at samples. Run tests. Check with the printer.

22. To create peaceful unity, choose colors that relate to each other in one or more of the following color attributes: hue, value or chroma. Hue means "red," "blue" or "green." Value means lightness or darkness. Chroma means intensity of hue: its dullness or brilliance.

23. Use aggressive colors to identify what is important; use conservative colors as background. Make charts and graphs vivid with color. Make the main point the brightest and subdue supporting information with quiet colors.

24. Set type that will be run in color larger and/or bolder. Any color, no matter how "bright," has lower contrast with white paper than black does. Compensate for that by enlarging the area to be covered with colored ink.

25. Bright colors tire the eye. Avoid using them in text because nobody will read more than the first sentence.

26. When printing colored

Electronic Art World

type on top of another color, make sure it is legible: use a larger size and simple, straight-forward typefaces. Don't use faces that are exaggerated in boldness or condensation, which are hard to read to begin with. In color-on-color, the counters (the little areas inside the lowercase "e," for instance) can fill up and become illegible.

27. Be sure that there is at least a 30 percent difference in tone value between type and its background color, especially when dropping out type in white. In any case, use larger or bolder type. Always compare your color to a gray scale. Don't be fooled by brightness.

28. Watch out for the unpleasant vibrating effect that complementary colors of equal tonality produce. Orange/blue, red/green and yellow/purple are unpleasant next to each other unless there is a distinct difference in lightness and darkness between them.

29. Black-and-white photos look best in black-and-white. Avoid weakening their impact by running them in color, which is bound to make them look washed out. Only the darkest browns, greens and blues can retain halftone quality.

30. Run black-and-white pictures on a screen of color only when what you gain by colorfulness outweighs what you lose in the detail of the image. You sacrifice sparkle because the color fills the highlights. That is a cheap way of producing color, and it looks it.

31. Colorful versions of black-and-white photos are best made as duotones. The two superimposed halftones reinforce the dark areas, keeping the highlights light. Tritones use three superimposed halftones. Subtle and/or surprising variations can be created by unexpected combinations of hues. This is where you should experiment and be bold.

32. Use special effects with discrimination. Mezzotints, posterizations, polarizations and other technical manipulations are fun. New technology makes them easier to produce. Use them when their specialness adds the graphic dash you

Identify a recurring theme with color. This helps establish the flowing relationship of visual to visual. A presentation is a sequence of correlated impressions over time, so its components must have visual continuity.

need; but remember, too much spice spoils the stew.

33. Use color responsibly and with great circumspection, because readers are at the mercy of the presenter's pace. They cannot study the information at their leisure. Furthermore, the presentation is a flowing sequence of impressions that must be coordinated to support the speaker's words. It must not draw undue attention to itself.

34. Treat color as a valuable tool. Don't use it merely because it is available. Don't simply decorate with it. Use it to sharpen the message:

explain, highlight, emphasize, lead the eye. Yes, this is the same as the first point, but it demands repetition in the context of presentations.

35. Use color to prioritize information. Readers will look at the brightest area first. Control their response by assigning the brightest color to the most important material.

36. Emphasize one point per visual in color. Draw attention to one segment of a pie chart, one graph trend, one row of figures, one verbal point, one bar or column, the bottom line, the focal point of the equipment under discussion.

37. Identify a recurring theme with color. This helps establish the flowing relationship of visual to visual. A presentation is a sequence of correlated impressions over time, so its components must have visual continuity.

38. Use color to help the audience sort out the material. Use it to distinguish one set of elements from another. If, for instance, the purpose of a series of visuals is to demonstrate some projected results, then pop out that aspect of the recurring reasoning in brilliant color and contrast it against the dull-colored current condition.

39. Use color sequencing to build the presentation to a climax. Start with cool green, then change the color imperceptibly over the series until you wind up with brilliant orange. Or start out with a dark shade and gradually become lighter until the final slide is white. Presentations are

Electronic Art World

also serious showbiz.

40. Use color to signal changes in the direction of the presentation. Think of a color interruption as symbolizing an announcement that "here begins something new." This might well be the place to change the orientation of the slide from the normal horizontal to a vertical, so the shape and color combine into a strong message.

41. Use as many visuals as there are points to be made. Make only one point per visual, and make it vivid and fast. Think of visuals as though they were billboards. It is impossible to make them too simple.

42. Make an impact by concentrating only on the nub of the message in the visual. Avoid long blocks of copy. Explain the supporting evidence in the speech or in handouts.

43. Break up a complex thought into a sequence of components and build them to a climax in a series of related visuals.

44. For formal presentations in dark rooms to large audiences, make the backgrounds of the slides dark. You want the words and images to shine, not the screen, so reverse the type out from the background.

45. On dark backgrounds (black, blue, grey, brown, green), drop out type in light colors. Though we don't think of white as being particularly valuable because it has "no color," it is brighter on a dark background than other colors. It is therefore the most valuable

and should be reserved for vital points. Yellow is equally visible but a little aggressive, so it should be used with great circumspection, and red should also be used sparingly. Supporting matter can be shown in any other color.

46. For informal presentations in light rooms to small groups, make the background of overheads or slides pale. The best color is no color at all, so the screen blends into the surroundings. The type should be dark on a light background.

47. On light backgrounds (white, pale green, pale blue, pale yellow, pale grey), use dark colors for the type. Black, dark blue, brown, etc., read best. Pick out words or phrases worthy of emphasis and assign them strong, bright colors and bold type.

48. Use simple type for slides and overheads, paying special attention to legibility. Slides and overheads are a different medium from hardcopy printouts. What works in print does not necessarily work in presentations.
- It is probably wisest to use Gothic (sans-serif) type, if it is not too exaggerated in boldness or too condensed.
- Avoid mixing more than two type families. Less is more.
- Avoid more than three sizes of type. Big for the heading, medium for bulk, tiny for asides.
Avoid type smaller than 18 points for overheads. It is hard to read from the third row.
- Avoid all capitals except for a word or two. They are harder to read than lowercase.
- Avoid italics except for short sentences. They are hard-

er to read than roman.
- Avoid fancy type unless it improves understanding. It draws attention to itself.
- Avoid long bulleted lists if you possibly can. They can be a bore. Like this one.

49. Leave generous margins around the outer edges of the visuals.

50. Start at the top-left corner, because that is where we all start reading. Align text beneath that, flush-left, which is easier to scan than centered copy.

51. Place the most important information at the top, where we are used to looking for it. That is why headlines are placed at the top of the page.

52. Be sure that the placement of elements follows a consistent pattern from slide to slide. The placement of the top line should be the same throughout the series, as should the space between recurring elements. It gives the presentation a thought-through quality.

53. Consistency in patterning, shape and color creates a good foundation on which deliberately placed inconsistencies can become that much more surprising. That is why, paradoxically, planned consistency can help to make a presentation more lively. ∎

This list is assembled from points made in Jan V. White's eighth book, Color for the Electronic Age, *which will be published in April 1990 by Watson-Guptill, New York City.*

LOOKING GOOD IN PUBLIC

Overheads, slides, on-screen, and videotape: how to present yourself successfully in any crowd

BY ROGER C. PARKER

Desktop presentation software can do more for your persuasive powers than just provide you with good-looking visuals. Today's programs provide speakers' notes for rehearsing your presentation and prompting you through your remarks. You can also supplement your slides with handouts, so your audience has something tangible to take home with them when the show is over.

Using such programs contributes to the success of a presentation in less tangible ways as well. With Aldus Persuasion and Symantec's More III, for instance, you start by using a built-in outliner to organize your ideas. These and other programs then can automatically generate graphics, like organization charts, from your outline. The act of preparing your visuals forces you to think about the goals of your presentation and increases your familiarity with your material.

Even more importantly, when you go into a presentation armed with clean, attractive graphics, you feel more confident about yourself and your presentation. Your audience picks up on this confidence and is predisposed to accept your ideas.

To succeed, your presentation—including the format in which it's delivered—needs to fit the audience. You can present your ideas with overhead transparencies, on slides, on a computer screen, or on video: Each format is appropriate to a particular range of situations. Choose your format according to the size and expectations of your audience, the room where your presentation will take place, your budget, and how much group interaction you seek.

These concerns also influence your choice of presentation software. This article explains when each type of presentation works best and what program features are most helpful. The table directly following this article will help you determine which programs have the features you need.

OVERHEAD TRANSPARENCIES

Overhead transparencies are not only inexpensive and simple to create, they offer several advantages over more advanced presentation visuals. You typically show them in normal or only slightly reduced room lighting, so you maintain eye contact with your audience and respond more readily to questions. This makes overheads most appropriate for small groups in conference rooms, where personal rapport is especially important. Leaving the lights on also enables your audience to take notes easily.

Overhead transparencies are lightweight and take up little space, so they travel well. They also make a flexible presentation, because you can add, remove, rearrange, or write on your visuals at the last minute, or even during the presentation.

Hardware

Black-and-white overhead transparencies are the easiest presentation material of all to produce because you print them right on your laser printer, using any font you want. (Make sure that the brand of transparency material you're using is approved for use in your laser printer.) Or, you can produce black-and-white paper originals on your laser printer and prepare the overheads on an office copier. The presentation you begin working on at 8:00 a.m. can be ready an hour later.

New color printers also make it easier and more affordable than ever to use color on overhead transparencies. Color ink-jet printers, such as Hewlett-Packard's PaintJet ($1,520) and Sharp's JX-730 ($2,395), cost less than many black-and-white laser printers; their range of color and resolution are limited, but they're fine for bar graphs, pie charts, and so on.

For the best possible overheads with the cleanest type and graphics and smoothest, most saturated color, send your files to service bureaus—such as Autographix or Genigraphics, that have offices nationwide. Microsoft PowerPoint, Micrografx Charisma, and similar programs come with software that lets you send files via modem directly to a service bureau.

Software

Since they're printed on standard desktop printers, you can prepare overheads using a wide variety of desktop publishing, word processing, or desktop-presentation programs. Programs designed specifically for presentations typically offer templates that automate layout decisions, and a slide show feature that permits you to preview your presentation in order, one slide at a time. This helps you check both the flow and the content of your presentation.

Tips for Overheads

• Use landscape (horizontal) orientation rather than portrait (vertical). Vertical overheads tend to resemble blown-up typewritten pages and tempt you to put too many words on a line, and too many lines on a page. Horizontal overheads restrict you to fewer lines set in larger type. Horizontal orientation also lets you project your overheads on the top half of the screen, so people in the back can see them.

• Light background colors with dark type remains legible in normal lighting and works best for color overheads.

• Remember that colors generally appear more washed out when projected than they do on your computer screen. Especially avoid using gradations of red and yellow. (See April and May's Showtime columns for more on getting the overhead advantage.)

35MM SLIDES

Choose 35mm slides for short, high-impact presentations. Slides offer the best color available: Their saturated, vibrant colors add impact to charts and improve text readability. Once the projector is aimed correctly, color slides are easy to present and they always appear properly positioned on the screen—unlike overheads, which must be carefully positioned by hand. Remote-control slide projectors permit you to concentrate on your words rather than on shuffling overheads.

Slides also travel well. Just about every hotel or conference room has a 35mm slide projector, and once your slides are loaded properly in a carousel, there's little chance that they'll get out of order.

You can easily incorporate builds—whereby each new image adds to the previous one—into 35mm slides. When presenting a list, you'd use a build to introduce one item at a time; on a chart, you can introduce each data series as you discuss it. By introducing information step-by-step, builds keep your audience from reading ahead of you and help to intensify your point. Persuasion offers the most comprehensive builds feature, automatically creating builds from text outlines, organization charts, or graphs.

The chief disadvantage of slides is that they require a darkened room, which disrupts your audience rapport. The audience is less likely to ask questions, and you'll find it harder to respond to them. Darkened rooms also make it difficult for your audience to take notes. Make your slide presentations short and to the point: Most audiences don't want to sit for too long in the dark.

Furthermore, although the cost of producing slides is going down, they cost more than overheads, and typically require at least a day to produce.

Hardware

Recent advances have made color slides more cost effective and practical than ever before. Mac programs such as Persuasion and Microsoft's PowerPoint, and PC programs such as Software Publishing's Harvard Graphics or PowerPoint for Windows come with utilities that enable you to prepare and proof your slides in your office and then send the files to a slide production service bureau. If you have a modem and you're willing to pay for overnight service, the slides you prepare today can be in your hands as quickly as tomorrow.

You can also use a slide recorder—a device containing a camera to record your graphics—and send the exposed film out for processing. Slide recorder prices start in the same range as a laser printer and climb sharply from there, so you may need to make a large volume of slides to justify the purchase. Whether you use a service bureau or a slide recorder, you have to be concerned with type and graphics compatibility. Most slide recorders offer only a limited number of fonts, and few can reproduce Encapsulated PostScript files. Without PostScript you must adapt your presentation to avoid unsightly jagged edges (see Showtime, September).

Software

All of the programs listed in the accompanying table offer features helpful for producing 35mm slides. PowerPoint, for instance, comes with 5,000 professionally designed color palettes. Each palette contains a set of background, foreground, and accent colors chosen to work well together. All you do is choose a background; the program then prompts you to decide on a foreground color and, on the basis of those two choices, offers a selection of appropriate accent colors. In addition to sample palettes for 35mm slides, PowerPoint includes recommended color schemes for overheads and on-screen presentations.

PowerPoint also lets you shade the background colors from dark to light, a common technique for dramatic slides. Persuasion goes further and lets you shade one color into another.

A slide sorter that lets you order your presentation by moving thumbnails of your slides around on screen is another useful feature found in most presentation programs. In slide-making programs you can create graphs, charts, and tables with a built-in graphing capability. Software Publishing's Harvard Graphics for the PC and CA-Cricket Presents for the Macintosh (a Windows version has been announced), can also produce graphs from data imported from a database.

Tips for 35mm Slides
• Color slides work best in darkened rooms, so choose dark backgrounds with light type to make your slides easy to read.
• Shaded backgrounds add interest without distracting your audience.
• Avoid including too many bulleted points and too many words on each slide. Three lines of three or four words each is about right.
• Don't put too much detail in your charts and graphs. You should use no more than six slices in a pie chart, for instance.
• Turn up the lights during your question-and-answer sessions.

ON-SCREEN PRESENTATIONS

On-screen presentations, for which your computer generates images while you speak, represent the most advanced form of presentation technology. They demand a lot from your hardware, but they offer the most flexibility and the most opportunities for creative expression. On-screen presentations work best when you have a great deal of control over the conditions under which you work: lighting, equipment, and so on.

If you're talking to a small group—say, in a product demonstration—you can give a presentation on your computer's regular monitor.

At this level, on-screen presentations are relatively inexpensive because you use your existing hardware. If the group is small enough, a laptop computer makes a handy, portable presentation tool.

For a slightly larger group, use a projector pad, which connects to your computer and sits on an overhead projector, allowing you to project what your computer displays. Only the most expensive projector pads offer color reproduction, though; most replace colors with patterns or shades of gray, which can clutter your carefully prepared charts and graphs.

For formal presentations to a large group, special big-screen, high-resolution monitors and video-projection devices are available. You can rent them, and they're often built into conference rooms and convention facilities. Mouse-controlled, on-screen pointers available in many programs are useful devices for leading your audience across the big screen.

Using on-screen presentations introduces the possibility of some dramatic special effects. With Persuasion, for example, one image can replace another by pushing it off screen, dissolving into it, or by any of several other options. These sorts of transitions work well not only between slides but between the successive elements of a build.

Another advantage you gain with on-screen presentations is flexibility. With both Persuasion and PowerPoint, for instance, you can advance from slide to slide with a click of the mouse, move backwards through your presentation, or go to a specific slide by simply typing in its number. Persuasion even lets you jump to a slide to answer a question, and then immediately return to where you left off.

You can also modify the presentation spontaneously to reflect new information or audience input. The built-in charting facility of Mac programs such as Persuasion and CA-Cricket Presents or PC programs such as Harvard Graphics and

PowerPoint for Windows means you can update existing charts with new data during your presentation.

"Hypermedia" capabilities, such as Harvard Graphics' HyperShow feature, make it even easier to customize a presentation. With hypermedia, you weave several series of images together, linking them with "buttons" on the screen. By clicking on a button, you can switch to another series, detouring from the main sequence, to amplify specific points. You could click on the axis of a graph, say, to bring up information about the source of the graph's data. Hypermedia is ideal for self-running presentations because viewers can take their own, independent detours.

At their most complex, on-screen presentations blend with multimedia to combine sound and moving images, either generated by the computer or stored on laser discs, videocassette recorders, or CD-ROM disks. For example, if you're describing a new product, you can insert an animated sequence showing how its parts are assembled. Then, to introduce your new sales force, you could cut to a videotaped segment of your staff. Multimedia presentations require special software beyond the programs discussed here, but the advantages of incorporating movement and sound into a presentation are obvious.

The primary drawbacks to any on-screen presentation are image quality and reliability. On a monitor, you don't get the same smooth type and graphics that you get on a slide. You also have to make sure that your display is bright and clear enough for the room size and lighting conditions in which you make your presentation.

Big-screen monitors are expensive to buy or rent. They're also heavy to move around, so they're most useful in corporate conference rooms or other situations where you can leave them for long periods of time.

Reliability is an important issue with on-screen presentations. Make sure to have backup computers and monitors available and the right hardware to use them; your entire presentation could be sabotaged by the wrong connecting cable. Even then, a presentation prepared on one computer or monitor may not work smoothly on another one. Complex color graphics make heavy demands on computer speed and memory, so make sure that the computer you use to deliver your presentation is as powerful as the one you created it on.

Hardware and Software

With the exception of multimedia presentations, the same programs that you use to prepare slides will work for making your on-screen presentations. However, there are certain products that can improve the appearance of your presentations.

Advanced video boards, for one, improve the color quality of both Macs and PCs. They're especially recommended for use with PCs that otherwise are limited in the number of on-screen colors possible. The Video 7 board enables you to display 256 colors on a VGA monitor without dithering—and this pays off in improved color saturation and smoothly shaded backgrounds.

Adobe Type Manager improves the display of PostScript Type 1 fonts in either the Mac or Windows environment, reproducing characters smoothly at any size.

Tips for On-screen Presentations
• Use builds to introduce information as you discuss it.
• Avoid dramatic effects that overwhelm your message: It's the message you want your audience to remember, not the medium.
• Always preview your presentation on the screen that you'll use to present it. Screen size influences the dimensions of your visuals, and you may have to resize and reorganize your images if you're showing them on a different monitor

than the one on which they were created.

VIDEOTAPE
Once you have your on-screen presentation prepared, you can save it to videotape using new hardware add-ons such as the Willow VGA-TV card for the PC and the Truevision NuVista Plus for the Mac.

With a video camera, a microphone, and some editing, you can even add announcers or voice-overs, producing what amounts to a television show that you can send to anyone who has a VCR. These self-running presentations are great for free-standing public displays.

The most serious disadvantages of videotaped presentations lie in their high production costs and relative inflexibility. Once you've prepared your presentation, there's no real way to modify it. You can't easily change your charts or graphs without rerecording the whole presentation, either. Likewise, there is no way for the audience to "jump around" from topic to topic.

Tips for Videotaped Presentations
• Avoid "talking head" videos, which are characterized by too much time spent with the camera focused on the narrator's face.

• Avoid abrupt movements and changes. Always provide smooth transitions, or segues, from segment to segment.

LIGHTS UP
Regardless of which presentation media you choose, planning is crucial for success. Ask yourself the following questions to help you plan your presentation:
1. Who is your audience? How much convincing do they require? How much do they already know about the subject?
2. Which type(s) of information are likely to be most persuasive for your particular message—lists, charts, maps, or illustrations?
3. When will your presentation take place? Do you have enough time to prepare color slides and overheads, or do you have to make do with black-and-white laser-printed overheads?
4. Where will your presentation take place? What's the lighting like? How familiar are you with the hardware and software?
5. How much do you want to "show" versus "tell"? Presentation visuals should provide a guide, but success depends on your confidence, enthusiasm, and knowledge. Remember that you've been asked to deliver a pre-

---

RESOURCES

The following books offer useful advice on the psychology and dynamics of presentation planning and delivery, as well as on designing high-impact graphics:

• *I Can See You Naked: A Fearless Guide to Making Great Presentations*, Ron Hoff (Andrews and McMeel, Kansas City, MO, 1988).
• *Presentations Plus: David Peoples' Proven Techniques*, David A. Peoples (John Wiley, New York, NY, 1988).
• *Dynamics of Presentation Graphics*, Dona Z. Meilach (Dow Jones-Irwin, Homewood, IL, revised 1990).
• *The Presentation Design Book*, Margaret Rabb, editor (Ventana Press, Chapel Hill, NC, 1990).
• *The Overnight Guide to Public Speaking*, Ed Wohlmuth (Running Press, Philadelphia, PA, 1983).

sentation because you possess knowledge that the audience wants. Go into it with the attitude that you have the answer to the audience's problems—or problems the audience may not even be aware of. Smile, present your visuals with pride, speak as if you were discussing a project with friends, and let your enthusiasm and knowledge emerge. You can't lose! ∎

Roger C. Parker is president of The Write Word, an advertising and consulting firm located in Dover, New Hampshire.

DEAN FITCH

Producing Foreign Language Presentations

Foreign languages and cultures present another chance to differentiate your product or service from your competitors' by identifying the language and cultural needs of your audience. Solid language skills will provide an edge to the well-prepared presenter.

It has been said that the only thing that remains constant in life is change. Today, the world is changing at an incredible pace. Economic and political barriers are coming down as surely as the Berlin Wall. The demand for presentation materials in foreign languages is expanding rapidly. The needs and ability to communicate effectively in foreign languages will affect the way AV materials are prepared and presented.

Economists are predicting the domestic market as having a flat or slightly negative growth period for the first half of the '90s. The expansion in the export sector is expected to continue increasing through the year 2000 and beyond. US firms are achieving record breaking export levels with continued annual increases (13 percent last year alone). Export will be the growth sector of the future. The proposed unified European Economic Community (EEC) was the cause of the $16.6 billion investments by US corporations throughout Europe in 1989. These investments are to establish a US presence in what will be the world's single largest market (320 million consumers).

The computerization of Eastern Europe and opening of free market systems will create a long-range growth potential. Likewise, the expansion of trade with the Soviet Union is expected to increase 200 to 300 percent before the year 2000. These and other areas such as the Pacific Rim and lesser developed nations will provide additional opportunities for expanded US economic growth.

This increase in exports will affect the presenter by requiring that he or she possess the language skills and address cultural differences when preparing the presentations necessary to tap this growing market. It was not many years ago that, when asked about presentations for foreign audiences, many US firms responded, "We only do business with English-speaking buyers."

With the prospect of flat growth in the domestic market and evidence of expansion in international markets,

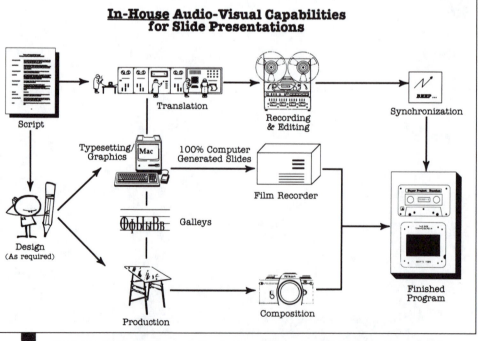

Producing foreign slide presentations is no small task as this chart from Omni demonstrates.

Reprinted with permission from *Audio Visual Communications,* pp. 17–20, Aug. 1990.

growth may depend on the presenter adapting to these opportunities.

The objective of a presentation is to address the needs and requirements of the audience while presenting an image of the company or organization and its ability to satisfy these needs. Foreign languages and cultures present an additional opportunity to differentiate your product or service from your competitors' by identifying the language and cultural needs of your audience. Proper language skills and knowledge of the specific market will give the competitive edge to those presenters who are properly prepared.

Some of the possible applications include corporate presentations to foreign trade ministries, foreign financial institutions, prospective joint venture partners and presentations to ecological and other special interest groups. Other areas to develop after the startup of foreign ventures include safety, training and marketing presentations. The cost of creating safety and training programs for foreign employees is small in comparison to the overall exposure to liability that exists in the event of an industrial accident. Trade show and promotional material may also be required and must be targeted toward specific languages and markets.

DOS AND DON'TS

The "dos and don'ts" involved in preparing foreign language materials would cover volumes on each language and culture involved. The most common practice is to convert existing presentations into their foreign language counterparts using existing materials and adapting them to the cultural requirements of the audience. Critical considerations include:

• Allocating adequate time to prepare the English presentation. A poorly written presentation translates poorly.

• English is a very compressed language. In video presentations compensation for language expansion is a concern. Each language may expand the length of the presentation at different rates. As an example French may expand 20 percent while the same presentation in Russian may expand as much as 35 to 50 percent.

• Avoid the use of slang, colloquialisms

Audio Visual Communications in Cyrillic.

or jargon that's not a recognized industry specific term. Abbreviations must be avoided or spelled out for the translator.

• Allow enough time for the translation and a review of the translation by a second source to verify the accuracy of the material before and after production. The production schedule must also provide enough time for production of the foreign language artwork. Presentations in foreign languages often require the same amount of time as the original English version.

• Never translate the text on the slides without the accompanying script. The wording on slides is very compressed and each word may have more than one meaning. A minor error in translating can be a major embarrassment.

As a slide show producer once told us, "Slide shows are never finished; they are just given up on at a certain point." That point is usually determined by the amount of time remaining before the presentation.

Production of foreign language presentations requires specific skills and equipment. An evaluation of the alternatives to accomplish this to provides the presenter with a way to produce cost effective and meaningful programs to the foreign audience. Remember: not giving a presentation is preferable to giving a bad one.

• In-house productions are the most cost effective, but many firms lack the required equipment and/or the foreign language competency necessary to create effective presentations. Just as owning a 35mm camera does not make one a professional photographer, speaking more than one language does not make one a translator.

• Foreign subsidiaries or agents are often used to produce programs for a localized market. This adds to the time and cost, while giving up control of the corporate image. These agents

may also lack the English language skills required to fully understand the material and effectively convey the intended message.

• Technical translating services are best prepared to translate the material, but some may lack either the industry specific terms or the equipment to produce the artwork. On the other hand, university language departments and language schools are like going to a barber's school to get a hair cut: you never know who will do the work or what the end result will look like until it is finished, too late to correct any mistake.

• All of the above: The combined resources of the in-house department, agents and qualified technical translating services will result in the best final product.

For firms that want to retain control of the production or send out the created art work to a service bureau, many options are available. Those firms with dedicated work stations and film recorders to create slides should check with the manufacturer for the availability of foreign language fonts. The evaluation of your options begins with the equipment available.

We tested software for both MS-DOS and Macintosh computers for this article.

The computer used may define the approach taken to create slides in foreign languages. One of the primary differences between MS-DOS and Macintosh computers is the way in which fonts are handled.

In the MS-DOS environment, the fonts must be resident in the software program or downloadable fonts that programs can use. In the Mac the software is font independent and you can import other fonts into the application.

The fonts used in all languages are Roman, which includes ANSI character sets, or non-Roman. Many software applications use the Roman/ANSI fonts but require keyboard manipulation. Most film recorders and slide service bureaus support these fonts. ❑

Omni Intercommunications Inc., Houston, TX, is a technical translation firm specializing in graphic arts for foreign language presentations. It provides the

foreign language skills and production experience necessary to support clients' internal production departments or create turnkey presentations. In-house facilities include desktop publishing using Mac equipment, an audio recording studio and an off-line video editing suite.

Dean Fitch, a graphic consultant experienced in multimedia presenations, provides the interface between clients and Omni's various production departments to create foreign language catalogs, brochures and other printed materials, as well as slide and video productions. Omni translated US Summit Committee documents recently.

Foreign Language Slides & Video

BY DEAN FITCH

We tested software and service bureaus that support those software. The presentation software test was limited to fonts included in the extended ASCII files, which make up the most common foreign languages that use Roman characters.

This limitation was necessary due to the lack of available non-Roman foreign language fonts for film recorders and MS-DOS software. Not all of the fonts required are directly available as extended ASCII characters.

The MS-DOS programs tested are object independent, and many of the missing characters could be made up through some keyboard manipulation to reproduce the font, but one of the benefits of presentation software is the ability to import text files into a created template to automate the slide production process. The results were as follows:

Harvard Graphics (version 2.13), a text-based program, requires a great amount of input and provided unwanted surprises. Not all of the extended ASCII characters were available.

Zenographics Mirage (version 5.2) is menu driven and the most powerful of the presentation software programs we tested in the MS-DOS environment. All of the extended ASCII characters were available in a limited number of fonts.

Neither program, however, offers foreign language versions.

The Macintosh programs offer the most easy to use applications and that the programs are font independent and the fonts are available limited the test only to a search for film recorders whose installed fonts support the extended characters used in creating foreign language slides.

We tested Power Point (version 2.01), Persuasion (version 2.0) and CA-Cricket Presents (version 2.0), all of which handle the extended ASCII character and come in foreign language versions.

The benefit of these foreign language versions is they have spell checking dictionaries to avoid embarrassing errors. These dictionaries can be installed in your English presentation program by "dragging" the existing dictionary to a separate folder and installing the foreign dictionary renamed as the same dictionary it replaces. Check with the individual manufacturers on how to buy these foreign language versions.

To test the production of slides created in non-Roman characters, using RIP (raster image processing), we called on the talents of Cal Stanley of Executive Presentation Graphics, a Houston vendor of desktop presentation systems. After testing combinations of presentation and art programs, he recommended creating artwork in Canvas (2.0) and saving it in an Encapsulated PostScript (EPS) file. The RIP conversion was done using VBS Professional Output Manager software and output to a Matrix PCR film recorder.

Further testing will be required to determine the full capabilities of this method of producing slides, but it does work.

The ChromaScript from AGFA Matrix is a hardware solution to the vector to raster image conversion necessary to produce non-Roman foreign language slides for a service bureau or in a multiple workstation environment using recently released AGFA film recorders and other devices. The JMT Group used a ChromaScript to output the test slide we created in Pagemaker (version 4.0) using PostScript Cyrillic fonts saved as an EPS file. This file was then processed through a download program to the ChromaScript and output to a Matrix PRC film recorder.

Photo Composition of slides uses multiple exposures of lithographic negatives and projection of a specified color for each color to appear in the completed slide. FrameWorks Production produced our test slide using a triple exposure of our pin-registered artwork. The time and cost to produce slides in this manner increases as the number of colors to be incorporated increases. The benefit of this method of production is that many of the languages not available in computer programs or on film recorders are available from Omni and other typesetters specialized in the production of foreign language graphics.

Foreign language character generation for overlay onto existing video can be done through dedicated character generators on-line or through off-line desktop video systems. The off-line systems we tested include video overlay boards and boards capable of frame grabbing for other applications. To provide an objective evaluation of the MS-DOS boards and programs, we enlisted the aid of Bruce Meltzer of Desktop Systems, a Houston vendor of desktop video systems.

"We tested three boards for the ability to produce overlays of foreign language fonts with an IBM compatible PC. The Willow GEO and Questel CVG are combination VGA adapters and NTSC encoders. Willow Peripherals VGA-TV GEO is designed for an AT class machine and is controlled through its software setup. The Questel CVG 10 will work in an XT class machine. The control software is memory resident enabling the user to change from VGA to composite and Genlock from within other applications. Both boards performed adequately for in-house production.

"The character generation software program used in the VGA tests was Video Titler [release 4] from Enthropy Engineering. This program includes rolling titles and other effects for the creative user. The fonts appear flicker free and come in a wide range of point sizes. At the time of preparing this evaluation, Enthropy Engineering had just released its Spanish and French fonts for the VGA boards with other languages to come. A TARGA version of the fonts will be out soon.

"The TARGA 16 board is designed as a video adapter for the PC. The Truevisons' TIPS TypeWrite fonts designed for use with this board are scalable in different sizes and include the extended ASCII character sets as well as selected foreign language fonts to create Greek, Hebrew, Cyrillic and other characters. This combination of hardware and software is capable of producing results acceptable for in-house corporate or industrial quality video."

Omni tested Colorspace Iii video board on a Mac II. We generated video overlay in a wide variety of languages (all the Roman character languages, Polish, Rumanian, Russian even Chinese). Zoom and scrolling titles were created using Macro-Mind Director (version 2.0) which uses the computer's screen fonts and those that run under Adobe's Type Manager. ATM gives the sharpest line edges. This combination produced the greatest number of foreign languages and at high quality. ❑

Part 4
Delivery Techniques and Tools

THE BEST planned and prepared technical presentation can be ruined if it is not properly delivered. An engineer's speaking ability and poise can significantly influence the effectiveness of his or her presentation. Other factors that can impact presentation delivery include presenter attitude, personality, and readiness; physical arrangements and environment; and supporting equipment capabilities and reliability. This part contains articles that describe various techniques and tools for delivering presentations with clarity and confidence.

A presentation can be either dry or dynamic. The result is most often a function of the presenter's ability to speak clearly and authoritatively, and the development of individual ability requires training and practice. In his article, "The clinical approach to creative speech," J. Campbell Connelly defines the attributes of a good speaker, and identifies the problems that must be overcome to achieve this capability. In addition, he describes techniques for developing professional poise and exercises for improving speaking ability. Connelly also stresses the importance of developing self-awareness and minimizing self-consciousness.

In her paper, "At ease: Five simple cures for public-speaking panic," Mary Jane Mapes indicates that nervousness is natural, and can be a blessing. Without a rush of adrenaline, a presenter could seem bored with the topic and/or uninterested in communicating with the audience. To harness the energy that comes naturally in presenting a technical paper, she describes five techniques involving preparation, self-assurance, practice, body language, and audience feedback. She also includes a checklist for writing a speech.

Should a presentation be read from a formal manuscript or delivered informally using only notes? This question has been the subject of much debate over the years without achieving a meaningful consensus of opinion. In her paper, "It's all right to read your speech—and here's how," Joan Dornbusch explains that engineers may need to read their presentations, especially when they must convey precise technical material or comply with a specified time limit. Success in giving effective presentations of written material begins at the writing stage, and Dornbusch gives suggestions for accomplishing this. She also offers some practical techniques for reading and includes an example of a short speech written for reading aloud.

Humor is a characteristic not usually associated with technical presentations. In his paper, "Humor can improve your technical presentations," however, James Gleason indicates the advantages of using humor in text and illustrations and analyzes several examples. This analysis addresses three points. First, humor can make technical material easier to understand. Second, the effectiveness of humor is a function of the author's freedom from external constraints. Third, the audience must be sufficiently knowledgeable of the subject to recognize and appreciate the humor. If humor is suitable for both the material and the audience, it can significantly enhance the effectiveness of an engineer's message.

Because most engineers are not experts in delivering presentations, audiences are often forced to sit through dull paper after dull paper, fighting drowsiness and distraction. In his paper, "Presenting the peer paper," Robert Garver describes some conventional public-speaking methods that address the major deficiencies observed in the presentation of technical papers. The specific problems addressed are the lack of presenter excitement, poor use and control of the available delivery time, distractions from unreadable visuals, unnecessary complexity, and apathetic responses to questions. He also addresses the special situation of the invited presentation.

Dale Neeck presents a list of tips to improve impromptu skills in the paper, "Off-the-cuff presentations." The specific points he covers include developing the right attitude, reflecting confidence while speaking, drawing on personal experience, and practicing impromptu speaking. He also includes a series of steps to follow when called upon to speak on the spur of the moment.

In his paper, "Performance guide for oral communication," Gerald Ratliff describes an orderly and consistent approach for organizing, supporting, and delivering technical presentations. The key elements of this approach are an analysis of audience interest in and knowledge of the topic, the role of speech in presenting complex technical ideas, and the principles of dynamic communication as they relate to audience perception and information retention. His paper also includes a comprehensive survey of the public-speaking situation, suggestions for scripting and vocal delivery, and a checklist of common errors encountered in presenting factual material.

In his paper, "Staged right," Mard Naman presents advice for arranging and controlling equipment to assure a smooth presentation. He stresses that, with proper planning and technical support, the chances of encountering an embarrassing glitch can be minimized. He specifically addresses procedures for microphones, lighting, equipment backup, crew communication, and teamwork. He also emphasizes the importance of detailed planning.

In her paper, "How to display an image," Dona Meilach describes the wide choice of projection hardware available to support the increasing needs of today's technical presenters.

She indicates that, while some hardware can serve more than one application, no one system can meet all requirements. The specific types of equipment addressed include overhead projectors, slide projectors, small-screen CRT and video projectors, and large-screen projectors. The impact of resolution, scan rate, and brightness is also addressed.

Christopher Rauen, in his paper "LCD panels bring presentations down to size," describes a relatively new accessory for overhead projectors that allows a presenter to take advantage of computer capabilities. LCD panels, which are about the size of a pizza box, mount on the platen of an overhead projector in place of the normal transparencies. Using information from a separate computer, an array of pixels in the LCD panel forms the desired image, which is then projected via the overhead projector onto a screen. Although the image quality is less than that provided by a normal transparency, slide, or video projector, it allows a presenter nearly real-time control of the image content to accommodate last-minute changes and/or react to audience feedback.

Peter Covino, in his paper, "Multi-image presentations: Still powerful after all these years," indicates that multiple screens were initially used at the 1900 Paris Exposition. Although new, more sophisticated technologies are replacing multi-image presentations for many applications, they are still the best approach for certain situations. He notes that, in today's budget-conscious business climate, the multi-image approach continues to prove its worth as an effective communication tool, and he cites several considerations for using it correctly.

Giving impromptu presentations while traveling merits special attention. In his article, "Visual presentations—How to survive giving one on the road," George Beiswinger addresses the required physical preparations and coordination. He also describes a "visuals first-aid kit" that can be handy for supporting out-of-town presentations.

In his paper, "Presentations on the road," John Callender notes that giving a presentation away from the author's relatively friendly and controlled home office environment requires careful planning. While mobile presenters can often secure the necessary equipment and facilities at a remote presentation site, unpleasant last-minute surprises are not uncommon. To avoid this situation, he describes a wide range of relatively new presentation tools that are both portable and easy to use. Although these tools often lack the functionality of their nonportable counterparts, the ability to assure control and reliability is usually well worth the quality tradeoffs.

In Chapter 10 of his book, *Presentations Plus*, David Peoples addresses the handling of questions, answers, and troublemakers encountered in presentations. He indicates that questions are an excellent tool for getting attention, keeping interest, and receiving feedback, and he provides suggestions for handling different types of questions. He also underscores the importance of establishing a stress-free environment that encourages an audience to ask questions. In addition, he identifies different types of troublemakers and how to handle each diplomatically while enhancing the effectiveness of the presentation.

THE CLINICAL APPROACH TO CREATIVE SPEECH

by

J. Campbell Connelly

Summary

The clinical approach to Creative Speech will:

1. Make you a sought-after speaker
2. Enable you to enjoy delivering presentations
3. Improve your poise and self-confidence
4. Give you new stature with your colleagues
5. Enable you to create a good impression for your company
6. "Make you conscious of yourself, but not self-conscious"

This paper is designed for your use in organizing speech groups within the PGEWS or within your respective companies.

Introduction

Being a professional person will give you some status with your company, but it will not guarantee you automatic advancement. You must be able to communicate your professional ideas. The ability to present them orally in a creative, clear, interesting, and motivating manner provides an excellent means of putting them, and you, before your company and the customer.

A professional person well trained in his profession, with an imagination, and the ability to clearly express his ideas has every advantage over his tongue-tied colleague.

Some of the best engineers are extremely interested in Stereo and HiFi. They will buy nothing but the finest matched components for their HiFi Consoles. They spend hundreds - and thousands - of dollars to get the "just right" combination of components to achieve the exact tonal qualities desired. Yet these uniquely trained and highly motivated people seem to ignore the best "built-in HiFi" set in the world, the human voice.

The training of that human voice to near perfection costs far less than a HiFi Console, and this training could result in raising you to a higher plane in your chosen profession.

A technical presentation can be either dry or dramatically dynamic. This depends, to a great degree, on the formal speech training of the person presenting the information. You may have a natural flare for speaking before groups, or you may be timid in such situations. In either case, you can be a more effective speaker – with training.

Pick up several good books on effective speaking, and they will give you some keys for improving your speech. They will suggest that you:

1. develop a keen mental attitude

2. know and understand your audience

3. develop your speech outline

4. make your information clear and interesting.

5. enunciate your words clearly

6. be persuasive

7. be forceful

8. be direct

9. be pleasant

Some books tell you to do these and much more, but most books fail to tell you how. The "how" of improving your speech is the area which will be developed in this paper.

You may believe that a speaker's first problem is that of being "overly nervous." I use the phrase "overly nervous" to distinguish that condition from the natural nervousness which should be a result of your desire to succeed in your presentation. This "overly nervous" state should occur only when your subject matter is not developed as close to perfection as possible, you do not know your audience, or you feel vocally untrained.

Reprinted from *IEEE EWS Nat. Symp. – Eng. Writing and Speech: An Art or a Science?*, Sept. 13–14, 1962, pp. 17–32.

I assume that a professional person knows his subject matter well and will know something about the audience he is addressing. Therefore, the actual oral presentation would be his major problem. One way to overcome this problem is through the Clinical Approach to Creative Speech. So, let's get on to some constructive work.

Remember, how much you improve will depend upon how much effort you put forth, how much time you spend, and your degree of interest and motivation. No one can do the job for you. It is your practice that will make you perfect — if you practice perfectly.

Poise

The first thing that the audience notices about the speaker is his stage presence. First impressions are formed by the audience (either consciously or subconsciously) about the speaker while he merely sits, waiting to be introduced. If the speaker appears uncomfortable and nervous, the audience receives the impression that the presentation will be pretty much the same. Conversely, if you look confident, well-poised, and alert, you will project this feeling to the audience.

Once you have arrived on stage, do not adjust your clothing. This should have been done while you were in front of a mirror prior to making your stage entrance. Next, if you lumber out of your chair, unsure of yourself, fumbling for notes, and straightening your clothes, you are conveying an impression that your presentation will also be nervous and disorganized. Sit, rather, in a relaxed comfortable position, your legs crossed at the knee or at the ankle. (The latter position is the only acceptable one for women.) When you have been introduced, do not slouch from your chair. Always use your leg muscles when rising.

Many speakers are bothered by stage deportment. They have no idea how to stand before an audience, and they become self-conscious and nervous in the absence of a lectern. You can be a far more effective speaker without a lectern, for you will have more freedom for physical flexibility. You should be capable of speaking to an audience without hiding behind a lectern. A lectern usually becomes a crutch for most speakers. You can capture and hold an audience merely by the manner in which you move about on the stage.

Your presentations will also be much more effective if you hold your 3" by 5" cards in your hand and get out in front of the lectern. This, at first, seems more difficult and next to impossible.

Most speakers put their hands either in their pockets, in back of them, or folded in front of them. These positions, too, are crutches. They take away the poise and dignity of the speaker. Simply stand on the stage with your hands by your side and talk. The only time your hands should move is when you are making gestures, or consciously desire to do so. When practicing, stand as I have just mentioned; and put that energy that you use making nervous gestures into vocal emphasis.

The following positions will be uncomfortable for you for quite some time, but only because they are foreign. However, after you have integrated them into your personality, you will feel so good, you'll wonder why you didn't always stand in this position.

1. Stand with your feet approximately shoulder-width apart and with your weight on the balls of your feet. Let the front portion (just before the sole) of your heel touch the floor.

2. Bend your knees slightly. This takes away strain and helps prevent the knees from shaking. It also gives a more relaxed appearance.

3. Pull your stomach in and up.

4. Pull or tuck your derriere in.

5. Keep your shoulders down and loose.

6. Keep your arms loose.

7. Keep your hands down by your side — forget them.

8. Acquire the habit of standing exactly in this manner, looking your audience directly in the eyes. This is an interesting exercise all by itself. You wonder what they are thinking about you, and they wonder what you are thinking about them.

9. Do not put your hands in your pockets, in back of you, or crossed in front of you.

10. Become consciously aware of any subconscious hand, body, or head movements. Have one of your colleagues bring any physical movement to your attention. Only when you become consciously aware of these unintentional

Figure 1   Poise

movements, can you begin to correct them.

11. This is perhaps one of the most important exercises of all. Do all of the above exercises during your everday conversations, whether it be during formal business hours or informal conversations -all of the exercises, that is, except numbers 8 and 10. At first you will feel stiff and uncomfortable; and then soon – but not too soon – you will feel uncomfortable if you revert to your old stance habits.

After you master these physical exercises, you will have developed much more poise and self-assurance. You will now begin to radiate and project these feelings and qualities to your audience.

Another aspect of poise which should be mentioned here is that many speakers give themselves away when they make verbal, or grammatical, mistakes and when they cannot instantly recall their next thought phrase. Minor verbal and grammatical errors may go uncorrected. They are usually just a slip of the lip and will, in all probability, be corrected when you say them again. The gravest error is that of letting the audience become aware that you have forgotten what you were going to say next. Some speakers give themselves away by becoming nervous and upset. Others do it by saying uh-uh, or m-m. Still others make all sorts of physical and facial gestures. You, the speaker, are the only one who can give yourself away in such a situation.

However, if you just stand on the stage and look at your audience with an intelligent (not blank or helpless) look they will never know what is going on in your mind. You may even move about the stage in a pensive mood. By the time you have done this once or twice, either your thought phrase will have come back to you or you will be able to improvise if you stay poised. If you allow yourself to panic, then the only thing that can happen is that your panic will grow.

If you do panic, it should be only an extremely temporary moment due to the fact that, as a well-prepared speaker, you have notes for just such a moment.

Considerable time has been devoted to poise because it is one aspect of speech that will develop your capability of delivering a creative presentation.

Before we become involved in the following exercises, I would like to relate an incident. This episode specifically points out that a few brief words criticizing a "dry run" can confuse a speaker more than it will help him. The help should begin a year or two before one ventures to represent his company as a speaker. One day a person asked if I would help him with a speech. I agreed and asked when he would like to begin. His reply nearly left me speechless. He said that he had to present a paper the next day and he would like some pointers. I offered to listen to his "dry run" and was again staggered by his reply. "Oh, I've already given that. Someone suggested that I should see you to obtain suggestions for highlighting the presentation."

I asked how long he had studied for his professional degree. He looked at me quizzically but answered, "Four years." I reminded him that it had taken him four years to learn something completely new. Somewhat confused, he asked what I was trying to tell him. Cautiously, but firmly, I told him that he couldn't expect to improve either his speech or his presentation in a few short hours. It would be grossly unfair and unkind of me to hear his presentation and then evaluate it without his having enough time to do any constructive work with it. I would only confuse him; and, while he was making the presentation, he would be thinking about all the things I said instead of concentrating on the subject and evaluating his audience.

Upon returning from his presentation, he asked when my next Basic Speech class would begin. When told the class wouldn't begin for six weeks, he was quite disappointed. However, he not only studied Basic Speech but went on to the Intermediate and Advanced classes. Now he wants to repeat the Intermediate class each year. He feels that this specific area of speech has been of the greatest value to him.

One evening during a break in class, he said, "Camp, now I know what you were talking about when you mentioned that it is impossible to improve your speech in a few short hours. First, we must break old habits and rid ourselves of unnecessary inhibitions. Only then can we really begin to improve."

He had learned his lesson well. It takes just as much training to be able to speak on a professional level as it does to become professionally proficient in any other endeavor. It is imperative that you continue studying, in order to remain at the peak of efficiency.

You can actually enjoy the following exercises by doing them at home with the whole family, at parties, or on your way to and from work. I feel that these exercises have been of the greatest value to students in my Speech classes. Just a word in passing. Some instructors criticize or constructively criticize a student after a presentation. If a person knows that he is going to be criticized in front of a group, he might be reluctant to join. I prefer to constructively evaluate a presentation. This puts the entire class on a higher plane, and no one's pride or dignity is injured.

Exercise No. 1. Diaphragmatic Breathing

You cannot speak correctly unless you breathe correctly. Diaphragmatic breathing is the key to a pleasant sounding voice and a well projected voice. Without diaphragmatic breathing, it is impossible to color your voice or to utilize its full range.

Now think of the air going to the bottom of your stomach. Take a deep breath when you inhale from the bottom up, not from the top down. Incorrect breathing is the origin of most speakers' troubles. Some schools of speech call this stomach breathing. The important thing is to get the air where it belongs – in the bottom of the lungs first – not the top. Do not let your shoulders or chest rise when you are taking a breath.

The following steps will aid you in this exercise.

1. Put your hand on your stomach.

2. Breathe deeply. Your stomach and your hand should move forward.

3. Now exhale by pulling in your diaphragm. The diaphragm pushes against the lungs and forces out the air.

Do not think of the biological make-up of the voice, the throat, or the palates. Imagine the air coming through a tube from the lungs to the mouth.

Repeat this exercise several times a day.

Exercise No. 2. Facial Exercises

One reason for our difficulty with some words is that our facial muscles are tight. If we are to be effective speakers our facial muscles and articulating organs must be flexible.

For good facial muscular flexibility perform the following exercises:

1. Close mouth and move lower jaw from side to side.

2. Repeat above exercise with mouth open.

3. Loosen your upper lip muscle by messaging it.

4. Make a wide grin (with lips closed) and then form your lips as though you were going to whistle. Do this several times until your facial muscles become tired.

5. Flutter your lips (like a horse) to relax the facial muscles.

6. Make all the foolish looking faces you can by stretching your mouth and face in all directions.

7. Flutter your tongue (like a Scotchman rolling his R's).

Be consciously aware of keeping your teeth apart while talking.

The above exercises are also interesting at parties. I call them, "Facial Fractures."

Exercise No. 3. To Lower Pitch of Voice

In our busy days most of us speak in our middle or high register. For greater effectiveness we should endeavor to lower the pitch of our voice by three full tones.

A good exercise for this is to open the mouth wide, and say, "Ah! ---- Ah" in your lowest tone.

If you have a piano available, try this exercise by going up and down the scale. Remember to put the emphasis on the lower tones.

This exercise is especially designed for those persons who tend to speak in a monotone, but it develops voice range for all who practice. This exercise also develops voice variety and gives the speaker greater speech authority.

Exercise No. 4 The Uvula

The uvula can and does give trouble to a great many speakers. This organ is located at the extreme end of the soft palate. The normal position of the palate is closed, which is one

Figure 2.   Diaphragmatic Breathing

Figure 3.   Facial Exercises

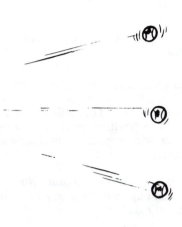

Figure 4.   Pitch

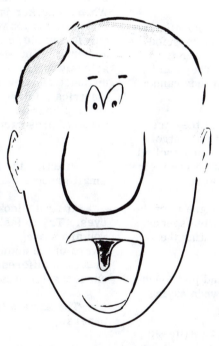
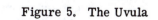

Figure 5.   The Uvula

reason why some people speak with a harsh nasal voice. A lazy uvula can make an otherwise potentially beautiful sounding voice sound absolutely ugly.

The following exercise will aid in keeping your uvula limber.

1. Open your mouth wide.

2. Say "Ah -- ah" (This will put your uvula in an open position so the sound can be emitted with minimum interference from the uvula.

3. Say "Ah -- ah" once again. Now force the sound up into the nasal regions. (Forced nasal tone)

4. Alternate the forced nasal tone and the open tone. You should be able to feel the uvula alternate. This exercise should be done several times a day in order to keep your uvula flexible.

This exercise designed specifically for those people who speak with a nasal tone. Remember – the uvula must be open for good, clear tones.

Exercise No. 5. Tongue

Many potentially fine speakers are not effective speakers simply because they are not consciously aware of the position of their tongue when they are talking. Their projection is good. They can be heard in the back of the theatre, but they cannot be understood.

One major reason for this is that they are swallowing their words. They have no control over their tongue. It is floundering uncontrolled all around their mouth. This is what gives their speech a muffled and indistinct sound.

The correct placement of the tongue is of the utmost importance for clear, unmuffled speech. The following exercise aids in preventing the speaker from swallowing his words:

1. Place the tongue behind, and just touching, the lower front teeth when saying any vowel.

2. Say the vowels several times daily with all their various sounds - long - short, etc.

3. Be consciously aware of the various

positions of the tongue. Also be aware of the various pressures of your teeth on your tongue while saying the different vowels.

Exercise No. 6. Eye Contact

Some schools of speech suggest that, if you develop stage fright or become nervous, you look just over the head of your audience. To me, this is like talking to a blank wall. I fail to see how this over-the-head-approach aids in overcoming nervousness or can be of any worthwhile benefit.

The only way you can judge the reaction of the people in your audience is to look them in the eye. There is no substitute for eye to eye contact with your audience. Having direct eye contact with your audience is one of the most important determining attributes of delivering a creative presentation. You can have no conception of your audience's reception unless you can see what reactions your words inspire. Only by direct eye contact with your audience can you intelligently vary your presentation. This is one great advantage that a speaker has over an actor. The actor has to rely on his developed intuitiveness. He must be able to feel the reaction of his audience. A speaker, with training, can both see and feel this intangible electrical flow from audience to speaker. When a speaker is narrating a presentation with slides or motion pictures (in a dark room), he has only intangible electrical reactions from the audience. You should train yourselves to be consciously aware of the visual reactions to your audience and develop a keen sensitivity to its electrical, intangible reactions. If you train and develop yourselves to this degree, you will deliver creative presentations.

Practice the following eye to eye exercise in small groups.

At first, look directly into each individual's eyes. Try to feel and see his reaction to what you are saying. Being keenly and consciously aware of the audience's reactions can sometimes make the difference between a successful or unsuccessful presentation.

The same pitfall awaits those who read their papers.

If you do the following exercises in groups, get accustomed to looking at your colleagues while doing them. This also aids in eliminating some of your unneeded inhibitions.

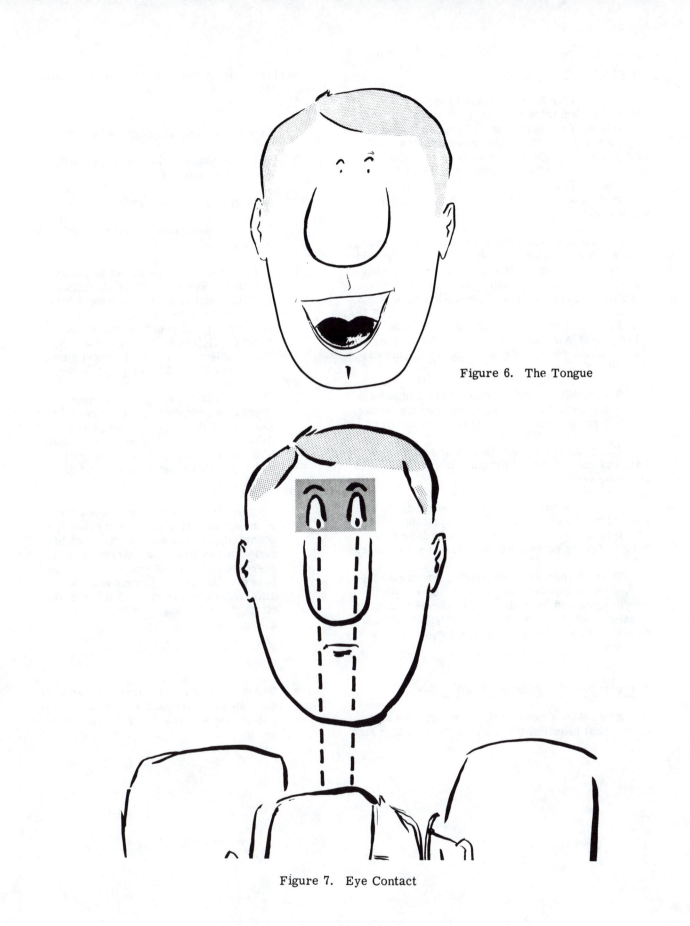

Figure 6. The Tongue

Figure 7. Eye Contact

Exercise No. 7. Projection

Now that we have conquered the previous exercises, we are now ready to develop our vocal projection. The correct use of your articulating organs will take care of the understanding - but now you must project those well articulated words to the back of the room. Some people do this by shouting. These people can be heard but not understood, and we want to be both heard and understood.

Before describing the exercise on projection, I believe that a confused semantic difference should be clarified. That is, the meaning between shouting and projection. Shouting is unrestrained speech, while projection is controlled speech.

First. Take a deep diaphragmatic breath. When you exhale, say, "One" and think of your voice leaving your mouth and actually hitting the back of the room.

Repeat this exercise, taking a new diaphragmatic breath for each count. Use all of your air on each of these counts. Do this several times daily.

For controlled projecting, stand close to a wall and say, "One" as though you were talking to someone very near.

Take one step backward and repeat the process. Each time you step back, your projection should be correspondingly greater. The main problem here is that of controlling the pitch of your voice.

Your awareness of this difficult exercise becomes more apparent as you increase the distance between you and the wall. This exercise takes a great deal of practice; consequently, conscious awareness and frequent conscious repetitions are necessary.

Exercise No. 8. Forming Nasal Tones

A nasal tone can be one of the ugliest or one of the most beautiful tones in the English language.

Perform the following steps to form nasal tones correctly.

1. Touch your lips together (delicately)

2. Put your tongue behind and just touching your lower front teeth.

3. Hum in your medium range.

4. Think of your voice as being out in front of your mouth.

5. Project the humming sound as far as possible. You should feel vibrations on your lips and around your nose. You will feel these vibrations extending to your throat and chest as you become more proficient in this exercise. After you do this exercise well, then go on to the words in the following exercises.

6. Hang onto the M and N in all these exercises before saying the vowel or the rest of the word. Some people have difficulty saying the M at the beginning of a word, others, in the middle of a word and still others, at the end of a word.

7. This same exercise may be used for enunciation, articulation and pronounciation of the words. The combinations of words is an extremely useful exercise. You must emphasize the M in order for anyone in your audience to differentiate between the B and the M. Exaggeration is the key word in all these exercises.

The N is done in the same manner with the tongue placed on the front of the hard palate just above and behind the upper teeth.

Figure 8.   Projection

Figure 9. Nasal Tones

Articulation Drill - Letter M

I. Nonsense Syllables

(m-m-i)	(m-i)	(mi)	(i-m-m)	(i-m)	(im)	(i-m-m-i)	(i-m-i)	(imi)
(m-m-e)	(m-e)	(me)	(e-m-m)	(e-m)	(em)	(e-m-m-e)	(e-m-e)	(eme)
(m-m-a)	(m-a)	(ma)	(a-m-m)	(a-m)	(am)	(a-m-m-a)	(a-m-a)	(ama)
(m-m-o)	(m-o)	(mo)	(o-m-m)	(o-m)	(om)	(o-m-m-o)	(o-m-o)	(omo)
(m-m-u)	(m-u)	(mu)	(u-m-m)	(u-m)	(um)	(u-m-m-u)	(u-m-u)	(umu)

II. Single Words

made	map	came	am	beaming	anthem
make	match	come	climb	bombing	atom
man	mean	farm	dim	famous	balsam
may	meet	hime	dime	foamy	bedlam
me	mesh	home	dome	hammer	emblem
men	might	name	from	lemon	heroism
mine	mob	same	gem	memory	phantom
much	mood	some	gum	pommel	problem
must	moss	them	lame	rumor	random
my	mouse	time	seem	timber	symptom

III. B and M

bean-mean	batch-match	bob-bomb	sob-psalm
bit-mitt	bill-mill	rob-rum	lobe-loam
bake-make	beet-meet	mob-mom	robe-roam
boss-moss	bug-mug	hub-hum	rib-rim

IV. Sentences

1. The man's face wore an impenetrable mask.
2. The mummers came to Mary's home on Christmas Eve.
3. Robin Hood's messenger met the Merry Men under the massive elm.
4. The bumblebees and humming birds were murmuring melodiously.
5. You must memorize this poem in twenty minutes.
6. Marion sat meditating by the Mediterranean.
7. Tom grumbled because he had to come home when he wanted to see the new film at the movies.
8. My mother made a warm muff for Mary out of my mink coat.
9. The murmurous haunt of flies on summer eves.
10. "Eternal smiles his emptiness betray.
    As shallow streams run dimpling all the way."
11. "Through caverns measureless to man."
12. "The murmuring pines and the hemlocks."
13. The explosion of the bomb on the moor sounded like the coming of doom.
14. Moses was known as the meekest man.

Articulation Drill - Letter N

I. Nonsense Syllables

(n-n-i)	(n-i)	(ni)	(i-n-n)	(i-n)	(in)	(i-n-n-i)	(i-n-i)	(ini)
(n-n-e)	(n-e)	(ne)	(e-n-n)	(e-n)	(en)	(e-n-n-e)	(e-n-e)	(ene)
(n-n-a)	(n-a)	(na)	(a-n-n)	(a-n)	(an)	(a-n-n-a)	(a-n-a)	(ana)
(n-n-o)	(n-o)	(no)	(o-n-n)	(o-n)	(on)	(o-n-n-o)	(o-n-o)	(ono)
(n-n-u)	(n-u)	(nu)	(u-n-n)	(u-n)	(un)	(u-n-n-u)	(u-n-u)	(unu)

II. Single Words

knee	knot	seen	lone	bonny	nonage
knit	nine	in	gone	contemplate	pony
not	name	then	lawn	dinner	rainy
gnat	noun	cairn	rain	funny	shiny
nurse	noise	an	vine	gunning	sunny
nut	near	can't	own	honey	tiny
noon	Noah	earn	town	incongruous	tonnage
known	newer	sun	coin	linen	vanish
gnaw	nude	moon	tune	money	whinny

III. Sentences

1. Anita knits industriously morning, noon, and night.
2. Nan can't earn enough money to buy all the new novels.
3. Nora lives at the corner of Ninetieth Street and Ninth Avenue.
4. Nonresident students often travel late at night and early in the morning.
5. Vachel Lindsay wrote a poem called "The Congo."
6. The knights were confident of victory in the impending conflict.
7. "A little nonsense now and then,
   Is relished by the wisest men."
8. "The wild November comes at last.
   Beneath a veil of rain."
9. "Now the hungry lion roars,
   And the wolf behowls the moon,
   Whilst the heavy plowman snores,
   All with weary task fordone.
10. "The nightingale, if she should sing by day,
    When every goose is cackling, would be thought
    No better a musician than the wren.
    Now many things by season seasoned are
    To their right phase, and true perfection."
11. "Be noble! And the nobleness that lies
    In other men, sleeping, but never dead,
    Will rise in majesty to meet thine own:
    Then wilt thou see it gleam in many eyes,
    Then will pure light around they path be shed,
    And thou wilt nevermore be sad and lone."

Exercise No. 9. Speed

We can perform all of the preceding exercises to perfection while delivering a presentation, and still not have an effective presentation because we spoke too rapidly. This effect is usually caused by nervousness and/or a desire to finish the presentation as quickly as possible. A speaker should be extremely cautious about this, or in all probability the speech will be finished – before it starts.

One of the more practical methods of slowing your speech is to say the Articulation exercise words and sentences every slowly. If you are consciously aware of your diaphragmatic breathing, articulation, facial muscles, uvula, the pitch of your voice, nasal tones, the correct placement of your tongue and projection, you <u>will</u> do these exercises slowly. It is only by executing the exercises slowly and correctly that you will be able to integrate these speech attributes into your own improved speech pattern.

Do not become discouraged because of the way you sound while developing your speech. You will sound horrible. You will sound even worse than you look while doing your "Facial Fracture Exercises."

Remember, you have many old habits to break before you will improve to any noticeable degree. You also have many unneeded inhibitions which you must eliminate. All of these exercises will aid you in overcoming both of these creative speech roadblocks.

Conclusions

The probability of your delivering a creative and interesting presentation will be greatly increased if you follow these five steps:

1. Desire to inform and interest your audience.

2. Be consciously aware of yourself but not self-conscious.

3. Develop the clinical approach to creative speech.

4. Project your personality.

5. Desire to be of service and value to your company and your audience.

You must motivate not only yourself but you must also motivate your audience. Let your personality speak, allowing your listeners to be aware of your interest in the subject.

If you develop the individuality of your voice, integrate this individuality into your personality, project this trained accumulated personality to your audience, and combine these attributes with the previous exercises; you will deliver a creative presentation.

Remember - your success as a professional may depend not only on your intricate and creative abilities but also in your capabilities of expressing your ideas and expressing them creatively.

Remember, too, that within two weeks, your audience, which is composed of persons with no formal listening training, will retain approximately only 25%* of what you said! Keep this in mind and be determined to increase that percentage by using the Clinical Approach to Creative Speech, thereby making your presentations come alive.

If you will forgive a personal reference, I will borrow a unique phrase which my good wife, Beverly, coined while guiding a creative development session. It is this: "One important aspect of a creative development program is to develop the individual to be conscious of himself but not self-conscious."

And how true this statement is! You must be consciously aware of the correct manner of performing these exercises before you can integrate them into your everyday speech and your presentations.

The impressions a speaker leaves with his audience concerning his company are lasting! It is with this thought in mind that we venture forth for better and more creative presentations at future symposiums. You are a professional man! Why not sound like one?

---

* Ralph G. Nichols and Leonard A. Stevens. "Are You Listening?" McGraw-Hill, 1957.

At Ease: Five Simple Cures for Public-Speaking Panic

MARY JANE MAPES

It's a waking nightmare. As the stranger announces your name, you rise, clutching sweat-stained notes, and begin the terrifying walk to the podium. Your heart pounds, your eyelids twitch and your tongue feels twice its normal size. You doubt you will make it to the lectern, let alone to the end of your speech. When it's over, you vow never again to accept a speaking engagement.

Does this scenario sound familiar? It does to millions of businesspeople who must make presentations that could be crucial to their careers. According to national surveys, public speaking is the number-one fear of most executives. Every year, I coach hundreds of men and women who suffer from stage fright that ranges from butterflies to utter panic.

Every time Chris Halpert* had to make a presentation, her heart beat so hard and fast that she wanted to bolt from the room. She would agonize for days over an upcoming presentation, finding it difficult to eat or sleep. Yet, she had no trouble selling herself in one-to-one situations, even in job interviews. After graduating from college with a degree in anthropology, Chris had become a secretary for a large aerospace company. A few years later, she decided to apply for an interesting job that was posted on the company bulletin board, even though she didn't have the background in computers, graphics and English required for the position. Chris not only talked her way into the spot, but she also convinced the director to upgrade her title from proposal/graphics coordinator to marketing supervisor, a change that meant a sizable pay hike.

It also meant that she would regularly have to make presentations to higher-ups from corporate headquarters. Fearing that her public-speaking phobia would prevent her from advancing in her company, Chris sought my help. "How," she asked, "can I control my nervousness so it doesn't block my message?"

*pseudonym

Reprinted with permission from *The Executive Female*, the bimonthly publication of the National Association of Female Executives, 127 West 24th, N.Y., NY 10011, (212) 645-0770.

I told her what I tell all my clients: that it's normal to feel uncomfortable if public speaking isn't something that you do all the time. Human beings are creatures of habit. We like to stay where we feel comfortable—seated in a chair. The moment you stand to speak, the dynamics change. It's no longer one-to-one—it's one to a group. You become the focus of attention, the leader who is expected to perform. Even the distance changes in this new arena: The larger the group and the more formal the presentation, the greater the distance between you and your audience. When you're in this unfamiliar situation, adrenaline floods your body as you gear up to handle the perceived challenge—the threat of audience disapproval. Your rapid pulse, quickened breathing, sweaty palms and shaking knees are physical signs that your fighting instincts are aroused.

Nervousness is natural—and it can be a blessing. Without that rush of adrenaline, that tension "high," you could seem like a passionless speaker, bored with your topic and uninterested in communicating with your audience. It's up to you to harness that energy, so it works for you instead of against you. The next time you have to give a speech or make a presentation, use these techniques to turn panic into power.

1. Test your fears against reality.

To get to the bottom of her public-speaking panic, I asked Chris what she feared most about making presentations. She told me, "I'm afraid I'll make a mistake and look foolish." Next, I asked her to name the worst thing that could happen if she did slip up. "People wouldn't consider me as competent as I would like them to," she replied. "Would you be fired?" I asked. She admitted that this was unlikely. "I'm good at what I do and my boss knows it," she said.

When Chris considered the "worst-possible" outcome, she realized that many of her fears were groundless. They stemmed from imaginary, "what-if . . ." scenarios that, when analyzed, didn't carry the dire consequences she had been conjuring up. Because Chris' worst "what-if" worry—"What if I make a mistake?"—was something she could live with, she began to feel more positive about public speaking.

Reprinted from *IEEE Professional Communication Society Newsletter*, vol. 33, no. 1, Jan. 1989.

2. Reassure yourself.

Behavioral research has revealed that 77 percent of our thoughts are negative. You're more likely to dwell on the worst that could happen than to anticipate a happy outcome. These negative messages sap your energy and sabotage your efforts. Because it's hard to change thought patterns, I told Chris to prepare enough positive messages to counteract any negative thoughts that might enter her mind before, during and after her presentation. Days before she had to speak, she began telling herself: "I am well-prepared. I have been asked to make this presentation because I am knowledgeable in this area. I am excited about my topic and delighted to have the opportunity to share what I know with my audience. I am the expert; my audience wants to hear what I have to say."

You should cleanse your mind daily. For every negative message that can undermine you, you should have a positive one to wash it away.

3. Practice, practice, practice.

You should rehearse your speech many times. Someone once said that "preparation is the greatest substitute for talent that you can find." It's also great insurance against panic attacks.

For over a week, Chris rehearsed her presentation out loud while driving to and from work. Because she had recited her speech so many times, when it came time to give it, Chris found that her tongue had developed "muscle memory." The right words to express her ideas were on the tip of her tongue. She didn't flounder, trying to remember what she wanted to say.

Without adequate preparation and practice, you waste your energy trying to recall the order of your ideas. You focus on your speech—not on the people listening to you. It's only when you know your material "cold" that you can concentrate on connecting with your audience.

4. Watch your body language.

To command people's attention, you must appear self-assured and in control. Studies have shown that your body language can carry more weight than your words. Another client, Trish Winters,* had to sell her coworkers on a new computer system, and was afraid that her nervousness would result in a lack of confidence in her and in the project. It was imperative that she look and sound confident, even if she was not yet feeling it, in order to convince those who opposed the new computers that the system was best for everyone. An audience usually responds to the attitude that a speaker conveys through her stance, gestures, mannerisms and tone of voice.

Trish's body language broadcast how uneasy she was. She tended to stand with her weight on her left foot and her hip elevated, an awkward position that caused her right leg to shake and, in turn, triggered other nervous mannerisms. Her shaking leg shook her confidence, causing her to worry, "I must look like a fool," a thought that completely unnerved her. To stop her leg from jiggling, Trish would shift her weight from foot to foot, a noticeable movement that drew attention to her jitters.

To prevent this self-defeating dance, I told Trish to stand with her feet about 10 to 12 inches apart, with her weight evenly distributed on the balls of both feet. This reduced the tension in her legs. Next, I had her square her shoulders, raise her rib cage and hold her head up, so that she looked like someone who had a right to speak on her subject. Trish discovered that when she stood tall in a "planted" position and energized her body from the waist up by using gestures, she spoke with more authority. As a result, people responded to her favorably, and she gained confidence.

Also, once she stood up straight, Trish found it was easier to breathe deeply, which relaxed her. This increased the oxygen supply to her lungs and gave her the breath support necessary to project her voice. Lack of tension and fuller lung capacity made her voice stronger and more resonant.

5. Ask for feedback.

Ken Blanchard, the author of *The One-Minute Manager,* calls feedback the "Breakfast of Champions." After your speech, ask colleagues whom you trust to critique your presentation. Most likely, you'll discover that other people are less demanding critics of your performance than you are. Many of the things that worry you are invisible to your audience. Did your knees shake? Did you lose your place? In most cases, you'll find people didn't notice.

If you do hear criticism, don't torture yourself with your mistakes. Learn from them so that your next presentation will be first-rate.

After you've perfected my five techniques, public speaking still may not be your favorite pasttime, but it won't be your worst nightmare either. And it's one of the best ways to gain visibility in the business community. When you become more confident center stage, you should seek out opportunities to speak, even if it's only to members of your network or to the PTA. As the renowned orator Demosthenes once said, "Small opportunities are often the beginning of great enterprises." So, why not begin!

It's All Right to Read Your Speech—and Here's How

JOAN F. DORNBUSCH

Abstract—When speakers must convey precise technical material or keep within a specified time limit, they may need to read their speeches. Unfortunately, those who read speeches often read them badly. Oral presenters can, however, read their speeches effectively by following a few simple steps in writing the material and preparing the manuscript and by observing some principles of good oral delivery. This paper offers practical suggestions to those who want to read their speeches and, except for typesize and spacing, provides an example of a speech written for reading aloud.

MY favorite story about reading a speech concerns Alben Barkley. When he was Vice-President in the early 1950s he once asked a friend how his speech went. "Well," his friend replied, "in the first place you read it. In the second place, you read it badly; and in the third place, Alben, it wasn't worth reading!"

If you have ever been to a conference or a meeting where speakers read their papers—and read them poorly—you can empathize with Mr. Barkley's friend. As you listened to the speeches, you probably wondered if there was any point at all in your being there—or any point at all to the papers. No doubt it also occurred to you that you could have saved yourself and the speakers a lot of time and trouble by simply reading the papers to yourself—in the comfort of your own home.

Although it is generally better, especially if you are a beginning speaker, to deliver your speech extemporaneously from notes or an outline instead of reading it word for word, there are occasions when reading a speech aloud is necessary. Such occasions often arise for engineers and other presenters of technical material. For example, if you must deliver precisely worded information, then you may *have* to read your speech—just as the President of the United States must *always* read policy speeches. Because President Reagan is especially skilled at reading a speech, he often seems to be talking from memory, but he's not. His position is so important that he cannot risk making a wrong statement or being misquoted. Thus his prepared manuscript is read and, as you probably know, copies distributed to the press. So if you want to read your speech, you are in the company of many important and public figures who must carefully word their information.

There are, of course, other good reasons for reading a speech: to keep within a specified time limit, to save time

Reprinted with minor changes from the *Conference Record* of the IEEE Professional Communication Society Conference held in Boston, MA, October 13–15, 1982; Cat. 82CH1830-9, pp. 76–80, IEEE Service Center, 445 Hoes Lane, Piscataway, NJ 08854; copyright 1982 by the Institute of Electrical and Electronics Engineers, New York.

The author is a lecturer in speech and a professional assistant with the Communication Skills Center at Ohio Northern University, Ada, OH 45810, (419) 772-2000.

practicing an extemporaneous delivery, and, last but not least, to overcome the terror of being separated from your manuscript!

If you read your speech, however, you must master a few basic skills so that your material won't seem—even to your friends—not worth reading at all. These skills sharpen not only your oral presentations but also your competitive edge in the job market. All you have to do is (1) follow a few steps in writing and preparing your manuscript and (2) master a few techniques of good oral delivery. Then you will be able to speak more in the style of "the great communicator" than in the style of Alben Barkley. Political parties and candidates aside, the fact that Barkley became Vice-President, whereas Reagan became President, may say something about the significance of communicating well.

WRITING THE MATERIAL

First you must understand that writing a paper for oral presentation differs from writing a paper for publication. A published paper can be read and reread until it is fully understood. Its tone tends to be more formal, its sentence structure more complex, its vocabulary more learned, and its content more technical.

When you present a paper orally, however, there is no second chance for your audience to grasp its meaning. That's why your opening remarks have to grab the attention of your listeners and make them want to hear what you have to say. Though the kinds of devices I am about to suggest may be out of place in the introduction to a written technical report, they are not only appropriate but also necessary in an oral one.

Getting Attention

You can arouse interest in your paper in a number of ways:

1.   A *story* about your topic is sure to get the attention of your listeners. One carry-over from our childhood is that we all like to hear stories. If you remember, I began this talk with a story about Alben Barkley.

2.   A *startling statement* is another way to capture audience interest. Make sure, though, that you can back it up. You shouldn't shock your audience to attention by beginning with a startling remark that is not true; audiences don't like or trust speakers who trick them. Don't say, for example, "Listen carefully! I want to assure you that this manufacturing process will double our profits. I want to show you how we are going to make money with this process. I want to, but I can't." Such an opening for your speech may cost you your credibility. In fact, you should

Reprinted from *IEEE Trans. Professional Commun.*, vol. PC-26, no. 1, pp. 25–29, March 1983.

always be candid, never devious—even to the point of admitting that you don't have all the answers. Then your listeners will trust you, identify with you (they don't have all the answers either), and support you. So beware of a startling statement to get attention—unless, of course, the startling remark is true. Then it is a good way to start your speech.

3. A *question* at the beginning of your talk is a third way of arousing audience interest. You may ask a rhetorical question or one for which you expect audience reaction, such as a show of hands. Asking a question or a series of questions is an informal approach that immediately involves your listeners.

4. A *quotation* will capture the attention of your audience. Perhaps someone has expressed an idea about your subject better than you could say it. Although you probably wouldn't begin a technical written report with a quotation, this device is appropriate for an oral presentation.

5. An *example* or an *illustration* is another interesting way to begin a speech. The closer to the experience of your listeners it is, the more effective it will be. I followed my story at the beginning of this paper with an illustration about listening to papers being read at conferences, an experience with which I felt you, as listeners, could identify.

6. Finally, *humor*—if you are good at handling it—is especially effective in capturing audience attention. It must, however, be in good taste; you don't want to offend your listeners. Also, your humor must be about your topic. A joke, for example, will always get the attention of your listeners, but if it is not about the subject of your paper, it will not concentrate their interest on your topic. If your subject is robotics, don't open your speech with a joke about a management problem; make sure that your attention-getting device, whatever is is, is related to your subject.

If you can also show your listeners that what you have to say will benefit them, you will not only get but also keep their attention. You can make them want to listen to you by showing them how what you say will affect their lives and by appealing to their desires for economy, profit, security, esteem, and other motivating forces. If your listeners believe that your paper concerns something that will help them in some way, they will want to hear what you have to say. For example, if you are trying to explain a new piece of equipment to a company, show your audience (if you can) how this machinery will be more economical to maintain, more efficient to operate, and safer for workers—all of which mean greater savings and profits for the industry. At the beginning of my talk, I explained that following my steps for reading a paper aloud will sharpen your competitive edge. That is, if you listen carefully to what I tell you, you may earn a promotion and make more money!

Providing Purpose

After you have the attention of your listeners focused on your topic, you must make clear the purpose of your talk. Are you going to *explain* a word processor or do you want to persuade someone to *buy* a word processor? Because it is difficult to follow a speaker whose purpose is not clear, be sure that your listeners understand your reason for speaking. Another way you can help them follow your ideas is to let them in on the orgaization of your speech. Tell them the main points you intend to cover so that they know from the start where your talk is headed. Then they can listen more intelligently to what you have to say. You might state, for example, that there are four major parts to the electronic device you are explaining—and then *name* those parts.

Explaining Ideas

In addition to writing an interesting introduction that immediately captures audience interest, uses motive appeals to relate your topic to the needs of your listeners, states your subject and purpose, and previews your main points, you must express your ideas more simply than you would if your paper were meant to be read silently. That is, your paper should *sound* both informal and conversational. Although the degree of technicality depends, of course, on whether you are addressing a lay audience or professionals, it's wise to simplify your language and content to some degree even when you are speaking to those knowledgeable in your field. Because audiences can grasp more complex language and ideas when reading than when listening, use small rather than large words and express your ideas in sentences that come quickly to the point.

Personal pronouns, such as *you* and *I*, and contractions, such as *it's* and *there's*, also help to maintain a conversational tone. A good rule is to use the same language in writing your paper that you would use if you were speaking to individual audience members in your office. In other words, write the way you would talk in ordinary conversations to your listeners. The only difference is that your speech—unlike your conversations—must be well organized. Maintaining a conversational tone does not give you license to wander as you would in talking to friends. So be sure that the body of your paper follows the organization that you promised in your introduction.

Besides simplifying your language and content so that your listeners can understand your ideas more easily and so that your paper sounds more as though you are speaking than reading, you can help your audience grasp and remember what you say if you amplify and clarify your main points. Quotations from authorities, examples and illustrations, definitions, comparisons and analogies, and humorous anecdotes all help to reinforce your ideas. Such material also enlivens your speech and makes it more interesting. As you can see, some of the same devices that you might use to capture audience interest in your introduction help to main-

HE INSISTS HE READS ALL HIS SPEECHES, BUT
NO ONE CAN TELL WHAT HE HAS THEM WRITTEN ON.

tain that interest throughout your talk. After all, a professional report or technical paper does not have to be boring! In fact, if it is being *read* to an audience, it must be even more interesting than a talk delivered extemporaneously.

Using Visual Aids

Visual aids also sharpen your presentation and help your audience understand and remember what you say, particularly the more technical parts. Because you are reading your speech, you will probably find it more convenient to prepare your charts and transparencies ahead of time. If you use visual aids, be sure to

- Keep them *very simple*—one idea to a chart or transparency.

- Print or draw large enough so that your material can be seen from 25 feet away.

- Show your aids only when you need them in your talk and then cover them or put them away or turn off the projector when you are finished explaining them. If you're worried that switching the projector on and off will distract your listeners or cause bulb failure, cover the transparency table of the projector with black paper whenever you're not using it.

- Use your aids to emphasize key ideas or difficult material, not for every point you make. Your visual aids should support what you say, not give your talk for you. If you plan to read your entire presentation from charts or transparencies, it would be better to give your listeners a copy of your material to read for themselves.

- Have your materials and equipment ready, handy, and in order before you speak. Be sure, for instance, that your transparencies will be in focus and right side up when you turn on the projector. If you have a helper, practice with that person so that your timing is synchronized. It's awkward to interrupt your talk with, "Let's have the first slide, please." While we are speaking of slides, I advise you to use them sparingly. A slide show with dim or no light may put some of your listeners to sleep. In addition, it costs you eye contact with your listeners.

Another visual aid that you may find especially helpful is the handout. To reinforce technical points, you can give your listeners material like drawings and graphs to study at their leisure. You can also hand out questionnaires to get feedback. By including your name, business address, and phone number, you make it easier for interested persons to contact you later. But a word of caution: It is better not to pass out anything before you speak or while you speak because the material as well as its distribution will distract your listeners. Unless they need your handout to follow a technical point, avoid distributing material during your speech.

Reinforcing Key Points

Finally, when making any oral presentation, but especially when reading a paper, you must often summarize your main ideas. By repeating what you have to say, you can make sure that your listeners not only follow your important points but also remember them. You can reinforce your points in five ways:

1. As I have already suggested, preview your main ideas in your introduction so that your listeners know how your speech is organized and can follow your ideas more easily. Say, for instance, "Today I am going to consider three points: first..., second..., and third...."

2. Use the PREP system: State your *point* ("A circuit breaker will not protect you against electrocution"); back up your point with a sound *reason* ("Common household circuit breakers are 20 amps") and an effective *example* ("Since it takes only 1/100 of an amp to kill you, you could be dead before the circuit breaker trips"); and then restate your *point* ("A circuit breaker will not protect you against electrocution"). As its name implies, the PREP system is a way of *preparing* your audience to accept, understand, and remember your ideas.

3. Use transitional words, phrases, and sentences to help your listeners grasp the relationship between your ideas. "In the first place," "in the second place," "third," "finally," "however," "therefore," "on the other hand," "moreover," and "furthermore" are words and phrases that help tie together points for your audience—a service that is especially necessary when people are not reading for themselves.

4. Provide summaries in the body of your talk. Indicate
    that "So far we have covered this point and this point
and now we are going to consider this point." If some of
your listeners have fallen asleep, they might wake up again
and pay attention if you bring them up to date and let them
know you're beginning a new point.

5. Summarize your main ideas in the conclusion. Even if
    the attention of some of your audience wandered during
your speech, they will at least go away with your main ideas
if you tell them once again what you said. An old speaker's
rule is to tell your audience what you are going to say in
your introduction, say it in the body of your talk, and tell
them what you said in your conclusion. After summing up
your points, end with a memorable statement that will again
reinforce your central idea and give a sense of finality to
your paper.

PREPARING THE MANUSCRIPT

After you have written your speech in a conversational style
that captures and keeps audience interest and reinforces
your points, you must prepare your manuscript so that it is
easy to read aloud. Follow these steps in typing the final
draft of your speech:

- Use large, orator's type so that you can easily see your
  manuscript.

- Alternate lines of black and red type, if you like, to help
  keep your place as you read. Or try going over certain
  key words in red.

- Leave wide margins so that you can write notes to yourself concerning pauses, rate, emphasis, or other aspects
  of delivery. You can, for example, use single slash lines for
  short pauses and double slash lines for longer pauses or
  write "pause" or "slow" in the margin. You can also
  underline words you want to emphasize and note that you
  want to hold up your fingers as you enumerate points.

- Number your pages at the top and check their order just
  before you speak.

- Tape each page of your speech to black construction
  paper to achieve a more professional looking manuscript
  and to cut down on the rustling of paper, especially if you
  are using a microphone.

- You also might want to construct a "speech box" to
  hold your papers and keep them from sliding off the
lectern. A typing paper box serves this purpose if you flatten
one long side from the lid of the box and one long side from
the bottom. Then you can tape the two parts of the box
together, the lid edge over the bottom edge, and cover the
outside with dark, plastic-coated paper. The box can be shut
to carry your speech to the lectern.

© 1983 KENNINGS

MY TALK TODAY IS ABOUT HOW TO
KEEP YOUR AUDIENCE FROM REALIZING
THAT YOU ARE READING YOUR SPEECH.

DEVELOPING THE DELIVERY

Even though you are going to read your speech, you still
must practice it. It is a mistake to believe that your speech is
ready once you have typed the final copy and penciled in
some notes on delivery. You must read your manuscript
aloud and *often* until you are thoroughly familiar with the
material. Be sure that you practice only with the final copy
of your speech so that you know where you have each word
and idea. Then, if you lose your place, you'll be able to find
it more quickly. Also, if you can, use a tape recorder to
listen to how you sound. You may find that you want to
reword some passages so that you can read your speech
more smoothly.

Memorize as much of your speech as possible. It is especially important to know well your introduction and conclusion so that you can establish eye contact during these
important parts of your talk. You want the attention of your
audience as you begin your speech, and you want to make
an impact as you finish it. A good way to establish eye
contact even before you begin talking is to come to the
lectern, look around at your audience, and count to yourself
1,000; 2,000; 3,000; 4,000. That way you will get the eyes
of your listeners on you. If you have their eye contact, you
will have their attention.

You must also follow the principles of good oral delivery.
Some of these techniques are to

1. Read your material at the proper rate for its complex-
    ity—slower for more technical material, faster for ma-
terial such as illustrations and examples. A varied rate helps

keep your listeners interested. You want to avoid a monotonous rhythm—a sing-song effect that dulls the perception of your audience.

2.   Look at your audience when shifting from one page to another and move your pages aside instead of turning them over (to keep your manuscript inconspicuous). Never wave your pages in front of your listeners. Although you need a manuscript to assure yourself and your audience that your technical material is accurate, you do not want to distract your listeners in any way.

3.   Most important, become involved in your speech and let your audience see your enthusiasm for your material by the way you read it. If you are really excited about your topic, you'll show that enthusiasm to your audience by

• Inflection—varying your pitch to emphasize points and avoid a monotone.

• Appropriate facial expressions—showing your involvement in what you are reading by smiling when relating a humorous anecdote and furrowing your brow when discussing a matter of deep concern.

• Gestures—using your hands to show the size of objects, pounding your fist to emphasize an idea, pointing your finger in accusation—all as you do naturally in conversation when you are interested in and excited about a subject. If, however, any gestures seem awkward to you, it's better not to use them but to hold lightly to the lectern instead. Awkward, planned, or poorly timed gestures detract from your speech as do annoying mannerisms—jingling coins in your pocket and playing with a pointer used for visual aids.

• Movement—putting your nervous energy to work by shifting weight or taking a few steps when moving from one major point to another, by coming in front of the lectern for an informal story, and by using head movement to punctuate your points. Avoid constantly swaying back and forth or side to side at the lectern—distracting motion that will bother your listeners.

• Eye contact—looking directly at individuals and around the room to include all your listeners. Because you cannot get or keep the attention of your audience without adequate eye contact, you should aim for eye contact 70 percent of the time as you read your paper aloud. Another hint: Try to look at your listeners at the ends of sentences and paragraphs where the climax of your ideas, the points you want to emphasize, will come.

By observing these guidelines you can make your speeches interesting and keep your audiences interested, even if you do read to them. Remember to

1.   Write your speech in an informal style that captures and keeps listener interest, makes your organization clear, and frequently amplifies and summarizes your ideas.

2.   Prepare your manuscript so that it is easy for you to read, and mark it for delivery cues, such as rate, pauses, and emphasis.

3.   Practice reading your speech so that you are very familiar with the material and can convey your enthusiasm through eye contact, inflection, facial expressions, movement, and possibly gestures.

You will find that these steps sharpen your oral presentations. And sharpening your oral presentation sharpens your competitive edge. Why be vice-president when you can be president?

Humor Can Improve Your Technical Presentations

JAMES P. GLEASON

Abstract—**Humor is a characteristic not usually associated with technical publications. In this paper I discuss humor and its usefulness to text and illustrations by analyzing several examples that use humor effectively. I demonstrate some successful uses of humor in technical publications by looking at three areas. First, though technical material can be dry, humor can make it easier to use. Second, whether this can be done is a function of the author's freedom from constraints. Third, the audience must be sophisticated enough (at least with regard to the subject) to appreciate the humor.**

THE HOW'S AND WHY'S OF HUMOR

HUMOR is a funny thing. A person's taste in humor is as personal as taste in clothes or entertainment. It can come from his or her background, experience, or mood. Yet humor can be the difference between a great manual and a merely good manual. More to the point, it can be the difference between a manual that is used and one that is not.

Whether consciously or not, every author starts writing with two basic assumptions. First, the author assumes that he or she has some information that is important to the reader. Without this, there is nothing to write about. Second, the author assumes that there is someone who is interested in obtaining that information. If not, why bother writing at all?

The technical author faces an additional problem in that he or she must "sell" information that is often boring or tedious. To accomplish this, the author must use any means available to keep the paper or manual from being crushed under the weight of its own data. Humor can help attract the attention of the intended audience, keep their interest, and sell the message. It is up to the author to motivate the reader to want to continue reading.

Within the context of technical writing, humor is anything that lightens the presentation of the technical material. Using witty phrases, asides, and slang in the text, with the addition of cartoons or humorous photos, can make a discussion of heavily technical material seem more human and less intimidating. In effect, the author is sugar-coating what, in some cases, is a difficult pill to swallow.

It should be noted, however, that using humor is a delicate matter. It is best applied to areas of lesser importance to keep from obscuring necessary data. In addition, humor must be used judiciously to maintain the professionalism of the work. Humor must never be used as a substitute for accuracy or clarity but, rather, must serve to reinforce those qualities. Humor is a tool that can help a presentation; just don't use a hammer when a feather is sufficient.

Humor can make technical information more memorable by offering the reader a psychological hook to grab. A humorous approach appeals to the reader's creativity and personal ex-

Manuscript received October 29, 1981; revised February 5, 1982.

The author was an Information Developer with the IBM Information Products Division. He is now Corporate Internal Communications Administrator with Lexmark International, Inc., Lexington, KY.

perience by adding elements of originality and surprise. This helps satisfy the reader on both technical and personal levels. In addition, humor can relax the reader, thus making him or her more receptive to the information and lessening some of the drudgery of obtaining it. Moreover, it may soften the reaction to any negative information that must be conveyed, as might be found in an engineering status report.

Humor lightens the tone of a paper by applying wit to less essential areas while leaving the technical portions intact and preserving their integrity. One way is to use funny or odd names for people, places, or companies in the text. Examples are "Fly-By-Night Air Freight" and "Acme Widget Company, Wingdale, East Virginia." This is one of the easiest and safest ways to use humor in technical publications. It also protects the author from being sued by some real person or company.

Illustrations can play a key role in affecting the mood of a paper. You might remember the simple equation,

$$1 \; picture = 1000 \; words.$$

Many papers are less than a few thousand words long; the impact of two or three pictures is obvious. An illustration is one of the first things to meet the reader's eye and so sets up the reader's preconceptions about the paper's content or the approach the author will take. The art should contribute to the author's purpose and complement the text. (Whereas this is true for all illustrations, it is especially important in technical works, which often rely heavily on graphics to convey technical data.)

Cartoons are becoming more acceptable for use in technical publications. An artist can exaggerate a particular part or tool, provide the reader with x-ray vision, and otherwise do things not possible with photos. They also allow the author to use color to stimulate the reader and to emphasize various aspects of the drawing. Obviously, cartoons aren't appropriate as an illustration in every instance. Their value is to increase the writer's options when trying to convey some information. In the end, cartoons are just one more tool.

EXAMPLES

Bill Chadbourne uses illustrations as an integral part of his text. In *What Every Editor Should Know* [1] he demonstrates principles of layout and typesetting by using humorous examples as part of the text. For example, he explains the formula for "Optimum Line Length" while not using it (Fig. 1). In this way he humorously demonstrates why the formula is effective.

Chadbourne found most of the photos in his book in old magazines and newspapers and adapted them to his subject by adding voice balloons or captions that humorously illustrate each point. In every case Chadbourne uses the appropriate layout technique (Fig. 2).

Many instructional manuals use examples containing extraneous information. I'm referring to things like text-entry exercises and word problems. This is an opportunity to use an interesting

Reprinted from *IEEE Trans. Professional Commun.*, vol. PC-25, no. 2, pp. 86–90, June 1982.

READ THIS IF YOU CAN

Edmund Arnold's "Optimum Line Length" formula provides a way to determine line length by math. The Richmond, Virginia teacher and consultant found that a measure one-and-a-half times the length of the lowercase alphabet of any font is the column width which is read most easily, most quickly, with minimum fatigue and with maximum comprehension. You need not be so scientific. Rely on your common sense. This paragraph is set on too wide a measure which has made it difficult to find the next line upon leaving the last. The space between the lines, called leading, (pronounced "Ledding", after the metal strips used by the hot-type compositor to separate lines of type) also should be in proportion to the line length. As the measure (line length) increases so should the leading. A sans serif face may require greater leading than a serifed face.

Fig. 1. This paragraph shows why an optimum line length should be considered. From [1]. Copyright 1977 by Bill N. Chadbourne; reprinted with permission.

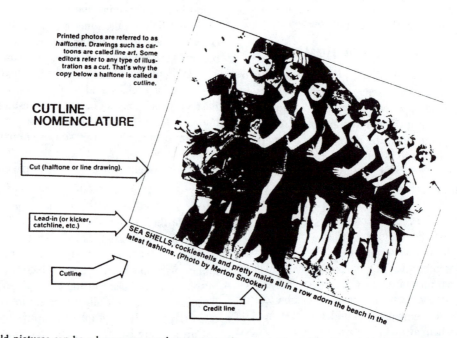

Fig. 2. Old pictures can be a humorous supplement to modern text. From [1]. Copyright 1977 by Bill N. Chadbourne; reprinted with permission.

or humorous piece of text rather than something dull, such as a requisition letter from the Highway Department. There is no reason the example can't be entertaining, as illustrated in Fig. 3.

This technique is used in the *IBM Virtual Machine/System Product: System Product Editor User's Guide* [2]. (Snappy title, huh?) The originality is unusual because the authors were able to be humorous and still keep within the guidelines of similar books produced by IBM. The authors used portions of *I Wouldn't Have Missed It* by Ogden Nash [3], as well as *The People's Almanac* [4], as the text to be manipulated in text-editing exercises (see Fig. 4). I found myself reading ahead just to see what the later quotes would be.

It is possible to present something as unfamiliar (and as potentially tedious) as a computer language in an interesting and even entertaining way by adopting a conversational tone. Many authors use slang, contractions, short asides, and, probably to the dismay of their editors, an occasional incomplete sentence in an attempt to humanize their presentations.

Another IBM book, the *Introduction to Interactive Computing with VM/CMS* [5], uses humorous examples to teach. The reader enters sentences like "Thus is my fisst experiense typing on tge VM systim..." and "Typing errors are easy to fiz whem using VM..." and then must correct the errors. Later, the reader is

asked to create a new file named PICNIC INGREDTS A1, which contains such items as "a loaf of bread," "a jug of wine," and "thou." The text continues, "You decide that a loaf of bread, while nourishing, isn't really all that appetizing by itself. You decide to include some ham, adding it after the bread entry. You could use the ADD command...."

Rather than just listing words to be entered, both manuals win points with the reader by using illustrations that show these entries as they appear on the screen. This helps the user feel comfortable with the terminal and the material to be learned and, in turn, helps increase the usability of the books. It is easy to believe that these exercises are more effective than the boring ones usually used.

One of the more humorous technical publications around is John Muir's *How To Keep Your Volkswagen Alive* (*A Manual for the Compleat Idiot*) [6]. This book is unusual in that it is meant to be a highly technical instruction manual for someone without technical experience. Muir states in his introduction, "This book contains...clear and accurate procedures to heal and keep well your Volkswagen. You supply the labor, the book will supply the direction, so we work as a team, you and I." He continues, "This book has been designed, in addition to making me some bread, to fill the gap between What To and How To. I'm trying to close

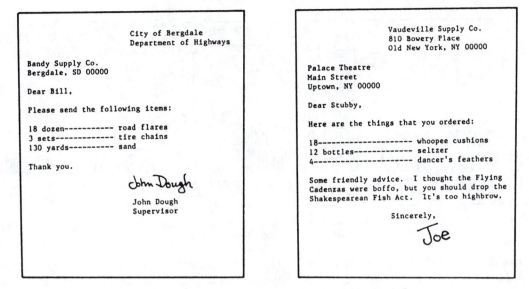

Fig. 3. Which letter is the more interesting typing exercise?

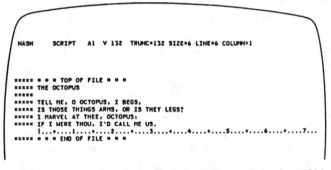

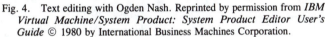

Fig. 4. Text editing with Ogden Nash. Reprinted by permission from *IBM Virtual Machine/System Product: System Product Editor User's Guide* © 1980 by International Business Machines Corporation.

this gap and have assumed that you know nothing, i.e., that you are an idiot mechanically."

Muir leaves nothing to chance. He gives all the information needed to perform whatever task is being done, from changing a taillight bulb to rebuilding the engine. Most important, Muir never condescends to his audience. He uses a conversational tone without ever sounding forced or quaint. In addition, the informality leaves plenty of room for useful asides. The first chapter describes the methods of using the procedures in the book. He tells us

> Take your time. Just do the job once and well. You have an eternity. Don't improvise.
> Just do it the way it says.
> Wear the right clothes. There's no better way to keep peace in [the] family than to wear car clothes to work on the car.

It seems important to him that we fix the car not only with a minimum of problems but also with a minimum of mental anguish. He always tries to keep morale high and this is one of his keys to success.

Peter Aschwanden did the art for Muir's book, as well as for *Poor Richard's Rabbit Book* [7], a similar manual for Volkswagen Rabbits. His illustrations of various parts and tools are useful, but the little asides offering service tips, and the little jokes or sight gags, such as in Fig. 5, make the job seem easier.

They reveal his enthusiasm for the project, and this enthusiasm is quite infectious. This book is one of the few technical manuals I recommend reading just for enjoyment, even by someone who doesn't own a Volkswagen.

FREEDOM FROM CONSTRAINTS

When producing technical manuals, most large corporations tend to have a general format for their books. They usually produce a large number of manuals and this approach ensures uniformity and consistency throughout their library. These companies usually do not credit the individual authors in each publication, and the authors do not have much latitude outside the company's guidelines. However, a little creativity can produce effective results even within these constraints, as in some of the examples I've described.

Lanier chose an interesting approach to writing an instruction manual for their dictation equipment line. They shunned the more traditional textbook in favor of a short, colorful booklet modeled after a cookbook.

Even the headings use cooking terms; for example, Ingredients, Start Dictation Cooking, Appetizer, and Dessert. Throughout the manual you come across lines like "Pens, pencils, or yellow pads…still play an important part in the daily routine of cooking up business ideas" and "…you will really be

Fig. 5. Making car servicing seem like fun. Drawn by Peter Aschwanden, from *How to Keep Your Volkswagen Alive*. Copyright 1990 by John Muir Publications; reprinted with permission.

cooking as a word originator." The author even lets a little humor unrelated to cooking slip in. He explains, "You need the ability to talk. Note: There are many brands of speech on the market and all are acceptable."

The *Executive Dictation Cookbook* [8] conveys the idea that using dictation equipment can be just as much fun as using the manual. And in spite of the technical material presented, this manual is fun to use. Lanier gives credit to both the author, Don Rogers, and the illustrator, Penny Perkins. This is unusual in an instruction manual and is probably due to its different appearance.

The *Cookbook* also uses cartoons to sell its message. For example, it depicts the leaders of industry in business attire and chef's hats (Fig. 6). Lanier makes excellent use of such drawings while maintaining a professional approach to the subject.

AUDIENCE SOPHISTICATION

Technical expertise is wasted if the presentation isn't suitable for the audience. This is also true with humor. The author must take care to tailor his humor to the level of the audience.

Fig. 6. A caricature helps imply that learning can be fun. From [8]. Copyright 1979 by Lanier Corporation; reprinted with permission.

For example, John Muir was looking to fill a particular gap in automotive repair manuals when he wrote *How to Keep Your Volkswagen Alive*. As the subtitle indicates, the book was designed to be "a Manual for the Compleat Idiot." In other words, he assumed the user would be a technical novice and wrote with that reader in mind.

The following excerpt is from a reader of a quarterly IBM publication: "The entire issue was eminently readable and informative. At the least point where it might have become tedious, there is an illustration to brighten it up…. The humor is splendid—all the more so because it is so unusual in an official publication." This high praise is for the *APL Jot Dot Times* [9], a publication about the APL computer language. This book incorporates illustrations originally used in newspapers, advertisements, and periodicals between 1850 and 1925. The authors added appropriate captions, voice balloons, and the like with interesting results.

The *Jot Dot Times* is an example of humor that is funny only to a particular audience. The illustrations refer to the APL language, the terminals, or the programmers themselves. If the reader has no experience with those, it is likely that much, or all, of the humor will be lost. Anyone who would have a reason to be reading the *Jot Dot Times* would have enough computing background to know what a circular function is and would appreciate the humor of illustrations like "the model for the Original APL Circular Functions" (Fig. 7).

Parody and satire are difficult forms of humor to master. They require a thorough knowledge of the subject matter, an equally thorough knowledge of the form to be satirized, and the wit to pull both together in a funny way. An example of parody at its best is *The Journal of Irreproducible Results*, the "Official Organ of the Society For Irreproducible Research" [10]. It is essentially a parody of those ponderous research journals that contain mountains of dry facts and references and serve as the main publication outlet for research data.

Articles in the *Journal* often produce lucid and insightful conclusions about areas of research no one ever thought needed research. Dinesh Mohan, of the Engineering Department at the University of Michigan, submitted "Interference of Labium Superious Oris Hair with Spherical Ice Cream Surfaces—A Theoretical Analysis" [11]. It offers this introduction: "Ever since the appearance of ice cream in spherical configurations a large number of men (some women, too) have been battling with the

Fig. 7. Specialized humor for a limited audience. Reprinted by permission from *APL Jot Dot Times* © 1980 by International Business Machines Corporation.

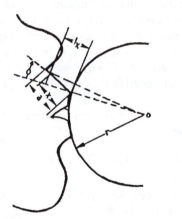

Fig. 8. Considering the interaction of ice cream cones and mustaches. From [11]. Copyright 1978 by the *Journal of Irreproducible Results;* reprinted with permission.

upper-lip-hair ice-cream interaction phenomenon." Please note that the author is talking about ice cream cones and mustaches (Fig. 8).

One *Journal* article of particular interest here is "How To Write Research Papers" [12]. The author, Andreas Jesse, believes people read research papers in an order other than starting at the beginning. She offers evidence that the reader goes to the acknowledgments first (presumably to look for his or her own name), then reads the conclusion, and finally the introduction. She feels that since few people read the entire paper anyway, the main text is almost irrelevant. She suggests that it would help the reader to arrange the paper in the observed reading order. (Incidentally, she wrote her paper using that sequence.)

CONCLUSION

The use of humor is not something to be taken lightly. If the humor is suitable to both the material and the audience, its use can significantly aid the author's real purpose, which is to transmit technical data to the reader.

A humorous approach to technical writing appeals to the reader's creativity and personal experience, and helps satisfy the reader on both technical and personal levels. In addition, humor can make the reader more receptive and may soften his or her reaction to any negative information that is presented.

While the examples I've used may provide some direction for the writer or illustrator trying to incorporate humor, the best advice I can offer is to keep your eyes open. Humorous text and illustrations are everywhere, especially in current advertising and magazines. Draw from as many influences as possible.

Also, remember that "Less is more." In other words, it is better to be too subtle and have the humor be missed than to try too hard for a laugh and have the whole message be missed.

ACKNOWLEDGMENT

This material was presented in poster form at the IEEE Professional Communication Society Conference in Arlington, VA, September 16–18, 1981. Another version of the text appears in the *Conference Record*, Cat. 81CH1706-1, IEEE Service Center, 445 Hoes Lane, Piscataway, NJ 08854.

REFERENCES

[1] B. N. Chadbourne, *What Every Editor Should Know.* Arlington, VA: Printing Industries of America, 1977.

[2] *IBM Virtual Machine/System Product: System Product Editor User's Guide.* White Plains, NY: IBM Data Systems Division, 1980.

[3] O. Nash, *I Wouldn't Have Missed It.* Curtis Brown, Ltd., 1971.

[4] D. Wallechinsky and I. Wallace, *The People's Almanac.* New York, NY: Doubleday, 1975.

[5] *Introduction to Interactive Computing with VM/CMS.* Poughkeepsie, NY: IBM Data Systems Division, 1980.

[6] J. Muir, *How To Keep Your Volkswagen Alive.* Santa Fe, NM: J. Muir Publications, 1979.

[7] R. Sealey, *Poor Richard's Rabbit Book.* Sante Fe, NM: John Muir Publications, 1980.

[8] D. Rogers, *Executive Dictation Cookbook.* Atlanta, GA: Lanier, 1979.

[9] *APL Jot Dot Times.* Kingston, NY: IBM System Products Division, Fall 1980.

[10] *The Journal of Irreproducible Results*, G. H. Scherr, Publisher, P.O. Box 234, Chicago Heights, IL 60411.

[11] D. Moran, "Interference of Labium Superius Oris Hairs With Spherical Ice Cream Surfaces—A Theoretical Analysis," *The Journal of Irreproducible Results*, vol. 24, no. 1, pp. 28–29, 1978.

[12] A. Jesse, "How To Write Research Papers," *The Journal of Irreproducible Results*, vol. 22, no. 4, pp. 14–16, 1977.

Presenting the Peer Paper

ROBERT V. GARVER, SENIOR MEMBER, IEEE

Abstract—Some methods beyond the conventional public speaking art are presented which address the major deficiencies observed in the presentation of technical papers. Problems addressed are lack of excitement, poor control of time, unreadable visuals, bewildering complexity, poor question conduct, and abusing an invitation.

A FEW engineers seem to be born with a knack for saying things in an interesting manner and some others have recognized the importance of speaking well and have developed it. Most engineers, however, are not experts at delivering papers, which causes most of us to suffer from their lack of expertise as we sit through dull paper after dull paper, fighting drowsiness and mental distractions. The conventional body of knowledge on public speaking does not place sufficient emphasis in areas where it is needed to solve the problems encountered in delivering technical papers at peer group conventions. This paper is written to focus on the most common problems observed in sitting through years of papers at IEEE technical conventions. Topic heads, terseness, and cartoons are not conventional for a published technical paper but are designed to capture and entrap the busy professional into reading the text and to illustrate methods that are helpful in delivering a paper.

Some of the more common crimes committed at the podium in technical meetings are these:

Hypnotizing the audience is often a problem for engineers presenting technical papers. The hypnotist Dreskin speaks in evenly paced monotone words at the beginning of his show to see who in his audience tends to become heavy-eyed. Those people he asks to volunteer as hypnotic subjects. Many of our technical speakers unconsciously use the same speaking method for their entire papers, leaving many of their listeners to fight mental battles to stay awake or sit in a stupor.

In committing the **time crime**, a speaker out of control of his time can lead to a powerful session chairman's cutting him off before he is through; or under a less powerful session chairman everyone's question periods are destroyed.

Unreadable visuals can lead to your listeners' exclaiming to themselves "I can't read the figure. The printing is too small. There is too much in it. It is too dark."

Bewildering complexity can lead to your listeners' thinking "The speaker lost me in the first minute. I can't understand what he is talking about. He just changed subjects. How is this related to what he has said already?"

Manuscript received Sep. 13, 1979; revised Nov. 30, 1979.

The author is a Supervisory Physicist and chairman of the Editorial Committee at the U.S. Army Electronics Research and Development Command, Harry Diamond Laboratories, 2800 Powder Mill Rd., Adelphi, MD 20783, (202) 394-3010.

"The print is too small."

A typical response to **poor question conduct** might be "All I asked was what his characteristic impedance was and he went into a full lecture on how to calculate characteristic impedance."

Abusing an invitation can lead to the following response: "What a letdown this invited paper was. I expected to hear an easy-to-understand theory of how Josephson junction devices work, but ended up hearing a paper that I couldn't hope to understand on materials that I couldn't appreciate the importance of."

The solutions to these common crimes may lie in the "secrets" disclosed here.

EXCITEMENT IN YOUR DELIVERY—HOW YOU GIVE IT

Rehearse.

Do you get bored when you rehearse? Are you afraid you will then bore your listeners?

The number of rehearsals is the secret. The interest conveyed in the delivery of a speech first decreases by number of rehearsals as boredom sets in and then increases as problems with delivery are worked out. This interest change can be shown as in Fig. 1. Numbers cannot be assigned to the coordinates because there is no convenient unit of measure of interest and the number of rehearsals required for the crossover varies from person to person. Indeed, you do become more boring when you practice the same speech—the first few times. But after more rehearsals, the speech stops getting worse and begins to get better. After enough rehearsals, it is better than it was originally and from there on you are way ahead.

When I give a paper, I begin rehearsing at home. I reduce my slides to 3-in. × 5-in. file-card size by repeatedly reducing

Reprinted from *IEEE Trans. Professional Commun.*, vol. PC-23, no. 1, pp. 18–22, March 1980.

them on a copying machine. I record the speech on a portable tape recorder as I rehearse it because recording forces me to keep going and not stop to think. I time the speech and listen to the tape. To observe the improvement in my speeches as the result of rehearsals, I save some earlier rehearsal tapes and compare them with final ones. I also take the tape recorder on the trip for further rehearsals: first thing in the morning and every time I enter my hotel room—three, four, five times a day. This method works. When I finally deliver the paper, I know it cold.

There is another way to give good speeches by making use of the interest curve. I use it when the anticipated audience is small and informal. I call it the minispeech. The complete speech is delivered only once—at the time of presentation. But it is divided into smaller parts called minispeeches. The minispeeches are developed by notes and mental rehearsals. It is essential that two of the minispeeches be the introduction and the conclusion. The minispeeches are mentally rehearsed one at a time until even the weakest one is satisfactory. Then when the speech is delivered, it has the zero-time enthusiasm shown in Fig. 1 because the novelty of hearing it all together

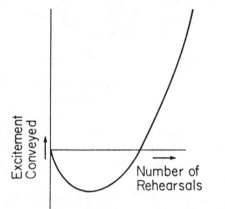

Fig. 1. How rehearsal affects the audience appeal of a speech.

fuels the speaker's own excitement. The minispeech is good for small groups when time is not critical. When time is critical, the speaker should build in flexibility. The points should be ranked in his own mind, especially in the later parts of the speech, so that he can drop or add them depending on how much time is left. Sometimes a secondary point can be interpreted as the answer to a question or otherwise conveyed in the question period. In any event, time should be saved for the conclusion, and it should be brief.

Nothing kills the excitement of a speech faster than a written or memorized introduction that is oblivious to questions or material in the preceding papers. Even though an introduction and a conclusion are planned and rehearsed, they should not be used in deference to ones based on the present state of humor and mood of the session. The speaker should be present in the sessions prior to his paper to sense this mood and be innovatively revising his introduction and conclusion based on what he hears.

There are also some hazards that can disarm a prepared speaker in front of a large audience and make him nervous and defensive, detracting from the excitement that he can convey.

For example, I had been working on X-band waveguide diode switches. The switch is about $1\frac{1}{2}$ in. on a side, and it was shown on my first slide. Projected, the switch was enlarged to 10 ft by 10 ft. I was overwhelmed by its appearance and delivered a good part of the paper in terror of what might happen as I showed the rest of my slides. It is always advisable to try the hall before you use it. Try to gain access to the hall when it is not being used and have a friend run the projector for you while you rehearse. Exercise the public address system and pointer to be sure you are in command of them.

Instead of "Lights on, . . . lights off," use a blank slide.

Another problem that detracts from the sparkle of a technical speech is the inability of audiovisual technicians to change the lighting quickly. Some speakers insist that the projector be turned off and lights be turned up in the middle of their paper. The usual long pause irritates the audience. The technical problem is that most symposiums are watching their costs and the projectionist or some volunteer operates the lights. No one is standing by the light switch. If a speaker wishes to have no slide on, he should make up a blank slide or a title or a cartoon that supports the change of subject he is conveying.

In general, the exciting speaker is active between the screen and the audience. He is not tied to the lectern. He uses the public address system and speaks up as if it were not there. And he bases his formality on the size of the group, leaning in favor of the formal.

THE TIME CRIME

Clocks seem to run faster when you are in front of a large audience. Don't deceive yourself into believing they will go slower. It helps if when you are rehearsing you glance periodically at your stopwatch. Then when you make your final delivery, glancing at your watch allows you to adjust your pace. (This is a good excuse to buy a new electronic wrist watch with stopwatch functions.) Allow two minutes per slide or chart. If you think you can cover 40 figures in 20 minutes, you are fooling yourself, and you are being lazy—not creative. Your listener has to be alert all day. If every speaker made the listener's mind run at top speed all day, how long do you think he would last? Rank the points you wish to convey and, depending on how time is going, give more or less supporting detail. Know how long your conclusion takes and start it at a prescribed time. Rehearse in less time than assigned, allowing for delays in the handling of slides or a generally truncated program because of other speakers' running overtime. Have points you can add in the question period in case there are few questions. Have the guts to leave out points if you just can't include them in the allotted time.

SLIDES

Black lines on a white background are best for slides. They are easy to read in a dimly lit room, and the screen itself gives off enough light to read by. Read what? Most conferences now provide their registrants with a digest of technical papers and the listener may be reading your paper or someone else's while you speak. Some people believe their presentations will be more dramatic if they use deep background colors with black lines and I have even seen black backgrounds with deep colors for the lines. The room becomes so dark that with the speaker's monotone it rates 99.5 out of 100 as conducive to sleeping. Sometimes it may be necessary to turn off the room lights. Sometimes a photomicrograph of an integrated circuit has such a wide range of color and such fine detail that it can be seen only in a dark room. But I believe the visual arts people can do a lot better than draw some of the dark shadows I have had to squint at on the wall. The speaker and his employer are being judged by the viewers and they should try to project a clear, bright, and cheerful image. Deep colored lines on a white background can introduce color without inducing sleep.

Slides should have a horizontal format.

All slides should be made to favor the horizontal format because this matches most screens. The speaker should project them as soon as he has them made so that problems can be corrected while there is still time. Seldom do I go to a symposium that some speaker does not have a slide that is too tall

for the screen and the poor projectionist is forced to manually tip the projector (often nervously) to get top and bottom alternately on the screen. Letters should be no smaller than four percent of the figure height. Remember that the people in the back of the hall would like to read your legends, too.

Tall slides create a problem for the projectionist.

Figures should be kept simple and familiar. Well known variables should be used. If an uncommon graph is to be used, its evolution from familiar graphs should be given. One welcome sight seldom seen is cartoons. Most illustrators are trained in art, not just graphic arts. They welcome the opportunity to interact with the speaker in livening up a dull technical paper.

Good visuals should have no equations, no long tables, and no long text.

DULL, HEAVY TECHNICAL PAPERS

The first step in composing your paper is audience analysis. Ask yourself "What can I assume that the listener already knows? What do I have to tell him that is new and useful to him? What are the problems in this area that people have had in the past? What is now in vogue? Can what I have to say be couched in the popular vogue or related to it? What will the impact of my paper be? Is there other similar information that the listeners already have?" To a novice writer, the period of audience analysis is very frustrating because it is seemingly not productive. No text or figure comes rolling out. There is nothing physical to show for all the mental agony of thoroughly analyzing your listener and your own position before beginning to compose or make notes.

If you don't do this, a dull paper is about to be born. Don't stop performing your mental analysis and start composing before you complete your preparation because once you have the paper composed, you will resist repeating the agony of giving birth and will treasure the "gem" you gave birth to in the first place. Little by little you will see the inadequacy of what you have done and become defensive about it. Little by little you will increase your own confidence in the value of your work, knowing that, after all, your paper was selected while others were rejected and therefore it should be enough for you to convey only the technical content of your work. Wrong! You may not be explaining your work in terms familiar to your listeners. You may be giving them fine details that only you consider important.

I define "dull" as when the speaker is going into fine details that are of no importance to me. "Heavy" is when the details may be important to me but I am not able to follow the development of the ideas. A heavy paper can be lightened by having a number of beginning points sprinkled around in the talk. This permits the listener to join you again after he has gotten lost. "So far I have explained X. If you do not understand it, you can take it on faith. I am going to explain Y, which makes use of X." Frequently, a new development comprises parallel derivations rather than series derivations. Clearly delineating these in the talk also provides the listener with new beginning points where he can begin to follow you anew.

Another helpful way to lighten a speech is to use an overview slide that shows by topics or by some physical representation all that is being discussed in the various sections of the paper. Each time a new section is begun, the overview slide can be shown again to keep the listener oriented. (Have copies of this slide in the right places in your stack or you will confuse the projectionist.)

QUESTIONS

Anticipate questions. Present your paper to your coworkers and ask them to ask you every conceivable question. Try your answers on them. Listen. The asker is probably more nervous at this point than you are. Repeat the question. If you think he meant to phrase it differently, check it out with him before you answer. Make your answer brief and to the point, especially if there are many questions. Be complimentary, acknowledging a good question or a good comment. Be honest. If you missed a reference or failed to measure something obvious to someone else, admit it and thank the questioner for his helpful suggestion.

Be honest. If you missed a point, admit it and thank your prompter.

THE INVITED PAPER

The acceptance of *contributed* papers is normally based on whether the subject is a new and useful device, whether it is a new application or a new problem, or whether it is a very good example (to be followed by others) of a known technique applied to a known problem. An *invitation* to give a paper is not a blank check to give the paper that was turned down last year. The purpose of the invited paper may be to bring everyone up to the same understanding of the state of the art, to set the stage for the remainder of the session following it. The speaker is encouraged to quote liberally from other people's works, giving credit. The speaker of the invited paper must be more thorough than normal in analyzing his listener. He should determine what his inviters had in mind when they selected him and try to live up to it.

ACKNOWLEDGMENT

The encouragement and suggestions of Della A. Whittaker are gratefully acknowledged.

Off-the-Cuff Presentations

DALE A. NEECK

Abstract—This article presents a list of tips to help improve impromptu speaking skills. The points discussed include developing the right attitude, having confidence while speaking, a series of steps to follow when called on to speak off-the-cuff, and a way to practice impromptu speaking.

JOHN SMITH, a mechanical engineer, was asked to participate in a discussion group that was formed to examine the merits of some equipment his company was interested in purchasing. The equipment was reported to have saved other companies hundreds of man-hours and thousands of dollars. Smith spent hours in the library investigating the history and value of the equipment so that he would be fully prepared when the time came for him to speak.

On the day of the discussion, Smith was ready with valuable information that would help in the analysis. He felt that it would take about one hour to present his material and he was confident that he would give an effective presentation.

At the start of the meeting, the director of the company noted that a problem had developed which would require immediate discussion and that the equipment analysis would have to be postponed. The director said that the topic requiring this urgent discussion was in Smith's field of specialization so he asked Smith to present a short history and analysis of the new problem.

What would you do if you were in Smith's shoes? Would you be able to give an effective presentation dealing with technical material when you had little or no time to prepare?

It's not surprising that many people are unable to do this. Throughout our lives we are conditioned to think that speaking, especially impromptu speaking, is the worst thing that can happen to us. Consequently, we approach impromptu speaking situations with hesitancy—with a fear that grabs our throats, chokes our voices, and strangles our thoughts.

Today's fast-paced business world demands that its participants have the ability to express themselves clearly and precisely. For every opportunity that we have to present a well-researched, carefully prepared, and thoughtfully constructed speech, there are many more situations that require us to speak "off-the-cuff," to be precise and clear with little or no preparation.

The ability to speak off-the-cuff can easily be learned if some effort is put into attaining this skill. What follows are five tips that will help you improve your impromptu speeches.

Manuscript received Sep. 17, 1979.

The author is a Graduate Teaching Assistant in the Dept. of Communication, University of Wisconsin-Milwaukee, WI 53201, (414) 963-4261.

Ann Jones

1. Develop the right attitude.

We have all been conditioned to dread impromptu speeches. This conditioning *can* be reversed. If you think of impromptu speaking as a situation where you will be addressing a group of people who are sincerely interested in what you have to say and who want to see you succeed (and they do), then your task will be much easier. No one takes pleasure in seeing a speaker "knocked off his feet" by nervousness and fright. Everyone wishes the best for a speaker.

2. Remember that you have done it before.

When you talk to your family at the dinner table, to your friends at a cocktail party, and to small groups of associates on the job, you are practicing impromptu speaking. Think of how many times each day you have done this and remember that you haven't prepared for one of these talks. You are already practiced in impromptu speaking.

3. Have confidence in your abilities.

You have spent many hours of study and training to get where you are today. Use your education and experience as subject matter for your impromptu speeches. Rest easy knowing that you are competent and capable in your profession.

4. Follow these steps when called on to speak off the cuff:

- Before doing anything else, breathe deeply three or four times. This will help make your breathing smooth and regular so you won't be gulping for air as you begin to talk.
- Begin your speech slowly. You can think much faster than you can speak. If you start slowly, your mind will be able to speed ahead and prepare a mental outline of what you wish to say.

Reprinted from *IEEE Trans. Professional Commun.*, vol. PC-23, no. 1, pp. 17–18, March 1980.

- Jot down the mental outline while you're speaking. You should not let your writing interfere with your speaking but nothing prevents you from jotting down a few key words or phrases that will help jog your memory as the talk progresses. With these notes recorded, your mind will be free to expand and develop each topic.
- Concentrate on having the audience *understand* what you wish to say. If you concentrate on developing understanding, you need not concern yourself with word choice, vocal variety, or any of the mechanics of speaking. The words and thoughts will come if you let them.
- Don't be alarmed if your mind draws a blank. Remember that the audience doesn't know that you have temporarily lost your train of thought. They will think that you are pausing for effect. When you lose your train of thought, try to summarize your previous points. This will help you regain your lost thoughts.

5. Practice.

Although you are already practiced in impromptu speaking on your professional specialty, the time will come when you are asked to speak on some topic with which you are not completely familiar. A good way to practice for that occasion is to speak extemporaneously during spare moments in the day. For example, after you hear a bit of controversial news on the radio while you're driving to work, turn off the radio and imagine that you have just been asked to comment on the topic. Talk out loud to yourself, trying to organize the material so that it will be easily understood by others. If you can't think of anything to say, try to use past experience as subject matter for your presentation. If the news you heard was about increased federal taxes, you might not be familiar with all the details of federal taxes but you might know quite a lot about your state tax policies. Try to draw a comparison or analogy between the knowledge you have and the topic you wish to talk about.

Impromptu speaking skills are easily developed through constant practice. Anyone who has given an effective off-the-cuff presentation knows that there is a tremendous sense of satisfaction and accomplishment that comes from giving such a speech.

Performance Guide for Oral Communication

GERALD LEE RATLIFF

Abstract—Public speakers whose primary responsibility is the giving of technical or scientific information need to acquaint themselves with an orderly and consistent design for organizing, supporting, and delivering an oral presentation so that they more effectively engage in dynamic communication with an audience. The elements of such a design are an analysis of audience interest in and knowledge of the topic, the role of speech in presenting highly complex technical or scientific language, and the principles of dynamic communication as they relate to audience perception and retention of information. This guide includes a comprehensive survey of the public speaking situation, suggestions for vocal delivery and scripting, and a checklist of common errors made in presenting factual material.

TWENTIETH-CENTURY man lives in an age in which communication is probably more essential to his professional and social well being than it was at any other time in recorded history. With the continued exploration of space, the rapid advances in both social and applied sciences, and the ever-inciting, ever-changing political and economic problems which beset this modern age of technological wonder, it is imperative that those engaged in worthwhile practical and scientific activity communicate the narrow world of their study to the great world outside.

Yet it is alarming to discover that people in the best position to share their meaningful discoveries and theories more often than not find themselves literally at a loss for words when faced with a public speaking situation. Indeed, a great many otherwise qualified and gifted professionals appear more like G. K. Chesterson's imaginary traveller, wandering aimlessly and incoherently through a maze of self-constructed obstacles in search of an avenue to express themselves and to share a new understanding of what they have observed, than like the effective public speakers an audience has come to expect from those engaged in scientific and natural discovery.

What is needed is a simple "performance design" for communication to make it possible for the professionally trained artist, craftsman, or scientist to assess his own public speaking skills and to better present his ideas and technical information to an audience.

A Primer for Performance

A convenient starting point in developing effective public speaking skills is for the speaker to remember that every public situation, whether it is an oral report, lecture, address, planning session, or business meeting, is an occasion that affords audience response. The speaker, through his appearance,

Manuscript received June 27, 1979; revised Nov. 26, 1979.

The author is a member of the Speech and Theatre Faculty at Montclair State College, Upper Montclair, NJ 07043, (201) 893-4313.

personality, manner of presentation, preparation, and attitude, is able to contribute greatly to the listener's response if he is aware of the following performance concepts. Each of these concepts is based on the assumption that the more interestingly the information is presented, the more easily it will be perceived by an audience. The speaker should also remember that he has an obligation to present both himself and his information in the best possible manner.

- Listen to yourself on a tape recording and observe yourself in a full mirror, noting any peculiar physical or vocal mannerism that might detract from the presentation.

- Select your wardrobe with care and taste, avoiding extremely bright colors and the admixture of contrasting hues. For many in the audience this will be their first glimpse of you, and what you choose to wear reflects upon the seriousness and the content of your presentation.

- Have a good mental attitude toward your presentation and don't be unduly alarmed by nervousness.

- Come before your audience fully prepared and confident, having rehearsed the presentation aloud.

- Maintain a conversational tone in speaking, and feel free to pause for emphasis or for audience participation.

- Look directly at your audience and maintain good eye contact so that the listeners sense an attitude of communication and concern for the importance of the material being presented.

- Be enthusiastic and animated, demonstrating your confidence in the material and suggesting your poise.

Receiving a favorable response is the first concern of any public speaker but it is equally important to promote good will and respect toward yourself and your topic. Although a speaker should be concerned with how well he does, inordinate fear, tension, inexperience, and a host of other physical and vocal distractions commonly referred to as "stage fright" can interfere with effective and meaningful communication.

To develop confidence it is necessary to confront these basic weaknesses in public performance and recognize and correct some of the flaws typical in a beginning public speaker's visual and oral portrait, particularly those which do not contribute to audience empathy or identification with the material being presented.

- Remember that the speech begins the moment you leave your seat or are introduced. The audience's first impression of you is the manner in which you approach the speaker's platform.

- Do not lean on the podium.

Reprinted from *IEEE Trans. Professional Commun.*, vol. PC-23, no. 1, pp. 11–14, March 1980.

Ann Jones

• Avoid a stiff and rigid appearance and refrain from shifting weight from leg to leg. You should stand comfortably with the limbs properly aligned and the hands relaxed at the side.

• Avoid reading notes; glance down at prepared information and then up at the audience to maintain continuity of delivery.

• Avoid unpleasant and distracting vocal phrases such as "uh" and "ah."

• Avoid slang and vernacular.

DYNAMIC COMMUNICATION

The manner in which a speaker uses his voice may well mean the difference between presenting information and dynamic communication. To formulate ideas and concepts worth communicating, to arrange them in a logical and meaningful sequence, and to phrase them accurately and interestingly are just the prelude to effective oral communication; there must also be "vocal orchestration" that suggests dynamic sound and action to the audience.

An audience does not immediately "hear" ideas, concepts, supporting materials, language or organization; rather, an audience's initial response, favorable or not, is to "sound." This sound is the result of voice quality, pitch, rate, volume, articulation, and inflection which distinguish one speaker from another and, when used properly, reveal the message or content of the presentation. The effective public speaker must be aware that the vocal mechanism does much more than merely transform a word or symbol into an audible transmission; it also permits a speaker to highlight meaning, focus attention, and direct impressions. For these reasons it is important for the speaker presenting technical information to pay careful attention to vocal attributes and to make a personal checklist of vocal weaknesses.

A speaker's vocal performance contributes to dynamic communication if it has the following basic characteristics:

• The volume is loud enough and the rate slow enough for the listener to hear without difficulty.

• The technical phrases and definitions are pronounced distinctly so that they are readily recognized and understood.

• The delivery is natural, animated, and forceful.

• The pitch and inflection have sufficient variety to create a high level of interest.

• The quality is resonant and free of nasality and hoarseness.

Although some of a public speaker's effectiveness in presenting information to an audience depends upon natural gifts or professional training, frequent practice and experience are still necessary to develop communication skills:

• Learn to hear voice as the audience hears it. The first step in learning one's voice is to record it and develop awareness of its range.

• Discover the pitch range best suited to your natural voice. To determine range, hum from your lowest possible pitch to the highest; then count on the musical scale the full-step tones in the complete range and divide by three. The lowest third of the range usually proves to be the most effective range of pitch for spoken communication.

• Maintain an adequate supply of air to achieve proper volume. To do this, learn to breathe deeply and develop either abdominal or thoracic breathing.

• Suit your voice to the occasion and the size of the audience and modify vocal characteristics accordingly, especially volume to overcome distractions or disruptions.

• Be aware that proper tongue placement is essential for good articulation and enunciation, especially for consonants like t, d, k, and g. A test of good pronunciation is to recite the alphabet through clenched teeth.

AUDIENCE PARTICIPATION

Once the principles of the vocal mechanism have been specifically charted and carefully explored in rehearsal, the speaker can then direct his attention to the informative material. When the subject involves scientific terminology, professional jargon, definitions, or so-called "technical talk," there are special problems that must be addressed so the listener can adapt to the documentary nature of the presentation.

"If it will make you feel any better, I don't understand half of this gobbledygook myself."

Reprinted from *Better Homes and Gardens*® magazine. © Meredith Corporation 1969. All rights reserved.

The speaker should *not* announce his topic and then develop the main points of the speech. "Telling" is not enough to ensure that the listener will receive the information and decode it into a meaningful and accurate understanding. Technical speeches in particular are usually of such complexity and demand such illustration that care must be taken to make possible a full appreciation of the subject matter.

Not infrequently a speaker engaged in communicating technical information of a numerical or statistical nature may discover a resistance to learning because he has not properly laid a foundation of curiosity about the subject or a motivation to comprehend it. He may even have forgotten that the evaluative tool for determining the success of a technically oriented speech is not how much information is *presented* but how much information is *understood* and *retained* by the audience.

The effectiveness of communicating highly technical information to an audience and the excitement of interest in the subject can be greatly enhanced if the speaker acquaints himself with some of the devices commonly used in speech making to gain attention and to promote understanding.

In those instances when the audience has a limited understanding of the topic, it is the responsibility of the speaker to offer them a reason for listening. Selected approaches include (1) adapting the topic to the interests of the audience and focusing on the significance of the material to the well being of each member; (2) suggesting the uniqueness of utilitarian nature of the topic; (3) posing problems presented by the topic and exploring possible solutions; (4) providing analogies between the topic and current events; (5) relating the topic to something that is familiar to the audience; (6) demonstrating the visual aspects of the topic; and (7) associating the topic with either a penalty or a reward.

It may be necessary on some occasions, particularly those in which the audience has no knowledge of the topic, for the speaker to "draw" the boundaries and parameters of the presentation and to develop the topic so that it appears less technical than it really is. Some popular devices that can be used effectively to communicate technical information in this manner include films, charts and graphs, models, photographs, posters, slides, tapes, and cartoons.

Remember, too, that the nontechnical listener may have a short retention span when confronted with complex information of a scientific or graphic nature, and communication cannot occur if he loses track of the speaker's development of the topic. It may be necessary to organize the material in either a chronological or a particular sequential pattern so that the audience senses clearly the verbal development of the speaker's subject.

Elementary compositional devices that can aid a speaker in structuring main ideas and supporting materials easily and precisely are

frequent summaries following main ideas,

restatement and repetition of supporting materials,

simile, metaphor, and personification to promote parallel relationships,

partition of points or support material to demonstrate isolation of thought,

description, "for instance," and example to stimulate recognition,

quotation and rhetorical questions to reinforce critical commentary,

comparison and contrast to suggest casual dependence and interrelationship, and

analogy or parable to elaborate relevance and significance.

Although motivation of the audience to learn, clarity of presentation, and illustration of the topic to promote audience understanding and comprehension remain the key elements of any speech that promises to convey technical subject matter, an effective speaker must also be aware of common mistakes made in the information-giving situation. Many speakers . . .

. . . attempt to read from a prepared manuscript and discover that they stumble over words, appear mechanical and artifical, or cannot maintain eye contact with the audience.

. . . forget to modify or simplify the language or terminology of the subject in terms of audience familiarity or knowledge.

. . . present such elaborate and detailed visual aids that the audience cannot distinguish the more important elements.

. . . neglect to impose order or form on the subject and wander aimlessly among main points and supporting materials.

. . . omit documentation of the material and the audience is led to question both the speaker's credibility and the subject's importance.

. . . attempt too much in an informative speech and communicate too little.

. . . lack vivid and descritptive word choices and figures of speech to help the listener receive a visual impression of the topic.

. . . conclude their presentations without any thought of provoking an audience to action.

SUMMARY

As a member of a profession that must use speech as a primary means of communicating technical information and theoretical strategies, the modern technologist must acquaint himself with an orderly and comprehensive approach to organizing, supporting, and delivering an oral presentation that will effectively communicate ideas and concepts to an audience.

For the conscientious professional who regards the informative speech as a practical matter, as a challenging activity to educate and enlighten his listeners, this preformance guide should serve as a basic introduction to the demands and expectations of the speaking situation, the role of the voice in sustaining interest in the topic and in highlighting significant ideas, and the principles of dynamic communication as they are related to audience perception and retention. This design should also provide the direction and the selected devices, in both the preparation and the performance of the speech, that will help the public speaker approach his speech with poise and confidence.

Although the scientist or engineer cannot expect to discover an easy formula for predicting the degree of success of an informative speaking situation, these guidelines support the proposition that public speaking is a *learned* art and that the best training for effective communication is to use daily conversation as a laboratory for developing a relaxed, natural, and enthusiastic approach to speaking.

Staged Right: Backstage Advice for Giving a Smooth Presentation

MARD NAMAN

The situation had all the makings of a speaker's worst nightmare come true. It was the 1988 Macworld Expo in San Francisco, and Apple Computer's Chairman John Sculley was in the middle of making the conference's keynote address. Sculley was presenting a sneak preview of an unreleased, prototype Macintosh; his keystrokes and mouse movements were displayed to the audience on a large overhead screen. All went well until suddenly—in front of the full house—the computer froze.

Realizing what had happened, an audio/visual technician working behind the scenes immediately hit a button to switch the screen projection from Sculley's computer to an identical Mac backstage; a cable and a scanner installed before the show made it possible for Sculley's keyboard and mouse commands to control the backstage Mac. For the rest of Sculley's presentation, the backup computer projected what was seen on the large auditorium screen; the audience never realized there had been a technical glitch, and the technician with the backup Mac saved Sculley from embarrassment in front of thousands of people.

It doesn't matter whether you're the chairman of a large corporation, like Sculley, or a marketing manager introducing a new product, or a district sales manager rallying your sales teams together—the fact is, you can't always foresee technical snafus. But with the proper planning and technical support, you can at least minimize your chances of encountering an embarrassing glitch or of not being equipped to handle one. After all, presenting ideas to an audience is scary enough; who wants to worry about "technical difficulties," too?

So, to help those who must stand up in front of a group and run an a/v show, either for the first or 50th time, we asked some of the best a/v technicians in the country for advice on how to run a smooth show. These experts stage dozens of large shows each year, ranging from Comdex to the Academy Awards; they've seen all the mistakes a presenter can make, and they know how they can be avoided. Most a/v experts work on large-scale productions, but their advice is applicable to even one-person presentations.

"The most common technical gaffes are microphones that are improperly handled or wired, incorrect lighting, and computer programs that freeze," says Richard Heller, vice president of operations for the Interface Group in Boston (which produces the annual Comdex trade show).

Here are the masters' tips on how to prepare for and deliver a compelling a/v presentation that won't degenerate into a comedy of technical errors.

1. KNOW HOW TO HANDLE A MICROPHONE

Your microphone is your primary communication vehicle. Remembering some simple pointers will make your presentations clearer and more professional.

Speak clearly into the microphone. When using a mike, presenters need to keep the proper distance. "With a standard podium mike, talk about six inches from the microphone, or a little closer," suggests Wayne Vincent, of Presentation Video Services. But don't get too close, he warns; if you do, you'll start "popping your *p*s," meaning that the *p*s become over-emphasized in your speech—an annoying sound for audiences.

If you're planning to move around as you speak—to gesture dramatically or to take questions and answers—tell the a/v suppliers in advance, so they'll be sure to have a clip-on Lavalier mike available. The Lavalier should be attached to a jacket lapel, collar, neckline, or tie above the mid-chest level, but not against the larynx; when the latter happens, your voice becomes muffled.

Repeat questions from the audience into the mike. This helps everyone hear the questions from the audience, and it's essential if your session is being recorded on audio or videotape. Sometimes audio/visual technicians place a standing microphone in the audience for questions, but that can be unsatisfactory. "We've found a lot of people are intimidated by having to get out of their seats and walk to a microphone," says Heller. "They'll readily yell their questions from their seats, but a lot of people won't get up and speak in public."

2. PUT SOME LIGHT ON THE SUBJECT

Lighting is particularly essential to the success of a presentation, Heller says. One thing to ascertain in advance is the heat generated by spotlights or other lighting; the lighting used in an auditorium can sometimes broil a speaker.

To avoid third-degree burns, try a dress rehearsal of your presentation from the podium; you'll experience how the lights will feel shining on you while there's still time for making adjustments. If that's impossible, ask one of the technicians on stage to stand in for you, Heller advises.

Some a/v experts believe that a speaker doesn't need

a spotlight (and the problems that such lighting can create). Adds Walter Silverberg, President of Silver Mountain Productions in Dallas, Texas, "My solution is to not turn the room lights all the way off during a presentation. The main thing that bothers speakers is not the spotlights shining in their eyes, but the visual contrast" between the speaker (in the spotlight) and his or her audience (in the dark). "Of course, you don't want too much room light, because the screen will be washed out."

In addition, turning off the lights for a prolonged period has been known to induce drowsiness among audience members, particularly after a meal. For that reason alone, it's a good idea to soften the contrast between the house lights and the light shining on the speaker.

Another consideration is the podium lighting, which can't always be relied upon. If you're making a slide or video presentation and you need to read your notes during the presentation, it's a good idea to take a pen light with you, several a/v experts suggest. If the podium lighting proves inadequate, a pen light will enable you to read your notes without having to turn up the house lights.

3. HAVE BACKUP PROCEDURES IN PLACE FOR COMPUTERS, SLIDE MACHINES, AND VIDEO EQUIPMENT

Computers are notorious for freezing up during a presentation—it's a matter of Murphy's Law—so be prepared. "For every piece of equipment we use in a presentation, we have a backup machine or a backup procedure in place," says Doug Hunt, executive vice president of Audio/Visual Headquarters (A.V.H.Q.) in Inglewood, California, which produces shows for Apple, IBM, and Sun, as well as the Emmy and Oscar presentations. When John Sculley's demo crashed at Macworld, for example, the technicians at A.V.H.Q. had a backup plan ready.

Contingency plans for computer equipment can be expensive, though. In addition to the duplicated equipment, the A.V.H.Q. technicians used a $50,000 routing box to save Sculley's Macworld demo. The routing box redirected computer projection signals so that the backup Macintosh offstage could replace the Macintosh that Sculley used onstage.

In the case of a presentation using rear-projection slides or video, an equipment backup plan isn't always practical. That's because rear-projection equipment must be precisely aligned, a task that no one would want to handle in a hurry in front of an impatient audience. As an alternative, plan to have a projector on hand with a 35mm slide of your company's logo; if the rear-projection system should fail, have the slide projector turned on to shine the logo onto the screen, suggests Heller. "That way, at least you're not standing up there beside a blank screen."

In fact, slides can often play the role of stand-in for computers *and* rear-projection devices. For example, if you don't have backup computers for your presentation, it's a good idea to have 35mm slides made of the highlights from your computer presentation; that way, you've still got

a visual aid, and your audience will at least get a flavor for the subject of your talk. This is also a good alternative if your budget prohibits use of a sophisticated computer or rear-projection display.

4. DON'T RELY TOO HEAVILY ON EQUIPMENT

The key to a strong presentation isn't the equipment you use. Knowing the message you want to get across and understanding that the message itself is what's important, says David Elliott, A.V.H.Q.'s senior engineer. "You should be able to do your presentation on a blank stage, with no props, and have it work on its own," he adds. "The equipment is only there to support you; if the computer crashes, it shouldn't stop your presentation."

According to Elliott, the worst presenters are "those who get on and just read the slides." The whole idea of slides and other visuals is to *enhance* your presentation, Elliott says, not to be the *purpose* of it.

5. PLAN AHEAD AND COMMUNICATE WITH YOUR CREW

The most common headaches in an a/v show could be avoided with proper planning and communication. To prevent last-minute worries, rehearse your presentation and communicate with the a/v providers as much in advance as possible, advises Wayne Vincent, technical coordinator for Presentation Video Services in Boston.

"Give the a/v technicians more information than you think they'll need," Vincent says. "If there's anything the least bit peculiar about what you're doing, such as using multiple screens or specially-synchronized sounds, make sure they realize that and understand how to deal with it. Too often, presenters assume you'll be able to do what they want you to do on site, and it's not always the case."

Says Lee Sterbens, sales manager of Greyhound Exposition Services in Las Vegas, "The biggest problem I have is speakers walking in 15 minutes before their presentation and wanting to rehearse their speech or run through their videotape. Or if we're providing teleprompter service, they'll walk up minutes before the speech with pages of changes they want to make. They don't understand that, in most cases, somebody else is in the process of making a presentation, and we're using the stage and the system for them. You can't just stop someone else's speech to make changes."

The a/v staff is almost always under tight time constraints, so the more time you give them to prepare for your presentation, the better you'll look in front of the audience. "The one key to success or failure on a big presentation is time, and often we aren't given enough to prepare the show properly," says A.V.H.Q.'s Hunt.

"For example, if the show's budget doesn't pay the $4,000 to rent the auditorium an extra day, we have only one day to put the equipment in and rehearse the presentation, instead of two," Hunt says. As a result, "mistakes get made, because you don't have the proper amount of time to tune up the system."

6. Find a Good Team and Stick with It

If you plan to make a/v presentations on a regular basis to sizeable audiences, you'll need to have a consistent group of technical people to support you, just as a concert singer travels with the same band. With a reliable a/v team in your corner, you'll be able to carry on without missing a beat—*a la* John Sculley—and the audience will enjoy your show with little or no distraction.

"There are many variables at work in a typical audio-visual presentation," concludes Hunt. "The key to pulling it off is having a team that works together on a consistent basis and has been through enough shows to know what to do, and when to do it."

Even if you don't have a regular a/v team, you can still plan ahead, communicate in advance with the a/v suppliers, and, most important of all, rehearse. Richard Heller's philosophy for averting disaster is a simple one: "Check it, check it, and re-check it. And when you think you've checked it enough, check it one more time."

You don't have to do any of these things, of course, but wouldn't it be better to have the audience remember your presentation for its relevant points and not for its embarrassing glitches? The more technical your presentation, the more likely you are to have something go wrong.

Whether you're making a presentation in a large auditorium with extensive high-tech equipment or in the conference room of your office with a few basic audio/visual aids, the experts agree: plan ahead. ∎

MARD NAMAN is a freelance writer in Santa Cruz, California.

How to Display an Image: Let Us Count the Ways

DONA Z. MEILACH

AT A RECENT presentation of Harvard Graphics by Software Publishing Corp, a 37-inch Mitsubishi monitor held center stage for an audience of about 50 people. Colors were incredibly bright and luminescent as the text scrolled, built up or revealed the topics. What could have been a relatively ordinary, static presentation with slides or overheads became an animated sequence of images spilled directly from the computer to the big picture monitor which served as a large-screen projection device.

It used to be that a presentation setup required an overhead projector or a 35mm projector with properly prepared visuals. No more.

There's a wide choice of projection hardware to support the increasing needs of today's presenters. There's no one answer as to which system to use for what; some do double duty and overlap. It's a market expected to grow to $8.5 billion by 1993. That represents a compound annual growth rate of nearly 22%, or $3.2 billion since 1988, according to Desktop Presentations Inc., a San Jose firm that analyzes the presentation market.

OVERHEAD PROJECTORS

Manufacturers are giving a new cachet to the traditional overhead transparency units. Some are portable and small enough to fit beneath an airplane seat. Larger units are compact, lightweight and transportable with heads that fold down for easy storage. Overhead projection presentations are best for small audiences in an informal situation. Prices range from about $220 to $600.

Polaroid has developed a new line of sleek overhead projectors. Several incorporate an optional unit for displaying Polaroid's 3-1/4 × 4-1/4 inch instant transparency, called the Small-Format Colorgraph or a "mini-transparency." A mini-transparency can be taken with a special Polaroid camera, and the peel-apart film is processed and cardboard-mounted in four minutes.

A mini-transparency can be projected using a traditional projector with a special template placed on the light bed. The new Polaroid projectors accommodate the standard transparency and have a platform for the mini-format. Polaroid and other companies market an attachable platform that will hold the mini-transparency above the light stage, so no templates are needed. These have a magnifying glass, so the image will project large.

LCDs, COMPUTERS AND OVERHEAD PROJECTORS

Trainers and presenters who need to work interactively with a computer and a large-screen projector are embracing the new, inexpensive and very portable Liquid Crystal Display (LCD) units. A unit consists of a panel about 2-1/2 inches thick that sits on the light stage of an overhead projector and is connected to the projector and to a computer. With this system, whatever is displayed on the computer monitor is simultaneously projected on a large screen.

There are about 50 vendors that bounce in and out of this market. But several companies have had staying power. The newest units are compatible with high-resolution EGA-VGA IBM PC systems or with Macintosh computers. Generally, they project in different shades of gray or in a narrow range of pastel shades. They're not yet perfected to project color as it's seen on a color monitor. Only the InFocus System offers memory on the LCD, so you can opt to project a show without interfacing the LCD with a computer. LCD units range in price from about $800 to $1800.

35mm SLIDE PROJECTORS

The original big-screen projection device, the 35mm slide projector, is usually recommended for large audiences in a relatively formal presentation. But with brighter light units, projection rooms need not be so dark, and presenters can choose a formal or informal style.

New features are built into many models. Dukane, for example, has a single projector that will fade one slide into another. Navitar's new LX slide projector lets you plug another projector into it for automatic dissolves. Traditionally, one needed a special controller box to fade or dissolve between two projectors.

The Dukane projectors let you randomly access slides rather than serially—a boon for jumping from slide 5 to slide 15, then to 20 or back again. Slide projectors run $350 to $1200.

SMALL-SCREEN CRT AND VIDEO PROJECTORS

When it's necessary to project images from a computer screen or via video, devices fall into a small-screen category.

Small-screen projection would be a cathode ray tube (CRT) such as your computer screen or a video terminal such as those used for VCRs. An average 12- or 14-inch computer screen can be viewed efficiently by a few people at one time. A few more can gather around a 19-inch computer or VCR screen. Often, several small video units are placed around an auditorium, so groups of 15 or 20 can view each unit. Monitors require coordinating and interfacing to a host computer or VCR. Even then, it's likely one or more aren't perfectly coordinated or that color isn't true. Wiring can pose a foot-traffic hazard.

Reprinted with permission from *Audio Visual Communications,* pp. 26–28, Nov. 1988.

But even these are expanding in size. The new XC3710 large Intelligent Display Monitor from Mitsubishi Electronics America is 37 inches with a viewing surface equivalent to a two-page spread of a tabloid-sized newspaper. With screen resolutions up to 800 pixels by 560 lines, the display quality is outstanding.

A Stand-Alone Monochrome Projector

The Vivid Systems' Limelight monochrome projector also takes images from the computer and instantly outputs them on a large screen. It works off standard composite video signals generated by an Apple II, Macintosh or IBM color card, or from a VCR. The projected image is green. The unit must be placed close to the screen. It's quiet because it doesn't have a fan-cooling system as do conventional slide projectors.

The Kodak Video Projector

The Kodak LC500 is a breakthrough in large-screen video projection because it can plug directly into a VCR, laser disc player or computer. When the unit is interfaced with a computer, on-disc images can be played or the presenter can work interactively with a program. Videotape and computer graphics can be combined in a single presentation.

Computer/Video Color Projectors

Large-screen color projectors aren't all aimed at vast stadium audiences. Businesses and organizations are becoming increasingly aware of the value of video as a communications medium. And combining output from video and computers is becoming the norm in a presentation environment. Expect to spend from $3000 to $25,000 for an average range system. Larger systems can cost much more—up to $450,000.

Where are such projectors used? In the growing number of corporate presentation rooms, large and small auditoriums and meeting rooms, at conventions and trade shows, and in training centers. They're applicable wherever it's more practical to run a program directly from a computer screen, a video tape or camera, rather than use 35mm slides.

Using a video projector system to display computer output makes good sense; one unit can replace several 19-inch monitors placed around an auditorium for audience viewing.

The image size possible from a video projector ranges from about five to 15 feet diagonally, a vast improvement over 19-inch and smaller monitors. The units are large and heavy, but they can be placed on a rolling cart, table or permanently mounted on a ceiling. One disadvantage is units must be placed close to a screen.

WHAT TO CONSIDER IN COMPUTER/VIDEO DISPLAYS

Most units have separate lenses for each of three primary colors, red, green and blue (RGB). The early units required careful adjustment to get all three lenses aligned and focused, called "convergence." Set-up time could take up to a half-hour if the units were transported. Poor setups caused blurred lines or spots and failure to resolve fine details.

Manufacturers have now incorporated microprocessor control to automate conversion and diagnostics. One-time setups have improved image quality and reduced dependency on an operator. Other improvements include color-coded remote control systems, so the projectors are easier for the presenter to use. They are lighter weight and more portable than early units.

The Electrohome 2000, for example, has only a single exterior lens to focus rather than the usual three lenses that must be synchronized. The unit's three internal color lenses are pre-focused, which aims for a set-up time of about 15 minutes.

Units are designed for different room sizes and distances from projector to screen. Because light intensity drops with the square of the distance, the farther away and bigger the picture, the more throw light is needed. Don't buy a low-brightness unit thinking you'll place it farther from the screen. That will cause sharpness and brightness to suffer.

RESOLUTION, SCAN RATE AND BRIGHTNESS

Resolution and image quality depend on the horizontal scan rate of the host CRT. The higher the operating frequency of the display, the higher the resolution and the quality of its signal. Video projectors aim for high horizontal scanning rates designed to support the range of resolutions that exist in graphics systems.

Resolution rating is usually based on a 12-inch screen. When you blow the image up to 10 feet, the resolution will be spread over 10 or more feet, so clarity is reduced. A crystal sharp image on a computer EGA or VGA monitor will still look good blown up, but not nearly as sharp as it appears on the computer screen. Some units will support EGA or VGA only if they have a scan rate of up to 25 MHz.

Most computer systems can generate images greater than the standard number of 512 scan lines. That resolution depends on the graphics board used such as the CGA (Computer Graphics Adapter), EGA (Enhanced Graphics Adapter) or VGA (Video Graphics Array), the Number 9 Board, AT&T Targa, and high-performance graphics workstations. High-resolution boards can produce images with as many as 1024 lines.

Projectors generally have the industry standard scan rate of 15 to 27 kHz: 15.75 kHz is the low end; high-end systems have 50 to 60 or 100 kHz. Most units are between 15 and 25 kHz which covers about 75% of users' needs.

Higher scan rate units are more expensive and indicated for CAD applications. A system developed to cater to high-resolution output is Electrohome's ECP CAD/CAM with horizontal scan frequencies as high as 71 kHz. It will support a 1280 × 1024, 60 kHz, non-interlaced display for CAD workstations. Chromalux Technology has a "CADD" video graphics projector with a 15- to 90-kHz horizontal sweep that will accept input up to 1280 × 1600 pixels.

One problem among large organizations that invest in a video display is they may have more than one generation of computer graphics in use. If they use a computer-to-video display, the units may not be compatible. The answer is a system with multi-scanning capabilities or an investment in a converter between projector and computer equipment.

Multi-scanning projectors are coming on the market quickly. Sony's VPH 1270Q was recently introduced. It responds to a wide range of scanning frequencies. An optional signal interface switcher lets you control various sources such as a VTR, video camera or varied-res computers.

All customers don't require high resolution, so many companies offer models with slower scan rates to achieve lower prices. Electrohome's model ECP 2000 with a 15 to 33 scan rate is about half the price of its high-resolution ECP CAD/CAM unit, which has a 30- to 71-kHz scan rate.

Scan rate is one measure; brightness is another. Brightness differs by unit and the number of lumens output. (Lumens is a measure of light intensity.) Watch advertising claims of brightness carefully. Some manufacturers advertise brightness at peak level, but rarely will you operate a unit at peak all the time because that would shorten the life of the tube. Also, as brightness is increased, contrast and sharpness decrease.

Some manufacturers cite specifications that are more conservative than others in their measure of brightness. A unit advertised as 400 lumens is not necessarily brighter than one advertised at 300. The same image viewed from these two units side by side may appear the same. Some ads claim "500 useable" lumens, but all may not be used at all scan rates.

There are other aspects of a machine and perceived brightness. GE's Talaria "Multi-Standard Projector" uses a light valve technology that transmits or reflects light depending on the video signal. It can generate brightness of 3800 lumens compared to less than 600 lumens in other systems. That means Talaria images can be seen well even in brightly lit rooms. The Talaria projector also uses a single lens.

When you consider buying a large-screen computer/video projector, research the market. Renting may make sense and look at the company's upgrade policies.

MULTI-MEDIA PROJECTION

The definition of "multi-media" is in transition. A multi-media show was built around multiple 35mm projectors and several screens—maybe with sound added.

Regardless of the definition, projection methods are required. This may mean a variety, or multiples, of projection devices. Also, there's the video wall, a bank of monitors arranged as a single unit. (See *AVC*, November, p. 22).

What's the best projection device for your needs? With so many choices, the answer is look at your presentation requirements for now and the future. Chances are you'll need several systems for different applications.

Dona Z. Meilach is a contributing editor to AVC. She is the author of "Dynamics of Presentation Graphics" and other books.

LCD Panels Bring Presentations Down To Size

by Christopher Rauen

A FEW YEARS AGO, desktop presentations required bulky CRT projectors to transmit images directly from the computer onto the big screen. Today, compact LCD projection panels serve the same function. Their convenience and low cost make them viable alternatives for many educational, training and business meeting applications.

About as big as a medium-size pizza box, these panels are especially popular for conducting presentations on the go. A typical panel weighs less than eight pounds and fits over most standard overhead projectors. After receiving the image information from the computer, an array of pixels in the LCD panel either blocks or transmits light to form the image. The image is then projected via the overhead onto a projection screen.

LCD panels have evolved from exclusively low-resolution monochrome models to true-color projection in less than three years. Introduced in 1986, the first generation products were compatible with CGA/Apple II graphics. By 1987, products appeared offering 640-by-480 resolution and compatibility with VGA and Macintosh graphics. These panels improved to deliver eight to 16 levels of grayscale, then *pseudocolor* with up to 16 shades of yellow or magenta. In 1989, the first true-color models appeared.

In Living Color

Today, color LCD panels, like the 5000CX PC Viewer from In Focus Systems in Tualatin, Ore., offer nearly 5,000-color projection capability. For Stephen Rogers, who formed Beyond Software, a multimedia production company, in 1989, that's a significant development. He calls that panel "the missing link" in a low-cost, portable system for projecting Super VGA graphics to small- and medium-size audiences.

At Visualon, a Cleveland, Ohio-based mailorder company specializing in presentation graphics systems, product manager Sylvia Povirk reports that color LCD panels are taking off. About 70 percent of all panel sales are for color panels. "We're selling as many as we can get our hands on," she says.

The appeal is to a middle ground that, according to Rogers, was never before addressed in presentation technology. "These panels give you everything you could ask for—portability, ease of setup, ability to connect to a laptop computer—and they handle as many colors as a computer can output at a reasonable resolution," he says.

Meantime, at In Focus Systems, the refinement of color LCD panels is attracting a new class of user. Marketing manager Scott Niesen says that more than 40 percent of color LCD panel sales involve corporations with more than 50,000 employees. Computer companies like Microsoft, Borland, WordPerfect, Apple, IBM and Toshiba are finding these panels remarkably cost-effective in sales training. In most cases, these companies had previously relied on light-valve projectors that cost nearly $1,000 a day to rent.

For those willing to accept the compromise in image quality when choosing an LCD panel over a CRT-based or light-valve video projector or over 35mm slides, the return on investment can be pronounced. "Once you've bought into the technology and you're past the learning curve, the payback is fast," says Rogers.

Toward Higher Resolution

I MPROVEMENTS IN RESOLUTION continue. The next advance for these panels will be to 1280 by 1024. Already, In Focus Systems holds a patent for a 1280-by-1024 monochrome LCD display system, and a color version is in the works. That advance would make LCD panels even more competitive with CRT projection systems.

Of equal significance is LCD technology's contribution to the projection of animation. Currently, thin-film transistor (TFT) technology offers the only practical method of presenting full-motion animation graphics. These panels use the same LCDs as those found in hand-held televisions. Telex Communications in Minneapolis, Minn., and Sharp Electronics in Mahwah, N.J., offer TFT-type (also called active matrix) panels.

Sharp is hoping its recently introduced QA-1000 active matrix LCD panel will deliver a level of animation quality suitable for interactive presentations. Previously, this level of quality had been available only in CRT projectors selling for twice the price, contends Bruce Pollack, national marketing manager for Sharp's Professional Products Division. Its 64-color capability, he adds, will satisfy the bulk of the market for corporate presentations and education. "Others offer more colors, but at a lower resolution," says Pollack. "We find most people want the 640-by-480 resolution."

At $4,995, the QA-1000 falls between Proxima's 512-color panel, which retails for $4,795, and the In Focus PC Viewer 5000 CX at $5,995. Both Proxima and In Focus rely on a subtractive color process similar to that found in photographic technology. Also referred to as passive matrix technology, this design places three LCD panels (cyan, magenta and yellow) on top of each other. Each pixel area appears white. The display then selectively subtracts primary colors from white to produce a spectrum of hues. This contrasts with active matrix design, which takes an additive approach. It creates color by illuminating three sub-pixels (each a primary color) associated with each pixel on the panel.

Problems in manufacturing originally limited shipments of active matrix panels. According to In Focus Systems, TFT manufacturing yield rates were down to 20 percent. However, both Telex and Sharp contend that the yields have risen since the beginning of the year, and shipments are beginning to meet a backlog of orders. In Focus, meanwhile, plans to refine its proprietary subtractive technology to boost both color resolution and refresh

Reprinted with permission from *AV Video*, pp. 50–57, May 1991.

> *For those willing to accept the compromise in image quality when choosing an LCD panel over a CRT-based or light-valve video projector, or over 35mm slides, the return on investment can be pronounced.*

rates, the latter as a means of improving the projection of motion.

Boon to Training

CURRENTLY, SOFTWARE TRAINING represents the most widely accepted application for LCD panels. Creagh Computers Systems in Solana Beach, Calif., has been using them for several years to demonstrate its commercial real estate and contract management software packages. According to chief executive officer Bill Berkley, LCD panels are a vast improvement over slides. "With slides, you're always fumbling back and forth. It's easy to get lost in a presentation," he says. "The LCD panels, though, allow you to see the functioning of the software right before your eyes."

For that reason, LCD panels may become standard equipment in classrooms and training centers across the country. Povirk recalls taking a computer class in which the teacher walked around the room responding to student inquiries about PageMaker. "It would have been so much easier if she could have projected the software onto a large screen," she points out.

LCD panels also allow for production of what Bernard DeKoven, founder of the Institute For Better Meetings in Palo Alto, Calif., calls "random access presentations," during which the audience defines the direction of the presentation. With the computer acting as an electronic flip chart, the LCD panels can project any computer screen. "That creates a more responsive environment," says DeKoven, "perfect for brainstorming or strategic planning. The results are incomparable in terms of increasing productivity in meetings."

Weighing Presentation Alternatives

DESPITE ADVANCES IN LCD TECHnology, resolution from color LCD panels remains their greatest drawback. Typically, LCD panel projection requires trade-offs between image contrast and brightness, as well as compromises between the number of colors projected and color saturation. When high-resolution

and saturated colors are both presentation priorities, slides and CRT projection remain superior methods of display.

A lack of applications software that takes advantage of presentation capabilities, along with a dearth of users skilled in PC-based presentations, presents other obstacles. Alan Ayars, director of Dataquest's multimedia industry services, claims that sophisticated presentations, particularly those involving multimedia, are not for "the faint of heart."

"For the most part," he says, "it's too difficult for the non-intuitive user to do the kinds of things that are available with the technology today."

> *It's quite clear that acceptance of LCD panels isn't happening as rapidly as are advances in the technology, but the problem could be that people just aren't used to authoring presentations on the spot.*

According to DeKoven, LCD panels are as much restricted by entrenched business habits as by any limitations in design. While he applauds their ability to transform meetings into dynamic experiences offering a high level of audience participation, it's quite clear that acceptance of LCD panels isn't happening as rapidly as are advances in the technology. The problem could be that people just aren't used to authoring presentations on the spot.

"I still find that people aren't catching on [as] they should simply because they must learn how to see the meeting experience in a different light," he says. "It will require a new way of thinking about meetings."

These deficiencies might explain a glitch in demand reported by the Rochester, N.Y.-based market research firm Hope Reports. Hope Reports chairman Tom Hope says that LCD panel sales actually declined about 10 percent from 1988 to

1989 after experiencing a 44 percent increase in sales during the 1987-1988 period.

Tony Janicki, marketing manager for Eiki International of Laguna Hills, Calif., believes LCD panels will be short-lived. "They're just a stopgap measure," he says. As prices decline and data projection capabilities improve in LCD projection systems, Janicki expects that LCD panels will be relegated strictly to low-end applications.

However, at Desktop Presentations Inc., in Mountain View, Calif., president William Coggshall sees a brighter future for the product category. He shows worldwide shipments of LCD panels nearly doubling from 1989 to 1995, reaching sales of 250,000 units and $770 million in revenues by 1995. By 1993, Coggshall predicts that electronic presentations will for the first time overtake slides in actual quantity produced.

New Class of Products

WHAT'S MORE, NEW PRODUCT hybrids are beginning to reach the market. Companies like Sayett Technology of Rochester, N.Y., are incorporating LCD panels into compact projection units of their own design to control heat and image brightness. The Sayett Media Show weighs 20 pounds, features active matrix technology capable of projecting animation and is expected to retail for about $7,000. Others, like the LiteShow II from In Focus Systems, can project images without the use of a computer. At $1,995 suggested retail price, the LiteShow II weighs less than four pounds and features a single 3.5-inch disk that stores 30 or more graphic files designed on a PC.

Priced between those two products at $3,995 is Visualon's PC 9800 Memory LCD Panel. Weighing under seven pounds, it stores as many as 100 images on a two-inch disk. Povirk suggests that corporations use the projector to produce presentations at headquarters and then ship the presentation out on disk to regional managers. They, in turn, can conduct presentations in the field "without having to lug a computer around."

Prices on these augmented LCD panels start at about $1,000 for monochrome models. Ease of use, hardware compatibility with VGA and Macintosh graphics, system reliability and heat tolerance are all factors to consider when examining alternatives.

Further, while a variety of colors for projection may be important, not everyone agrees on the ideal palette. In Focus Systems touts the nearly 5,000-color projection capability in its 5000CX PC Viewer as a potential that will influence future applications.

Coggshall of Desktop Presentations suggests that 500 may be sufficient in most instances. "Nothing you generate on a computer requires more than that," he argues. "You only need more when you have essentially a photograph."

Whenever possible, those considering LCD panels should compare products projected under identical room lighting conditions and with identical overhead projectors — and not just any projector will do. Coggshall observes that many just don't exploit the potential of these panels. "The average projector is regularly out of focus and runs too hot," he says. "I'm surprised these panels are selling as well as they are." He cautions that buyers should test these panels on a standard overhead before buying.

Some LCD panels may achieve comparable image quality, but only when projected by specialized projectors such as metal halide models. That not only raises the cost of a complete system but removes the advantage of portability that LCD panels provide when projected from standard overheads. Povirk asserts that a bright projector is mandatory to guarantee image quality. She prefers Dukane projectors outputting 3,000 lumens of light. "It's amazing what a difference that makes," she says.

Other features that differentiate competing LCD panels include the manner in which these products interface to the computer and what type of control panels and accessories are available. Jeffrey Belk, Proxima marketing manager, claims that options can quickly raise the cost of a complete projection system. Proxima includes EGA and VGA cables, remote control and software control as standard with its color LCD projection panel. "Most other companies unbundle that," he asserts.

Emerging Presentation Options

OTHER PRODUCTS ARE GROWING up around LCD panels to deliver more complete presentation solutions. One example is Proxima's Cyclops interactive presentation pointer system. Retailing for $1,195, it turns a projected wall or screen into a touch screen. A pointer with an LED sensor at the tip functions like a mouse, remotely controlling the presentation from anywhere in the room. The proliferation of Windows in the DOS world and the library of presentation software that will grow up around it should also help promote the LCD cause. Al-

ready, software packages like Quatro and Harvard Graphics incorporate slide-show features for assembling charts and presenting detailed sales analysis. This introduces elements to a presentation that enhance audience interactivity. "That's something you can't do with a static medium such as foils and 35mm slides," observes Belk.

LCD-based presentations may mimic the phenomenon of desktop publishing in which a new application gradually captured the fancy of a broad user base. If so, many projects which were once designated to outside service firms and design professionals may be assigned in-house.

So, too, may LCD-based presentations transform the department manager into a bona fide presentation specialist. Beyond Software's Stephen Rogers is especially enthusiastic, particularly regarding the potential for adding sound, synchronizing interactive video and using a PC as central control. "Once you have a computer in the meeting room," says Rogers, "you'll find all sorts of new things to do with it." **AVV**

Christopher Rauen is a West Coast-based freelance writer who has covered video and presentation technologies for several publications.

Multi-Image Presentations: Still Powerful After All These Years

BY PETER COVINO

At the 1900 Paris Exposition, a mock balloon ride was simulated on hand-tinted 70mm film projected onto 10 screens in a circular configuration. After three days, however, the presentation was closed down as a fire hazard because of the extreme heat generated by the projectors.

This preempted show marked the origin of the multi-image presentation—a slide show that uses two or more projectors in a programmed presentation. Over the years, multi-image has become more sophisticated and has survived the emergence of competitive technologies, maintaining a stronghold as a powerful presentation technique.

A Grand Beginning

The multi-image presentation of today is a far cry from the Paris Exposition slide show. Multi-image, you might say, grew up on the stages of the world's fairs and expositions. With each production it grew more majestic and better orchestrated.

In one of its earliest forms, multi-image was a part of the epic 1927 film "Napoleon," by Abel Gance, which used three semi-synchronized projectors in some sequences to display both panoramic and multi-image combinations.

The first self-contained multi-image presentation was unveiled in 1939 at the New York World's Fair. "The Calvacade of Color" used 11 screens, each 22 feet high and 17 feet long, arranged in a semi-circular format. The projectors were capable of dissolve transitions, and each contained 96 slides on precision-ring gears. The show also included a synchronized-sound film. It was produced, appropriately enough, by Eastman Kodak.

The multi-image presentation was a hit at subsequent world's fairs and exhibitions. "The Creation of the World," a 15-minute show at Expo 67 in Montreal, for example, dazzled the audience with its display of more than 12,000 slides. Fourteen new images appeared every second, projected on a 22- by 32-foot screen area. The projectors were hidden in 112 cubes, each containing two carousel projectors mounted one above the other.

Multi-image had come of age.

Video Enters the Picture

Ironically, at the same time that the first automated multi-image program debuted, so did the first experimental live television broadcast.

"It seems that we've been in competition with video from the very beginning," says Dr. Ken Burke, president of the Tampa, Fla.–based Association for Multi-Image International.

The struggle for multi-image to survive alongside video has been a constant one, with speculation about the industry's future usually a major topic at trade meetings and conferences. Its imminent demise has been predicted by both its supporters and its detractors for the last few years. Some say that eventually, as video, laserdiscs and other media forms continue to grow and become less costly, multi-image will be considered only as an art form.

Doug Spitznagel, president of Multi-Media Presentations, a division of The MultiMedia Group, Inc., says he has found that many multi-image producers have switched over to video. "We have certain standards, and we find it is getting harder and harder to get someone who can do multi-image as well as we'd like it done," he says.

MultiMedia, a Culver City, Calif.–based firm that specializes in audiovisual presentations, sees the use of multi-image becoming more limited as the century comes to a close.

"We are using less and less multi-image per se," says Spitznagel. "So many of our shows are video/film compilations now. A lot of our multi-image has been speaker-support slides and short-feature product modules, but more of our speaker support is moving toward electronic modules. It is really nipping at our heels in terms of what it can do."

Along with that, Macintosh and other computer graphic systems—as well as videodisc and other new technologies—are starting to edge into multi-image's turf.

Spitznagel says the medium may be becoming outdated as other technologies get closer to achieving some of the same end goals at a comparable or cheaper cost. "Multi-image was initially developed because film and video were so expensive," he explains, "but video has come down in price, and the cost of multi-image has gone up. And because multi-image is a collaborative medium, you need to invest in a whole team of people " he adds.

He observes that aside from the cost factor, there isn't as much interest in the art of slide-making as there once was. "Young, creative people are going into computer graphics. I don't think they see any future in slides," he adds.

The Best Defense

Despite these circumstances, multi-image still has its ardent supporters. A substantial amount of money is continuing to be spent on slide shows, as indicated in a 1988 study by Hope Reports. It cites an estimate of $5.5 billion in annual commercial revenue generated by slides, with approximately $500 million from multi-image production and the balance from speaker support. The report also indicated that about 2,000 U.S. companies produce slides for commercial purposes, of which 64 percent also produce multi-image programs.

Washington, D.C.–based Todd Gipstein, who has created dozens of award-winning multi-image presentations over the past decade, is a firm believer in the enduring value of multi-projector/multi-screen productions.

"I've worked with many different kinds of media through the years, but nothing has a more powerful effect than multi-image," he states. "Its power is

Reprinted with permission from *Presentation Products Magazine*, pp. 28–34, May 1990.

derived from so many different technical factors. Multi-image taps into that natural process of how we see the world. No other medium has the same capacity for pacing and dissolves."

Seattle-based producer Charlie Watts of Watts/Silverstein, an award-winning production house that does about half of all of its media work in the slide format, also has a number of reasons why he favors this medium.

"Multi-image is our medium of choice in certain parameters," says Watts. "I like to use it for a controlled environment. In a fixed-installation theater or a sit-down meeting, the light and the mood can be controlled as part of the presentation."

According to Watts, the decision to use multi-image comes down to cost versus visual impression. "You can do a cheap video or an inexpensive film, but to make it impressive, it becomes more expensive. You can impress your audience less expensively with multi-image," the producer says.

A good example is Watts/Silverstein's Mount St. Helens show. Installed permanently at the Mount St. Helens Visitor Center, the presentation powerfully displays the greatest volcanic explosion in the continental U.S. of this century. The 15-projector show ended up costing only about $65,000.

"I could have done the Mount St. Helens show in film, wide-screen Panavision or multi-screen video, but at what cost?" asks Watts. "It would have been impressive but expensive to produce."

Peter Ryan, former president of the Association for Multi-Image and a producer who owns Peter K. Ryan Associates in Foxboro, Mass., agrees that multi-image is often the most economical way to go. "In today's budget-conscious business climate, the bottom line is that multi-image continues to prove its worth as an effective communication tool," he says. "A typical video presentation is produced in editing suites, furnished at either high capital cost or rented at hourly rates starting at $250. However, the best programmers in the multi-image industry, with all equipment at hand, can edit, program and create those magical visual effects on-screen for an hourly rate of $65."

Another benefit, say its supporters, is that multi-image gets messages across more easily. This was certainly the case with the Watts/Silverstein multi-image show "Elephants." The production was meant to raise funds for a renovation project at the Woodlands Park Zoo in Seattle. Government cutbacks threatened the building of a new house for the elephant population, so the zoo turned to the private sector. With the help of this two-projector award-winning show, the public became aware of the zoo's plight, and the elephants' new home became a reality.

Quality and Control

The keepers of the faith in multi-image maintain that there will always be a place for the medium's sharp, highly-choreographed effect.

"I keep hearing about what high-definition television is going to do for video," says Gipstein, "but it requires use of heavy equipment, and you just can't project video as well either. The slide is still the way to go. Multi-image holds up very well, and the productions are of very high quality."

Watts concurs. "I think multi-image will continue to be a strong medium until such time that video equipment can match the quality of the 35mm slide." He believes, however, that continued technological improvements in film and film processing may mean that electronic media will never catch up with the visual quality of a slide.

"It's hard to compare the beauty of the projected image with any other form," says Spitznagel. "Electronic delivery systems are great, but they don't provide the same pristine density or allow the same visual choreography that multi-image does."

Ryan has his own theory why multi-image will be around longer than the doomsayers forecast. It will succeed, he believes, simply because it is not "the same old stuff."

"Multi-image is a unique medium because of its use of screens, bold images and choreography," Ryan says. "Whereas film and video operate at continuous frame rates, multi-image has its own meter—images can appear according to musical tempo or any desired pulse. The programmer has discrete control over each frame, with a single slide appearing on the screen for tenths of a second or several seconds depending on the desired effect."

Despite all the advances and changes in video and related media, Ryan is convinced that multi-image's place remains secure. "Nothing approaches the image quality of a projected 35mm slide. This quality difference is critical for end users because it has a direct relationship to the impact of their presentation on the audience." Nor will high-definition television supplant multi-image, he argues. "It is clear that this technology is not a panacea, and it is at least several years from reality."

Multi-image will continue to grow and endure, according to Ryan, not only because of its high quality but also because of its flexibility.

"The readily adaptable nature of the multi-image production process provides another reason for the medium's continuing popularity," he says. "Instead of being subject to the complexities of video editing, a multi-image show can be edited by spreading the slides out on a light table and substituting individual images. A single slide can be reused in any number of presentations. In fact, many companies send a more cohesive message by incorporating the photography and graphics from their multi-image shows into collateral materials."

Video, Ryan contends, will not spell the end to multi-image presentations. "It is a phenomenon of today's lifestyle that many people have a television on for hours at a time but don't watch it. Instead television is treated as a background source; it has become the aquarium of the late 20th century. When audiences see the slide-based medium, however, it's a different experience for them, and they have a positive reaction."

For communicators of the 1990s who want to provide a memorable visual experience, multi-image will continue to provide a convenient, flexible and economical solution to business communication needs.

Multi-Image Glossary

Animation—Simulating movement with sequenced still images.

Digital audio—Recording technique that stores sound as a train of numbers for improved fidelity.

Digital data—Numerically storable data that can be computer-maintained and processed.

Dissolve—Gradual fading of one image into another, done by changing brightness of two superimposed projectors.

Dissolve control—Device that may be used alone or as part of a system to fade lamp brightness between two or more superimposed pictures.

Dupes—Duplicate slides made from original transparencies.

Fade—Increasing or decreasing projection

lamp light level to bring up or dim down an image or dissolve images.

Filmstrip–Still images in a continuous strip of film with a horizontal positioning of an 18mm x 24mm image.

Masking–Special-effects procedure involving high-contrast film and multi-exposure techniques.

Matte white screen–Smooth-surfaced screen with high sharpness and relatively low reflectance for wide-angle viewing.

Multi-image–Concept of presenting the viewer with several images simultaneously.

Multimedia–Technique of incorporating a variety of visual or other media in any presentation, e.g., film and slides.

Multi-screen–Projecting images onto several image areas.

Random accessing–Projector-control system allowing the user to call up any slide in any order.

Real-time programming–Encoding cues onto the track at show speed.

Rear projection–Projection of an image onto a translucent screen material for viewing from the opposite side.

Single screen–Presentation in which all images are superimposed on the same screen area.

Staging–The process of setting up and operating a multi-image show.

Transfer—Reproduction of a slide show on film or videotape.

Xenon projector—Very-high-brightness projector using a non-adjusting arc in a xenon gas-filled lamp.

Visual Presentations: How to Survive Giving One on the Road

GEORGE L. BEISWINGER

IT'S ONE THING to make visual presentations at your home base, but traveling with them, well, that's an entirely different ballgame.

You have become quite adept at making visual presentations, whether utilizing a simple chalk board or riding herd on a 15 projector A/V presentation. You know your equipment and your facilities, have adequate supplies, and there is generally always someone nearby who will lend a friendly hand if a problem develops. But not so if you are many miles from your office, in a strange hotel or convention center. Then, making such presentations involves far more than just bringing along your equipment.

It is true that not much can go wrong with a flip chart, felt-tip marker or a simple overhead projector. But almost anything more requires advanced planning. If your presentation is to take place in a large room and involves the use of such equipment as a 16mm projector, one or more slide projectors, video tape and large-screen TVs, or a PC equipped with an overhead projection device, you need to visit the site at least a full day in advance of your meeting. There are many things to nail down.

First of all, it is vitally important to meet the facility's personnel and ascertain the nature and extent of services.

If you will be at the location for more than one day, carefully record the telephone numbers of all applicable personnel, from the manager on down. Also find out the hours that each will be working on the day of your presentation.

WHO PROVIDES WHAT?

On your advance visit, reach an understanding with the facility manager on what equipment the house will provide, what you will need to bring, and what must be rented locally. If you are renting a VCR, large-screen TVs, projectors or sound equipment, it is a good idea to have a sales or technical representative from the rental company present.

Find out if the house will provide podiums or risers, projection stands, scaffolding, if required, as well as such things as tables for handout materials. Will the facility's personnel perform all necessary labor, will you or your staff be required or permitted to perform certain setup duties? Maybe, maybe not.

Will the meeting facility drape the podium, provide background draping, if needed, and also drape your screens?

George L. Beiswinger is a Philadelphia-based writer who has done complex visual presentations for several major corporations.

Probable answers: the house will drape the podium, an outside convention exhibit service will be required for background draping, and your A/V rental agency will drape its screens. Perhaps.

CHECK THE ROOM

Now it's time to check your meeting room thoroughly. If you're using front projection, do you have a clear "throw" to your screen? What about obstructing chandeliers? When you visit a facility in advance, you can often arrange to have large lighting fixtures moved, but seldom is this possible when the request is made the day of the presentation. You will need to know who will perform such services and who will bear the expense, which could be considerable. Several years ago, I found it necessary to have a large crystal ballroom chandelier removed for a three-screen, rear-projection setup. I had to engage an independent millwright at a cost of $700.

Will dedicated circuits be required for computerized equipment, such as a PC with a projection device? Will a surge and spike suppressor be required? Electrical devices elsewhere in the facility could play havoc with your sensitive solid-state equipment. Are circuit-breakers and fuses adequate for your intended uses? Will podium lights be required and will they interfere with your projected visuals? If your room is part of a larger area, separated by folding doors, can the sound system be properly isolated? Is the tape recorder you brought compatible with this system? Do you have the right jacks and connecting cables?

And what about those mirrored posts often found in many facilities, which can play havoc with front projection. Will they need to be draped? Few large rooms with windows are equipped with blackout shades. You may need to cover these openings. Who will do this? When? Suggestion: most building or garden supply stores sell large, precut sheets of black builder's plastic. This material does an excellent job of screening out even the brightest light. The sheets are easy to hang with thumbtacks, and economical enough for one-time, throw-away use.

The A/V and computer visuals field has exploded during the past decade, but many meeting facilities have failed to keep pace; even new halls are not always constructed with A/V in mind. That's why it's essential to do your own measuring, rather than assume that the facility's spec sheet is adequate and accurate. The latter may show a 10-ft. ceiling, but fail to indicate that a bisecting I-beam or heating duct lowers it to 9 ft. for projection purposes. The plat may

Reprinted with permission from *The Office*, pp. 70–73, Oct. 1988.

indicate a 24-ft. stage—plenty of room for utilizing three, eight-ft. rear projection screens—but fail to indicate that it narrows to 18 ft. at the rear, making projector placement impossible. Check to determine if room length (for front projection) or stage depth (for possible rear projection) meets your lens requirements. Perhaps a stage extension will be required. Again, who will provide it, and who will pay the bill?

SET A TIMETABLE

Determine the exact time you will be permitted in the room for setup and final preparations. A multiprojector slide show, for example, can require the services of several persons for four hours or longer. You must consider such things as how much wire is to be run and taped down, and how much prior testing, focusing and the like is required. Allow enough time to test-run your entire presentation at least once. Gremlins have a way of sneaking into carousels.

After your equipment has been installed, focused and checked out, do not leave it unattended for one second prior to your presentation—even if you have to bring a box lunch and stay with it yourself. For some reason or another, even the most simple electronic setup acts as a magnet for everyone from the house electrician's helper to the electric guitar player in the lounge. If you aren't there, they may "fool around with it a bit," all with the best of intentions. Later, you could find that your show has been partially "reprogrammed." Or, upon returning, you may find that your carefully placed equipment happened to be in the only available path for removing the grand piano, which the room next door needed immediately. You're back where you started. Stay with your setup until you go on!

It is also a good idea to turn on your equipment 15 to 30 minutes prior to show time. Most projection lamps seem to burn out when they are first energized. Allow ample time to replace them, and keep an insulated glove handy to handle hot lamps. Don't forget to check the slide retaining rings on your carousels before you turn a projector upside down to change a lamp. After I tighten these rings, I secure each one with a small piece of masking tape. This serves two functions: it keeps the rings from working loose and provides a quick visual assurance that all is well.

Don't be like the A/V man who thoughtlessly inverted a projector and its slide-filled ringless carousel just five minutes before an annual meeting presentation. He was still surveying the random collage of slides on the floor, transfixed in horror, when he heard the chairman of the board say, "And now, a pictorial review of the year's events."

THE COMFORT FACTOR

Visual presentation people are often concerned only with their equipment, to the unfortunate exclusion of the things that affect the other senses. Temperature and room air quality are important, too, as are the nerve endings in one's posterior. Just ask those who have had to endure a presentation, sitting on metal folding chairs in a hot, smoke-filled room. If your meeting facility doesn't have padded chairs, arrange to rent them during your advance visit. The audience will love you.

Prior to the meeting, pick an able monitor—two if possible. They should occupy the two rear corners of your room. On prearranged cues, they can monitor the sound, check air quality, and raise and lower room lights. Nothing is more unprofessional than for the speaker to begin by asking, "Will someone please get the lights?" Room monitors are important, so use them.

FIRST-AID NEEDED

If you are going to be responsible for road presentations on a regular basis, it will pay you to invest in a "visuals first-aid kit." This lifesaver, which will fit into an average-size briefcase, should include a claw hammer, several regular and Phillips-head screw drivers, pliers, adjustable wire strippers, electrical tape, a box of miscellaneous nuts, bolts and washers, a small roll of utility wire, several different-gauge nails, heavy rubber bands, an assortment of audio plugs and jacks, AC socket adapters, an extension cord, finger bandages and aspirin. Anyone who has ever handled more than one out-of-town A/V presentation will not have to ask what these things are for—especially the latter two.

Presentations on the Road

NEW

....

TOOLS AND TECHNOLOGIES ARE

....

MAKING IT EASIER TO DELIVER POWERFUL PRESENTATIONS

....

AWAY FROM THE OFFICE.

....

In the relatively friendly environment of the office conference room, a presenter can be reasonably sure that equipment will be where it is supposed to be and will function as it is designed to function. Such considerations as lighting, viewing distance, power supplies and room acoustics, as well as the size and composition of the audience, are known variables. But when a presentation is taken on the road, all bets are off.

BY JOHN B. CALLENDER

In the past, mobile presenters often had to count on those at the remote presentation site to provide such presentation equipment as overhead and slide projectors, video projectors and monitors, VCRs and computers. The main advantage to this approach, of course, was that the presenter did not have to transport these items. The main disadvantage was the possibility of an unpleasant surprise at the last minute—promised equipment not being available or not being suitable for the presenter's needs, for example.

Recent advances in presentation technology have done much to alleviate this problem. A wide range of tools are currently available, tools that are both highly portable and easy to use, including everything from compact overhead projectors to laptop computers to dedicated presentation systems. Although these products sometimes lack some of the functionality of their nonportable counterparts, for mobile presenters the trade-off is usually well worth it.

Understated Overheads

For the presenter who wants to use colorful visuals while reducing the risk of technical problems to the absolute minimum, it's hard to beat the combination of overhead transparencies and an overhead projector. Recent advances in presentation-graphics software and color printers have made it easier than ever to produce stunning transparencies on the desktop. If one can count on finding a suitable overhead projector in the presentation location (and overhead projectors are probably the most common projection devices in existence), all that needs to be transported is a thin sheaf of transparencies.

But why chance it? The latest generation of compact overhead projectors represents an attractive alternative. Typically using a light source that is mounted in the projector's overhead arm, these projectors are reflective, rather than transmissive, making them unsuitable for certain applications. On the plus side, they fold into a compact package that is only a few inches high, making them perfect for carrying in the trunk of a car or aboard a plane.

Larry Venable, a management consultant based in Minneapolis, Minn., who is on the road 100 to 150 days each year, is firmly committed to overhead transparencies. "There are a number of advantages to using transparencies," he says. "Overhead projectors are so common that I rarely have to bring one with me. The transparencies fit in my briefcase, and because they are normal size, I can review them and make notes on the frames while I'm traveling. In use, they are more practical than slides; the lights are on in the room,

so I can see the people in the audience and interact with them."

Personal interaction is crucial to an effective presentation, says Venable. "Charts and graphs and other kinds of visual aids are important tools, but that's all they are—tools. They can't make the presentation for me. I make the presentation, showing the audience the material, interpreting it for them, and pointing out trends and highlights. That's why I prefer the overhead projector and transparencies; they offer a comfortable, natural way for me to interact with the audience."

The Safety of Slides

While slide projection represents a significant step up in image sophistication and quality, it sacrifices some of the ease of interaction and on-the-spot editing inherent in overhead transparencies. Capable of being projected for a large audience, and featuring unsurpassed color resolution, slides have long been and continue to be one of the most popular choices for displaying presentation visuals on the road. As with the process of creating overhead transparencies, slidemaking has been revolutionized in recent years by advances in personal-computer technology, as film recorders (the devices used for producing slides from computer graphics) have become more capable and less expensive.

On the delivery side, slide projectors have also been making advances. New random-access 35mm slide projectors eliminate one of the chief objections to slide-based presentations: being locked into a serial sequence of images. With a random-access projector

Reprinted with permission from *Presentation Products Magazine,* pp. 22–30, June 1991.

and a numbered list of the presentation's slides, a presenter can quickly jump to any image in response to a sudden thought or an audience question. More and more hotels and conventions sites now have random-access machines available on request.

Dona Meilach, president of CompuWrite in Carlsbad, Calif., who spends much of her time on the road teaching presentation techniques, prefers 35mm slides for their unmatched quality and reliability. In her presentations, she offers this advice regarding the use of slides: "If people are going to be using slides more than three or four times, I strongly recommend they put them in glass mounts. That prevents the slides from buckling due to the heat of the projector, which can make the images go out of focus."

Meilach also warns presenters to make sure slides are oriented properly in their carousels. "Always double-check one last time before the presentation," she emphasizes, "because you always miss one."

Technology on Trial

Any business presenter who has watched television courtroom dramas may envy the fictional attorneys' abilities to hold a jury's attention while making verbal arguments. In reality, though, real-life attorneys sometimes turn to the business world for examples of how to present information in an effective and interesting manner. J. Ric Gass, a trial attorney with the law firm of Kasdorf, Lewis & Swietlik, S.C. in Milwaukee, Wis., speaks frequently to defense-lawyer organizations on how to use presentation technology to lend visual interest to their courtroom arguments.

Gass travels the coun-

Apollo **Dukane** **Medium**

Portable overhead projectors are of two types: transmissive (center) and reflective. Reflective models have the edge on portability, but cannot be used with LCD projection panels.

try delivering his presentation, which he calls, "Trying Cases Visually." The equipment he brings with him includes transparency makers, overhead projectors, the Varitronics PosterPrinter and Kodak Carousel slide projectors. "Any audience, whether it is a jury or a group of businesspeople, has a limited attention span," he says. "If you have important information to tell them, they will listen to you, but only for so long. Show them interesting pictures, though, and they'll watch forever."

One of the more eye-catching pieces of equipment that Gass uses is Elmo's EV-308 Video Presenter. Consisting of a video camera mounted on a stand, the Video Presenter enables Gass to project live images through a large-screen video projector or color monitor. "Anything

you place under the camera is projected for the audience," he says. The device is capable of projecting photographic prints, slides, blueprints and even three-dimensional objects.

Using presentation technology is relatively simple, says Gass, but one must be willing to spend a certain amount of time becoming familiar with the equipment before facing an audience.

"I've seen very experienced, talented attorneys stand before a jury and speak with no fear at all, but those same people can get near an overhead projector for the first time and become intimidated and go to pieces," he says. "If they would just take five or 10 minutes before the presentation and walk through it, putting their transparencies on the projector and jotting down notes on the frames, they could get over that anxiety easily."

Like many other presenters experienced with the uncertainties of the road, Gass carries an "emergency kit" that includes an extra extension cord, backup bulbs for overhead projectors, transparency marking pens, and numerous other small items that can make the difference between disaster and a graceful recovery when unexpected difficulties crop up. "Mental attitude is important," he advises. "When problems occur, and they will, I'm ready with a couple of quips to put the audience at ease while I work on solving the problem. You can't let that kind of thing fluster you. Instead, just go about fixing things, or if that isn't possible, switch to a backup plan."

Other forms of video technology are also finding their way onto the mobile

Hitachi **Panasonic**

Compact VHS and 8mm videocassette players with built-in LCD screens can be used for one- or two-person presentations. For larger groups, the video signal from the players can be output to a video monitor or projector.

presentation circuit. The latest developments in Liquid Crystal Display (LCD) technology have fostered a host of new display options for portable presenting. Tiny units combining VHS and 8mm videotape players, speakers and small liquid crystal monitors are available from several manufacturers, and are rapidly gaining popularity for small-scale, one-on-one presentations.

At M.D. Buyline, a Dallas firm that helps hospitals and other healthcare institutions purchase major items of medical equipment, the company's 18 marketing directors spend much of their time in the field, helping existing clients and pitching the company's services to prospective ones. Each marketing director carries a Sony Video Watchman, a hand-held device with an 8mm videocassette player and a 4-inch monitor that can be used to view the company's marketing videos. "Our marketing people also carry VHS-format videos

Video presentation stands, such as the Video Presenter from Elmo (above), allow presenters to display 2-D and 3-D objects.

to show on the customer's equipment," says Larry Malcolmson, the company's president, "but the Video Watchman is perfect for small presentations involving two or three people."

LCD technology has also helped reduce the size and weight of video-projection devices. Where large-screen monitors

and high-resolution projectors still require forklifts or extensive set-up time, those who can live with lower resolution can take advantage of a new generation of portable projectors from companies such as Sharp, Eiki, Sanyo, GE, Telex, Panasonic and Hitachi.

The projectors, which are about the size of small suitcases, use small, internal LCDs to project images through a single lens to as large as 120 inches. They typically weigh in at about 25 to 30 pounds and accept a standard video signal. While the best resolution of these devices is far from that of most CRT video/data projectors, some manufacturers have been showing prototypes which offer brighter and sharper images that rival tube-based, three-lens products.

Compact Computing

LCD projection panels are yet another application of LCD technology helping to put wheels under traveling presentations.

Movable Multimedia

The prediction, by now, is a familiar one. Computer-based multimedia, say its proponents, will revolutionize the way people communicate with one another. The use of computer-integrated and computer-controlled video, audio, graphics and text will overshadow more traditional media, paving the way for a new era in which multimedia tools play a major role in the presentation of information. While some may argue about the precise timing of these coming changes, few who have experienced the power of multimedia firsthand will disagree that change is coming.

One area of multimedia development that holds particular promise for mobile presenters is that of compact disc/read-only memory (CD-ROM). Combining small size with the ability to store vast amounts of computer data (up to 600 megabytes) on a single disc, CD-ROM technology seems tailor-made for the computer-based presenter on the go.

Indeed, some mobile presenters are already using CD-ROM technology.

David Kuehn, field sales manager for special products at Warner, Elektra, Atlantic Corp. (WEA), Los Angeles, Calif., says his company, more commonly known for music distribution, has recently become involved in distributing multimedia CD-ROM titles for the Warner New Media division. "One thing we're trying to do is expand the distribution of CD-ROM products beyond traditional software outlets," explains Kuehn. As part of that effort, he recently went on a three-city demonstration tour.

Transporting the equipment needed for the demonstrations was a huge task, admits Kuehn. "I had a Macintosh IIci with a large monitor and an external hard disk, the CD-ROM drive, speakers, extra batteries and a lot of product literature. Getting all that

through the airports was a real headache; I was always worried about things getting broken and, by the end of the trip, the shipping boxes were pretty beat up."

More-portable approaches to CD-ROM presenting are on the horizon, however. A new product from Hitachi, currently selling in Japan but not yet available in this country, is the "CD2," a device that combines a CD-ROM/XA drive and a 16-bit microprocessor with a 5-inch LCD monitor, all in a single compact unit. A similar product from Sony, called the "Data Discman," is also being sold in Japan, although Sony has no plans to market it in the United States until more CD-ROM titles are available here. Toshiba also has a similar product in development.

Currently, those who wish to take CD-ROM technology on the road in this country must lug along a separate computer and display device, although it is possible to put together a fairly portable solution with a notebook or laptop PC, an LCD projection panel and a battery-driven CD-ROM drive, such as the new XM-3300 series from Toshiba.

David Thornburg of the Thornburg Center for Professional Development in San Carlos, Calif., uses this type of system to address an estimated 20,000 educators and businesspeople each year, discussing multimedia technology and methods of learning. "I currently use existing CD-ROM titles in my presentations," says Thornburg, "mostly to demonstrate their power as reference tools. I'm considering having all of my own presentation materials pressed onto a CD-ROM, along with some additional reference materials. Then I could leave that behind after a presentation, in order to reinforce the points I had made while speaking."

—J.B.C.

When used in conjunction with an overhead projector and a portable personal computer, these notebook-size products allow text and graphics to go directly from the central processing unit to the screen and weigh less than 10 pounds.

Personal computers also have shrunk beyond the "portable" or "laptop" designation, down to the point that they now are referred to as "notebook" or even "palmtop" PCs. Notebook PCs are available with large-capacity hard drives and high-resolution monitors; they can also be equipped with onboard modems so that the presenter has access to the very latest company information, a real boon when working in highly competitive, time-sensitive businesses.

Because much of the information that M.D. Buyline provides to its customers is delivered through an online computer system, the company's LTE/286 notebook computers from Compaq are an important part of the marketing directors' presentation equipment. To display the computer information, the marketing directors use a Proxima Data Display LCD panel in conjunction with a standard overhead projector.

"There's no question that our field people appreciate having this kind of technology," says Cecil Kraft, head of M.D. Buyline's Systems Department. Because the company's marketing directors have online access to a constantly updated database, they can be sure

Liquid Crystal Display (LCD) panels provide a convenient, cost-effective (and portable) means of displaying computer-generated data. The LCD panel spectrum ranges from economy monochrome models to high-end, active-matrix units that generate up to 5,000 true colors.

that the information they are giving clients is as timely as possible.

"Those marketing people are some of the most important factors in our company's success," says Malcolmson. "We have people in the home office who are available around the clock to help them deal with problems. It would be ridiculous to hire these incredibly talented field agents, and then have them sit around unable to do their

jobs because of a minor technical problem."

M.D. Buyline is not alone in taking advantage of the power of computer-based presentations. Having the ability to project a live computer image in a remote location has distinct advantages especially in the case of companies that sell software or other computer-related products.

Edward Keeley, marketing manager with Unisys Corporation's UNIX Systems Group in San Jose, Calif., uses the technology on a regular basis. "I do a lot of presentations on the road, showing people what we're working on in terms of new-product development, and I wanted to put a lot of pizzazz into my presentations," recalls Keeley. "Overhead transparencies were nice, but they just didn't deliver the kind of impact I wanted." His solution? Live, computer-based presentations.

With the help of an overhead projector, a portable computer and an LCD projection panel, Keeley now can deliver live, full-color presentations anywhere that has an electrical outlet. In addition, he can transport all his equipment with him as carry-on baggage when flying. "I don't trust airline baggage-handling," he admits. "I feel much more comfortable having everything with me."

Careless baggage handling is not the only way one can lose a sophisticated piece of presentation equipment, as Keeley discovered. "I first put the system together

While they cannot yet produce the same levels of resolution and brightness, portable LCD projectors offer a lightweight and affordable alternative to three-lens, CRT-based projection systems. The smallest of the lot is the Panasonic AG-505 (second from right).

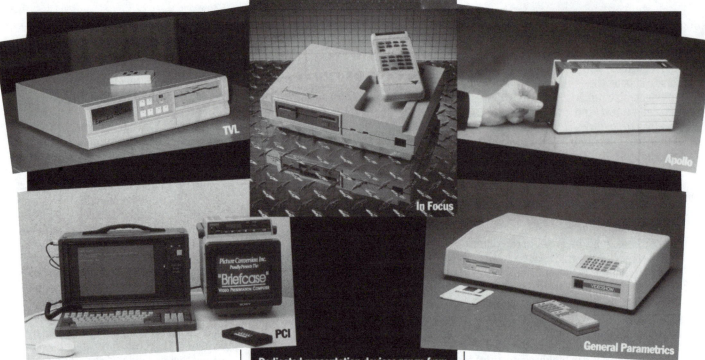

in January of 1990. I did an internal presentation for several Unisys sales reps,
and one of them was so taken with the
panel he talked me into letting him borrow it for a trip to New Zealand. When he
used it there, our New Zealand people
were so impressed that they wouldn't let
him bring it back."

Stephen Rogers, president of Beyond
Software, Portland, Ore., also uses portable
computer projection to communicate complex technical information. "We're a systems integrator," says Rogers, "and we do
a lot of work helping clients take advantage
of computer-based multimedia. Given the
nature of our business, it really wouldn't be
appropriate for me to go out [into the mar-
ketplace] with overhead transparencies
or slides."

Rogers' projection solution features
an In Focus 5000CX LCD panel and an
overhead projector. "Nearly everyone I
give a presentation to is blown away by
the technology," he says. "Most have never seen it before, and if they have seen
it, they didn't realize it could be this
portable."

Traveling Smart

Some mobile presenters may balk at
the idea of giving computer-based presentations for fear of being overwhelmed by
the technology. Such presentations are
certainly becoming easier to create and deliver, but for those who still react with a
shudder when the word "computer" is

mentioned, a number of dedicated presentation systems are available that offer
many of the advantages of computer-based
presenting, while reducing the associated
complexity.

The Briefcase Video Computer from
Picture Conversion, Inc., for example,
combines the graphics and video-pro-
cessing power of a portable computer with
all the necessary features for presentation playback. Composite video, RGB and
S-Video outputs allow users to display
through external computer monitors,
standard television monitors or through
data/video projectors using a compact
remote-control unit. In addition, the Briefcase can be directly linked to videoconferencing systems for use as a graphics
generator.

Another option for on-the-road presentations is to create the presentation
in the home office and store the data on
a dedicated presentation device for later
delivery in the form of an electronic
"slideshow."

Among the products designed for
this approach are the Liteshow II from
In Focus Systems, the Memory Smart
Data Presentation System PC-9950 from
Apollo Audio Visual, the General Para-

metrics VideoShow and the TVL Executive Presenter from TeleVision Laboratories. Wireless remote controls, optional
keyboards and a variety of built-in transition effects allow users of these devices to
be every bit as effective as their laptop-tot-
ing competitors.

Bob Pike, president of Creative
Training Techniques in Eden Prairie,
Minn., has logged an estimated 600,000
miles with his General Parametrics
VideoShow equipment. Pike begins each
presentation by projecting a colorful, animated VideoShow graphic to make his
audiences sit up and take notice. "[The
graphic] sets a tone before I've even said
anything," he says. "The audience will
assume that the rest of my presentation
will be as up to date as that beginning.
I've won some credibility there. I can
still lose that credibility in the course
of the presentation, but that's a better
situation to be in than having to gain
credibility.

"We live in the age of entertainment.
Thirty years ago, people would watch one
channel of black-and-white TV all evening
long without changing stations. Today,
they push a button on their remote controls and say, 'You've got six seconds to
impress me.' They flip through 56 channels and say, 'There's nothing on.' They
go to a video store with 5,000 titles and
walk out saying, 'There's nothing to
watch.' That's the audience we face when
delivering a presentation today." *PPM*

Questions, Answers, And Troublemakers

D. A. Peoples

QUESTIONS

What would you guess is the single most important weapon to use in getting attention, keeping interest, and receiving feedback on how you're doing?

You just read the correct answer. It's the question. Nothing can do so much for so many as the question. And nothing is as effective as the question to give you immediate feedback on the comprehension, understanding, and agreement of the audience. Questions are an essential and integral part of an effective presentation. Not just any question, but well-thought-out, preplanned and prepositioned questions. And the planned question is an important item on your cheat sheet.

You should consider a question for either the introduction of a key point or as a way of finalizing the key point, or both.

For example, in presenting my company's corporate strategy, I introduce a key point by asking the audience whether they think our company is primarily a technology-driven company or primarily a market-driven company. Well, let me tell you, that really gets the juices flowing. Most people had never thought about the company in those terms. Not only do they voluntarily start responding, they start arguing with each other. Nobody ever went to sleep during that question, or the presentation of the material that followed on that subject.

The key to the question is to make it stimulating and thought provoking: questions that call on experience, views, or opinions. Questions that start with phrases like, "What is your opinion of . . .?" or, "What is the first thing you would do if . . .?" or, "What do you think is the cause of . . .?"

What we do *not* want are mundane questions with a self-evident answer of yes or no.

You do want to cause early success with the audience, so you'll want the first few questions to be easy to answer.

The most dramatic and tongue-in-cheek example of causing early success I ever saw was a presenter who asked a member of the audience to pick a number between 1 and 10. He responded with "four." The presenter said, "That's the correct answer." Of course, any answer would have been the correct answer.

Reprinted from *Presentations Plus: David Peoples' Proven Techniques*, D. A. Peoples, Copyright © 1988. Reprinted by permission of John Wiley & Sons, Inc.

On a more serious note, you should ask questions that:

- Relate to the key point you are presenting
- Are clear and concise
- Emphasize one point only
- Reveal the audience's understanding

There are different types of questions and different questioning techniques. Here are some types of questions that are suited for an audience size of fewer than 50 people.

The Rifle Shot Question

This is where you make eye contact with a specific individual, call him by name, then ask the question. This is the most common type of question.

The Time Bomb Question

This is where you ask the question of the audience as a group, then call on a specific person to answer only after you have finished the question. This is an effective technique for getting attention and keeping interest. After a few of these types of questions, the audience will really perk up since they don't know who is going to be called on to answer—and nobody wants to be embarrassed by not knowing what the question was.

The Ricochet Question

This is where you redirect a question that has been asked of you to another member of the audience. This is a good technique if you want a little more time to think about your answer. It is also helpful in assessing the understanding of the audience about the subject. It's effective for audience participation and especially good for handling some types of troublemakers, as we shall see shortly.

The Rebound Question

This is where the presenter rephrases the question and directs it back to the person who asked it. This technique is also good for certain types of troublemakers and tends to reduce or eliminate frivolous questions.

Which type of question is the best?

No single one, but *all* of the above. It is best to have a mix of questions, with the inherent elements of suspense and surprise.

These types of questioning techniques presuppose that you either know the people or they have tent cards to facilitate calling them by name. But what if you don't know the people, and they don't have tent cards? The answer is a seating chart. A seating chart is an $8\frac{1}{2} \times 11$ piece of paper with the names of the people and their relative position in the room.

The simplest and easiest way to get a seating chart is to ask the host or person running the meeting if they have one. Sometimes, but not often, they will.

The next best way is to sit in on the meeting kick-off if the attendees are asked to introduce themselves. Have the room arrangement roughed out with the boxes for each position. Then all you have to do is write their names in the boxes. This has the added advantage that you can also make brief notes about special backgrounds, experiences, skills, and so forth. This will allow you to personalize and tailor your questions to specific individuals.

Another way of getting a seating chart with names is to sit in the back of the room and ask a staff member, or another presenter who is familiar with the audience, to help you fill in the names.

If none of the above works you can still sit in on other presentations and listen to names as they are used by other presenters, or by attendees among themselves before the meeting and at coffee breaks. Using this technique, you can get at least 25 percent of the names. That's all you need to personalize your questions.

In any event, get some names. You want to direct your question to John by name, not by, "Hey you."

Well, all that Q and A is fine for a small audience, but what if it's a larger group? You can't really have questions and answers for a large group, right? Wrong. You sure can have questions for a large group—you just change your style and use a different technique. The key is to phrase the question in such a way that it causes a short answer and provides a hint to the type of answer we are looking for.

For example, we have the sing-a-long.

The Sing-Along Question

Here the question is directed not to an individual, but to the entire audience. Examples: "The bottom line reason we're in business is to make_____" (What?). Or, if I were talking about the relationship between unit cost and volume, I might say, "If I increase the volume the unit cost will go_____" (Which way?).

Another version of the sing-along question is to ask for a show of hands on a subject, an opinion, an experience, and so on.

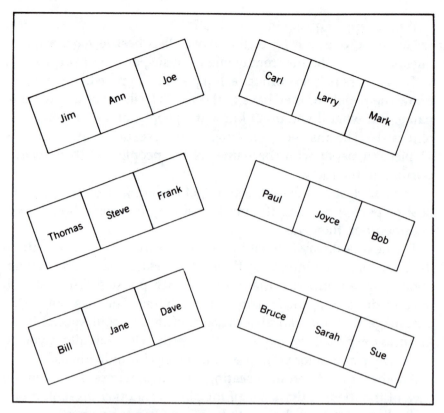

FIGURE 10A A seating chart will allow us to call on people by name.

The Multiple Choice Question

Here we give the audience a hint by suggesting some answers. For example, in talking about lawyers I make the following statement: "You would probably guess that we have more lawyers per 1000 of population than the Japanese. . . . But how many more? . . . Twice as many? . . . Three times as many? . . . Ten times as many?" The key to this technique is to make brief eye contact with multiple people as you call out the multiple choices. At the same time extend your arm in a gesture of directing the question to multiple people.

The Ask and Answer Question

This is a technique where you ask a question, pause, and then answer it yourself. For example, "Of all the stockbrokers in the country, what percent would you guess don't own any stock?" (Pause) "Would you believe 70 percent of all stockbrokers don't own any stock?"

Please, let's not use the mundane "Are there any questions?" If we have done our homework, we will have planned provocative questions that will be of interest to the audience. Answers to the questions will let you know whether or not your audience understands and agrees.

Don't let your mind trick you into false thinking about the value and importance of questions. For example:

- They're more trouble than they're worth.

  Answer: Not true. It's not a lot of trouble. It is a lot of planning. As for its value—there is no known substitute.

- I just don't have the time. There's too much material to cover.

  Answer: You do have the time. You will do much better to cover less material with questions and discussion than more material without.

- I'm afraid I'll lose control.

  Answer: Not if you do your homework and plan the content and position of each question. What we're looking for is a controlled discussion using questions.

- I'm afraid I'll get into arguments and personality conflicts with loudmouth know-it-alls.

  Answer: Not if you follow the rules of how to handle the troublemaker.

- I'm more comfortable just giving a straight presentation.

  Answer: You may be. However, the primary objective of your presentation says nothing about your comfort level. You are more likely to achieve your objective if you have interaction with the audience.

There is no better way to get attention and keep interest than a well-thought-out and well-planned question. It is natural for people to want to know the answer to a good question. Questions will also enhance the comprehension and retention of the material.

Questions are just as important to you the presenter. The quality of the answers let you know how effectively you are communicating your message. Areas of misunderstanding are quickly identified.

If the answers to questions in a particular area are consistently vague, or wide off the mark, then you need to either spend more time in that area or, more likely, think of a different conceptual way of presenting the information—such as using an analogy.

Here is something you will run into when giving a presentation to different levels of people within the same organization. People at the bottom of the organizational chart are sometimes reluctant to express an opinion until they know which way the wind is blowing at the top of the organization. So direct a few early questions to the top of the mountain. Make sure they are opinion questions. There are no wrong answers to an opinion question. That will clear the air for the rest of the climbers. Even better is to get the top man to introduce you and the subject and announce his support and endorsement. That guarantees a successful presentation.

ANSWERS

So far we have talked only about questions you would ask of the audience. But what about the other side of the coin? How do we handle questions that are asked of us?

The first thing you must do is create a penalty-free environment for asking questions. The audience must know that it is safe to ask a question. That there will be no ridicule, no rudeness, and no sarcasm. Never say or do anything that will make the questioner feel stupid or foolish. If in fact they are, let the audience come to that conclusion—not you.

In creating an open and positive environment for questions, our body language speaks louder than our words. For example, if you say, "Are there any questions?" while looking at your notes or down at the floor, your body language says, "I'm not interested in your questions."

If you genuinely want questions it will be evident to the audience. If you don't, it will be just as evident.

To a large degree you can control the number of questions and the content of the questions. Here's how. At the end of a presentation, when you have some quiet time to yourself, flip back through your cheat sheets and make a note of what questions were asked and where. After you have given the presentation a few times, consistent patterns of questions will begin to develop. It's a simple matter then to incorporate the answers into your presentation. That way, those questions will never be asked.

On the other hand, you might plan on not answering some specific common questions. Then, when the question is asked, you appear to be giving a spontaneous answer. But it is actually a carefully rehearsed response. This is a particularly impressive technique if the answer involves statistics, dates, names, places, and so on, which you have rehearsed and noted on your cheat sheet.

Now to some Do's and Don'ts on answering questions.

Listen to the Question

Carefully. Have you ever been in a meeting where someone asks a question and half way through the question the presenter interrupts the questioner and says, "The answer is so-and-so"? Then the questioner replies, "That was not my question."

What happened here is that the presenter assumed that this question was going to be the same as other questions that had been asked before.

Do not make that mistake. Do not interrupt. Even if you are correct about what the question is, the other members of the audience want and need to hear the complete question.

Repeat the Question

If you don't the audience will ask, "What was the question?" If they don't ask it, they will be thinking it. This problem is most common when the question comes from the front of the room. Typically, those in the back of the room can't hear the questions from the front of the room.

In my company when we get a new instructor on board, an experienced instructor will sit in the back of the room for their first few presentations. The experienced instructor will have three pre-printed signs which can be held up as a flag to the new presenter. They say:

"Louder"

"5 Minutes"

"Repeat the Question"

(That tells you the three most common problems with our presenters.)

You want to do more than simply repeat the question. You want to rephrase the question and direct your answer, not to the person who asked it, but to the entire audience. This will enhance the interest of the entire audience in the subject and your answer.

There is a side benefit in this technique. It gives you more thinking time to formulate your answer. We talk slow but the brain is fast. When we rephrase and repeat the question there are a lot of surplus nanoseconds for the brain to organize its answer. So give yourself that extra time. You'll have a better answer.

Eye Contact

Look directly at the person while the question is being asked, but when you answer, break eye contact with that person and direct your attention to the entire audience.

Answering Mistakes

The three most common mistakes in answering questions are:

1. Answering too much. Keep your answers brief and to the point. If you answer a question with a speech, you sure put a damper on other questions being asked.

Long-winded answers are boring, do not help your case, and will cause the audience to tune out.

2. Answering too soon. This is the interrupt problem we talked about that causes you to answer the wrong question.
3. Dialogue with one person. Don't allow yourself to fall into a dialogue with one person. Offer to speak with the individual at the end of the program, break eye contact, and move on.

Don't Bluff

If you don't know the answer or aren't sure—don't bluff. And don't hesitate, either. If you do, your body language says, "I'm not sure of this answer, but I'm going to try it." Instead be prompt to admit that you do not know the answer. If it's important to the subject at hand then volunteer to find out. Make a written note to yourself (that's important). One effective technique is to reserve a page of a flip chart for questions you are going to follow up. That demonstrates interest and sincerity to the audience.

Don't be embarrassed or apologetic about not having all the answers. If you did, you wouldn't be where you are, and you wouldn't be doing what you're doing.

Don't Insult the Questioner

You will if you say, "Your question isn't clear." A better way is ask the person to repeat the question.

TROUBLEMAKERS

Let's Talk about Troublemakers

The first thing we need to realize is that out of every 100 people, there's at least one nut. We need to recognize that we simply are not going to convert a hostile attitude about our company, our product, our service, or a new idea in a one-hour presentation. What we can hope to do is to neutralize the participation and the effect of the hostile troublemaker.

Further, even among fair-minded people, not everyone will want to sing word for word out of our songbook. It's important for us to have a realistic attitude about our audience and our expectations. If we anticipate the troublemaker, we are less likely to come unglued when someone says, "It'll never work." The best way to anticipate the audience is to take the time to get background information on the audience. This will allow you to better understand the opinions, feelings, and biases of those who are likely to disagree with you.

If on the basis of your homework you anticipate disagreement, or if the nature of your subject is somewhat controversial and likely to arouse strong feelings, you should address the disagreement before it addresses you. The best way to ward off trouble is to head it off at the pass. The way we do that is to recognize in our early remarks that there are some other points of view on this subject. Then—and this is important—you state in summary form the other point of view. If you are the one who brings it up, you can explain it in your words and within the context of the view you are going to present. That will burst the balloon of the hostile troublemaker. You have stated his case for him and thereby taken the sting out of his comments. The audience will appreciate the fairness and evenhandedness of recognizing other points of view. The bottom line effect of this is to lend credibility and strength to your case.

We will now categorize the troublemakers and talk further about techniques of neutralizing, defusing, and minimizing their effect.

The Hostile Troublemaker

He is the worst of all. He or she is the one who'll burst out with statements like, "That'll never happen"; "It'll never work"; "I don't agree." His remarks may even take the form of a personal attack on you or what you represent.

One strategy for handling the hostile troublemaker is to persuade the rest of the audience to your way of thinking before the troublemaker can do his damage. The way we do this is to preface our presentation with a remark like the following: "For the next 30 minutes I am going to present a new concept. I would like to ask that we just have an open mind and hold our comments or questions until I finish. Is that all right?" Of course the group will agree that it's all right. Then if the troublemaker tries to interrupt we can merely reference the agreement of the group to hold comments until we finish the presentation of the new concept. If you have followed the Blueprint for Success you will have persuaded the audience to your view. If the troublemaker now makes a statement like, "It'll never work," he will tend to be put down by the audience and viewed as not having an open mind. And rather than you responding to the troublemaker's comment, it's more effective if you let another member of the audience respond. Now his or her disagreement is not with you, but with the rest of the audience. That's like a hammer to the head. The message to the troublemaker is, "Shut up."

Another way of defusing the troublemaker is to use the weight-of-evidence technique. Since you know what you are going to present, you will also know the more common objections. That being the case, you can prepare yourself in advance with facts, figures, references, quotes, and so forth for the common objections. The strategy is to drown the troublemaker with the weight of the evidence you have prepared or collected. He is disadvantaged since he is not as prepared as you are for an intelligent discussion. Your dialogue might sound like this: "You may be right, but let me review the facts and the evidence that supports my position." You can now, once again, turn to the audience for support.

Hostile questions sometimes have hostile words imbedded in the question. Words like rip-off, sneaky, hedging. You can defuse these words by asking for clarification. Do not repeat hostile words when you rephrase the questions. In fact, a truly hostile question that is loaded with emotion should not be repeated. Approach it instead as follows: "I can't answer your entire question. If, however, what you mean by _____ is _____ then my answer is _____." Or, "If what you would like to know is _____, then my answer is _____." Clearly state your position but do not let the interrogator goad you into a debate or an emotional argument. Again, it's far better to ask for input from someone in the audience whom you know does not agree with the opinion of the troublemaker.

Another way of handling troublemakers is to let them destroy themselves. You do this by answering a hostile question with a question. "If you feel that way about the situation, then what do you think should be done to correct it?" The answers to these kinds of questions tend to be recognized by the audience as more emotional than well-thought-out, logical answers. The longer he talks the more the troublemaker hurts himself. It starts to become apparent that he has an ax to grind.

Negative comments or questions are not always hostile. Some people just like to argue or play the devil's advocate. You know the type. You say it's hot, they say it's cold. Their comments or questions tend to be nit-picking, directing attention away from your central point. The strategy here is to get their agreement on the larger point. You can then respond with, "Although we have a difference on the detail, we're in agreement on the concept."

Finally, don't lose your cool. Avoid eye contact with the troublemaker. The more visual contact you have with the troublemaker, the more irritated you will become. If all else fails, just say, "It looks like we have different views on this subject. Why don't we discuss it in more detail after the meeting?" Strange thing. Rarely do they want to discuss the subject after the meeting. They seem to be more interested in a verbal interchange in front of an audience. Guess that tells us something about the troublemaker.

The Know-It-All Troublemaker

This one has a club used to intimidate people. Some types of clubs are:

- Length of service
- Advanced degree
- Experience
- Title
- Professional status

Their remarks are prefaced with:

- "I have a Ph.D. in Economics and. . . ."
- "I have worked on this project more than anyone in the room and. . . ."
- "In my 20 years of experience. . . ."
- "As a senior systems analyst my opinion is. . . ."

The unstated assumption here is that he knows more than you do, hence he is right and you are wrong.

The key to handling the know-it-all is to stick to the facts. Do not theorize or speculate. Stick to your own experiences and well-documented evidence. People can legitimately question and disagree with your theory or your speculation. But they cannot question your experience or documented facts.

Another way of handling the know-it-all troublemaker is to use quotes of other experts whose credentials are even greater than those of the troublemaker.

Let me tell you another approach I stumbled on by accident. Often you know in advance or can find out in advance if you are going to have any know-it-alls in the audience. Arrange a meeting with them in advance. Acknowledge their credentials. Tell them what you are going to present and ask for their support. You will be amazed more often than not to find that they will support and endorse your program. So what started out to be a problem has now become a reference.

Let's take the worst case scenario. Suppose the know-it-all will not support you. You can still take the sting out of his punch by announcing in advance that you and he do not agree and here's why. You are stating his case for him and thereby defusing him.

The Loudmouth Troublemaker

This is the person who talks too much, too loud, dominates the meeting, and seems impossible to shut up.

The most subtle techniques for coping with loudmouths involve your physical position in relation to them. Try moving closer and closer to them while they are talking and maintain eye contact until you are standing right in front of them. Your physical presence—you are standing, they are sitting—will often make them aware of their behavior and they will stop talking.

Here are some other techniques for dealing with loudmouths:

- Interrupt them with the question, "What would you say is your main point?"

- Make eye contact with the loudmouth and say, "I appreciate your comments, but we would like to also hear from other people."

- After a reasonable amount of time ask the loudmouth, "What is your question?"

- Questions that are vague, open-ended, or not relevant can be answered as follows:

  — "I'm not qualified to give you an intelligent answer to that question."

  — "That's a good question, but in the short time we have I would like to stick to the subject of _____."

  — "Interesting point, but how does it relate to the subject of _____?"

- Avoid eye contact and conveniently don't see their hands.

- Ask them to record, take notes, or list questions and "to-do's" for follow-up. (That will keep them busy.)

- Finally, during a coffee break you can recognize their interest in the subject, but tell them you are running behind because of the open discussion, and ask for their support in keeping the discussion down. You could even suggest they jot down questions they would like to discuss with you after the meeting.

The Interrupter Troublemaker

This type starts talking before others are finished. Often, the interrupter doesn't mean to be rude, but becomes impatient and overly excited. Like the loudmouth, the interrupter is afraid that a new, red-hot idea will be lost if it isn't blurted out immediately.

There is a simple and easy solution to the interrupter. Every time they start doing it, jump in and say, "Wait a minute Jim, let's let John finish what he was saying." After you do this a few times the interrupter will get the message.

The Interpreter Troublemaker

They continually want to speak for other people. They can always be recognized by the phrase "What John is really trying to say is . . . ," or "What I hear John saying is. . . ."

The first part of our solution is the same as for the interrupter. If John is still in the middle of talking we want to jump in quickly and say, "Wait a minute, let's let John speak for himself. Go ahead, John, finish what you were saying."

If John has already finished talking, then turn to him and ask, "John, do you think Jim correctly understood what you said? Was his restatement an accurate representation of what you were saying?"

A couple of these will cure the interpreter real quick.

The Gossiper Troublemakers

They introduce gossip, rumors, and hearsay into the discussion. Valuable time can be wasted arguing over whether something is true or not.

"Isn't there a regulation that you can't . . . ?"

"It seems like I remember that. . . ."

"I thought I heard so and so say. . . ."

Immediately ask if anyone can confirm or verify the accuracy of the statement. If they cannot, then give the ball back to the gossiper with the statement, "Let's not take the time of the audience until we can verify the accuracy of the information."

The Whisperer Troublemaker

Nothing is more irritating to a presenter than two people whispering while you are presenting. There are two solutions.

One is to walk up close to the whisperers and make eye contact with them. The other is to stop talking and establish dead silence. When you do, what was a whisper becomes a roar—and an embarrassment to the whisperers.

The Silent Troublemaker

They sit in the back of the room, don't say anything, may be reading a newspaper, rolling their eyes, shaking their heads, crossing and uncrossing legs, pushing their chair back from the table, and so forth. In many ways they are the most difficult of all. At least with the overtalkative participants, you know where you stand. With the silent treatment, you don't know if they understand what you're talking about, aren't interested, are thinking about something else, are shy and unassertive, aren't interested in the subject, don't like you, or what.

The only real weapon you have for silent troublemakers is the Rifleshot Question. Call them by name, then follow with an open question that calls for an opinion, an experience, an example, and so on.

The other thing you can do is talk to them at the break on a personal basis about the subject, their understanding, their questions, their agreement or disagreement, or what have you.

In presentations, as in life, the silent treatment can be the worst. And as in life, there are no magic answers, just a need for patience and tolerance.

The Busy-Busy Troublemakers

They are always ducking in and out of the meeting, constantly receiving messages or rushing out to take a phone call, or deal with a crisis. What's worse, the busy-busy is often the manager or senior person in the meeting. That's why he or she feels so free to come and go. But by doing so, the busy-busy ends up wasting his or her time, and the time of the rest of the participants. During each departure, the meeting may come to a standstill. Or the busy-busy has to be briefed upon reentry. Often there is no point in continuing a meeting if a key person is absent.

There are four ways of dealing with the busy-busy troublemaker.

1. The simplest and most effective way is to hold the presentation on your turf or neutral turf, and not his home ground. That way you remove him from his support systems, and you will be in control of the messages.

2. Another solution is to schedule the presentation either before or after normal business hours.

3. If you have to give the presentation on his turf, then you can announce in advance the time and duration of the break for coffee and phone calls. He will usually get the message.

4. Sometimes you will know in advance that you will have this problem because you know the individual. If that's the case then go to Mr. or Ms. Busy-Busy and tell him or her that you want to schedule the presentation on a date and time they will be able to attend with minimum interruption. Again, they will usually get the message.

The Latecomer Troublemaker

This is a tough one, but here are some thoughts:

- Pick an odd time for the meeting or presentation to start. Don't pick 8:30 or 9:00. Have it announced and publicized that the presentation will start at 8:47. That kind of a start time is a tip-off to the attendees that this meeting is probably really going to start at 8:47.

- You can also have it announced and publicized that a door prize will be awarded at 8:48. The value of the door prize is not important. It can even be a novelty item.

- Make remarks to the latecomer only if it feels natural and comfortable for your body chemistry. And smile when you make them, such as

  "I'm sorry, I must have started early."

  "Are you the one who's giving the lecture on time management?"

- Stop talking and establish dead silence while the latecomer makes his way to a seat.

- Establish a late kitty. Anyone late for a meeting has to put a quarter or a dollar in the late kitty (used for coffee and rolls).

- Announce to the latecomer that she has been volunteered to do some follow-up staff work.

The Early-Leaver Troublemaker

Few things are more disconcerting to a presenter than someone standing up and walking out in the middle of the presentation.

The best way to stop this is to get the audience to agree in advance that it will not happen. You can do this by announcing the time that the presentation will be over, and then asking if anyone has a problem with that schedule. If no one says anything, then we have established a gentlemen's agreement. A potential early-leaver would be pretty embarrassed to walk out after that.

If anyone does state they cannot stay for the entire presentation, he or she will usually volunteer a legitimate reason.

Part 5
Computer Graphics

PRIOR TO the availability of personal computers, the production of graphics for technical presentations was the domain of dedicated art departments. With the proliferation of personal computers and the introduction of graphics software packages, however, the task of producing presentation graphics has been distributed among three levels of users:

1. A relatively small group of specialists with an art/design background, expensive equipment, and extensive professional experience in producing graphics for sophisticated applications, such as sales meetings, stockholder reports, customer briefings, professional society conferences, and video presentations;
2. A somewhat larger group of creative people with some art/design capabilities who lack expensive equipment but are required to produce professional-looking graphics for routine technical applications, such as proposals, reports, design reviews, and presentations; and
3. A broad spectrum of authors who are not artists but have personal computers and presentation software, and want to produce their own visual aids for routine applications, such as technical status reports, internal management and coordination briefings, training sessions, and project proposals.

The capabilities at all three levels are continually being expanded and enhanced, usually starting with the first level and then being refined to make them user friendly for the other two levels. Sufficient computer power and software tools are currently available to allow an engineer with minimal artistic capability to produce usable visual aids. Simply providing the required tools, however, does not turn an engineer into an accomplished artist any more than providing a brush and palette to a house painter makes him or her a Michelangelo. In addition, the cost-effectiveness of using skilled engineering personnel to perform tasks that could be performed more efficiently by talented nontechnical people is questionable.

While some nominal savings may be realized in the initial preparation of graphics at any of the three levels, most of the advantages of computer graphics derive from the ease of incorporating subsequent changes, corrections, and updates. In addition, as a data base is assembled, all or parts of existing graphics can be used as the basis for developing new graphics, resulting in additional time and cost savings. Some computer graphics software also includes the capability to automatically convert numerical data into corresponding graphical representations of various types (e.g., pie, bar, line, and stack charts), resulting in greater accuracy and efficiency than corresponding manual conversions.

The cost-effectiveness of computer graphics for technical presentations is also enhanced by the fact that most current engineering and scientific tasks already involve use of related technologies. For example, many engineering firms routinely use computer-aided design and manufacturing (CAD/CAM) for tasks formerly performed by drafting groups. In addition, engineers, technicians, and researchers in most fields routinely use computers to prepare graphics in reports, proposals, and other similar documents. New applications, such as image processing, pattern recognition, artificial intelligence, robotics, and micromedicine, also depend heavily on computer graphics for success. As a result, many of the graphics originally produced for job-related tasks can easily be adapted for use in technical presentations.

The hardware available to produce computer graphics ranges from powerful mainframes to desktop personal computers. The capabilities of these computers can be enhanced and expanded by a variety of input devices such as special keyboards, mice, light pens, digitizer tablets, image and text scanners, and video cameras, as well as a variety of output devices such as displays of various types and sizes, impact and nonimpact printers, plotters, screen cameras, slide film and microfilm recorders, facsimile devices, and various types of projectors. The associated software ranges from relatively simple packages offering only basic graphic capability to highly sophisticated packages providing three-dimensional, animation, and/or interactive video capabilities. The specific hardware/software configuration selected to produce visual aids for technical presentations should be based on the anticipated graphic requirements, user capabilities, and available funds. These and other issues for using computer graphics to prepare visual aids for technical presentations are addressed in the papers in this part.

The raw data used to produce charts and graphs may express pinpoint accuracy, but the way in which they are presented greatly influences how they are interpreted. To be effective, they must be designed to not only present facts but also to help make a decision, communicate information, or influence someone else's decision. In his paper, "Presentation graphics primer," Steve Lambert defines the basic types of charts used, their primary components, related standards, and the application of computers in producing them. Although the information is oriented toward Macintosh equipment, it is pertinent to most applications involving charts.

To provide a basis for hardware/software selection, the methods of displaying computer-generated graphic images are examined and several alternative technologies are evaluated in the paper "Visual communication technologies for computer graphics" by J. O. Hamblen and A. Parker. The principles of operation for various color display technologies, including screen cameras, film recorders, and various projection devices are described. In addition, several equipment comparison charts are included.

In their paper, "The audiovisual capabilities of computers," Linda Shapiro and Steve Rubin compare the use of computers for technical presentations with film, video, and slide/tape alternatives. The specific applications addressed include graphics, speech, music, and animation, individually and in various combinations. They emphasize the interactive and rapid updating capabilities of computers and describe techniques for using these capabilities effectively.

In his paper, "Driving your point home," Christopher O'Malley describes how to use computer graphics to stimulate interest and foster understanding in technical presentations. While his message is primarily sales oriented, he offers many ideas and techniques that are readily adaptable to technical presentations that are often focused on selling an idea, concept, process, or product. He describes several procedures and tasks for using computer graphics effectively. Although the particular tools described are constantly evolving, his suggestions can significantly enhance the effectiveness of virtually any technical presentation.

In his paper, "Getting started with desktop presentations," Jim Heid presents an overview of graphics software and hardware. He indicates that, while desktop publishing and desktop presentations are similar, some significant differences exist. He reviews the ways software addresses the individual visual elements, the process of sorting and showing visual aids, and various output alternatives. He also compares the technologies used for overheads and slides, and describes software features available to accommodate presentation variations.

While animation is still generally considered to be marketing-oriented and expensive, it is now being used for some technical presentations. In their paper, "Presentations with punch," Salvatore Parascandolo and Kristi Coale describe two software programs that can be used to produce relatively inexpensive animation for technical presentations. These two programs are used in much the same way as draw, paint, and presentation programs. In addition to describing the features of these programs, the paper defines the two basic types of animation, includes a glossary of related terms, and addresses the use of interactive animation.

The specialized topic of animation is further addressed by Jim Heid in his article "Getting started with animation." He defines the concept of animation and presents reasons for using it. He also presents a step-by-step description of animation production using the Macintosh computer and available software. The related procedures and tools can be easily adapted for technical presentation applications.

In his paper, "Authoring, modeling, and animation," Lon McQuillin describes three common computer graphics approaches that are used to produce multimedia presentations. One uses an authoring system, such as HyperCard, SuperCard, or Authorware, as its base for specifying relationships among segments or user-controlled events. The second approach involves the use of basic animation to achieve and control the desired dynamic effects. The third approach extends the use of animation to three dimensions. In addition to presenting the history and characteristics of each approach, McQuillin describes how all three techniques are currently combined in desktop systems to produce presentations with a compelling sense of reality.

Presentation Graphics Primer

STEVE LAMBERT

Abstract—**Programs that enable you to create charts rapidly from raw data have increased in power and ease of use. The Macintosh uses its sophisticated operating system to make complex application programs, such as *Microsoft Chart*, easy to learn and use.**

A CHART is a mixture of art and science. In Bertrand Russell's words, it is a geometric metaphor. The numbers you have to chart may express pinpoint accuracy, but the way they are displayed greatly influences how the numbers are interpreted. An effective business chart is an editorial comment—designed not only to present facts but also to help make a decision, communicate information, or influence someone else's decision. As such, a business chart can express your opinions and point of view.

In more specific terms, a chart is a visual representation of numeric information. When you are using *Microsoft Chart* on the Macintosh, for example, information can be entered manually for each chart or gathered automatically from outside sources such as the Dow Jones News/Retrieval service. Once stored in the computer, numbers can be manipulated in *Multiplan* and passed to *Chart* for final distillation.

BASIC COMPONENTS

A chart used to display numeric information can take many forms. The most common types of charts are column, bar, line, pie, area, and scatter (see "Types of Charts"). Although they differ in appearance, each expresses the relationship between two variables: independent and dependent. *Chart* refers to these variables as the *category* (independent) and the *value* (dependent). A chart is created by plotting the points corresponding to the category/value combinations, called *data points*. This process is demonstrated in Figure 1, which identifies the components of a basic line graph. The box on the left contains all the data points to be plotted; the category is in the left column and the value in the right column. The set of data points that describes the specific event you are plotting is called a *series*. Most charts can display more than one series, which means that they will have several lines or sets of bars.

Column, line, and area charts measure the category along the horizontal axis and the value along the vertical axis (see Fig. 1). Bar charts reverse this relationship, measuring the category along the vertical axis and the value along the horizontal axis. Pie and scatter charts have their own methods of expressing the relationship.

Any chart's usefulness as a communication tool is due largely to the fact that the human mind, through a process known as preattention perception, comprehends a visual representation much faster and retains it far longer than words and numbers explaining the same relationship. This concept is more pleasantly expressed in the familiar proverb, "One picture is worth a thousand words."

CHARTING HISTORY

Just as it is difficult for a ten-year-old of today to imagine a world in which video arcades did not exist, most adults assume that business charts have been with us since the first enterprising ancient philosophers offered their services as business consultants. Graphic representation of numeric information is hardly a new concept, but it is not as old as one might think.

The logic behind this method of communication can be

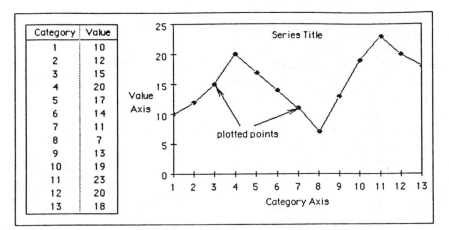

Fig. 1. Components of a line graph. The box on the left contains the series of data points, which are formed by the category/value combinations.

traced to Rene Descartes, the seventeenth-century mathematician and philosopher who lent his latinized name to the Cartesian coordinate system we currently use to plot charts. The idea wasn't exactly a box-office hit at the time—the civilized world was still dogmatically caught up in Aristotelian reasoning, and proponents of "new" ideas were often called upon to support them with their lives.

Not until more than a hundred years later, in the late 1700s, did references to the use of charts in the study of history, genealogy, chronology, and finance start appearing in books. By the early 1900s graphs were in common use. Unfortunately, everyone had ideas as to how they should be created and used; the results were often confusing or misleading. In an attempt to solve this problem, a Joint Committee on Standards for Graphic Presentation was formed in the United States in 1915, in hopes of promoting universal acceptance of graphic methods by establishing standards.

STANDARDS VS. DISTORTIONS

The passage of time has eroded many of the standards established in 1915, and changes in taste and methods of production have further modified them into less stringent guidelines. *Chart* takes these guidelines into account as it automatically offers numeric information in a variety of graphic formats. Whether you choose to plot a basic bar chart or a sophisticated scatter graph, the information you enter will be presented in an acceptable form. You are then free to modify it to meet your special needs.

Chart cannot force you to adhere to graphic standards when you modify the charts you create, but you will find that doing so beautifies your charts and adds credibility by discouraging some of the more blatant editorial tricks. Here are a few examples of what can happen when a person tries a little too hard to prove a point.

It is possible to change completely the apparent value of a graph by varying the length of one axis relative to the other. The three graphs in Fig. 2 plot the same series of numbers: 2a follows accepted standards, but 2b and 2c have been modified to the point of distortion.

A similar distortion is caused by changing the scale on the value axis. This may be even more deceptive because the axes ratio is correct, so the shape of the chart does not warn you to look closely. Figure 3 shows another set of numbers pushed to extremes; the chart titles help perpetuate the distorted point of view that is being passed along.

Of these charts, only Fig. 2a honestly portrays the facts by presenting information in a standard format. The vertical axis is about three-fourths the length of the horizontal, and the scale extends from zero to a value just greater than the highest point plotted.

Sometimes the connotative value of a chart can be changed without actually distorting the truth. By carefully selecting the time period to be plotted, you can give the impression that business is soaring to a new height (see Fig. 4a), rather than climbing out of a pit (see Fig. 4b). There is a fine line between charting the facts in a manner that emphasizes your

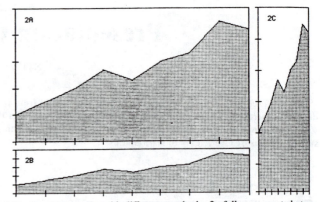

Fig. 2. Line graphs with different standards; 2a follows accepted standards, and 2b and 2c have distorted them.

point of view and distorting the facts to imply something untrue.

CREATING A CHART

The process of creating an effective business graph can be divided into four stages. You must identify the purpose, select a format, gather and plot the data, and check for accuracy.

Every graph should have a well-defined purpose. The same information can be presented in a variety of ways, with implications from positive to negative to nonsense. If you are

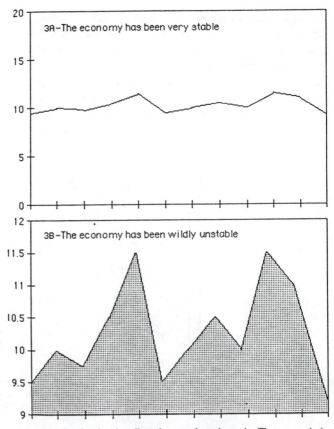

Fig. 3. Graphs showing distortions on the value axis. The axes ratio is deceptively correct, and the chart titles add to the distortion.

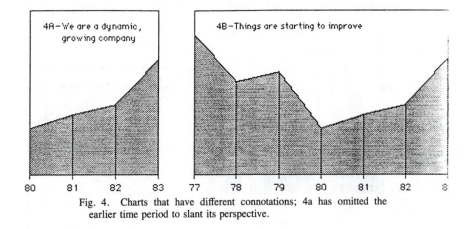

Fig. 4. Charts that have different connotations; 4a has omitted the earlier time period to slant its perspective.

going to take the time to create a graph, make sure that it works for you.

Many graphic formats are available, but usually only one is appropriate for both the type of data you want to present and the audience to whom you are presenting it. Choose both carefully.

Although gathering and plotting data would seem to be the most significant aspect of creating a chart, it really isn't. Once the other decisions are made, this part is fairly mechanical—the Mac does the difficult work for you.

It is amazing how often people will go to the effort of gathering and plotting information and then present their chart without checking to see if the numbers actually got on it correctly. Although this procedure takes only a few minutes, it can help avoid a lot of embarrassment.

Good Taste in Graphics

Many charting standards deal more with taste than with honesty. People judge you, your company, and its products by the quality of your letters and charts. It is to your advantage to stick to the standards in the areas of shading, labels, scales, and legends.

Shading and hatching can be used to attract attention, add interest, emphasize a point, or differentiate between chart segments. *Chart* and *MacPaint* together offer countless patterns, but don't try to use them all on the same chart. If you think you need more than five or six patterns, it's probably time to redesign the chart. Multiple patterns should be arranged in order from dark to light (see Fig. 5).

Titles and labels should be descriptive and short enough to be printed on one line. If you are charting sex in the suburbs, the title of your graph should be "Sex in the Suburbs," not "Courting and Reproductive Rituals Practiced in Outlying Residential Communities."

Your choice of scales should make it easy for readers to interpret the plotted points. Measurements should start from zero, if possible, and extend to a value just greater than the maximum (or minimum if negative) number plotted.

When a graph displays more than one series, each is represented by a different pattern. A legend, as shown in Fig. 6, can be used to relate each pattern to the series it repre-

sents. *Chart* will provide a legend at your request. If you change its size or location, consider the overall weight and balance of the chart.

Fortunately, *Chart* consistently follows these standards. You have to be concerned with them only if you make modifications. Some creative people have become rich and famous by disregarding standards and adding their own imaginative flair to everything they do, while others remain unemployed due to the same flair. Unless you have absolute confidence in your creative genius, you should follow the established guidelines.

Purposeful Business Charts

Charts created for business purposes fall into three broad categories: personal decision-making, peer information, and presentation. The methods used and the standards applied differ for each aim.

Personal decision graphics help you make decisions based on the rapid manipulation of data. A practical method of doing this is to link a chart created with *Chart* to a model created in *Multiplan*. Each time you instruct the Mac to plot a linked chart, it reads the current values of the variables from the *Multiplan* file to which the chart is linked. This process allows you to use the powerful "what if" capabilities of *Multiplan* and to express the results in a sophisticated

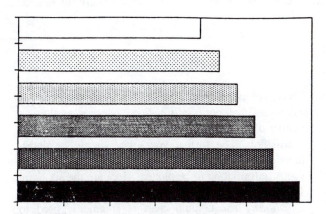

Fig. 5. When you use several patterns, it's best to arrange them in order from dark to light.

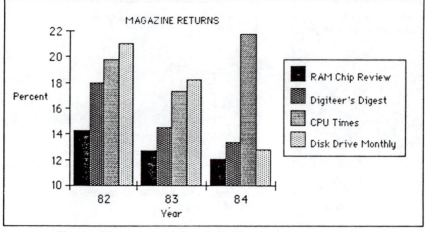

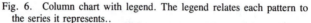

Fig. 6. Column chart with legend. The legend relates each pattern to the series it represents..

graphic format for easy evaluation. Since the chart probably will never leave the Mac (it is only for your personal enlightenment), an unmodified standard format from *Chart's* Gallery menu should suit your needs.

Peer information graphics communicate your decisions and the information they were based on. Charts of this nature are more sophisticated than decision graphics, requiring explanatory text and perhaps legends, arrows, and other enhancements to emphasize the most important points. Although these charts will ultimately be reproduced on the Imagewriter, you can detour them to a word processing program such as *MacWrite* for inclusion in the body of a report. (Because the Mac enables you to cut and paste parts of one document into another, you can easily include charts and illustrations in your reports.)

Presentation graphics are charts of the highest level, designed to prove your point emphatically and impress your audience. With a little practice and some artistry, you can create charts of this quality using the Mac, and for some occasions you can use them just as they come out of the printer. Often, however, when you need illustrations for an important sales presentation or inclusion in an annual report, you will use the Mac to produce rough drafts in a variety of sizes, shapes, and styles.

Special symbols, a logo, or an illustration can be added in *MacPaint,* and once you decide which charts to include in

your presentation, the rough drafts can be given to a graphic artist to use in producing the final version by hand. Better yet, they can be telecommunicated to an agency that can reproduce them in minutes on a high-resolution (over 2000 lines) graphic terminal. Colors, three-dimensional views, or pictorial overlays can be created from your draft, and slides, transparencies, or color photos can be produced and returned to you almost immediately.

KNOW YOUR AUDIENCE

The audience for which your charts are intended has a definite bearing on which format you select and how complex you let the chart become. An audience familiar with graphing techniques and with the information you are plotting can be expected to appreciate complex relationships expressed in a sophisticated fashion, such as a multiple-line graph with logarithmic scales. The same information presented at the stockholders' meeting might be understood more easily if displayed as a series of separate line graphs on different scales.

In today's fast-paced business world, in which busy executives are called upon to make split-second decisions that have long-term effects, the picture generated by a computerized charting program may be worth far more than a thousand words.

TYPES OF CHARTS

Whichever category your chart falls into—personal decision-making, peer information, or presentation—a number of graphic formats are available to present your information. The format you select should be determined by the audience the chart is intended for and the type of information to be plotted. Here are the six basic chart types and a few standards that apply to each.

Column. The column chart is commonly used to show variations in the value of an item at equal time

intervals. You can compare the variations of several items by plotting the values of each on the same chart, either by stacking the resulting columns on top of each other or by placing them side by side. The figure labeled ''Column Chart'' displays the same information in different column formats.

When you plot multiple columns, the order of plotting, the amount of overlap, and the spacing between groups affect the balance of the chart and therefore its effectiveness.

Bar. Although a bar chart looks like a column chart

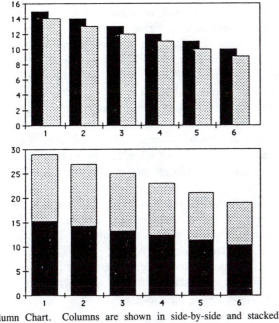

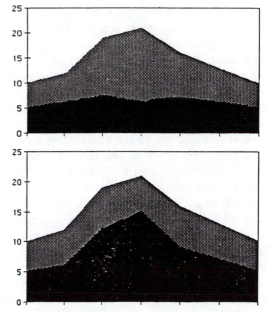

Column Chart. Columns are shown in side-by-side and stacked formats.

Area Chart. The two charts show the same information with the bands arranged differently.

rotated 90 degrees, there are other significant differences. The main difference is that bar charts usually compare the values of different items at a specific point in time. The figure labeled "Bar Chart" shows a multiple-series bar chart in the 100-percent stacked format. In this format the segmented bars extend the entire width of the plot area; each segment's length is proportional to its contribution to the whole.

Line. The line graph effectively presents large amounts of quantitative information in a form that enables readers to recognize trends and relationships quickly. The fluctuation of the line indicates variations in the trend, while the distance of the line from the horizontal axis at any given point indicates a quantity. By plotting multiple lines on the same axes or placing several graphs on the same page, you can evaluate information rapidly and form opinions about cause-and-effect relationships. This type of chart is particularly popular with financial forecasters and other prophets; they often put as many as 20 small charts on

one page, tracking all aspects of a company's financial condition.

Pie. The pie chart is universally recognized and understood. It is the least intimidating of all formats, probably because it is impossible to express a complex relationship with it. Pie charts are used to compare relative proportions of parts that make up a whole. A circular pie represents the totality; and the size of each sector shows its share. The arrangement of sectors allows you to compare them to each other as well as to the whole.

Area. You can create a simple area chart by filling in the space below the line on a line chart—perhaps

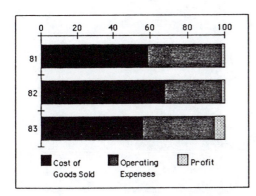

Bar Chart. The multiple-series chart is shown in 100 percent stacked format.

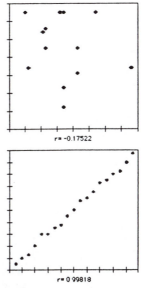

Scatter Chart. The two charts show a low and a high correlation coefficient.

with a pattern from *Microsoft Chart* or an illustration created in *MacPaint*. More complex area charts are created by plotting several series on the same axes. Each series is represented by one band—the thickness of the band at any point indicates its value at that point. The bands are stacked one on top of another, so the distance from the horizontal axis to the top of the upper band indicates the sum of all the bands at that point. The order in which you choose to plot these bands can have a significant effect on the finished product, as shown in the figure labeled "Area Chart," which charts the same information twice with the bands arranged differently.

Scatter. The scatter chart is used to determine the relationship between two variables, which can vary from none at all to very substantial. This relationship is measured by computing the correlation coefficient (referred to as r) for the graph; *Chart* will perform this task for you. The value of r will always fall between -1 and $+1$. A positive value indicates that as one variable increases in value, so should the other. If r has a negative value, you can expect that one variable will decrease as the other increases. The closer r is to plus or minus one, the more direct the relationship between the variables; a value of zero indicates a random association. A typical use of scatter charts is to discover an unusual condition buried in a mass of similar conditions, such as a deviation in the expected return rate of a product when certain production parameters change. The figure labeled "Scatter Chart" shows two scatter charts and their correlation coefficients.

Chart allows you to select a chart format from a gallery of stock variations by pointing to a picture. Your data will be plotted automatically in that format, and you can modify it to suit your specific needs. Combinations of these formats are also possible, both in *Chart* and by sending the individual charts to *MacPaint* and combining them there.

Visual Communication Technologies for Computer Graphics

J. O. HAMBLEN, MEMBER, IEEE

A. PARKER, MEMBER, IEEE

Abstract—Methods of displaying computer-generated graphic images are examined. Several alternative technologies are evaluated. Principles of operation are given for each of these advanced technologies. These color display technologies include screen cameras, film recorders, LCD projection devices, and RGB projectors. A brief survey of commercially available devices is presented.

A PICTURE is worth a thousand words. Studies have shown that people using visual aids in their presentations are perceived as being more professional, persuasive, and effective [1–12]. It is now quite commonplace to develop presentation graphics on a personal computer using a word processor, graphics, paint, or CAD program. With personal computers, it is now possible to produce high-quality images. However, there is more to a successful presentation than large numbers of graphics. The quality of presentation graphics has both esthetic aspects [13] and empirical ones [14, 15]. The need for sound design still remains.

Traditionally, the graphic image has been printed on a dot matrix printer, a multipen plotter, or a laser printer. The color choices and resolution of standard hardcopy technologies are inadequate to display color graphics developed on personal computers [16–18]. Assuming that one has developed effective visual aids using a computer, a number of more advanced technologies can be used to display computer-generated graphics in a presentation. These specialized technologies include screen cameras, film recorders, large monitors, and video projectors. After a brief examination of the common graphics standards for personal computers, a short discussion of each of the display technologies is presented, followed by a survey of commercially available products.

COMMON GRAPHICS STANDARDS

There are several types of video signals (figure 1). Low-resolution color graphics may use composite video, which is based on conventional color TV technology. In composite video, a single signal wire contains all of the color, brightness, and synchronization (sync) information.

Medium- and high-resolution color graphics use RGB signals. In RGB signals, separate red, green, and blue signal

Dr. Hamblen and Dr. Parker are both assistant professors in the School of Engineering at Georgia Institute of Technology, teaching courses in the area of computer engineering and doing research in computer engineering parallel processing. Dr. Hamblen's Ph.D. in electrical engineering is from Georgia Tech; Dr. Parker received his Ph.D., also in electrical engineering, from North Carolina State University.

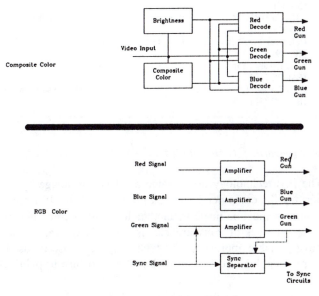

Figure 1. Color Video Signals

wires are used to achieve more precise control of color than with composite video. RGB signals can be digital or analog. Digital RGB signals use TTL (transistor transistor logic) with signal levels of zero and 5 volts. A fourth connection may be required for sync, or this signal may be supplied in the green signal. RGB signals also have different scan rates. For increased resolution, a faster scan rate is required [19].

Several of the common computer graphics display standards are shown in table 1. Currently, on IBM PCs the enhanced graphics adapter (EGA) is the most popular high-resolution graphics standard. The recently announced IBM VGA display standard offers higher resolution than the EGA and resolution equivalent to the Macintosh II.

The AT&T Targa and Vista boards display a large number of colors simultaneously. This feature allows the AT&T graphics boards to be used to display images with three-dimensional appearance similar to standard video images. Systems are available which allow the AT&T graphics boards to capture, display, and modify standard video images [30].

DISPLAY TECHNOLOGIES

Display technologies available to those who want to present computer-generated material are described in the following paragraphs.

Reprinted from *IEEE Trans. Professional Commun.*, vol. 31, no. 3, pp. 135–141, Sept. 1988.

TABLE 1
COMMON GRAPHICS STANDARDS

Graphics Standard	Resolution	Colors	Video
HGA for IBM PC	720 × 348	2	Monochrome
CGA for IBM PC [20, 21]	640 × 200	4	Digital RGB
EGA for IBM PC [22–24]	640 × 350	16	Digital RGB
VGA for IBM PC [25–27]	640 × 480	16	Analog RGB
Color for MAC II [28, 29]	640 × 480	16	Analog RGB
AT&T Targa for IBM PC [30]	512 × 480	256	Analog RGB
AT&T Vista for IBM PC [30]	2048 × 2048	16,000,000	Analog RGB

TABLE 2
CRT CAMERA MANUFACTURERS

Company	Model	Price
NPC Photo Division 1238 Chestnut St. Newton, MA 02164 617–969–4522	Screenshooter [12]	$189
Photographic Sciences 770 Basket Rd. Webster, NY 14580 800–828–6489	DATACAM I [31, 32] DATACAM 35 mm	$545 $875

Photographic Technologies

The first technology investigated produces an image by recording the computer-generated graphic on film. If a presentation must be made in another location, a container of slides or stack of transparencies is certainly the safest and most portable approach. This technology can also be used to produce high-quality artwork suitable for use in publications.

Color slides can be made without additional equipment by placing a camera on a tripod about one foot in front of the center of the computer monitor. Long exposure times (one second or more) require a darkened room to eliminate reflections and a tripod to hold the camera steady. Focus the camera carefully and set the f-stop to provide adequate depth of field so that the entire screen is in focus. If graphics are required on short notice, Polaroid makes a 35mm instant color slide film that can be processed without a darkroom. After some experimentation with exposure time and settings, satisfactory results can be achieved.

Screen cameras are also available which perform this function. These devices have a hood which fits over the front of the monitor to block reflections from ambient light. The hood also holds the camera so that a tripod is not required. Screen cameras are available in several sizes for different screens and film formats (table 2).

The only disadvantages of these photographic techniques are that the sharpness of the image is limited by the computer's monitor, and images are contorted because of the curvature of the monitor screen. This effect is most pronounced near the edges of the screen.

Film recorders can overcome these problems; however, they are expensive. A diagram of a typical film recorder is shown in figure 2. A film recorder contains a flat, small, high-resolution monitor and an attached camera. Three color filters are normally used to produce color images from a monochrome CRT. In these devices, the film in the camera is exposed to three different images, red, green

and blue, to produce a single-color photograph. A film recorder produces these three images by placing the appropriate filter in front of the CRT.

A film recorder requires fewer adjustments than a camera and tripod. Detailed exposure settings for different film types are normally supplied with the film recorder. Many film recorders attach directly to the computer's video output connection to the monitor.

Table 3 lists commercially available film recorders. Before you purchase a film recorder, you must check with the manufacturer to ensure compatibility with the computer's video signals and software.

The Focus Autocam film recorder automatically adjusts to a number of different video signals just like the new multi-sync computer monitors. Extremely high-resolution images are obtained using film recorders. High-end film recorders, such as the Genigraphics Masterpiece, are usually faster and can also use 4- by 5-inch and 8- by 10-inch film to obtain sharper images. The Hewlett Packard 7510 film recorder attaches to the computer's serial port and functions like a plotter. It has a resolution of 16344 by 10986. This resolution is much higher than standard computer video signals; however, several minutes are required to transfer the image data to the film recorder.

Film recorders produce the highest-quality graphic images of all display technologies. High-resolution color images can be recorded on slides, photographs, or transparencies. Photographic technologies allow presentations to utilize standard film projection equipment. The major disadvantage with photographic technologies is that they cannot present interactive or live demonstrations.

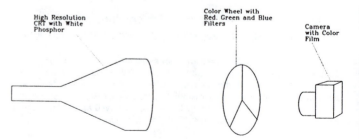

Figure 2. Film Recorder

TABLE 3
FILM RECORDER MANUFACTURERS

Company	Model	Price
Focus Graphics 10 Twin Dolphin Drive Redwood City, CA 94065 415-595-6920	AutoCam Multisync [33] Imagecoder [32]	$2,999 $6,000
Polaroid Corporation 549 Technology Square Suite 25 Cambridge, MA 02139 617-577-2000	Polaroid Palette [31, 35]	$1,799
Lasergraphics 17671 Cowan Ave. Irvine, CA 92714 714-660-9497	Personal Film Recorder	$4,995
Matrix Instrument One Ramland Rd. Orangeburg, NY 10962 914-365-0190	Model 3000 [32] Matrix QCR2	$9,865 $24,995
Presentation Technology 743 N Pastoria Ave. Sunnyvale, CA 94086 408-749-1959	ImageMaker [34]	$4,995
Hewlett-Packard 3000 Hanover St. Palo Alto, CA 94304 415-857-1501	HP7510	$13,900
Dunn Instrument P.O. Box 917 Springfield, VA 22150 703-922-4600	Micro Color 35MM [32] MultiColor 35MM CompactColor	$5,000 $7,500 $14,000
Genigraphics Corp. 4806 W. Taft Rd. Liverpool, NY 13088 315-451-6600	Masterpiece 8740	$55,300

REAL-TIME DISPLAY OF COMPUTER VIDEO SIGNALS

Two additional technologies can be used to display live video signals generated by the computer. A number of software packages that use the computer to simulate a slide projector are available. These technologies can also be used for presentations containing interactive software demonstrations or animation to large groups of people. A computer system with the required graphic software must be located within a few feet of these devices.

A large RGB monitor works well for presentations to groups of less than 20 people. Displaying the computer's video signal allows interactive demonstrations. In addition to a variety of video signal inputs, monitors have different resolution or bandwidth. A monitor can only display as many lines as are generated by the computer's video output. For a clear, sharp image, a monitor must have the same or higher resolution than the computer's video signal. The major component inside the monitor is the color CRT (cathode ray tube) shown in figure 3. The electron beam is scanned over the viewing screen in a sequence of lines. The deflection yoke uses magnetic or electrostatic fields to deflect the electron beam to the appropriate position on the face of the CRT. The information in the video signal is used to control the strength of the electron beam. Light is generated when the beam is turned on by a video signal and it strikes a color phosphor dot or line on the face of the CRT. The face of a color CRT contains three different phosphors: one for each of the primary colors, red, green, and blue. The entire screen is normally refreshed 60 times a second to avoid the sensation of flicker.

The resolution or bandwidth of a CRT or monitor depends on a number of factors. Resolution is normally described as X by Y. X is the maximum number of distinct horizontal lines displayed and Y is the number of vertical lines displayed. Conventional TV signals contain approximately 525 lines. High-resolution graphics may approach 1000 lines. The limit of human vision has been estimated to be in excess of 3000 lines [36, 37].

Resolution may also be described in terms of bandwidth. Bandwidth is a measure of the information content of a signal. A composite video monitor can display a video signal with a maximum bandwidth of less than 7 MHz. RGB monitors typically have bandwidths in the range of 15–20 MHz. Ultra-high-resolution RGB monitors can have bandwidths greater than 40 MHz [38].

Another way to compare the resolution of color monitors is to compare the dot pitch. As shown in figure 4, the dot pitch indicates how closely the color phosphor dots or lines are placed together on the face of the CRT [40]. A dot pitch of 0.60 mm or greater is normally described as low resolution and a dot pitch of 0.40 mm or less is termed high resolution. The next generation of high-resolution color monitors will have average dot pitches of 0.25 mm.

Low-resolution color graphics, such as the IBM CGA, can use large-screen, low-cost TV monitor/receivers with RGB inputs. For higher-resolution color graphics, such as the IBM EGA, PGA, or VGA, special high-bandwidth RGB monitors are required. These monitors are expensive and they typically have smaller screens. Special options, cables, and possibly a video signal converter may be required. Several commercially available large RGB monitors are listed in table 4.

PROJECTION TECHNOLOGIES

Video projection is another technology available to display computer graphics. It uses the computer's video signal to project an image on a large screen. As shown in figure 5, low-end systems use a special projection CRT and a reflective or refractive optical system to project the image [39]. To produce brighter images, many projection CRTs contain a fluid layer to help dissipate the heat generated. In this technology the use of a CRT limits the intensity and size of the projected image.

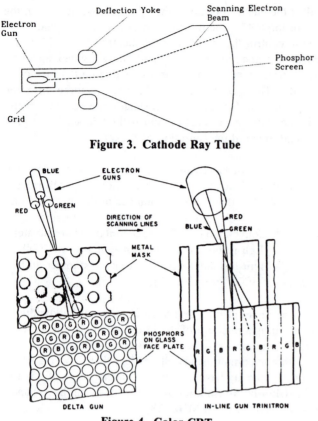

Figure 3. Cathode Ray Tube

Figure 4. Color CRT

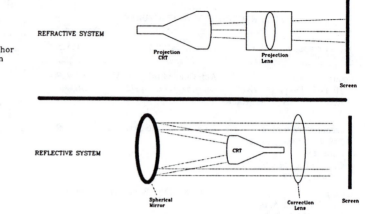

Figure 5. CRT Projection Systems

TABLE 4
LARGE RGB MONITOR MANUFACTURERS

Company	Model	Video	Price
Sony One Sony Drive Park Ridge, NJ 07656 800–222–0878	25XDR	IBM CGA	$1,199
Mitshubishi 991 Knox St. Torrance, CA 90502 213–515–3993	CM-2501	IBM CGA	$1,100
Hitachi 41 W. Artesia Blvd. Compton, CA 90220 213–537–8383	CT2250B	IBM CGA	$700
Microvitec 1943 Providence Ct. Airport Perimeter Pk. College Park, GA 30337 404–991–2246	AUTO-SYNC	IBM CGA/EGA and PGA	$2,195
Electrohome 809 Wellington St. Kitchener, Ontario Canada N2G 4J6 519–744–7111	ECM1910	IBM CGA/EGA	$2,795
Aydin Controls 401 Commerce Dr. Ft. Washington, PA 19034 215–542–7800	8888	IBM EGA	$2,530
Conrac 600 N. Rimsdal Ave. Covina, CA 91722 818–966–3511	7111, 7211, 7241	IBM CGA/EGA	$2,200

Larger and brighter images can be produced using light valves, which are built using an oil fluid layer or a liquid crystal. Light can pass through this layer or be blocked, hence the name light valve. The transmissive properties of this layer are then controlled by using a scanning electron gun, a laser, or small electrodes. As shown in figure 6, a high-intensity light source passes through the light valve and a reflective or refractive optical system is used to project the image [39].

Two portable monochrome video projection systems are Kodak's DataShow and the Vivid Systems' Limelight. The DataShow system is about the size of a notebook. It contains a 640 by 200 array of LCDs (liquid crystal displays), shutters that are driven by the computer's video signals. It is placed on a standard overhead projector to display a blue and white image. The Limelight projector, about the size of a 16mm projector, displays a green image up to 10 feet diagonally. A new color projection system, Telex's MagnaByte II, uses an overhead projector and an array of LCDs. Similar to Kodak's DataShow, this device is small and portable.

Superior to monochrome projection systems are RGB projectors which display large color images on screens using the computer's video output. These systems are unsurpassed for interactive software demonstrations to large

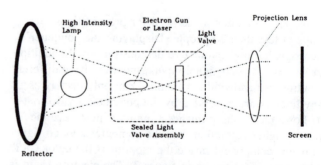

Figure 6. RGB Light Valve Projection System

groups of people; however, they are expensive. The RGB projectors in table 5 range in price from $3000 to $100,000. Typically, RGB projectors with larger, brighter images and higher bandwidths are more expensive.

A number of complicated adjustments are required to focus and align the red, green, and blue images of RGB projec-

TABLE 5
LOW-COST PORTABLE PROJECTORS

Company	Model	Compatibility	Price
Eastman Kodak 343 State Street Rochester, NY 14650 800-445-6325	DataShow [44–46]	IBM CGA RGB or Composite Video (monochrome)	$1,270
Vivid Systems 41752 Christy St. Fremont, CA 94538 800-331-2834	Limelight [41–43]	Composite Video (monochrome)	$3,950
Telex Communications 9600 Aldrich Ave. Minneapolis, MN 55420 800-328-5727	MagnaByte II	IBM CGA RGB (color)	$1,580

tors. However, permanent installations minimize the need for adjustments. Many facilities mount the projector on the ceiling and use a remote control device. If needed, the projector can be placed on a cart and moved to different locations. In a new location, however, focus and convergence adjustments will be required.

The images produced by most of these projection devices are not as bright as those of a standard film projector. Brighter images are obtained with special reflective screens which cost several hundred dollars. These special screens produce an optical gain of 10 or more directly in front of the screen. If the room cannot be darkened, these special screens are very useful.

Video projection technologies provide real-time display of computer-generated graphics. RGB projectors produce the highest quality real-time images; however, they are expensive. If moved, they will require a number of adjustments and a controlled lighting environment.

CONCLUSIONS

A number of alternative technologies are available to display computer-generated graphics. Each technology is supported by several products and manufacturers. There are advantages and disadvantages to each of these technologies.

Presentations or demonstrations using computer-generated graphics can be made more effective by selecting the appropriate technology. Photographic technologies are porta-

ble. Film recorders produce the sharpest and brightest displays. RGB video projection technologies are superior when used for live demonstrations to large groups of people.

The hardware required to support these technologies is costly; however, it can be shared by a number of users. This additional investment is required to take full advantage of the color and graphic capabilities provided by the current generation of personal computers. Ideally, personal computer users who are developing graphics should have access to both a film recorder and a RGB projector.

In the near term, for IBM personal computers, VGA is likely to establish itself as the common graphics standard. As a result, multisync monitors which support analog RGB inputs will become popular. Complete systems for PC-based presentation graphics will appear which support film recorders, projectors, and standard video formats.

At present, graphics produced on personal computers have a flat, two-dimensional appearance. In the future, however, computer graphics will have increased resolution and color selection. Personal computer applications software supporting animation or three-dimensional graphic images with color shading will then appear. The extensive computations required for higher-resolution images and color shading are too time-consuming on the current generation of personal computers. With advances in VLSI technology, the future generation of personal computers will have the computing power required to support the increased graphics requirements.

Thus, as long as esthetic and empirical standards for graphic display are maintained, present and future display technologies will make our presentations more effective than ever.

REFERENCES

1. Prasad, S., "Effective Presentations: A Bibliography," *IEEE Transactions on Professional Communication 30*, 1 (1987), pp. 39–40.
2. Wilcox, R., *Oral Reporting in Business and Industry*, Englewood Cliffs, NJ: Prentice-Hall, 1967.
3. Smith, T., *Making Successful Presentations*, New York: John Wiley and Sons, 1984.
4. Morrisey, G., *Effective Business and Technical Presentation*, Reading, MA: Addison-Wesley, 1975.
5. Mambert, W., *Effective Presentation—A Short Course for Professionals*, New York: John Wiley and Sons, 1976.
6. Kemp, J., and Dayton, D., *Planning and Producing Instructional Media*, New York: Harper and Row, 1985.
7. Howell, W., *Presentational Speaking for Business and the Professions*, New York: Harper and Row, 1971.
8. *How to Make Successful Presentations*, New York: The Research Institute of America, 1986.
9. Holcombe, M., and Stein, J., *Presentations for Decision Makers*, Belmont, CA: Lifetime Learning, 1983.
10. Coulter, C., et al., *Winning Words: A New Approach to Developing Effective Speaking Skills*, Boston, MA: CPI, 1982.

11. Eagleson, D., "Getting Your Graphics Seen," *Systems User 6*, 5 (1985), pp. 5–6.

12. Stahr, L., "Presentation Graphics," *Personal Computing 8*, 9 (1984), pp. 117–121.

13. Tufte, E. R., *The Visual Display of Quantitative Information*, Cheshire, CT: Graphics Press, 1983.

14. Houghton, H. A., and Willows, D. M., *The Psychology of Illustration: Basic Research*, Vol. I, New York: Springer Verlag, 1987.

15. Houghton, H. A., and Willows, D. M., *The Psychology of Illustration: Instructional Issues*, Vol. II, New York: Springer Verlag, 1987.

16. Foley, J. D., and Van Dam, A., *Fundamentals of Interactive Computer Graphics*, Reading, MA: Addison-Wesley, 1982.

17. Enderle, G., Kansy, K., and Pfaff, G., *Computer Graphics Programming: GKS, the Graphics Standard*, New York: Springer Verlag, 1984.

18. Giloi, W., *Interactive Computer Graphics: Data Structures, Algorithms, Languages*, Englewood Cliffs, NJ: Prentice Hall, 1978.

19. Porter, M., "Looking Into Monitors," *Computer & Electronics 22*, 12 (1984), pp. 44–51.

20. Reed, S. R., "Color and Graphics: The CGA Compromise," *Personal Computing 10*, 3 (1986), p. 252.

21. Bellamah, P., "New PCs From IBM Will Feature Built-in Graphics," *PC Week 3*, 2, December 9, 1986.

22. Honan, P., "EGA Boards and Monitors," *Personal Computing 10*, 5 (1986), p. 115.

23. Petzold, C., "Exploring the EGA, Part 1," *PC Magazine 5*, 9 (1986), p. 367.

24. Petzold, C., "Exploring the EGA," *PC Magazine 5*, 13 (1986), p. 287.

25. Cummings, S., "VGA Moves Graphics to the Forefront," *PC Week 4*, 4, July 28, 1987.

26. Rosch, W., "Makers of Compatible Video Boards Set to Confront IBM's VGA Standard," *PC Week 4*, 2, May 12, 1987.

27. Cummings, S., "IBM's VGA Should Increase Market for High-End Graphics Products," *PC Week 4*, 2, July 28, 1987.

28. Brennan, L., "Graphics Software Developers Embrace Mac II," *PC Week 4*, 1, August 4, 1987.

29. Poole, D., et al., "Macintosh II: Opening to the Future," *Macworld 7*, 4 (1987), p. 126.

30. Hansen, R., "Merging Video and Business Graphics," *Computer Graphics World*, November 1987, pp. 52–56.

31. Lincoln, J., "Turn Your Screens Into Slides: Four Screen-Imaging Systems," *A+: The Independent Guide for Apple Computing 3*, 2 (1985), pp. 35–40.

32. Kunkel, G., and Luchak, H., "Screen Cameras: An Output Essential," *PC (Independent Guide to IBM Personal Computers) 4*, 20 (1985), pp. 130–131.

33. McGrath, R., "Innovative Recording," *Computer Graphics World 10*, 7 (1987), pp. 149–150.

34. Rosch, W., "Imagemaker," *PC Magazine 6*, 5 (1987), pp. 298–291.

35. Rosch, W., "Polaroid Palette Merits a Plus for Adding EGA Compatibility," *PC Magazine 6*, 4 (1987), p. 54.

36. Cornsweet, T. N., *Visual Perception*, New York: Academic Press, 1970.

37. McCann, J. J., "Human Color Perception," in *Color: Theory and Imaging Systems*, R. A. Enyard (ed), Washington: Society of Photographic Scientists and Engineers, 1973.

38. Porter, M., "Looking Into Monitors," *Computers and Electronics*, December 1984, pp. 44–49.

39. Todd, L., and Sherr, S., "Projection Display Devices," *Proceedings of the SID 27*, 4 (1986), pp. 261–268.

40. Herold, E., "History and Development of the Color Picture Tube," *Proceedings of the SID 15* (1974), pp. 141–149.

41. Edge, L., "Product Profile: Lightweight Unit Projects CRT Image," *T H E Journal 12*, 2 (1984), p. 12.

42. McClain, L., "Limelight Squeezes Onto Center Stage," *Hardcopy 14*, 1 (1985), p. 98.

43. Alperson, J., "Limelight," *PC World 4*, 2 (1986), pp. 258–259.

44. Rowinsky, W., "Device Lets PC Output Be Viewed via Projector," *PC Week 3*, 34 (1986), p. 128.

45. Poor, A., "LCD for Overhead Projectors Offers Clear Computer Images: DataShow Display System," *PC Week 4*, 4 (1987), p. 59.

46. Alperson, J., "Kodak DataShow Projection Pad," *PC World 5*, 3 (1987), pp. 170–172.

47. Frazier, D., "Larger Than Life," *Digital Review 3*, 2 (1985), pp. 80–84.

48. Birkhead, E., "Movies in the Sky with Inflight," *Hardcopy 13*, 9 (1984), p. 124.

49. Shipley, C., "Large-Screen Video Projectors, PCs Team Up," *PC Week 2*, 23 (1985), pp. 90–91.

50. "Projector Brings Interactive Video to Computer Classes," *T H E Journal 12*, 5 (1985), p. 92.

51. "Large Screen Data Projection for Computer Classes," *T H E Journal 13*, 2 (1985), p. 74.

52. Edge, L., "Product Profile: Video Projection System Is Portable," *T H E Journal 12*, 2 (1984), p. 14.

53. "Projecting Into the Graphics Market," *Electronic Business 11*, 11 (1985), p. 68.

The Audiovisual Capabilities of Computers

Linda Shapiro
Steve Rubin

"LIGHTS! COMPUTER! ACTION!" . . . COMPUTER? The display terminals of computers offer audiovisual capabilities similar to those of film, video, and slide/tape. These capabilities include speech, music, graphics, and animation (moving graphics). Although the quality of sound and visuals on terminals is not as good as that of other audiovisual media, terminals compensate by offering several unique capabilities. These capabilities, which are described here, make display terminals an attractive option to consider when you are going to produce an audiovisual presentation.

The phrase *audiovisual* media brings to mind images of slide projectors, video recorders, and film projectors.

Add computer terminals to this list.

Granted, few people think of a computer terminal as an audiovisual medium. People often mistake today's terminals for their predecessors. Those old terminals had an audio capability of one-note beeps. Their visual capabilities were limited to green letters on a black background. Occasionally, these terminals displayed graphics composed of creatively arranged letters.

Today's display terminals offer a full range of audiovisual capabilities. Their audio capabilities include music and reproducing speech. Their visual capabilities include life-like color graphics, animation (moving graphics), and windows—a special device for emphasizing information. Many of these features are the same as those of video, film, and slide/tape.

Admittedly, today's terminals do not offer the full capabilities of an audiovisual medium. The sound on terminals is tinny and electronic and the quality of the visuals is not the same as that of visuals that were filmed or videotaped.

Terminals, however, offer capabilities not available with video, film, and slide/tape. Terminals offer two-way communication, real-time production, instant turn-around, and electronic distribution—features that make them an attractive alternative to film, video, and slide/tape. This article identifies the audiovisual capabilities of display terminals and describes how to use these capabilities to effectively communicate information.

WHAT ARE THE AUDIOVISUAL CAPABILITIES OF DISPLAY TERMINALS?

Terminals offer not only the standard audiovisual capabilities but a number of additional ones, such as touch and interaction and the ability to update easily. These capabilities are described below.

Audio Capabilities

The audio capabilities of a terminal range from a simple beep to synthesized speech and music.

Beeps. A beep—a simple buzzing noise—attracts the user's attention to the screen. Beeps are used by many application programs. They are audible signals, telling users a message is displayed, for example, or indicating that processes have started, such as printing.

Musical sounds. Musical sounds that can be played by many microcomputers include bee-bops, clicks, and music. These sounds can be used in tutorials and other applications for re-enforcement. For example, a short happy tune tells users of a tutorial that their response to a question is correct; a sad tune tells users their answer is incorrect.

Musical attachments available for microcomputers let you create a wider and more realistic range of sounds, sounds ranging from pouring water to symphonic music. These attachments are increasingly used in commercial television and film production for sound effects and background music.

Speech. A number of devices are available that let computers "talk." One device is a speech-recording and playback device. This device records speech digitally—that is, as bits and bytes. The sounds can be used by computers because computers store all information digitally. The speech-recording and playback device records sound with more precision than traditional recording devices, so the playback sounds just like the voice that was used to record it. Some computerized phone systems use this technology for recording phone messages.

Other devices that let computers talk are speech synthesizers. These devices let users command computers by speaking through a microphone. Computers respond in spoken words. However, problems with the way speech synthesizers handle both input and output operations have kept them from coming into wider use.

There are several problems associated with using speech synthesizers as input devices. For example, consider the variety of accents that need to be interpreted: people in Minnesota pronounce the word "dog" differently than do people in Georgia, who pronounce the word differently than peo-

Reprinted with permission from *Technical Communication*, published by the Society for Technical Communication, vol. 35, pp. 16–22, First Quarter 1988.

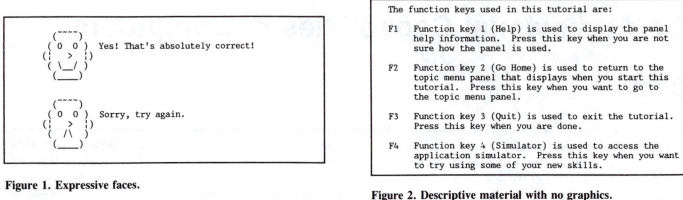

```
   (~~~~)
  ( 0  0 )  Yes! That's absolutely correct!
 (¦   >  ¦)
  ( \__/  )
   (_____)


   (~~~~)
  ( 0  0 )  Sorry, try again.
 (¦   >  ¦)
  (  /\   )
   (_____)
```

Figure 1. Expressive faces.

```
The function keys used in this tutorial are:

F1    Function key 1 (Help) is used to display the panel
      help information.  Press this key when you are not
      sure how the panel is used.

F2    Function key 2 (Go Home) is used to return to the
      topic menu panel that displays when you start this
      tutorial.  Press this key when you want to go to
      the topic menu panel.

F3    Function key 3 (Quit) is used to exit the tutorial.
      Press this key when you are done.

F4    Function key 4 (Simulator) is used to access the
      application simulator.  Press this key when you want
      to try using some of your new skills.
```

Figure 2. Descriptive material with no graphics.

ple in Pittsburgh. Consider, too, the variety of voice tones that needs to be interpreted. Words gain their meaning as much from tonal qualities, gestures, and facial expressions as from the words themselves. Computers don't recognize tones. Word recognition by a computer, therefore, is poorer than that by humans. The rate of word recognition by a computer varies from 67.2% to 96.7%, depending on the device used and the voice selected. The rate of word recognition by people varies from 96.1% to 99.4%.

Similar problems arise with the use of speech synthesizers as output devices. Computer-generated speech is limited to programmable spelling rules and phonetic implementation of those rules. Thus, speech synthesizers have a reputation for speaking in a tin-like, electronic sound, although more recent synthesizers speak in more human-like voices.

Speech synthesizers are already available for a wide range of terminals. As the problems with input and output are addressed, speech synthesizers will come into wider use.

Visual Capabilities

The visual capabilities of a terminal range from text-mode graphics (graphics composed of characters) to animated graphics-mode graphics.

Text-mode graphics. Text-mode graphics are created solely from characters entered at the keyboard. On microcomputers, these are called ASCII graphics, because the graphics are created with characters in the ASCII character set. ASCII graphics include special symbols in addition to letters

and numbers. Many terminals can display only text-mode graphics.

Although text-mode graphics are limited to characters and special symbols, they don't limit your creativity. You can create imaginative illustrations using the box characters, asterisks, underscores, and slashes available in text-mode. These characters can be used to create expressive faces, as shown in Figure 1, or to add interest to descriptive material, as shown in the examples in Figure 2 and Figure 3. You can also use reverse and blinking images, which can add a bit of movement.

Graphics-mode graphics. Graphics-mode graphics are created by coloring each tiny dot of light on the display. These dots are called *pixels*. Several pixels comprise a character. This type of graphics is sometimes called *all-points-addressable (APA)* graphics because you address (color) each dot on the display. Graphics-mode graphics provide more flexibility than text-mode graphics because you are not limited to characters entered at the keyboard. You can draw the graphics, using such devices as a light pen or a mouse.

Only certain terminals let you work with this mode. The terminals are usually called graphics terminals or displays. The quality of the graphics varies among terminals, depending on their resolution. Resolution is determined by the number of pixels on a display. The more pixels, the higher the resolution. Generally, the higher the resolution, the better the quality of the graphics.

You also need special software

for creating, storing, and displaying graphics. There are several application packages available, especially for microcomputers, to perform these tasks.

Graphics-mode graphics require more storage than text-mode graphics. A panel of graphics-mode graphics can use as many as 5000 bytes (characters) of storage. A panel of text-mode graphics uses no more than 1920 bytes of storage.

Last, some computers take a long time to display graphics-mode graphics on the screen. This can be irritating to users, especially impatient ones.

But you must weigh these technical considerations against the flexibility and quality of graphics-mode graphics.

Animation. Animated graphics can move while being displayed. Animation is created for a terminal as it is created for film: a sequence of similar pictures is created in which each successive picture is slightly changed. When shown together at a rapid pace, the changing pictures give the effect of movement. You can create animation with both text-mode and graphics-mode graphics. (See Fig. 4.)

Windows. Windows are displays of information that are smaller than the entire screen and that overlay information already displayed on the screen. Windows "pop up" on the display, so they are sometimes called *pop-ups*.

Windows are helpful in describing applications. You can use the entire screen to display a panel from the application, then overlay a window with a description on part of the panel, as shown in Figure 5.

F1	F2	Function key 1 (Help) is used to
F3	F4	display the panel help information. Press they key when you are not sure how the panel is used.
F5	F6	
F7	F8	
F9	F10	

F1	F2	Function key 2 (Go Home) is used to
F3	F4	return to the topic menu panel that displays when you start this tutorial. Press this key when you want to go to the topic menu panel.
F5	F6	
F7	F8	
F9	F10	

F1	F2	Function key 3 (Quit) is used to exit
F3	F4	the tutorial. Press this key when you are done.
F5	F6	
F7	F8	
F9	F10	

F1	F2	Function key 4 (Simulator) is used to
F3	F4	access the application simulator. Press this key when you want to try using some of your new skills.
F5	F6	
F7	F8	
F9	F10	

Figure 3. Descriptive material with graphics.

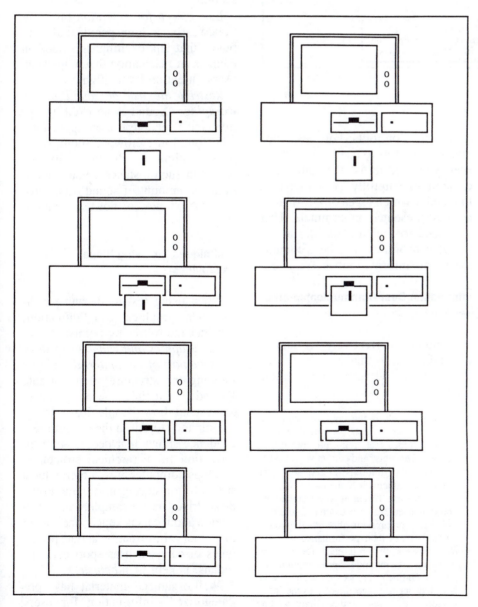

Figure 4. Example of animation using text characters.

Touch Capabilities

Unlike audiovisual media, which work with only two senses, some terminals work with a third sense: touch.

Terminals that offer this capability are called *touch terminals* and are relatively new. They allow users to make selections by touching the face of the screen. One use of a touch terminal is a description of a fair. Users first see a map of the fair grounds. To see the description of a particular exhibit, users find the location of the exhibit on the map and touch that spot. The computer automatically displays the description of the exhibit.

One benefit of touch terminals is that they don't use keyboards. People who don't know how to type aren't intimidated by this type of computer. Another benefit is that "hackers" can't make unwanted changes.

Additional Capabilities

To compensate for poorer quality audiovisual capabilities, display terminals offer many capabilities that aren't available with video, film, or slide/tape:

- *Interaction*. Interaction is two-way communication between users and the computer. Through interaction, users receive feedback to questions and can easily move within the material. When users interact with material, their attention span lengthens. The typical attention span for a computer-based tutorial is 30-45 minutes.
  In contrast, users can only listen and watch film, video, and slide/tape presentations. The attention span for these media is 12-18 minutes.

- *Bookmarks*. When leaving a computer-based presentation, users can set a bookmark to indicate where they last worked. When returning, users can start again where they left off. This is especially useful for lengthy presentations. In contrast, when returning to film, video, and slide/tape presentations, users must "hunt and seek" until they find the spot where they left off. (Viewers can turn off a videotape player without rewinding the tape, but failing to rewind the tape can cause damage.)

- *Modularization*. You can split the material into modules (small units of information), and give users the choice of selecting which module to view and in what order. In contrast, users must view film, video, and slide/tape presentations linearly—from beginning to end, with no breaks.

- *Ability to Update*. Technical material

```
                    Printer Options Panel


   Name of Input File .................. =>  _____

   ┌──────────────────────────────────────┐_____
   │ Enter the name of the file that contains │
   │ the data to be printed in the "Name of   │
   │ Input File" field.                       │ TP1
   └──────────────────────────────────────┘

   Line Spacing (1 or 2) .............. => 1

   Single Sheet Feed (Y or N) ......... => N

   Echo to Screen (Y or N) ............ => N
```

```
                    Printer Options Panel


   Name of Input File .................. =>  _____

   Name of Output File ................. =>  _____

   Printer Designation (LTP1 or LTP2) .. => LTP1

   Line Spacing (1 or 2) .............. => 1

   Single Sheet Feed (Y or N) ......... => N

   Echo to Screen (Y or N) ............ => N

   ┌──────────────────────────────────────┐
   │ Enter Y in the "Echo to Screen" field    │
   │ if you want to see the output displayed  │
   │ on the screen as it is being formatted.  │
   └──────────────────────────────────────┘
```

Figure 5. Examples of windowing.

changes quickly, and changes can be made quickly to information that is presented on a computer. In many instances, no special production capabilities are needed to make the change. In contrast, updating film, video, and slide/tape presentations can be costly and time consuming because entire productions may need to be reshot.

- *Inexpensive production and distribution.* Computer presentations are easy to reproduce on diskette. Computer presentations can also be distributed instantly through electronic mail. The cost of reproduction, in those cases, is limited to the cost of computer time plus diskettes or transmission costs.

  In contrast, film, videotape, and slide/tape production requires hours of studio time. Furthermore, film and video lose quality when reproduced; computer presentations lose no quality when diskettes are reproduced.

- *Self-paced, self-scheduled learning.* All that's needed to view information presented on a computer is the computer and the user. You don't need to worry about the availability of audiovisual equipment, trained technicians, and viewing rooms.

HOW CAN YOU EFFECTIVELY USE THESE AUDIOVISUAL CAPABILITIES?

When you use the audiovisual capabilities of display terminals, these devices become more than electronic page turners; they become powerful communication tools. For these communication tools to be effective, however, you need to use the audiovisual capabilities carefully. You first need to identify which types of information are best presented on terminals. Then you need to find the specific audiovisual capabilities that best communicate this information.

Information Best Communicated on a Terminal

Computers effectively communicate information that needs an interactive visual and verbal presentation. Examples:

- *Tutorials,* such as the ones we wrote. Information most effectively taught on a terminal includes facts requiring rote memorization, concepts, and certain verbal skills. For example, tutorials effectively teach grammar, mathematics, and the use of computer software. Teaching on a terminal lets you quiz users, ask for feedback, and control users' progression through the course on the basis of their performance.

- *References,* such as references for computer operating systems and for programming languages. Online references, as such material is called, let users go directly to the information they want rather than wading through hundreds of pages.

Techniques for Using Audio Capabilities

When using buzzers, music, and speech synthesizers, consider the following points:

Use music to provide feedback (as mentioned earlier). Music is especially popular with children. Be careful, however. Make the happy tunes used for positive feedback short and varied so they remain interesting. Use an unobtrusive, soft beep for the negative feedback—something that won't get on someone's nerves. The same sad tune played repeatedly could be irritating (so could an unfriendly beep or buzzer), especially when someone is trying to get through a difficult section.

Use speech for narration (as mentioned earlier). Some people find computers that talk exciting; this adds an element of fascination that helps keep users' attention levels high.

Restrict the use of sound in the workplace. Sound is an exciting option, but if people use an application with sound in a crowded room, other people might be distracted by the noise. In such instances, you need to limit the amount of sound used, provide ear phones, or not use sound at all.

Techniques for Using Visual Capabilities

When using text-mode and graphics-mode graphics and animation, consider the following points:

Use graphics and windows to show the structure of the material. Why is showing the structure so important? To understand this, you need first to understand how people learn information and store it in their memories. Evidence shows that people store information in hierarchical structures, much as computers use a hierarchical structure for storing information on a disk drive. For example, all information about cars is kept in one *cluster* in memory. Information about specific types of cars, such as sport cars and sedans, is kept in *subclusters.*

Well-organized material has "precategorized" information for users. Users can therefore retain such infor-

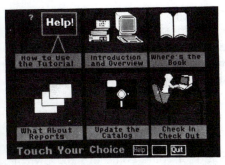

Figure 6. Windows establish information clusters.

mation easily. Audiovisual material—including that presented on terminals–has an inherent structure problem. Most audiovisual media present information sequentially. Users start at a beginning panel and work through to an ending panel. In the process of moving, users might not follow the structure of the information. The sequential presentation of information makes it difficult for users to figure out the hierarchy for themselves. Users can't thumb through an audiovisual presentation in an attempt to figure out its structure. They can't lay the information out before them to get a feel for it. They don't know how big it is or how many categories of information there are.

We used graphics and windows to solve this problem. The overview panel (Figure 6), one of the first in our tutorial, shows the structure of the presentation. The panel contains six information windows. The first window describes the entire presentation. The remaining five windows each depict a different function or work area discussed in the tutorial. This type of panel establishes the first level of information subclusters.

We used *zooming* to establish a second level of subclusters. Users first see an overview picture, then watch as the computer zooms in on a section of that picture. The section is enlarged so it fills either part or all of the screen. We used zooming to show individual processes taught within each section of the tutorial—and how these individual processes fit into the "big picture."

The examples in Figure 7 demonstrate our zooming techniques. This series of pictures begins when users select the "Where's the Book" infor-

Figure 7. Zooming forms information subclusters.

mation window from the overview panel in Figure 6. Then users see

1. A display of a work station
2. Same display with the screen highlighted, establishing a subcluster (panel) within a cluster (work station)
3. Enlargement of the screen display so that only the panel is shown
4. Highlighting of an option to show users which option to select
5. The work station, again, with the keyboard highlighted
6. An enlargement of the keyboard to indicate that the keyboard is used for the next step in the procedure.

This zooming example establishes two subclusters of information: data shown on the screen and information entered on the keyboard.

Use animation to explain abstract concepts. Concepts that involve movement of information are sometimes difficult to comprehend, especially when users must understand when specific elements of information are moved and where they are moved

to. Animation shows this movement. For example, animation can be used to show information moves from main storage to the processor to the display in a computer system. In some cases, you may need to include a verbal explanation of what the animation demonstrates.

We used animation similarly in our tutorial. Users commented that they had never understood these processes until they saw the animated sequence.

Ensure consistency in graphics. Once you choose a picture to display a concept, use the same picture throughout the presentation whenever you refer to that concept. Otherwise, users might think you are talking about something else.

Balance graphics with text. Graphics have certain uses. For example, graphics aid the learning process by giving users a visual image to associate with a spatial concept. Graphics show the relationship between processes (such as data networks) and the movement of information (such as reading in a file from a diskette). Graphics can also show diminishing and expanding elements, such as a stack of dissolving money to show cost savings.

There are certain times, however, when graphics should not be used: for lists, for steps, for command formats, for data protocol, and for items that require specific memory retention. For these types of information, use only text [1]; graphics may distract users from important reading.

Choose colors that can be effectively displayed. Sproull, Sutherland, and Ullner [2], offer some excellent advice to help with this difficult decision:

- Limit the number of colors you use—six is a reasonable maximum.
- Resist the tendency to use fully saturated colors.
- Don't use colors that draw attention to insignificant parts of the screen. Strong, bright colors draw attention.
- Avoid colors that cause problems for people with color deficiencies (color blindness). To avoid problems with users who have difficulty seeing red and green, avoid all but the brightest reds and greens. Be careful, too, about using blue and purple. Some eyes are not sensitive to these colors; for example, people who have trouble seeing red see purple as blue.
- Note that warm colors (red, orange, yellow) usually appear to move objects closer to users; cool colors (blue, purple, green) make objects appear to recede.

Consider, too, that the capabilities of the terminal change the effectiveness of the colors that you use. Color terminals use three electron guns that excite red, green, and blue phosphors on the inside screen surface. Other colors, such as magenta, cyan, and white, are displayed by converging the red, green, and blue beams on top of one another. Some color monitors are better at converging colors than others—for example, the color brown often displays differently on each different terminal.

This color problem is magnified on projector monitors (which project images displayed on a terminal onto a large screen, like large screen TVs). These devices have difficulty with color convergence. Some colors in the projected image may seem fuzzy.

If you are suspicious of the device that will be used for displaying your presentation, avoid colors that are displayed by means of color convergence.

Techniques for Using Touch Capabilities

When using touch, consider that touch screens are useful only for applications that do not require that information be entered from a keyboard. An example of this would be a description of how to program a computerized telephone. The touch screen could simulate the keypad from which information is entered to the telephone.

Touch screens would not be useful for applications that require users to enter information from a keyboard. For example, it would be inappropriate to use a touch screen tutorial to teach how to use a word processor.

Planning for Audiovisual Use

To communicate information effectively through display terminals, you must plan ahead of time how you will use the capabilities in a computer-based presentation to achieve specific goals of communication.

Use a *storyboard* for planning presentations. A storyboard is a tool commonly used for planning film, videotape, and slide/tape presentations. On a storyboard, you design information by pictorially laying it out—panel by panel, beginning to end. Each storyboard describes what graphics and text will be shown, what sounds, if any, will be used, and what interaction, if any, is requested from users.

Review storyboards by hanging them on a wall in the order in which the panels will be displayed. This method emphasizes movement through the presentation and can help you determine where to best place pieces of the information.

When reviewing storyboards, determine whether all topics are covered and covered in an appropriate order.

Correcting informational problems such as these at an early stage can eliminate a great deal of rework, rework that becomes costlier and more time-consuming as the project nears completion.

Prepare and review storyboards before writing a first draft. After completing the project, save the storyboard. It could come in handy when you update the presentation.

WHAT'S THE FUTURE FOR COMPUTERS AS AUDIOVISUAL MEDIA?

The future for computers as audiovisual media is bright. Each year, new products are developed that enhance the audiovisual capabilities of terminals. Products coming into increasing use include the following:

- *Interactive videodisc*, which lets you attach a videodisc player to a microcomputer. The computer controls the videodisc player. The computer can display video images or computer images.
- *Enhanced graphics displays*, which provide crisper pictures than previously possible and in a wider range of colors, as many as 256,000.
- *Scanners*, which let you scan printed images, such as photographs and drawings, into the memory of a computer. Later, you can use the scanned image as it is, or make changes to it.

SUMMARY

Modern display terminals are an alternative to audiovisual media such as film, video, and slide/tape. Computer terminals have a variety of audiovisual capabilities. Although these qualities are not as good as the audiovisual capabilities of film, video, and slide/sound, the reduction in quality is offset by the availability of features that enhance usability, such as touch and interaction. Material presented on a terminal is easy to update, cost-effective to produce, and simple to deliver (because it can be viewed independently), unlike material presented in film, videotape, and slide/sound.

The key to using these capabilities is careful planning. If sound, graphics, touch, and interaction are used to enhance communication and education, your finished product will be well received, as ours was.   Ω

REFERENCES

1. R. John Brockman, *Writing Better Computer User Documentation* (New York: John Wiley & Sons, 1986), 180-181.

2. Robert F. Sproull, W. R. Sutherland, and Michael K. Ullner, *Device-Independent Graphics* (New York: McGraw-Hill Book Company, 1985), 490-491.

BIBLIOGRAPHY

Anderson, Barry F. *Cognitive Psychology*. New York: Academic Press, 1975.

Anderson, John R. *Cognitive Psychology and Its Implications*. New York: W.H. Freemand and Company, 1985.

Bolt, Richard A. *The Human Interface: Where People and Computers Meet*. Belmont, California: Lifetime Learning Publications, 1984.

Blumenthal, Arthur L. *The Process of Cognition*. New Jersey: Prentice-Hall Inc., 1977.

Olsen, Leslie A., and Thomas N. Huckin. *Principles of Communication for Science and Technology*. McGraw-Hill Book Company, 1983.

Olson, Richard, Gregory Foltz, and Barbara Wise. "Reading Instruction and Remediation Using Voice Synthesis in Computer Interaction," in *Proceedings of the Human Factors Society—30th Annual Meeting—1986*.

Pisoni, David B., and Howard C. Nusbaum. "Developing Methods for Assessing the Performance of Speech Synthesis and Recognition Systems," in *Proceedings of the Human Factors Society—30th Annual Meeting—1986*.

LINDA SHAPIRO *is a Senior Associate Information Developer with IBM in Raleigh, North Carolina.*

STEVE RUBIN *is a Staff Programmer with IBM in Raleigh, North Carolina.*

Driving Your Point Home

CHRISTOPHER O'MALLEY

EVERY PICTURE tells a story, but some pictures tell it better than others. The distinction may prove entirely subjective in the realm of, say, impressionist painting, but it's an obvious one in the art of business presentations. Here, a picture can register a concrete return—as in sale or no sale. And it's the professional polish of sharp, lively images that brings the best response from your audience. With the help of high-quality graphics software and output hardware, you can create a presentation with your personal computer that has all the markings of commercial artistry save for the high cost and slow delivery. In the process, you can exercise greater control over form *and* content, making every picture tell precisely the story you intended.

The practice of communication by presentation has a place in many business activities. In preparing a sales pitch, for example, a visual presentation is almost a necessity since you must move your audience to action. If the bait takes and the prospective customer buys, your presentation is a success.

Even without a product or service to sell, putting on a good show can pay off. A polished presentation can generate interest and foster understanding where there was none before, providing feedback that is more subtle and often just as valuable as a sale. In fact, the communication of all but the simplest points can benefit from some visual amplification. Concepts can confuse and numbers can numb without the instant translation that pictures can relate. And in the course of communicating with clarity, you can focus attention on important information.

How can your personal computer help? With the right software and accessory hardware, your personal computer enables you to prepare a slick presentation—from start to finish—at your desk. Not surprisingly, most business presentation products have been geared to work with the IBM PC and its compatible workmates. A genre of graphics software has emerged for these computers that is tailored to the needs of the busy executive. These programs, unlike many "paint" and "draw" packages, supply more than an electronic canvas (with brushes), which can be intimidating without a knack for freehand sketching. They enable you to develop effective presentations even if you have little or no artistic talent. You simply choose, for example, to create a bar chart in red, plug in the required numbers and titles; and the software does the drawing for you. That approach not only makes creating visuals easier, but it makes the process considerably faster. Thus, you can lure clients, customers or colleagues with color and persuade them with graphics even if you're short on artistic talent *and* time. That is a welcome change.

Keeping involved parties appraised of weekly sales figures keeps Robert Eves, president of a real estate development firm called Venture Corp. in Mill Valley, Calif., plotting away at his personal computer. Eves uses bar graphs and pie charts to summarize the week's activities in a "sales status report" he issues every Monday to his staff managers and bank officials. To create the graphs, he sends data from his Multiplan spreadsheets into Microsoft Chart, selects a chart type, adds the needed labels, and then instructs Chart to draw the images on his Compaq Portable 286 (with a color monitor). When he's satisfied with the image on-screen, he puts his Hewlett-Packard HP 7475A color plotter to work drawing the graph. Once completed, the picture is placed amid spreadsheet printouts that spell out the details.

The charts that Eves produces for his weekly status report highlight important sales information that might otherwise be difficult to cull from a table of numbers. Simple bar charts relate the daily rise and fall of customer traffic and housing unit purchases at a specific development project. A line graph overlaying a bar chart compares traffic to actual sales. A pie chart depicts the percentage of housing units sold and unsold, while another pie chart supplies a "buyer profile" detailing the occupations of the purchasers. The high-quality color charts, says Eves, make it easier for his sales staff to identify trends and redress weak spots in their sales pitch. "It's one thing to look at the number of people that pass through one of our projects and then look at the number of people who buy," he notes, "but when you see it in color, jumping off the page at you, you can really see the difference. And it makes you begin to ask questions: 'Was the weather bad that week?' Or, 'Was our advertising approach wrong?' That's the kind of response these reports should encourage."

To encourage that kind of careful review, Eves carefully designs every bar graph and pie chart to emphasize certain aspects of the information he's communicating. The colors he uses, a greenish blue for some bar graphs and a medium gold for others, are hues he finds "optically pleasing" and easy to read. Thinner bars are sometimes outlined in red to "bring them out" of a busy picture. In creating pie charts, Eves says he is careful not to use similar colors or patterns in contiguous sections, so as to make clear the boundaries between areas. He scales graphs for maximum effect, too. By making the vertical axis of a bar graph only about as high as the tallest bar, for example, he can dramatize the differences in weekly figures. "There's a little psychology involved," he admits. "But we want to graphically demonstrate to our people exactly what's happening."

The professional look of his colorful weekly reports, in addition to effectively communicating sales information, also impresses the financial types who bankroll his housing projects, according to Eves. "Presenting ourselves as an orga-

Reprinted from *Personal Computing*, pp. 86–105, Aug. 1986.

nized, efficient operation can be just as important as communicating sales information,'' he explains.

The capability to generate and display graphics is a prerequisite for using presentation software. You'll need a graphics board in your computer. Most graphics programs can work in monochrome with a plug-in board like the Hercules Graphics Card or with the built-in green display of a Compaq portable. But the images that you form can be visibly enhanced with a color display. The standard color cards and monitors for IBM and compatible machines depict four-color graphics that are clear and bright, if a little rough around the edges at a resolution of 320 by 200 lines. The new crop of enhanced graphic adapter (EGA) cards and monitors can clothe the same images in as many as 16 colors with a resolution of up to 640 by 350 lines. Most presentation packages now include the necessary video ''driver'' routines to take advantage of EGA equipment.

Getting the images you create from the computer screen to a medium you can use for presentations is the task of output hardware. These devices range from the ordinary to the exotic. Dot matrix printers can grind out graphics to be included in paper presentations or photocopied onto transparent sheets for use with overhead projectors. Laser printers can do the same with a sharpness that rivals typeset characters and graphics. Color printers and plotters can cover paper graphs and overheads with a variety of hues. Photographic equipment gives you the ability to produce color slides from your screen images. With the proper video equipment, it's feasible for you to keep your presentation ''output'' on the screen. The display of large-screen monitors and televisions, and special video systems may suffice for smaller gatherings.

The ''best'' presentation format for you is the one that most clearly delivers your message within the boundaries of allotted time and money. Certainly, the nature of your audience should influence your decision. Presentations to fellow employees, for example, may demand less formality than a pitch to prospective clients. Portability may be a consideration, too. You can travel light with a set of transparencies, for instance, if there's always an overhead projector at your destination. Company image may be yet another factor. The quality of the visual medium is often associated with the company itself. Your company may have presentation standards (e.g., color slides only) already in place. You may find that a mix of presentation formats, say, overheads for internal meetings and color slides for client presentations, best accommodates varying situations.

Putting your presentation on paper is the easiest, quickest and least expensive way to deliver the picture. It is an especially effective medium in situations where it is inconvenient to gather all interested parties in the same room. Paper presentations can be hand-delivered or mailed to your audience. Paper images can also supplement a more visual medium like overhead transparencies or color slides. A paper graph, unlike a projected image, can be coupled with as much explanatory information (text and numbers) as needed to punctuate key points to staff members or clients. Though

unquestionably less visual than the projected pictures of overheads and slides, paper charts can enliven an otherwise tiresome compendium of characters.

The process of creating paper graphs is straightforward. You create an image with the help of presentation graphics software, like Chart-Master, Graph Writer or Microsoft Chart. You can type in the numbers to be charted or use the program's ''import'' facility to bring in data from files you've generated with other programs. (Many presentation packages can import data from Lotus 1-2-3 and dBase III, for example.) Once the numbers are set, you can label the graph. A printer or plotter handles the rest, etching the graph on paper. Black and white images produced by a matrix or laser printer can simply be photocopied for distribution. You can do the same for color charts drawn by an ink jet printer or plotter, but you'll lose the full impact that color provides. To keep the color intact, you'll have to make the needed copies directly from the color printer or plotter. A photocopier that copies in several colors can help you retain some color, too.

The quality of an image on paper, though sharper than the screen picture, depends on the output device. The graphs churned out by a standard dot matrix or color ink jet printer are adequate for modest paper presentations. A laser printer like the Hewlett-Packard Laserjet, on the other hand, produces exceptional text and graphics—both in black only, however. Color plotters are the best conventional tools for committing pictures to paper, as their color pens draw vivid reproductions of screen images.

Using color graphics to filter through a sea of numbers to get at the bottom line is a crucial task, particularly where numbers represent dollars. Van Spicer, an assistant vice-president at Commerce Union Bank in Nashville, Tenn., converts a pile of financial information on the bank's retail division into a dozen color charts each month. Working from ''a foot'' of fanfold paper emanating from the company mainframe computer, Spicer plugs the key numbers into a 1-2-3 spreadsheet on her IBM PC AT. She next moves these figures into Chart-Master, using the program's ''data grabber'' facility to import numbers from a 1-2-3 worksheet. The numbers are then graphed in Chart-Master and drawn on her HP plotter. The final product is a series of colorful bar charts that compares actual monetary performance of financial services in the division such as consumer credit or business loans to its budgeted targets.

The colorful paper presentation that Spicer creates goes into a ''staff notebook'' for general review, and is sent to the executive vice-president of the division. The latter audience of one, she says, appreciates the concise reporting of critical numbers that her monthly charts provide. ''I'm giving him information, not data,'' she explains. ''The computer printout doesn't mean much by itself. It doesn't tell you what's relevant in all those numbers.''

To speed things along, Spicer uses a number of preset graph formats, or templates, to which she can add specific numbers to produce the charts she needs. A simple line graph that compares actual earnings to budgeted earnings, for ex-

Presenting Graphs on Paper

Unlike the projected images of slides or overhead transparencies, a presentation on paper has only the light of day (or the fluorescent light of an office) to communicate information. But as a complement to slides and overheads, or as a primary medium for communicating to a remote audience, a paper presentation can put your message in the best possible light.

Preparing a paper presentation can be a quick and straightforward task when creating full-page graphics, but sometimes a tricky operation when merging text and graphics. The tools you'll need to get started include presentation software and at least one printer or plotter.

You can quickly create business graphics like bar graphs and pie charts with programs such as Microsoft Chart, Chart-Master and Graphwriter. Typically, you select a graph type and enter the appropriate numbers and labels. (The numbers to be graphed can often be imported from other programs, like Lotus 1-2-3.) The software then does the drawing for you.

The crucial step, of course, is getting the charts you've created from computer to paper. It's a fairly simple job if your charts are to appear on separate pages. A dot matrix printer will suffice for black-and-white graphics, though a laser printer may make the same monochromatic images appear sharper. Most graphics programs support popular matrix printers, like the Epson FX-80 and IBM Proprinter, and at least one laser, the Hewlett-Packard LaserJet.

Once the printing is done, you can duplicate the black-and-white images in your paper presentation with a photocopier. You can etch more vivid images with a color printer or plotter. Again, the most popular color printers and plotters are widely supported, including the IBM Color Jetprinter and plotters from Hewlett-Packard and Houston Instruments.

You can add text to your paper presentation in several ways. You can introduce or supplement full-page images with large text outlines produced with a graphics or "overhead" program. If you want to include more text, you can place full-page graphics before or after pages of copy you've created with your word processor, perhaps cueing the reader with phrases like "as you can see by the chart on the next page. . . ." Though it is more difficult to do so, you *can* keep text and graphics on the same page. You can manually cut and paste images onto a page partially filled with text, and then photocopy the full page.

Merging text and graphics electronically produces slicker results. A few word processors enable you to bring graphics into documents. PFS: Write, for example, enables you to import charts created with PFS:Graph or Harvard Presentation Graphics and print them with text. GEM Write enables you to do the same with GEM Graph or Draw. Some integrated programs, Framework II and Enable among them, have both word processing and graphing capabilities and can mix text and graphics on the same page. If your programs are without such features, you may be able to merge text and graphics with the help of a utility like Inset.

—C.O.

---

ample, is updated monthly. To create a new monthly chart, she adds the most recent figures to her existing line graph and replots the image. The basic framework of the line graph—a box outline with dollar amounts at left and calendar months at bottom—remains the same every month, as do the titles and legend she uses to key the graph.

Presentations to a gathered audience of more than one can have more visual impact when they are projected onto a large screen. The overhead projector is a tried-and-true method of propelling images onto a screen with a beam of light. Overheads are a convenient and inexpensive presentation tool for American business. For groups of four to forty, overhead transparencies are an effective vehicle for delivering a visual message. The overhead medium is also quite portable, as overhead projectors are standard equipment in many businesses, hotels and conference centers. Overhead projection is at its best when employed as an accompaniment to the spoken word, emphasizing important points and charting new trends. Here, the text outlines and translucent graphs can echo and amplify, and help keep an audience's attention focused on your presentation.

The modern technology of personal computers has given new life to the time-honored medium of overhead transparencies. No longer need you wrestle with grease pencils, typewriters or lettering machines (or pay a commercial art shop) to put your message on overheads. Presentation software like Overhead Express, Sign-Master and the GEM Wordchart specialize in enabling you to create slick—and sometimes imaginative—overheads in short order. These programs include a variety of type styles and bullet symbols, plus a handful of preset layouts, that make creating an overhead image as easy as choosing a font and typing in the text. The resultant text charts can be printed on paper with a dot matrix or laser printer, and then photocopied onto transparent sheets for a clean, dark overhead image. You can plot directly on transparencies.

Listening to Greg Moss conduct a session on stress management is much easier for managers at Ex-Cell-O Corp. in Troy, Mich., because he supplements his oral presentation with overhead outlines he prepares with Overhead Express on his IBM PC. Moss, the manager of human resources at the diversified manufacturing company, periodically makes in-house presentations to corporate managers on instructive topics like performance appraisal and interviewing. He says

Creating an Overhead Display

Using an overhead projector to spotlight major points of a discussion is not new. American businesses and schools have for decades been projecting words and images from transparent sheets to screens (or walls) to aid in getting the message across. What *is* new is that personal computer users who are rather short on artistic flair, and even shorter on time, can develop overheads that appear to have been professionally designed and produced.

Besides ordinary transparency sheets, you need only a printer or a plotter and some special software to get started. Any printer capable of etching graphic images will probably do, though popular models like Epson and IBM dot matrix printers or Hewlett-Packard laser printers will work with the least fuss. (Printers that emulate these are fine, too.) With most such printers, you'll need access to a photocopier to transfer the images from paper to transparency sheets. Color printers like the IBM Color Jetprinter, however, can draw directly on these clear sheets. So can color plotters like the HP 7475A model.

Graphics programs like Chart-Master and Graphwriter enable you to produce the lettering and symbols needed for most overheads, but a more specific genre of software for the IBM PC and compatible computers has emerged to expedite the task. Among the most popular of these are Overhead Express, Sign-Master and GEM WordChart. Typically, these programs sacrifice graphic versatility for speed and ease of use. They are not the right tools for creating bar graphs and pie charts, but they are ideal for generating a dozen word charts in under an hour. Such programs enable you to quickly design a page layout, fill it with oversized characters, and punctuate it with special symbols.

Overhead Express, though not as versatile as Sign-Master

and GEM WordChart in some respects, is the easiest of the three to learn and use. Once the program is loaded, you can create a word chart in minutes using its "express editor" function. You select one of the 12 predesigned overhead layouts, or templates, that the program includes to get you started. One of these templates, for instance, is a simple format consisting of a title, subtitle and several bulleted lines. To complete a bulleted chart of your own, you need only type in the text next to sections marked Title, Subtitle and Bullet. You can then save your creation to disk, preview the results on the screen if you have a graphics card in your system, and send the image to the printer. Once printed, you can transfer the word chart onto an overhead by replacing the paper in a photocopier with transparency sheets and copying the paper image in the normal manner. Overhead Express does not yet support color plotters, though Sign-Master and GEM Word-Chart do. The IBM Color Printer and Color Jetprinter are supported.

As with other "overhead" programs, Overhead Express enables you to create word charts from scratch (more or less). In its "custom editor" mode, you can fashion original page layouts or modify the supplied templates. There are four type styles in five sizes with which to fill your page design, as well as graphic symbols like arrows and check marks to denote key points. A single file can contain numerous word charts, which can be printed together or presented on-screen with the program's slide show facility.

—*C.O.*

he uses overheads partly to increase the retention of his message, but mostly to keep attention. "People simply cannot listen to another person talk for more than two or three minutes before their attention begins to wander," he asserts.

Until a year ago, Moss used a Kroy lettering machine to create overheads. Though the character quality was good, he found switching type wheels and repositioning pieces of tape often made producing overheads a tedious task. The characters produced by typewriters are too small, he contends, and commercial services would have cost about $35 per transparency. Having latched onto Overhead Express, he sometimes generates 15 original overheads in under two hours. He puts the word charts on paper with his Epson dot matrix printer and then transfers the images to transparencies on a Xerox copier, which darkens the lettering and fills in much of the dottiness left by a matrix printer.

The process of creating overheads is now faster and more flexible than before, according to Moss, and he maintains that the quality of the images in his presentation is better as well. Moss has ample evidence to support his claim. Evalua-

tion forms submitted by managers who attend his sessions show that his presentation materials rate a "4" on a 5-point scale—up from about "1.5" before he began using a personal computer to create his overheads. "People in the audience are sensitive to the visual aids that are used," he notes. "Good overheads make for more professional presentation overall, and that helps get the point across."

Adding color to overheads can help get the point across even more effectively. Moss alternates between colored transparency sheets to distinguish between screen images. When a color graph is needed, he prepares a grid (with titles) with Overhead Express and then colors in the chart with a felt pen. Some photocopiers can copy in several copies to add a touch of color. As with paper graphs, however, the most stunning use of color in overheads is with charts drawn by a plotter. Some of the "overhead" programs available can draw simple charts, and a few support plotters. If charts are a mainstay of your overhead presentations, though, you'll want a graphics program like Chart-Master or Graphwriter, too.

A full rainbow of colors is particularly advantageous in

The Oversized Presentation

Graphic images and words on standard sheets of paper do little to impress an audience in live presentations. But you can enlarge these images to produce graphs and text outlines that have larger-than-life impact on your audience.

The normal 8.5- by 11-inch graphic printouts generated by most dot matrix and laser printers can be enlarged with some office photocopiers. Such copier enlargements are far from spectacular in size, typically reproducing an image at 115 to 150 percent of its original size. But increasing the size of a bar chart by one-third, for example, may be sufficient for mounting the paper graph on a piece of cardboard or binding it into a flip chart and then making a presentation to a small group. Images drawn by a color plotter can also be enlarged this way, though you'll lose the colors.

In some cases, you may be able to enlarge the image right at your computer. A few business graphics programs enable you to draw larger images on wide-carriage (132-column) printers like the Epson FX-100 than on standard printers. Color plotters are a better bet in this regard, as many popular programs and plotters can hook up to produce vivid color graphs that are larger than 8.5 by 11 inches. The Hewlett-Packard 7475A plotter, for example, can draw both "A" size (8.5 by 11) and "B" size (11 by 17) charts in several colors. With Chart-Master, you need only specify "plotter limits" as the output dimensions to get the largest rendition possible.

Sharp color images of 11 by 17 proportions may prove sufficient for boardroom presentations. For bigger gatherings, you may want a bigger plotter. A select class of plotters can pen images in "C" (18 by 24), "D" (24 by 36) and "E" (36 by 48) sizes. The Houston Instrument DMP-56 plotter can draw images in sizes ranging from "A" to "E," as can the 7580 and 7585 models from Hewlett-Packard. The HP 7586 model can actually plot images of unlimited length (at 36 inches wide) by using a roll of paper for continuous feeding. Generally, computer-aided design (CAD) programs like AutoCAD enable you to drive these plotters to such lengths, but only a few presentation packages fully support these sizes. Chart-Master, for instance, is one of the few programs that supports the "C" and "D" sizes on larger HP models. Justifying the expense of a large plotter for presentations may be difficult, as prices for these devices are in the $9,000 to $17,000 neighborhood. But if expensive plotters can serve professional design needs as well, they're worth looking into.

Commercial art and photography shops offer an alternative to enlarging an image yourself. Color labs, as these shops are typically termed, can usually enlarge images from paper originals or slides. Often, you can have images enlarged up to four by twelve feet. Such labs can also mount the images on illustration board or a sturdier backing like "foam core" board, a Styrofoam-like material that is light but rigid. A mounted image of 16 by 20 inches, enlarged from paper or film, costs about $60—depending on the materials used and the turnaround time requested. Commercial services may be too expensive for a series of images, but a presentation that focuses on one or two graphics, like an intricate diagram or a flow chart, may be worth it.

—C.O.

communicating information on subjects that are less lucid than the transparency film of overheads. The world economy is certainly that, and Edward Asam of Chase Econometrics knows that fact better than most. As marketing director for international services at the Bala Cynwyd, Pa., firm, Asam pitches a wide range of economic planning and forecasting services to government agencies, major banks and large corporations. In presenting Chase's services to these prospective clients, he puts colorful graphs on overheads, as do many of the economic analysts at the company. Asam uses Sign-Master to create text outlines and Graphwriter to create color charts on his Compaq Portable 286. An HP plotter then draws the images on transparencies. The added emphasis of sharp color graphs helps him to visually identify the needs of a potential client and communicate the benefits that Chase's consulting services can provide.

Overhead presentations are a very flexible medium for live presentations, says Asam, since you can resequence transparencies as needed to address the concerns of a specific audience. Overheads also can be produced quickly, and at minimal expense. Nevertheless, Asam insists, the overheads at Chase Econometrics cannot appear slapdash. "We're a professional organization and we have to maintain an image of quality in making presentations to our clients," he says. "A hand-sketched chart or a dirty overhead doesn't lend any credibility to our services."

Though photocopiers can darken characters created by matrix printers, and color printers can pigment a pie chart, the resultant overheads are not as bright and smoothly rounded as those created with laser printers and color plotters. To keep overheads looking sharp, Asam and others at Chase routinely use color plotters to draw images directly onto the transparencies. If the overheads are for internal view only, or if a presentation to a client must be arranged hastily, Asam may use his HP LaserJet printer to produce sharp black-and-white images on paper and then photocopy them onto transparencies. A dot matrix printer may be useful to preview your work in some instances, but many "overhead" programs enable you to preview images on the screen of your monitor.

Applying Color for Impact

When communicating with color, *when* and *where* you apply certain hues can make all the difference between an effective and ineffective presentation.

Every presentation has its own requirements, but there are basic guidelines for using color for maximum impact. Andrew Corn, a partner at Admaster, Inc., a New York City design and production agency specializing in business presentations, suggests the following elements of design in creating a presentation:

Maintain good contrast. On paper, from a color printer or plotter, that means using dark text and brightly colored graphics set against a light background, probably white. A plotted chart with black text and blue bars, for example, will separate clearly from a white paper background. Through projected light, from a slide or overhead or monitor, text and background colors should be reversed. Use light text against a dark background, though black is often too dark. Again, use bright colors for graphic images. A color slide with white text and yellow bars set against a medium blue, for instance, will provide good contrast.

Choose appropriate colors. Certain colors can trigger associations that may not be desirable. Yellow is usually considered a positive color, while red is often considered negative—as in the red ink of financial losses. Blue and green are typically considered ''safe'' color choices. Don't mesh a lot of red and green, unless the topic is related to Christmas. Steer clear of pastels like pink and lavender, except for beauty products.

Fill charts with bright, solid colors. Patterns like diagonal stripes and crosshatch can be very effective in black-and-white graphs, but are generally unnecessary with colors. Bright, solid colors are generally more pleasing to the eye, and easier to distinguish. Fill the bars and pie slices in your charts with bright blues, greens, yellows and the like. Also, use shades of the same color when you want to denote that items are related but should be examined separately. A pie chart with four slices of graduated blue shades, for example, might better explain the operating costs of one department than might distinct colors. Make sure that shades are different enough to distinguish them from one another, however.

Spare the colors, don't spoil the picture. Color can begin to lose impact when many pigments appear in the same image. Use no more than five or six colors in one graphic for most business charts. If you use more, the image must be carefully arranged so as to avoid ''gum ball machine'' effect. An elaborate design with many colors may overshadow the message. Preserve clarity by working with a limited palette and distancing similar colors like yellow and gold.

Accentuate with color. Use bright colors like white and yellow to emphasize text or key portions of a graphic. Use darker or lighter colors to anchor the remaining pieces.

Adapt color to your environment. Factors like lighting and audience size should influence your use of color. Light colors are fine for darkened rooms, for example, but you should use bolder hues in brighter settings. Also, a small audience may pick up on subtleties in color, but larger groups require color differentiation that is very clear.

—C.O.

Perhaps the medium with the best mix of versatility and polish for live presentations is the color slide. Slides enable you to present text and graphics in the refined splendor of many colors. Slide images typically are suitable for large and small meetings alike. With a carousel projector, you can orchestrate an automated show from the set of slides you create, giving your presentation a professional air. And slides are very portable—even if you have to carry your own projector.

Customized color slides, however, have long been luxury items for many business people. Ordering color slides from a commercial art shop is an expensive, time-consuming procedure that sometimes yields inexact results. Commercially produced slides typically cost $30 to $40 apiece and take several days or more to generate from scratch. Elaborate images can cost more and take longer. And if the results aren't what you expected, then you'll have to shell out more money and wait again. An in-house art department might allay some cost and turn-around concerns, but even that fortunate circumstance doesn't always allow for last-minute additions and subtractions. Keeping presentations timely and under budget, therefore, often means opting for overhead transparencies or paper graphs instead of slides, or foregoing a visual presentation.

A personal computer and the right accessory hardware can change all that, enabling you to produce color slides quickly and at a reasonable cost. Special photographic equipment can be connected to your personal computer to capture screen pictures onto film, which can then be developed into color slides. The most popular of these devices is the Polaroid Palette, which uses a miniature display of its own and a 35-millimeter camera to enhance and then record images on film. Most presentation programs now have the necessary software drivers to work with the Palette. You can create ''instant'' color slides with the Palette by using Polarchrome film and the included accessories for developing and mounting the slides. Or you can use a more traditional 35mm film, like Ektachrome, to produce sharper slide images that can be developed commercially. In either case, the cost per slide is a small fraction of what art shops charge. Depending on the frequency and size of your presentations, a slide-maker like the $2,000 Palette can make an initial investment pay handsome returns—financially *and* visually.

As primary providers of statewide health care services, managers at Blue Cross/Blue Shield of Rhode Island feel a responsibility to keep costs down, and they offer (for a fee) to tell other companies how to do the same with medical insurance claims. Color slides they create with a personal computer and a Polaroid Palette help them do both. Terry Fleming, a supervisor in the company's "utilization review" department, uses the amplifying power of graphics on color slides to show large Rhode Island companies where medical monies are being spent and where costs might be contained. In the process, he is containing administrative costs within his own company by producing slides with the Palette, a color system he helped implement.

In developing a service called CUE (cost utilization evaluation), Fleming sought a presentation format that could convey to company executives recommendations on curtailing costs that are derived from a ream of statistics on insurance claims. Overheads were used previously, but he found the images less than impressive and the manual switching of images "too cumbersome." Color slides moving along a carousel projector promised slicker showmanship, but the services of commercial slide-making shops were "too expensive"—particularly for a company pushing cost containment. The Polaroid Palette proved to be the solution. Instead of paying $10 to $40 per slide to have the images produced by an outside slide service, Fleming can, with the Palette, create slides at a cost of about 30 cents apiece.

Fleming and others at Blue Cross headquarters in Providence channel their slide requests to a staff member who operates the Palette with Graphwriter, Sign-Master and Freelance software to recreate the images to be photographed. Slides can be developed and mounted in a matter of minutes with Polaroid instant film if necessary. More often, the company's slides are snapped with regular 35mm film (Ektachrome 100) and sent to a local photo lab for developing. A 24-exposure roll of film costs about $3.25 and developing the film as mounted slides costs only $3.80 per roll. A complete set of slides can be turned around within 24 hours.

The Palette's payoff at Blue Cross has been more than just monetary. Presentations are smoother and more concise than with overheads, notes Fleming. "It's the same information, but it's packaged much better," says Fleming of the multicolored bar charts, pie graphs and tables that help summarize the findings of Blue Cross analysts. "It's forced us to present our information more succinctly, and our clients come away with much clearer impressions of the measures we're proposing for containing costs."

Indeed, presentations with color slides can be a great deal more than cost-effective. They can elucidate intricate information. At St. John Hospital in Detroit, relating complex ideas and techniques to physicians, nurses and residents is a task that begs for visual amplification. Marilyn Wayland, the director of clinical research and curriculum planning at St.

John, coordinates much of the visual support given to hospital lectures and professional conferences with the help of her IBM XT and the Polaroid Palette. The cost per slide dropped from $40 to 50 cents; and a local photolab returns developed, mounted slides within 24 hours.

But the key difference, she notes, is that color slides are now available—instantly if necessary—to virtually anyone on staff who needs to present research data to medical colleagues. And ready access to slides, she explains, means that hospital presentations have more impact than before. "Color slides help an audience understand what the research and statistics mean," says Wayland. "They're not only an audio-visual aid, they can put information in a format that people can comprehend."

Despite the promise of cost reductions and slicker presentations, Wayland says she was hesitant at first about the purchase of slide-making equipment, fearing she would become the hospital's "audio-visual graphics technician." In fact, the task of preparing color slides for the entire hospital *does* rest with the computer and Palette in her office. But she finds the process of creating a graph with 35mm Express software and then transferring it to film with the Palette easy enough to teach others how to do it in about an hour. The system is also sufficiently versatile, says Wayland, to accommodate a wide range of presentation needs. Staff members often create simple text outlines and number tables to accompany a lecture, she explains, as well as bar graphs and pie charts to communicate the statistical findings of medical research and studies. Though it is a considerably slower process, staffers with an artistic bent can create "freehand" renditions of anatomical drawings to visually demonstrate medical techniques.

The informal, in-house meetings at St. John demand the bulk of the color slides produced with the Palette, and audiences seem more attentive now that slides have largely replaced paper charts and the blue-and-white diazo photography commonplace in medical institutions, Wayland reports. At the annual conferences of professional organizations like the American Medical Association and the American Public Health Association, the feedback on visual presentations is more direct. The findings of research projects conducted by Wayland and other hospital members are communicated to such groups with slides and are judged on content *and* presentation. Hospital members, she notes, consistently receive high marks on presentation with the color slides they create with the Palette. "Audiences today are more sophisticated than they used to be," Wayland observes. "They've come to expect colorful graphics. If they see a hand-drawn overhead or a sketchy chart, they're going to lose interest."

While the Polaroid Palette does a commendable job of rendering in-house slides, it is by no means the only such equipment available. Simple screen cameras like the Datacam 35 deliver less vivid images than the Palette, but also cost considerably less. At the other end of the cost and quality spectrum is Matrix PCR, a large and expensive ($11,000 and up) film recorder that can produce incredibly sharp color images.

In some situations, the very best color slides may be worth the investment. At U.S. Trust Company of New York, which oversees the considerable trust funds of many wealthy families, corporate image is an important asset. Presentations to clients and to the company's board of directors, says senior vice-president Peter Korndorfer, have to be first-class operations. Korndorfer went through stints with mainframe graphics systems, commercial artists and photo laboratories before investigating in-house slide makers. He examined the Polaroid Palette and other film recorders but found their picture resolution lacking—until he discovered the Matrix PCR unit. Impressed with its brilliant images, he recommended purchasing the Matrix recorder. The complete Matrix PER system costs about $15,000 but Korndorfer says he expects the unit to pay for itself in less than two years. More importantly to U.S. Trust executives, the company's slide shows are state-of-the-art presentations. "Nobody wanted to step down from the quality of professional services," explains Korndorfer. "And frankly, $15,000 didn't give anyone here great pause."

A relatively new method of producing color slides, one which combines commercial services and in-house preparation, may better suit many budgets. Some commercial graphics outfits allow you to send completed electronic images by modem to be made into color slides and then returned by mail. By doing half the work, creating the finished image, you can usually reduce the cost of commercially prepared slides by 50 percent.

Helen Anbinder, director of seminar services at Spar, Inc., in Tarrytown, N.Y., teams up with a company called Magi-Corp Slide Service Bureau in the nearby town of Elmsford to cut the cost of her presentations to consumer products companies. Anbinder creates graphs and text charts with 35mm Express software on her IBM AT to counsel clients on the best use of sales promotions aimed at retailers. She inputs text and numbers into 35mm Express to create the desired images, previews the results on her dot matrix printer, and then sends the graphics files to MagiCorp for processing. The 35mm Express makes sending images to MagiCorp as easy as selecting a menu choice. Normally, MagiCorp mails the final product to the sender the next day, but Anbinder is close enough to pick them up herself or arrange for someone to do so. The final cost is $17.50—exactly half of the $35 charge the company was paying for slides before the electronic messengering began. The turnaround time is half of what it was as well. Anbinder appreciates the cost and time savings but also relishes the newfound sense of precision that comes from creating the images herself.

"I have much more control over my presentations now," she says. "By doing it myself, I can make sure it's exactly what I want."

If methodically stepping through slides or overheads is still not as visually exciting as you'd like, a video presentation with moving pictures may be the answer. Piecing together an animated show with your personal computer is more difficult than creating a series of single images, but the right combination of software and hardware can make the process manageable for the non-artist. And while keeping your presentation on-screen involves more work up front, it eliminates the final step of putting your images on paper, transparencies or film.

Some graphics packages, Execuvision and Harvard Presentation Graphics among them, enable you to simulate a slide show on-screen by moving from one image to another. Artfully done, such a presentation can amount to more than an electronic flip-chart. Many programs with slide show capabilities enable you to exercise a variety of special effects, like moving images within a frame or fading pictures in and out of view to sequence the show. A few programs, including PC Storyboard and Show Partner, specialize in creating simulated slide shows. A normal computer monitor may be large enough to display these images for an audience of one or two; a large-screen color (RGB) monitor from the likes of Sony or Toshiba may be sufficient for as many as 10 to 20 people. If the presentation software you're using works well with composite video output, as PC Storyboard does, you can use a television as your display and even record the show on videotape for use with a VCR. Larger gatherings will require a screen at least as big as those used with overhead and slide projectors. A standard projection TV or special computer projection equipment like the Sony Videoscope can beam the images onto a bigger screen.

Using a color television (or projection TV) to display your presentation wares can yield bigger images, but the quality of your pictures may suffer. The composite video signals used with TVs generally produce pictures that are not as sharp as RGB circuitry. Display equipment for computers is your best bet.

A unique ensemble of equipment called VideoShow is perhaps the best purveyor of moving pictures and on-screen presentations. VideoShow is, essentially, a briefcase-size box of video circuitry that can create high-quality images from a palette of 1,000 colors when connected to a color monitor or video projector. You can add attachments to produce color slides and paper graphs, too. You use your personal computer to create the images with any presentation software that includes support for the VideoShow product (Microsoft Chart does, for example) and store the graphics on a floppy disk. You then put the disk into the VideoShow unit and control the show with its wireless remote control. Like the best slide makers, VideoShow is a considerable investment at $3,500 and up. But for frequent video presentations, it can pay off.

Van Spicer, in addition to plotting paper graphs, recommended that Commerce Union Bank purchase VideoShow for live presentations. The VideoShow system she wanted cost nearly $8,000 with software and slide-making accessories. Though she was "taken aback" by the cost at first, she figured the demand for high-quality visuals and the volume of presentations at the bank would make the system worthwhile. She convinced a "very conservative" chief financial officer that this would be so, and she now uses VideoShow to create

Creating Colorful Slide Shows

Color slides are fast becoming the presentation medium of choice in American business: Slides are—and have been for years—a portable, colorful, and generally vivid means of communicating images to an audience. But while vacation shots snapped with a hand-held camera can be easily developed into slides for home viewing, the process of capturing bar graphs and pie charts onto film for business presentations can be lengthy and costly. With graphics software and some special accessories, however, you can quickly and inexpensively put business graphics onto color slides.

Photographic equipment is now available that can be connected to your personal computer to transfer graphic images and text onto film, which can then be developed into color slides. Essentially, these devices are of two varieties: screen cameras and film recorders. Screen cameras capture images onto film exactly as they appear on a computer display. Film recorders draw images on film without regard to the display.

Screen cameras typically shroud a computer monitor with a fitted hood so a 35-millimeter camera at the far end of the hood can snap a picture of the display. The Datacam 35 ($875) is one such screen camera. It includes a Minolta 35mm camera and a connecting plastic hood that fits snugly on IBM color displays and similar monitors. Once it is so fitted, you merely snap a picture of the screen image and send the roll of film to a photo lab for developing and mounting. What you see on the screen is what you get on film, and that is a mixed blessing. The standard IBM color display (CGA) is not exactly vivid at 320 by 200 lines in three colors. You can improve the image on the screen, and thereby improve the image that appears on the slide, with newer EGA display cards and monitors. Even so, screen images often appear sketchy and slightly distorted on film.

Though the quality of color slides produced with a screen camera is sometimes mediocre, the process is fast and easy. The film and development costs for 24 slides is about $7.

Like screen cameras, film recorders capture images created with a personal computer to film with a 35mm camera. But film recorders process images electronically, instead of simply snapping a picture of the display, and can therefore surpass the resolution and color palette presented by a standard computer display. These recorders are linked to the computer through a serial port or a plug-in adapter card and, like a printer or plotter, can be made to produce images that are far more detailed than those on the screen.

A number of film recorders are available. The Matrix PCR is the most elaborate and expensive of these, a system costing nearly $14,000 that is capable of producing color slides of extraordinary quality. Bell and Howell's Color Digital Images occupies a middle ground at about $5,500. The Polaroid Palette may be the least expensive film recorder at $1,999 and is certainly one of the most popular. The Palette, a unit of shoe box proportions with a camera attached to one end, enables you to produce instant or processed color slides and prints. A Minolta 35mm camera is included, as is a Polaroid camera for instant prints. The unit is linked to the computer via a serial cable. Inside the Palette is a color wheel and, behind the wheel, a tiny black-and-white monitor. The images that the Palette photographs are drawn on this internal display with instructions from the computer and snapped through the staged filtering of the color wheel. As many as 64 colors from a palette of 72 are available. Using Polarchrome instant film in the 35mm camera and an included "auto processor" kit, you can develop and mount your own slides. The Palette's instant color slides are quite good; but conventional 35mm film will generally yield sharper results.

You can create the images and drive this inventive array of photographic hardware with most popular presentation programs. Chart-Master, Graphwriter, Microsoft Chart, 35mm Express, GEM Graph, and Harvard Presentation Graphics all have necessary routines to work with the Palette. To create a color slide with Chart-Master, for example, you choose Polaroid Palette from its menu of output devices.

The Palette can actually be used as a screen camera of sorts by cabling the unit to the composite (RCA-type) port of a color graphics card inside your computer. The Palette enhances the screen images before shooting them, but the slides are not as sharp as those recorded from software.

—C.O.

slides for, among other things, monthly meetings of branch managers. Spicer is set to move from slides to moving pictures, however. The bank is purchasing a Sony projection system with a 10-foot screen to do it all through animated video. "Slides have been our primary medium for presentations, and that was costing us $60,000 to $70,000 a year," she notes. "When you look at that cost, VideoShow begins to look cheap."

Clearly, all four of the major presentation formats—paper, overheads, slides and video—can communicate information in timely, lively fashion with a personal computer in your corner. Putting the medium you choose to its best use takes less artistic sense than you'd expect, but still requires a keen sense of your audience and the information you're attempting to communicate. With that perspective in hand, your personal computer empowers you to make presentations that put your message in the best light possible.

Producing a Video Presentation

In an age so influenced by the moving pictures of television and cinema, it's not surprising that an animated presentation can have maximum impact on an audience. But a conventional video presentation—as in lights, cameras, action—is beyond the scheduling and budgetary reach of many business people. With your personal computer and the right accessories, however, you can quickly create a fluid blend of moving images right at your desk.

Imparting the virtue of motion into a series of still images for an on-screen presentation is not as difficult as it sounds. Many business presentation programs include a facility for simulating a slide show on the screen. Harvard Presentation Graphics, for example, enables you to sequence the individual graphic images you've created and then display them in order, cueing screens manually (with a keystroke) or automatically (with a specified pause). Overhead Express can do the same for word charts. Many such features do little more than move from one image to another, but that may be enough to keep an audience attentive.

Not all on-screen slide shows are created equal. Some presentation programs are more amply geared to on-screen delivery, offering a variety of special effects that can enliven an otherwise unvarying sequence of images. ExecuVision, for instance, has long included a ''motion'' command that enpowers you to move text or graphics across the screen in different directions. Its new sister program, Concorde, has considerably more options, enabling you to employ a range of ''dissolve'' and ''fade'' techniques for moving from one screen to the next. Concorde also includes a ''library'' of predrawn images and animations, as well as a collection of canned music to accompany the presentation. Both Concorde and ExecuVision are fully capable of helping you prepare hard-copy items like slides and paper graphs, too.

At least two programs specialize in the art of on-screen presentations: PC Storyboard and Show Partner. They both include functions that enable you to draw an original image, capture a screen image from another program, edit these images into a flashy visual script, and run the show from a single disk (without loading the program). While the drawing functions in these programs may prove uninviting to stick-figure artists, the ability to capture screens created with other programs means you can use a product like Chart-Master to generate the business graphs you need for a video show. It's in the editing process that these images, regardless of their origin, become an animated presentation. Special effect commands allow you to make the transition from screen to screen quite inventive, and you can direct images on a single screen to move in various manners.

For small audiences, your computer monitor may serve adequately as a presentation screen. Show Partner, for example, supports standard color (CGA) and enhanced color (EGA) displays. For audiences of more than two or three, you'll need a larger display. Several manufacturers offer large color monitors, including Toshiba, Sony and Sanyo. There are video projectors available, too, which can be cabled to the computer to thrust images onto a big screen (or wall). For a more conventional large display, you might consider moving your video presentation to a television screen, and perhaps to a videocassette recorder for later viewing on a television set. To do so, you must convert the signals emanating from the composite video-out port (the round, RCA-type connector) on your graphics card to the radio frequency signals that a television or VCR can accept. By cabling the composite port to a radio frequency (RF) modulator—which is available at any electronics store—and then attaching the modulator to the television or VCR antenna screws, you can send converted video signals to a television screen. A projection-screen television can give your show a larger-than-life appearance.

VideoShow, a unique graphics system designed for personal computers, offers what is probably the most sophisticated set of video presentation tools. The briefcase size VideoShow box is a computer in itself, complete with an 8086 microprocessor and IBM-compatible disk drive. At $3,500 and up, it also has a computer-like price tag. You can create the images to be displayed with VideoShow's PictureIt software or with programs that support the VideoShow unit, like Microsoft Chart or Freelance. Then, by inserting the floppy disk with your creations into the VideoShow drive and attaching the unit to a standard color monitor or a video projector, you can display brilliant images in up to 1,000 simultaneous colors. Like slide show software, VideoShow includes special effects to move the presentation along, which you can pace with a wireless remote control. VideoShow's primary strength, however, is in the quality of its pictures. Its onscreen resolution is 2048 horizontal lines by 484 vertical lines—a near-picture-perfect setting for a presentation.

—C.O.

Getting Started with Desktop Presentations

Out to win friends and influence people? Presentation software can help.

by Jim Heid

Now and then, everyone dons a Willie Loman outfit and becomes a salesperson. Some sell products, others sell ideas and concepts. Whatever the wares, the steps are the same: you gather your facts, shine your shoes, and present your argument.

The Mac won't shine your shoes, but it can help with the rest of the process. *Desktop presentation* software can help you refine your ideas and create presentation aids such as slides, overhead transparencies, and audience handouts. The whole process smacks of desktop publishing, and indeed, there are many parallels between publishing and presenting, but there are also significant differences. This month, we take you on a tour of the world of presentation graphics software and hardware.

PRESENT OR PERISH

Desktop publishing implies a permanence to your work: you're preparing documents that will be printed and kept—at least for a while. With presentation graphics, however, your efforts are more transitory: each visual is seen just briefly, then it's gone. Because of their fleeting nature, it's important to create visuals with impact

and to plan your presentation so that your message sinks in.

Another key difference between desktop publishing and presentation graphics

lies in the output media. With desktop publishing, your efforts rest on paper. With presentation graphics, your results are usually projected onto a screen, the most common types of output media being 35mm slides and overhead transparencies. The Mac itself is another medium: using hardware I'll discuss later, you can project Mac screen images onto a large screen.

Combining text and graphics is something all Mac word processors and publishing programs do with ease. So do you really need a presentation program? The truth is, if you make only one or two presentations a year, you can probably get by with a word processor or publishing program, or even with a drawing program such as MacDraw II. But if presentations are a regular part of your job, you'll come to rely on the specialized features of a presentation program.

SOFTWARE TO PRESENT BY

Word processors and publishing programs are generalists; presentation programs are specialists: their text-editing and graphics-manipulation features are geared specifically toward producing presentation materials. Toward that end, most presentation programs play three primary roles: they help you develop and refine your ideas, create visuals, and structure and deliver your pitch. (For a comparison of the top programs, see "Picking the Best Presenter," *Macworld,* May 1989.)

When you're first developing a presentation, you need to organize and reorganize your ideas on the fly. Built-in *outlining* features, found in More II and Aldus Persuasion, help you do just that. If you use a presentation program that lacks built-in outlining, team it with Symmetry's Acta outliner desk accessory (included with Cricket Presents), or do your brainstorming with a word processor that has outlining features.

Presentation programs encroach on word processing territory in other ways. Many provide search-and-replace commands for making wholesale changes to your text. And because typos can turn a presentation into an embarrassment, you'll

Reprinted with permission from *Macworld,* pp. 209–218, July 1989.

find spelling checkers in PowerPoint, StandOut, Cricket Presents, Persuasion, More II, and MacDraw II version 1.1. But remember, spelling checkers aren't usage checkers. They don't know "capital" from "capitol" or "its" from "it's," so keep your dictionary handy.

VISUAL COMPONENTS

After you've refined your ideas, you're ready to produce visuals, which can contain three basic elements:

Text Usually short passages, often organized as *bullet charts* for fast reading (see "Bullet Charts"). More II, Persuasion,

and Cricket Presents let you turn outlines into bullet charts with one command or mouse click. Many programs also let you specify that the bulleted items in a list should appear one at a time on consecutive visuals, a technique called a *build*. By using builds, you can make your case and discuss it point by point.

Graphics A picture is often worth a thousand bullets. Graphs can visually depict trends or market shares, organization charts can spell out the corporate pecking order, and diagrams and drawings can illustrate complex concepts. Cricket Presents, Persuasion, and StandOut all provide built-in graphing features that let you create graphs from numerical data that you type in or

a dark background. All presentation programs let you specify such schemes. Many also provide special background effects that give visuals an elegant, professional look (see "Fancy Backgrounds").

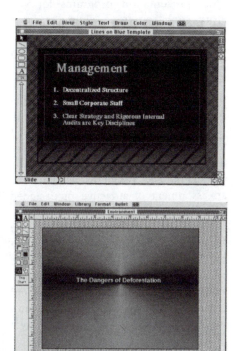

Fancy Backgrounds
A sampling of elegant backgrounds from PowerPoint (top) and More II (bottom).

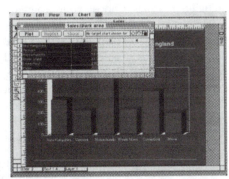

Built-In Graphing
Persuasion, Cricket Presents, and StandOut provide built-in graphing features that let you create graphs from numerical data that you type in or import from a spreadsheet. Shown here, a three-dimensional bar chart created in Persuasion.

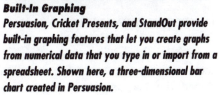

Bullet Charts
Bullet charts present information in short phrases, organized in bulleted list form. Many presentation programs let you specify that the bulleted items in a list appear one at a time on consecutive visuals, as shown here. This technique is called a build.

import from a spreadsheet (see "Built-In Graphing"). Other programs let you import graphs created in other applications, such as Excel or Cricket Graph, but that isn't as convenient. More II and Persuasion can automatically generate organization charts based on the indent levels in an outline. All presentation programs have drawing tools for making diagrams, and most can also import images created with a scanner or a drawing program.

Backgrounds On printed documents, text and graphics generally appear against the white background of paper. But presentation visuals are usually projected on a white screen in a darkened room, and in that setting dark text against a white background is hard on the eyes. It's better to use white or brightly colored text against

For assembling these components, presentation programs provide drawing and layout features that let you position text, create boxes and borders, paste graphics created in other applications, and draw various shapes. Most also provide on-screen rulers and alignment guides for precise positioning.

True, these features are all found in publishing and drawing programs. But presentation programs put a different spin on many of them. For example, publishing programs don't know overheads from slides, but presentation programs provide page-setup options for both. Thus, instead of having to calculate the proper dimensions for a slide or overhead and then type them into a Page Setup dialog box, you simply choose your output medium from a dialog box.

A presentation is more effective if its visuals are designed with care. Well-designed visuals use a consistent background and color scheme, and have a uniform layout to give the viewer's eye familiar points of reference. One way to achieve this continuity is to repeat a company logo or graphic on each visual. Most presentation programs let you store such repetitive elements on a *master page.*

For typographic consistency, many programs also provide word processor–like *style sheet* features that let you store and recall text formats. Cricket Presents, Persuasion, StandOut, More II, PowerPoint, and MacDraw II version 1.1 let you create *templates* to store your formatting preferences for future use. The aforementioned programs also include an assortment of attractive predesigned templates.

Most presentation programs let you assign descriptive titles to each visual in a presentation, so you can tell at a glance what the slide or transparency contains. You can also use these names to sort and organize the visuals using your program's sorting features.

SORTING AND SHOWING

What ultimately separates presentation programs from their publishing and drawing kinfolk are features that let you sort and arrange visuals and present them using the Mac's screen.

People who work with slides often use an illuminated stand called a *slide sorter* to view and organize their images. On-screen slide sorters—provided by PowerPoint, Cricket Presents, StandOut, and Persuasion—perform the same role by displaying numerous slides reduced to fit within a window (see "Sorting Slides").

Within the slide sorter window, you can change the sequence of slides in your presentation simply by dragging them to different positions. The slide sorter windows in PowerPoint, Cricket Presents, and StandOut also let you cut or copy slides to the Clipboard for pasting elsewhere in the same presentation or into a different one altogether. Most programs also provide *title sorters* that display only the slides' titles.

Finally, presentation programs provide *slide show* features to help you ex-

Sorting Slides
PowerPoint, Cricket Presents, StandOut, and Persuasion (shown here) provide on-screen slide sorters, which display numerous slides reduced to fit within a window. You can reorganize the sequence of the presentation visuals by dragging them within the sorter window.

hibit your visuals. Choose your program's Slide Show command, and the Mac becomes a projector, displaying each visual on the screen without the menu bar and tool palettes. In slide-show mode, the mouse becomes a remote control: click it to advance to the next visual. Cricket Presents, More II, MacBriefer, and StandOut also let you use visual effects such as a *wipe,* which causes one visual to "push" the previous one off the screen.

OUTPUT ALTERNATIVES

Now we come to the output: once you've created your visuals, how do you screen them for your audience? I've just mentioned the Mac-as-slide-projector technique. With this route, however, you'll probably need some additional hardware —unless you can convince a roomful of people to crowd around your Mac.

One option for Mac II and SE/30 owners is a 19-inch color monitor like those offered by SuperMac, RasterOps, Radius, and others. If you can spare about $8000, you might consider Mitsubishi's gargantuan XC-3710SS, whose 37-inch color video screen is guaranteed to get noticed (see "Attack of the Giant Monitor," *Reviews, Macworld,* October 1988).

But giant monitors weigh—and cost— a great deal. And in a large room even a 37-inch screen can seem small. If you can live without color, a more practical solution may be a *projection panel* or *pad,* which works with an overhead projector. You

attach the panel to your Mac and then la[y] the panel on the projector as though i[t] were a transparency. The Mac creates [a] video image on the panel's liquid crysta[l] display (LCD) screen, and the projecto[r] shows the image on the room screen. On[e] projection panel—In Focus Systems' P[C] Viewer—doesn't require your Mac to ta[g] along. PC Viewer contains a megabyte o[f] memory that, according to In Focus, ca[n] store up to 75 visuals.

Another way to project a video im[age]—this time in full color—is by using [a] *video projector* such as Kodak's LC500[.] The $3495 LC500 looks like a slide projec[tor] tor and, according to Kodak, project[s] images up to 12 feet wide. Unlike som[e] bulky video projectors, the LC500 is port[able] able (weighing about 13 pounds) an[d] doesn't require tricky setup and color convergence adjustments.

The biggest drawback to most video oriented presentation hardware is tha[t] you need to lug your Mac along. Wher[e] portability, color, and economy are im[portant, overhead transparencies and slide[s] are better alternatives.

OVERHEADS AND SLIDES

If you're like me, you probably slep[t] through a few overheads in high school[.] Ah, but old Mr. Crusty didn't have th[e] output options we enjoy today. By combining a laser printer with *transparency film* such as 3M Type 154 Transparency Film, you can produce overheads tha[t] would impress even Mr. Crusty. Need color? Use one of the new breed of affordable color printers, such as Hewlett-Packard's PaintJet or Tektronix's ColorQuick (see "Printing a Rainbow," *Macworld,* January 1989, and *Macworld News,* June 1989).

Overhead transparencies are eminentl[y] portable—dozens of them will fit in a binder—and overhead projectors are almost as ubiquitous as photocopiers. But overheads have drawbacks, too. You mus[t] manually flip from one transparency to the next, and that can make an otherwise sophisticated presentation seem amateurish—especially if the projector's fan blows half your visuals off the table. (Don't laugh, it's happened.) What's more, overheads can scratch and smudge, and the colors produced by inexpensive color printers can't approach the vividness of the ultimate presentation output medium—the color slide.

Slides are portable, too, and you can carry them in a tray, where they're always properly sequenced and ready to show. Slides are also inexpensive to duplicate, so it's easy to prepare duplicate sets as backups or for colleagues.

Slides also have the greatest dazzle potential. By combining two or more slide projectors under the control of a *dissolve unit,* you can create impressive presentations containing fancy visual effects such as dissolves and animation. And slides give you the vivid colors that only film can provide.

To create slides with a presentation program, you need a *film recorder,* a special kind of printer that provides images not on paper, but on film. Most film recorders contain a camera aimed at a video tube. Sandwiched between those components is a wheel containing red, green, and blue filters. The film recorder paints an image on the video tube, making separate passes for red, green, and blue light—three primary colors for video with which it can create a palette of over 16 million hues. A complete exposure usually takes a few minutes.

Not only do film recorders provide spectacular color, they offer tack-sharp resolution—usually in the ballpark of 4000 horizontal lines per slide. By comparison, a Mac II's screen display contains 480 horizontal lines; commercial television has 525-line resolution.

Alas, hardware with talents like these isn't cheap. Film recorder prices generally start at about $6000. A less expensive alternative is to use a slide-service bureau.

You send a disk to the bureau or transmit your visuals via modem, and your slides arrive from one to several days later, depending on the turnaround time you are willing to pay for. Service bureaus also provide other types of output, such as high-resolution overhead transparencies and color prints.

Most publishers of presentation software have cooperative arrangements with nationwide slide-service bureaus. Included with the software is a *driver* for the type of film recorder the bureau uses. You copy the driver into your System Folder, and use the Chooser desk accessory to select it prior to creating your visuals. Most publishers also include a special communications program that uses straightforward dialog boxes to automate communications with the service bureau.

PRESENTATION VARIATIONS

Most presentation programs are geared toward brainstorming and producing visuals, but some specialized ones also deserve mention. Programs such as Visual Information's Dimensions Presenter (still in development at this writing) and Dynaware's DynaPerspective allow you to create three-dimensional animated presentations. For example, architects can use these programs to take clients on simulated walk-throughs of their designs, and industrial designers or interior decorators can allow clients to view their proposals from literally any angle. Macro-Mind Director can produce animated presentations that incorporate sound effects and music. I'll be looking at these and other animation-oriented programs in a future column.

Then there's HyperCard. As a presentation program, Apple's software tool-kit has serious limitations, including no color and a fixed window size. But it does provide a selection of visual effects and it can play back recorded sounds. And by combining its built-in programming language with on-screen buttons, you can create dynamic presentations that people can use and navigate through on their own. (For more details, see "Getting Started with HyperCard," *Macworld,* May 1989.)

Before I began researching this column, I was somewhat skeptical of presentation software. The entire category struck me as being one created by software companies looking for another way to sell programs that provide drawing and layout features. But once I had worked with the programs a little and had seen what they're capable of, I thought of all the meetings I've snoozed through—where monotonal voices droned on and on in conference rooms, curing insomnia. Some interesting visuals and better planning certainly would have enlivened those proceedings.

It's true that a presentation program can spawn droves of uninteresting visuals when placed in the hands of a graphics philistine. And it's true that a presentation program's brainstorming features won't help someone suffering from a mental drought. But it's also true that every day thousands of people stand up and try to sell their ideas and products to others. That's hard work, and if I were in that position, I'm sure I would want all the help I could get. ▣

Jim Heid is a Macworld *contributing editor who focuses each month on a different aspect of Mac fundamentals.*

Presentations with Punch!

Banish Boardroom Boredom with Animation! Where still pictures take the place of many words, animated pictures speak volumes.

A city-planning director stands at the front of a boardroom explaining how the city will expand its residential and commercial zones. The audience sits, ready to hibernate. On the screen, a digitized photo of a building site fades in. As the planning director speaks, roads extend themselves and buildings sprout in a simulated downtown. The presentation's background music changes as statistics on city traffic appear on the screen in the form of a chart whose bars are composed of slowly accumulating cars. The audience watches intently for new changes as cars and buses move along the city streets, illustrating transportation bottlenecks. A subsequent scene shows new streets and freeways popping up around town. Inspired rather than expired by the presentation, the planners spend the next few hours hammering out the finer details of the city's plan.

Once considered only an art form, animation is now being used in business for presentations, training, and simulation. Adding animation to presentations can give clarity and punch to a visual explanation. It also works wonders to lengthen short attention spans. An ordinary presentation seems passé when compared with the message that animation packages can convey.

Me, Animate?

Consider whether the message of a presentation or demonstration can come across better with the aid of animation. If the subject is sleep-inducing, you should definitely try to get movement into the show. For complex subjects, animation supports explanations, reduces hand waving and verbal garble, and saves time for everyone concerned.

You don't have to be a highly trained professional to put together an animated presentation. With the programs we'll be examining, MacroMind Director and Studio/1, all you really need is some familiarity with paint, draw, and perhaps presentation programs. With a little practice, you can create a decent animated presentation.

Animations are not as simple as still graphics. You'll need to spend some time with samples, tutorials, and manuals. You'll find easy shortcuts for effects that could have taken hours to achieve by hacking and guesswork. Both MacroMind Director and Studio/1 come with sample animations and their components, such as cast members, graphic documents, and sound files. You can open them, examine their workings, and see what techniques were used. Experiment on your own, when not under production pressure, to see what you can create.

By Salvatore Parascandolo and Kristi Coale

PHOTOGRAPHY: MEL LINDSTROM

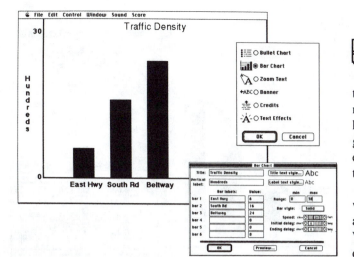

Director: AutoAnimate a Graph

Using Director's AutoAnimate dialog box, you can select predefined animated-text effects or self-rendering charts, and then fine-tune their action. The bars of the chart automatically grow from the minimum value to the value you specify. You can set the speed, start delay, and how long they linger after finishing. The Preview option is handy for making sure you have the effect you want.

A Seasoned Director

MacroMind Director is currently the highest-horsepowered animation tool for the Mac. Although it can't single-handedly create everything you need, it can combine or animate just about anything you have. With its ability to synchronize a wide variety of graphic elements, text, sounds, transitions, and visual effects, Director is the lone choice for creating Macintosh-based high-end animated presentations.

Director's still-thriving predecessor, VideoWorks II, was a breakthrough product in both its ambitious scope and its interface. Director is far more capable than Video-Works II, yet it's even easier to use. To really rev up its dormant animation engine, however, you must read its excellent user manual to understand its symbols, options, and window-related modes.

Director can create text and graphics, as well as import art in MacPaint, PICT, and PICT2 formats from such programs as Studio/8, PixelPaint, Canvas, SuperPaint, MacPaint, and Glue. It can also import color palettes and sounds from various sources and animations saved as Scrapbook pages or in PICS or VideoWorks format.

Interactive Animation

After you've given movement and sound to your presentation, you can go one step further. The best demos and training programs are interactive, drawing users into the action and letting them navigate freely by choosing areas of concentration and skipping over unnecessary parts. Neither Director nor Studio/1 offers much interactivity beyond starting and stopping. HyperCard, with its links, ease of use, and wide availability, is ideal for creating and using interactive training programs, but it offers no power animation tools. SuperCard builds on HyperCard's features with color, movable objects, multiple resizable windows, and other enhancements, including script-driven animation, but it lacks the rough-and-ready animation horsepower built into Director and Studio/1. Fortunately, with a little help from some friends, you can combine the two Cards' interactive abilities with the visual

ILLUSTRATION: MARK W. SWEENEY

powers of sophisticated animation.

Studio/1 comes with a (nearly) self-installing XCMD that, once added to any stack, lets it play native (S1AN) animations. VideoWorks II has a HyperCard driver that similarly accesses VideoWorks movies but with greater versatility. HyperAnimator and SuperAnimator create "talking actors" to bring lip-synched life to HyperCard and SuperCard presentations.

The HyperCard Connection

You install Studio/1's XCMD by opening the Studio/1 Demo/instruction stack, clicking once to go to the Installing card, and clicking again to copy the XCMD to any stack. Once the XCMD is installed, you create buttons to access the animation. In addition to the XCMD, Studio/1's stack includes sample scripts that offer several ways to play animations. They can be played in a specific rectangle anywhere on the card, fill the card, or fill the screen. Two animations can even be played simultane-

ously. There are options to control the animation mode and speed and to turn associated sounds on or off. The XCMD can be freely distributed.

VideoWorks II's animations can be played in HyperCard in a manner similar to Studio/1's, but the VideoWorks II HyperCard Driver must be bought separately. You install an XCMD into your stack, but it's not a stand-alone operator. The driver, which must be kept in the same folder as the subject stack, offers numerous options, such as playing specific frames and preloading a movie. In addition, up to 16 movies can be designated to play sequentially. Color movies play in full color on a Mac II. You can elect to play a movie at a specific card or screen location, with the background saved, and have it repeat until there is a mouse click. With features such as these, VideoWorks II gives far more control over the animation in stacks than does Studio/1. Scripts for these options are included in the driver's manual, which is also a HyperCard stack.

More options, including movies with different color palettes and resident sound resources, will be available in the MacroMind HyperCard Driver (release planned for July), an upgrade that will take advantage of the special features in MacroMind Director. It will sell for about the same price as the VideoWorks II HyperCard Driver ($99.95), with a special upgrade price.

The Director's Overview

Director has two easily switchable modes: Overview and Studio. The Overview mode is like a post-production facility in which all the elements of a show are assembled, tweaked, synchronized, and blended. This mode is the key to creating quick and slick presentations from existing art and built-in effects.

The AutoAnimate facility is a power tool that lets you create and control the timing and duration of animated effects with charts and text. It also serves as one of three direct links to the Studio portion of Director, allowing you to create custom film clips for presentations.

The Overview environment is much more than a slide sorter. It lets you easily sequence complex *sets* of events such as color fading in, music starting, and an animation sequence kicking off the presentation. If you were simply to lay in a sequence of scenes, your presentation would look a little rough and low-budget. Director includes strong editing capabilities for polishing a presentation,

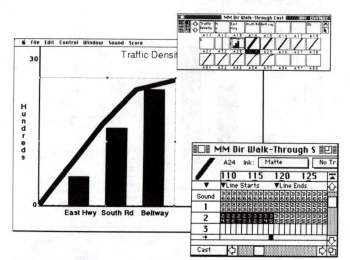

Director: Add a Red Line
A copy of the completed graph image was used as a background guide for drawing the red line. The effect of growth over ten steps was achieved by making ten duplicates of a whole line and trimming parts from each copy. Here Director's Cast window shows each segment, as a castmember scaled to fit its rectangle. The highlighted score segment animates the growing line.

SuperCard Animations

SuperCard (see "A Wilder Card," March '89) offers real windows and resizable, restackable, reshapable objects that you can animate with SuperCard's Move command. SuperCard's math, logic, and mouse-sensing functions let you calculate an object's intended path and speed on the fly, to suit the event, which means exceptional flexibility and power for interactive productions.

To create an animation entirely with Super-Card, you need objects and a script. The script identifies the object to be moved and the points it must visit. You can move the object from one point to another, along the perimeter of a polygon, or along a freehand curve. You don't even need a path object — any set of comma-separated x,y-coordinates will do. To simulate a car on a road, you create or import the car graphic, name it Car, and then draw a freehand or polygonal roadway and call it Road. Finally, write a short script containing the statement `Move graphic Car to the points of graphic Road.` That's it. You can optionally specify the speed of the motion and record the frames of action as a PICS file, which is exportable or playable by SuperCard.

Animated Agents

Beyond interaction, there's another way to enliven your stacks — HyperAnimator (see NewsLinks, May '89), from Bright Star, con-

tains an XCMD, RAVE, that lets you place talking characters into your stacks. Each actor has 16 images with eight facial expressions and seven lip positions. When the actors talk, their lips move to the correct positions and their expressions change. You create or import an actor in HyperAnimator's dressing room; save it to a stack; and create buttons to access and play it, a field into which to type what the actor will say, and a Talk button to cue the actor to speak.

Close on the heels of SuperCard comes a hyped-up HyperAnimator called SuperAnimator (release of which is planned for September). SuperAnimator is a stand-alone application written in SuperCard that provides color and gray-scale resizable actors. You can have a grand total of 127 images in an actor's file, with 32 speaking images. Bright Star promises better synchronization of lip movements with digitized sound and the ability to call up talking actors from as many authoring environments as are in common use on the Mac.

Claris and other software giants have included demos and help stacks with products

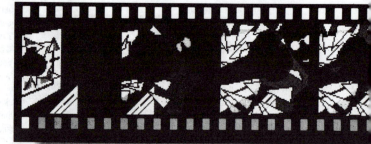

such as MacWrite II, but the same processes are available and affordable for small-business and even home users. Occasional users may stick with HyperCard. For relatively little money, more sophisticated animations can be added with Studio/1 or animation and color with Video-

Works II. Or you may find that SuperCard alone — with its color, animation, and programmable capabilities — is the ideal, inexpensive way to go.

SuperCard is priced at $199. Silicon Beach Software is located at 9770 Carrol Center Road, Suite J, San Diego, CA 92126; (619) 695-6956.

HyperAnimator costs $199.95. For information about it and SuperAnimator, contact Bright Star Technology at 14450 N.E. 29th Place, #220, Bellevue, WA 98007; (206) 885-5446.

— Laura Johnson

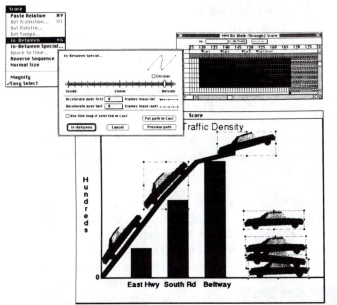

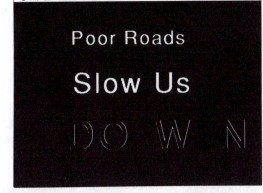

Director: Make a heap of cars
Cars will climb the line and leap off onto a growing stack. One car was hand-done, and rotated copies were created automatically by Director's Auto Transform feature. The set of rotated cars was dubbed a film loop, and it was tween-moved by definition of four key positions. After tweening, the developed event in the score was pasted into other channels, a few frames apart, producing multiple staggered climbing sequences.

Director: Add more with Overview
In Director's Overview mode, you can assemble a custom presentation from existing animations and stills, and then add audio, textual, color, and transition effects. Here, the graph movie produced earlier is preceded by the sound of decelerating machinery, a short delay, and a venetian-blind transition. The show ends with a built-in sliding-text effect on a blue background.

and they're relatively easy to learn and use.

The most common need is to rearrange the order of presentation. By dragging a document icon to a new position, you can change the order of slides or the order in which items appear on a given slide. To help smooth the presentation and highlight key slides, you may use one of the built-in, customizable transition effects, such as fade or dissolve.

Sounds, such as music and lifelike speech, can emphasize particular slides or change the tempo of a presentation, but they might also cause problems — a slide might advance before a playing sound ends, for example. To ensure that the current scene will be displayed until the end of a sound, you must adjust the length of time the slide will be on the screen. Director uses stopwatch icons to time events. After you install one, you can set and change its time to whatever duration is needed.

Call Your Editor

The Launch Editor is a valuable shortcut for changing documents in your show. It bypasses the Finder and takes you directly into the program that created the document. Highlight the document and select Launch Editor from the File menu. Once you've made the changes, save the document and quit the program — you'll return to the Overview window, where your changes will be automatically reflected.

Fortunately, an Overview document doesn't actually contain its component documents — it serves as a table of contents for the presentation and has links to the elements, wherever they are. One document, such as a background, logo, or animation, can be used by more than one Overview document. Even though the Overview remembers where its component documents are on any mounted disks, it's a good practice to put all the documents into one folder. Director can even do this for you with its Gather Documents option.

Universal Studio

Director's Studio mode puts you closer to the backroom details of animation. It's a stage and movable-character animator in which movie frames are built from image layers — similarly to conventional animation. You draw characters or parts of characters onto individual transparent overlays and then stack them in the appropriate order for viewing.

Director supplies a full-featured painting environment for creating and editing the elements of the animation. Each element is a free-floating, separately controllable object, which makes a movie much easier to manage and change.

Painting the Picture

Director paints in both black-and-white and 8-bit color. Although it's not immediately apparent, the program's bit-mapped–graphics power is on a par with that of the higher-end Macintosh color paint programs. It has the standard tools, including a widely configurable airbrush; special effects such as graduated fills, free rotation, perspective, and distortion; and a nearly unlimited variety of special paint effects such as lighten, darken, and blend, the degree of which can be user-specified.

Each painted object has an invisible registration point that helps line it up with preceding objects during playback. Director also has tools for rendering basic Quick-Draw shapes that can serve as memory-efficient background objects and animated elements with special object properties.

All the World's a Stage

The *stage, cast,* and *score* are the metaphors of a Director animation. The stage is where the action occurs. Any created or imported graphic is automatically placed in the cast and considered a *castmember.* Director gives each castmember a short identifier for referencing and managing purposes. The Cast window is like a set of dressing rooms, where all the actors wait, ready to go onstage. Unlike real theater, however, Director lets you have multiple copies of the same castmember onstage simultaneously, each in a different location, stacking order, and size.

The score is a script — a to-do list that controls the location, stacking order, apparent size, and other attributes of each castmember, plus the start of transitions and sound effects. Like a 24-channel tape deck, it can control up to 24 simultaneous activities in each movie frame. Objects controlled by higher-numbered channels pass in front of objects in lower-numbered channels. Four additional channels control sound, film speed, palette changes, and transitions such as fades and dissolves.

Events recorded in the score can be copied and pasted from one part to another like spreadsheet cells. This system makes it easy to set up complex motion, change the location or synchronization of events, or alter the stacking order of castmembers — all without error-prone manual tweaks to the graphics themselves. Hundreds of frames can be dealt with as easily as one. Double hats off to MacroMind on that one.

To include a castmember in the action, you select the channel and frame number where its action will begin and then drag the castmember from the cast and place it on the stage. Director automatically annotates the score. Once your object is on the stage, you can drag it around while Director records its movement or move it incrementally and advance one frame each shift, or have Director tween it to your specifications. Although its tweening function is strictly 2-D, it can produce straight or curved paths between two or more

Glossary

ILLUSTRATION: K. DANIEL CLARK

Tweening

Tweening is easily the most important capability in animation software. To move an object smoothly along a path through successive frames, you can either draw that object by hand in each frame or have your trusty animator automatically do it. You specify the starting position of the object in one frame and its ending position in a later frame. The program automatically creates a specified quantity of frames in be*tween* your two key frames, with the object precisely displaced in each. It's a priceless and essential power tool. Both Director and Studio/1 offer tweening, but each has its own methods, options, and extensions.

Cel or Cell

A cel is one layer of a frame of animation. The word has its roots in the clear, transparent sheets that are stacked to form a complete frame in conventional animation.

Film Loop

Any repetitive motion, such as the walking of legs, swinging of arms, or spinning of wheels, is like a length of film with the ends glued together to form a *loop.* Director and Studio/1 each offer this capability. Moving and tweening a loop are as easy as moving and tweening an object, but loops produce

complex motion, such as a bird flapping its wings and flying across the screen.

Frame

An instant in a movie. New things happen only when there's a change of frame. Characters appear, disappear, shift, or change shape, size, or color. You synchronize the actions of animated items by using the frame number as a reference.

Key Frame

A frame that is specified as the starting or ending point for a gradual movement or change.

PICS

A generic animation file format that contains sequential frames. Animation-capable programs that can read PICS files include VideoWorks II, Director, Studio/1, SuperCard, and some utility programs that are like play-only movie projectors.

Real-Time Animation

An animation mode in which an object is dragged with the mouse and its path automatically recorded for later playback.

Transition

A gradual change from one view to the next. Examples are fade-ins, fade-outs, dissolves, leftward slides, venetian blinds, and checkerboards. — Salvatore Parascandolo key frames. You can vary the type of path curvature, as well as the acceleration and deceleration of the tweened object.

If you don't mind the characters' having a Swedish accent, they can even talk, using MacInTalk speech, which is text that has been converted to spoken English. You can import and play digitized sounds and music or pass timing cues to a MIDI sequencer (which can be another Macintosh) that drives one or more musical instruments.

Occasionally in theater, one actor takes over a role from another. The role remains the same, but someone else goes through the motions. You can do the same with Director's castmembers. You select the piece of the score that describes the path of castmember X, select castmember Y from the cast, and choose Switch Castmember from the menu. Poof! Instant understudy. You can replace, for example, an older model of a demo product with a newer model.

To obtain such effects as revolving

Different Kinds of Animation

Painted Frame by Painted Frame

This is simple flip-book animation. The background, characters, props — everything you see in a frame — is one painting. Editing can involve slow and intricate work in repositioning items and patching holes. This type of animation can play back at high speeds, regardless of the complexity of the scene, but generally it uses up substantial disk space.

Stage and Movable-Character Animation

Everything floats and is not permanently attached to anything else. Even though most of the items can be bit maps, each bit map is treated as an individual object. So it's easier to do small positioning tweaks and major edits. The action slows as more animated elements are brought onstage. This type of animation produces relatively compact files. — Salvatore Parascandolo

titles, flashing signs, and shattering words, Director lets you create and animate bit-mapped text. Once set in place, it can't be edited as text. Permanent text is available. It can be animated or globally searched and replaced and have its attributes changed at any time. Special effects with permanent text are severely limited.

For some color magic, Director offers color cycling and palette transitions. Basically, cycling is a wave of color sweeping through an area. With it you can produce intricate

visuals such as shimmering oceans, flashing lights, and moving backgrounds, all in one frame. The palette transition is a gradual scenewide change from one color scheme to another. It can be used to mimic a sunset, sunrise, or weather changes or make other mood-altering shifts that can enhance a presentation.

To manage an animation project better, you can attach comments to any frame to note important changes in the action, characters' entrances and exits, names of sounds, plans for future action, or notes to other mem-

3-D Animation: Super3D and Swivel3D

In its true sense, 3-D animation is somewhat like clay animation. Currently, several modeling applications provide it. Some have features for simplifying both the creation and recording of frames. These applications manipulate objects in fully law-abiding 3-D space. You can rotate an item and see its previously hidden sides. There are true perspective and realistic lighting effects, with no need for mental calculation or guesswork.

Super3D from Silicon Beach and Swivel3D from Paracomp are two widely used world makers. They can create complex shaded renderings; execute tweening; and export frames as either Scrapbook pages, sets of individual files, or PICS files. When a movie is exported in any of those forms, it loses its 3-D nature and becomes a series of sequential snapshots of the 3-D world. Both programs work in monochrome and color and provide a wide range of hues. They offer options such as wire-frame or

fully shaded display; extensive abilities to create, combine, and edit objects; and the ability to change the viewpoint on the world in both angle and magnification. Each can also build, modify, and animate a 3-D scene from simple text instructions. There are some important differences between these two programs, however.

Super3D's interface looks like that of an object-oriented drawing program. Objects have handles that can be used to stretch, compress, and reshape them. An object can be dismantled into its component polygons and the parts removed or replaced. The program can use multiple copies of a master object and automatically reshapes all the copies when the master is changed. You can even make all copies of a given master object invisible when you don't need them, which lets you powerfully control the contents of a scene without manually removing and adding items.

Swivel3D's interface is less orthodox but more natural. Screen redrawing is exception-

ally fast. You create an object by entering the construction mode and drawing the object's cross-section, side, and top views. You can reposition, resize, and spin objects by dragging them with the mouse with a modifier key depressed. Objects can be combined into swiveling, sliding, or locked mechanisms while retaining their editable individuality. Each object's pivot point can be set for more-natural motion. Swivel3D objects can cast real shadows, and colored images can be wrapped onto the curved surface of a 3-D object. A painted rocklike texture on a 3-D lump can add striking realism to your world.

Swivel3D costs $395. Paracomp is located at 123 Townsend St., Suite 310, San Francisco, CA 94107; (415) 543-3848.

Super3D is priced at $495. Silicon Beach Software can be contacted at 9770 Carrol Center Road, Suite J, San Diego, CA 92126; (619) 695-6956.

— Salvatore Parascandolo

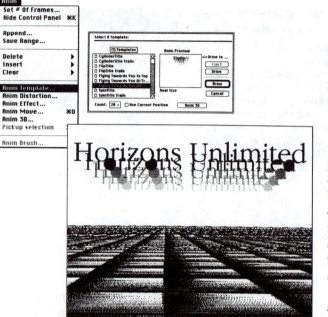

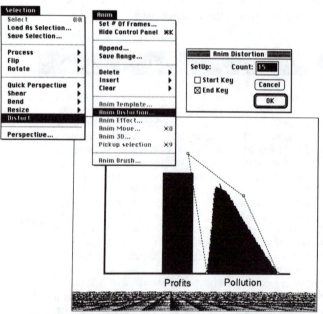

Studio/1: Set Up Flying Text
You select any text or image-area object and apply an animation effect to it. You can either specify starting and ending positions, using x,y,z-coordinates, or use one of the predefined animation templates that are included. As you browse for effects, the preview window shows an animated sample of the highlighted template. This one is a home brew. Trailing images are a built-in option.

Studio/1: Grow a Graph
Both the profits bar and the pollution pile were grown with tweened distortion. You select an area, opt to distort it, and then set up its beginning and ending shapes and the number of in-between frames. The profits bar was grown first, and then the pile began its growth after a two-second pause. After vertical growth, the pollution pile was further tween-distorted toward the profits bar.

bers of an animation team. Director offers a host of options for printing the elements of your animations. The printed forms can be used for such things as annotated presentation handouts, speaker notes, and storyboards.

Even on the fastest Mac, Director's playback slows when many characters are onstage. (Director itself takes up 1 megabyte in RAM; under MultiFinder, the default partition is 2 megabytes.) With Director's Accelerate option, you can convert your movie to a form that can play back at up to 30 frames per second. The conversion will cost you disk space — a 200K color movie can convert into a multimegabyte disk-blockbuster. Large movies also require substantial RAM for smooth playback, but the smoothness adds to the presentation's effectiveness and professional punch.

Out of the box, Director is not an interactive program as HyperCard is. You can't, for example, push a button to decide which animation sequence plays next. But Macro-Mind offers a way to join Director's talents with Hyper-Card's navigational nimbleness. This subject is discussed further in the "Interactive Animation" sidebar.

Direct from the Desktop

Director is a professional tool that lets you easily add interest and clarity to conventional graphics. With its Overview mode, you can hit the ground running — assembling a custom presentation in minutes from existing components. In the Studio mode, you can import or create simple or sophisticated graphics, and you have the means to create plain or complex motion. You won't instantly become Walt Disney or produce a space-shuttle simulation in your first five minutes, but with experience and experimentation, you can be a prime mover.

The Little Studio That Could

Studio/1 has two basic talents: It's a full-featured monochrome painting program and a painted-frame animator. Its tool sets are complete and rich with intelligent options, shortcuts, and special effects — and they're easy to use.

Studio/1 can be used for low- or high-resolution painting, scanning and editing 1-bit TIFF images at resolutions of up to 300 dpi, obtaining dynamic titling effects, illustrating mechanisms and processes, or producing action segments that can be played back from within Director or by HyperCard or SuperCard under interactive control (see "Interactive Animation" sidebar).

The program shows obvious efforts to overcome some inherent limitations of paint-only animation. For example, new objects and selections can temporarily float above the painted background. While floating, an object

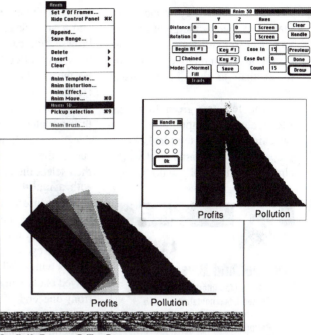

can be moved harmlessly, and it can be the subject of special effects and sophisticated tweened motion. With sound synchronization, variable playback speeds, acceleration and deceleration, live recording of mouse-dragged objects, and sophisticated tweening, you have an impressive package for only $150.

The Artist's Studio

Studio/1's painting tools are first-rate. In addition to the usual tools, the program features an editable Bezier curve and polygon, eight configurable airbrushes, and fully controllable gradient fills. It works at resolutions from 72 to 300 dpi and offers a separate layer for real text that remains editable and prints at the highest resolution of your printer. The ready access to high-quality text simplifies production of overhead slides, handouts, posters, and other mixed-media work.

Its selection tools include a free-form lasso, polygon, and rectangle. When combined with different selection modes, these tools let you quickly isolate parts of an image, which you can alter, using transformations such as rotation, perspective, bending, and free-form distortion. An area can even be masked to protect it from any editing action, including a full-document erase.

The *draft page,* a serious goof-reducer, is a spare work area that overlays the document and that can be painted like any other document. The draft page can be transparent so that you appear to be working over existing art without disturbing it. You can cut and paste items from the draft page or merge the entire draft page with the underlying document.

Your Anim Mate

Studio/1's extensive set of animation tools can help you pack plenty of movement, excitement, and entertainment into your frames. Studio/1 can tween motion, rotation, size changes, and distortions when supplied with the beginning and ending states of an item.

You may want a logo to revolve as if it were painted onto a slowly rotating cylinder, or you may need to send a propeller spinning off into the sunset along a diagonal path, going faster as it moves away. No sweat.

Although Studio/1's world is a flat, 2-D surface, Studio/1 can apply distortions, visual perspective, and scaling to objects as if they were moving in 3-D space. Because your object is only 2-D, it behaves like a paper cutout dancing in space. As it spins, you won't see its back faces come around. Even with this limitation, the effects can be quite striking.

You supply the data for tweened events through dialog boxes and by using the mouse. You then preview the

Studio/1: Tween a Falling Bar
The profits bar is going to be knocked over by the pollution pile. The bar's falling motion is a simple tween that's described numerically as a 90-degree z-axis rotation. Studio/1 lets you set the pivot point (handle) anywhere on screen. The lower left corner is appropriate in this case. You specify the number of tween frames and how many frames are to be used for acceleration. Here the Trails option produces a motion-blur effect.

Studio/1: Reuse an Animated Sequence
The animator constructed an animated brush earlier by drawing a bird at different stages of flapping in sequential frames that were saved as a named brush. In the Picture Preview window, the bird brush actually flaps. The brush was reconverted to a multiframe selection and 3-D–tweened many times. The key positions for each path were set numerically and by dragging. The result: independent birds, moving far and near at different speeds, in different directions.

tween and adjust key frames. You may even specify the pivot point of an object so that rotations and other operations occur relative to that point rather than to the object's visual center.

If you feel too green to tween, Electronic Arts supplies a set of tweening-effects templates with proper x, y, and z start/end values already filled in. All you do is select

MacroMind Director ♟♟♟♟♟

Follows Mac Interface	5
Printed Documentation	5
On-Screen Help	5
Performance	5
Support	5
Consumer Value	5

Comments: Clearly the best-equipped tool for creating color Macintosh animations for demos, presentations and entertainment. **Best Features:** It can import, sequence, and orchestrate truly impressive audiovisual events. Its self-contained, object-oriented animation capabilities provide virtually limitless editing freedom. **Worst Feature:** Large or numerous objects onstage slow the action significantly. **List Price:** $695. Published by MacroMind, Inc., 410 Townsend St., #408, San Francisco, CA 94107; (415) 442-0200. Version 1.0 reviewed. Requires Mac Plus or SE with two 800K drives for monochrome work, Mac II with hard disk and 2 megabytes or more of RAM for color productions, System 6.0.2 or later. Not copy-protected.

a part of the drawing as the subject and go shopping for an effect. As you browse, you can automatically preview each effect's action. You can use the templates as they are, customize them, or make your own.

Studio/1's version of a film loop is the animated brush. You first produce several frames of animation, such as the poses of a flying bird. You then select the bird *in all its frames* with a multiframe "cookie cutter" option and use the selection as a brush. As you move the brush across the screen in Recording mode, the frames play in a cycle, causing the bird to flap its wings. When the sequence is played back, you see a bird flapping from one end of the screen to the other.

Sounds can be associated with any frame. You simply go to the frame where you want sound to start; click on the Sound button in the animation-control panel; and choose a sound, which can come from any file that has SND resources, such as the System file or a MacRecorder sound document.

Studio/1 imports MacPaint, PICT, TIFF, EPSF, and PICS files from applications such as Swivel3D, Super3D, and VideoWorks. It exports files in MacPaint, PICT, TIFF, and PICS formats. Its Open dialog has a time-saving thumbnail preview facility that lets you see single illustrations or multiframe animation documents. You can preview MacPaint, PICT, TIFF, PICS, and Studio/1 files. Animation files actu-

Studio/1 ♟♟♟♟½

Follows Mac Interface	5
Printed Documentation	5
On-Screen Help	3
Performance	4
Support	5
Consumer Value	4.5

Comments: An excellent monochrome paint program with impressive flip-frame animation powers. **Best Features:** Top-notch painting tools with 300-dpi capability and real-text layer; uncluttered animation interface; includes HyperCard playback utility. **Worst Feature:** Painted animation is inherently more difficult to edit than object-oriented animation. **List Price:** $150. Published by Electronic Arts, 1820 Gateway Drive, San Mateo, CA 94404; (415) 571-7171. Version 1.0 reviewed. Requires Mac Plus or SE with two 800K drives, System 6.0.2 or later. Not copy-protected.

ally move during preview.

Final Frame

Studio/1 is an excellent and versatile black-and-white painting program and capable animator. The painting and animation interfaces mesh well. With Studio/1's special effects, applied artistry, and time, you can produce animations of biblical proportions, but when you plan to create big, complex projects, you'll want to migrate to MacroMind Director and use Studio/1 as a respected contributor to the production. ▦

Getting Started with Animation

Make it move with
this introduction to
Macintosh animation

by Jim Heid

JOHN CRAIG

Animation is everywhere—and it has been since newspaper cartoonist Winsor McCay thrilled turn-of-the-century theatergoers with his Gertie the Dinosaur cartoon. So novel were McCay's efforts that audiences "suspected some trick with wires," he later wrote.

Today's public is less naive but remains enchanted by watching inanimate objects take on lives of their own. Animation is hot, and computer animation is helping to fuel the fire. Enter the Mac, stage left. The Mac's sharp graphics and fast processor make it a natural platform for creating moving pictures. This month, I explore the concepts and challenges behind animating on the Mac and spotlight some animation products. For more details on the products discussed here, see "Move It!," *Macworld,* June 1989.

WHY ANIMATE?

You might think the only people who could benefit from Mac animation are in the entertainment business. There's no denying that Mac animation is making show biz inroads. Advertising agencies and production houses use Mac programs to mock up animations that will be created on broadcast animation equipment such as the Quantel Paintbox (a broadcast titling and animation workstation commonly used in the video industry). Filmmakers and video artists use the Mac to create animated art (see "Macintosh Masterpieces," *Macworld,* August 1989). Television stations use Mac animations for on-screen graphics.

But the truth is, you don't have to have delusions of Disney to use an animation program. Consider the following scenarios.

▪ To explain the benefits of solar water heaters at a trade show, a marketing manager for a solar heating firm uses an animation program to create a moving diagram in which blue (cold) water enters a solar collector, is heated (it turns red), then flows into a storage tank. This is followed by animated charts and graphs that show how much money and energy customers will save. The complete presentation runs continuously at the show booth.

▪ A history teacher wants to show how events combined to create the Cuban missile crisis. Instead of scrawling time lines on blackboards, he creates an animation that uses maps of Cuba and the United States to depict the key events: the missiles arriving on Cuban soil, the spy photos, the naval blockade. The result is more engaging than a lecture and reinforces the geography and timing behind the crisis.

▪ An architect bidding on a project creates an animated walk-through that lets clients look at the building from various angles and that simulates entering the lobby from the central courtyard. The clients can better visualize the final product than they could by examining blueprints.

FRAME BY FRAME

Smooth animation requires that the viewer see at least 16 images, or *frames,* per second. That's easy for any Mac, but creating the frames themselves is, for the most part, your job. Some programs can automatically create certain types of animation, but creating a realistic moving picture still takes time and effort.

The most basic animation package is Beck-Tech's MacMovies. You use a paint program such as MacPaint to create each frame, then you compile the frames into a *movie,* MacMovies' term for a single document file containing the images, compressed to use less disk space. You can play the results at up to 30 frames per second, and you can distribute movies along with a Beck-Tech projection program called, logically, Projector.

MacMovies is easy to learn and use, and at $99, it's an inexpensive way to test the animation waters. Note, however, that the program's current release, called Volume I, does not run on the Mac II or the

Reprinted with permission from *Macworld,* pp. 299–307, Sept. 1989.

257

SE/30 and has minor (that is, easily bypassed) problems with System 6.0. Beck-Tech says MacMovies II will fix the incompatibilities and add color support and additional features; look for it late this year or early next.

CEL MATES

More ambitious animation programs—such as MacroMind Director and its lower-priced cousin, VideoWorks II—electronically recreate the techniques behind *cel animation,* the conventional film animation technique patented in 1914 (by another newspaper cartoonist, Earl Hurd) and still used today. With cel animation, a background is drawn or painted on paper, over which are laid transparent sheets of celluloid containing only those components of the frame that have moved since the previous frame. The resulting composite is photographed, a new cel is laid in place, and the process is repeated. Cel animation saves animators time and effort by eliminating the need for them to redraw the background for every frame (as Winsor McCay did for all 10,000 frames of *Gertie the Dinosaur.*)

In Director (as well as in VideoWorks II), the components of an animation—backgrounds, animated objects, sound effects, and so on—are stored in the Cast

In the Director's Chair
MacroMind Director stores the components of an animated sequence—objects, sounds, backgrounds, and so on—in the Cast window (bottom right). Behind the Cast window, the Score window graphically depicts the events in the sequence. Behind the Score window is the Stage, where the animation takes place.

window. The Cast window is a cinematic database, an electronic casting office that holds the audio and visual cast of characters in an animation. You add a cast member to an animation by dragging it from the Cast window to the *stage,* or by using the Score window, which graphically depicts the events in the animation (see "In the Director's Chair").

If all this sounds complex, it can be. Fortunately, Director provides a host of labor-saving features for creating animated titles, bar charts, bullet charts, and other special effects (see "Automatic Animation").

Director also offers the computer animator's best friend, an *in betweening,* or *tweening,* feature. Tweening lets you animate without crafting every frame by hand. You create the first and last frames (the *key frames*) of a sequence, specifying the beginning and ending appearance of an object, as well as the total number of frames desired for the sequence. Director does the rest, creating the specified number of frames and moving the object from one frame to the next.

There's far more to Director than I've described here. For creating a computerized cast, the program provides color painting features that rival those of SuperMac's PixelPaint. It has impressive sound features, which I'll describe later. And it can exchange images and even entire animation sequences with other programs. That latter talent lets you combine Director with three-dimensional drawing programs to create 3-D animations.

NO SPECIAL GLASSES NEEDED

Much of today's computer-generated animation creates the illusion of three dimensions. A network logo spins toward you, its letters thick and glittering as though cast in platinum. A telephone company commercial takes you on a fantastic voyage through the circuits of a fiber-optic network. Seen on a big screen, 3-D animation can almost give you motion sickness.

Three-dimensional animation is also complex enough to give a computer headaches. Simulating depth, lighting, and

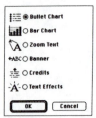

Automatic Animation
MacroMind Director provides labor-saving features that create animated titles, bar charts, and various text effects. After selecting the desired effect in the Auto Animate dialog box (top), you specify the text style, animation speed, and other options (bottom). Here, the From Right option, which causes each bulleted item to slide into place from the right edge of the screen, is selected.

shading requires time-consuming calculations that tax even the fastest Macs. For example, a Mac II takes over 10 seconds to display the image shown in "The Third Dimension." Multiply that figure by the number of frames needed for smooth animation (at least 16 per second), and the time factor behind 3-D animation becomes all too real. Consequently, no Mac 3-D graphics program provides real-time animation features. You can't draw a car, for example, and watch it move as you move the mouse.

But if you're patient, you can still create stunning 3-D animations. Many 3-D programs let you create animation sequences that you can save and play back or import into VideoWorks or Director for further embellishment. You'll find animation-oriented features in numerous 3-D programs, including Abvent's SpaceEdit, Dynaware's DynaPerspective, Visual Information's Dimensions Presenter, Paracomp's Swivel 3-D, Silicon Beach Software's Super 3D, and two new companion programs to MacroMind Director,

CREATING A HYPERCARD ANIMATION

T̶o̶ ̶d̶o̶ ̶t̶h̶i̶s̶ ̶t̶u̶t̶o̶r̶i̶a̶l̶, you must set your HyperCard user level to the Painting level or above. Use the Home stack's button to get to the User Preferences card, click the Painting button, then return to the Home stack's first card.

PHASE 1: CASTING CALL

First, you need a character. For this exercise, use the small dog in the Mac's pictorial Mobile font.

1. In HyperCard's Home stack, click the Art Ideas icon. When the Art Ideas index appears, click the entry that reads *Font—Mobile*.

2. Open the Tools menu and activate the selection tool.

3. Find the small black dog (it is on the top row, near the right edge of the card), and select it by enclosing it within the marquee. Don't select any extra space below the dog's feet.

4. Choose Copy Picture from the Edit menu.

PHASE 2: CREATE A NEW STACK

Now you're ready to create the stack that will contain the animation.

1. Choose New Stack from the File menu.

2. Uncheck the Copy Current Background box by clicking within it, then type a name such as *Animation Test*, and press Return.

Before pasting the dog into place, you must create a background.

3. Choose Background from the Edit menu (or ⌘-B).

(Stripes appear in the menu bar to indicate you're working on the background.)

4. Open the Tools menu and select the line tool.

5. Hold down the Shift key to draw a horizontal line from one end of the card to the other. This line will be the street upon which your silicon pooch will stroll.

6. You're finished with the background, so choose Background from the Edit menu again.

7. Add the dog to the scene by choosing Paste Picture.

8. Drag the dog (gently, please) until it's at the left edge of the card. *Do not deselect the dog; keep it within the marquee.*

PHASE 3: CREATE THE FRAMES

Now you're ready to create each frame. To do so, you will repeatedly move the dog, copy it to the Clipboard, create a new card, and paste the dog into the new card.

1. Choose Copy Picture from the Edit menu to put the dog and its new position on the Clipboard.

2. Choose New Card from the Edit menu or press ⌘-N.

3. Paste the dog into the new frame from the Edit menu or use ⌘-V.

4. Hold down the Shift key and drag the dog slightly to the right.

5. Copy the dog's position by pressing ⌘-C.

6. Repeat Steps 2 through 5 at least 20 times. The more frames you create, and the smaller the movements in each frame, the slower the dog will appear to move.

PHASE 4: ACTION!

1. Return to the first card in the movie by choosing First from the Go menu.

2. Display HyperCard's message box by choosing Message from the Go menu or by pressing ⌘-m.

3. In the message box, type *show all cards* and press Return.

Disney it isn't, but it's a start. From here, you might spruce up the background and add more characters. You might also create a background button whose script shows the movie for you (see "Roll 'Em").

Remote-Control Painting
You can use Hyper-Talk scripts to control HyperCard's painting tools. This script selects the dog (assuming it's near the left edge of the card and 1⅛ inches from the bottom), drags it to the right, reverses its direction, and drags it to the left.

```
on mouseUp
    choose lasso tool
    click at 25,270 with commandKey -- lasso the dog
    set dragSpeed to 200 -- adjust for desired speed
    drag from 25,270 to 490,270 -- move the dog
    doMenu "Flip Horizontal" -- turn the dog around
    drag from 490,270 to 25,270 -- move it back
    doMenu "Flip Horizontal" -- turn it around again
    click at 0,0 -- deselect the dog
    choose browse tool -- restore the browse tool
end mouseUp
```

```
on mouseUp
    show all cards
end mouseUp

on openStack
    lock screen -- turn screen updating off
    show all cards -- load images into memory
    unlock screen -- restore screen updating
end openStack
```

Roll 'Em
The top script, when attached to a button, will show all the card images when the button is clicked. The script on the bottom speeds̶...

currently in development and code-named 3DWorks and RenderWorks.

The most basic way to create frames for 3-D animation is to alternate between moving the objects (or *the camera,* the 3-D world's term for your vantage point) and issuing a command that saves the screen in a graphics format such as MacPaint or PICT. You can then move the resulting frames into Director, VideoWorks, or HyperCard.

An easier alternative to this move-and-save process is to use a program's tweening feature to create frames. Some programs provide a variation on the tweening theme by allowing you to specify not just the beginning and ending key frames but also *intermediate frames.* These let you use a program's tweening talents to create more complex animation sequences, such as entering a building, turning left, then climbing stairs.

You can save the frames generated by a tweening session in a number of ways. Some programs, including Super 3D and Dimensions Presenter, let you save frames in a document that you can play back using an accompanying projection program. Swivel 3-D lets you save frames in the Scrapbook file. Super 3D also supports an up-and-coming file format for exchanging animation sequences called *PICS.* MacroMind Director can import (and export) animations in numerous formats, including Scrapbook and PICS.

ANIMATING WITH HYPERCARD

You can get a feel for Mac animation techniques using HyperCard. By following the tutorial in "Creating a HyperCard Animation," you'll be using a method that closely mimics the process of traditional cel animation; it's also similar to the frame-by-frame animation techniques you can use in VideoWorks and Director. Another technique that usually provides faster animation than the frame-by-frame approach is to create HyperTalk scripts that select objects and move them on the screen. One such script is shown in "Remote-Control Painting."

The best way to add animation to stacks is to create it using VideoWorks and then use MacroMind's Play VideoWorks driver to play them from within a stack. A driver for playing Director animations is also in development.

The Third Dimension
With 3-D programs such as Silicon Beach's Super 3D (shown here), you can create animations that provide the illusion of three dimensions. Like most 3-D programs, Super 3D can save animated sequences in files that you can import into VideoWorks II or MacroMind Director.

WHAT ABOUT TALKIES?

The extent to which you can add sound to an animation depends on your animation program, on the nature of the sound, and on your Mac's memory and storage capacities. MacroMind Director has the most impressive audio capabilities of any Mac animation program. Director can use Apple's Macintalk speech driver to produce synthesized speech, and it includes a library of synthesized sound effects. Some are convincing, but most have a decidedly electronic sound to them.

A better way to sound off is to exploit Director's ability to play back sounds that you digitally record, or *sample,* using Farallon Computing's MacRecorder, the Impulse Audio Digitizer, or Digidesign's professionally oriented Sound Tools system. Director can also control music synthesizers by transmitting commands using the musical instrument digital interface (MIDI). If you have a sampling keyboard such as an Ensoniq Mirage or EPS, you can use the keyboard to record high-fidelity sound effects and narration and trigger the audio by having Director transmit MIDI commands to the keyboard. (For MIDI details, see "Music for Beginners," *Macworld,* April 1989.)

Animators using HyperCard or Silicon Beach's SuperCard can include synthesized and sampled sound in animations, too. And by combining HyperCard with Bright Star Technology's HyperAnimator, you can create animated talking heads whose mouth movements are synchronized with sampled or Macintalk-gener-

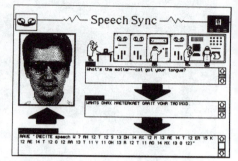

Read My Lips
You can use HyperAnimator with HyperCard to synchronize the mouth movements of this talking head with digitized or Macintalk-generated speech. HyperAnimator's Speech Sync mode converts text into phonetic codes and creates the HyperTalk commands needed to play back digitized sound. The numbers in the HyperTalk command represent timing values that control for how many sixtieths of a second each mouth and facial expression appears.

ated speech (see "Read My Lips"). Bright Star's SuperAnimator, which should be available by the time you read this, is an animation system designed to take advantage of SuperCard's color features.

NO BARGAIN MATINEES HERE

With its digital audio, color, and 3-D capabilities, the Mac makes a mighty impressive screening room. Unfortunately, the price of admission is high. Sophisticated animation and 3-D drawing programs are costly, and they have nearly insatiable hardware appetites. MacroMind Director, for example, retails for $695 and needs a minimum of 2MB of memory for color work; 4MB or 5MB is a more reasonable amount, especially if you use sampled sound. To transfer your work to videotape, add a specialized video board that outputs video signals in the NTSC format used by consumer and professional video gear. Throw in a hard disk for storing your movies, and you may not be able to afford popcorn.

But I don't want to discourage would-be animators on a budget. If you're content to create black-and-white animations and view them on screen (or even film them frame by frame from the screen), you can get by with a smaller system. Many of the programs I've discussed run—or at least trot—on a 1MB Mac Plus. And if you have HyperCard, you already have a serviceable animation program.

Some traditionalists in the animation

world decry computer animation, just as some musicians deride synthesizers and MIDI. I'd hate to see traditional cel animation disappear, but I also enjoy the distinctive, ethereal look that 3-D computer animation provides. Computer animation combines the magic of moving pictures with the art and science of mathematics and technology. Winsor McCay would have loved it. ▣

Jim Heid is a Macworld *contributing editor who focuses each month on a different aspect of Mac fundamentals.*

AUTHORING, MODELING, AND ANIMATION

BY LON MCQUILLIN

What do authoring, modeling, and animation have in common? Not much, except that the pioneers of multimedia authoring systems loved modeling and animation. The rest is history.

Figure 1: Animation may be recorded in either frame-by-frame or real-time mode, depending on the animation software being used and the quality of results desired. With simple images, real-time recording is possible, either directly from the 3-D program or via PICS files from an animation program such as Director. With complex images, frame-by-frame recording, using a VCR controller for single-frame animation, produces top-level results.

There are two common approaches to creating a multimedia extravaganza. One uses an authoring system as its base. An authoring system — such as HyperCard, SuperCard, Plus, or Authorware — is ideal for specifying relationships among segments or user-controlled events. The other approach uses an animation package as its starting point. Creating animation on the Mac can stretch your computer to the limit, however, and you must often use several programs together to get the desired results.

Logically speaking, animation and authoring go together about as well as oil and vinegar. But shake the bottle hard enough, and you get salad dressing. And historically, that's what has happened in the field of multimedia applications — mainly because products such as MacroMind Director have evolved from animation-creation packages to become the control centers of true multimedia presentations that combine audio, video, and interactive functionality.

Simultaneously, application-development or authoring systems such as HyperCard have

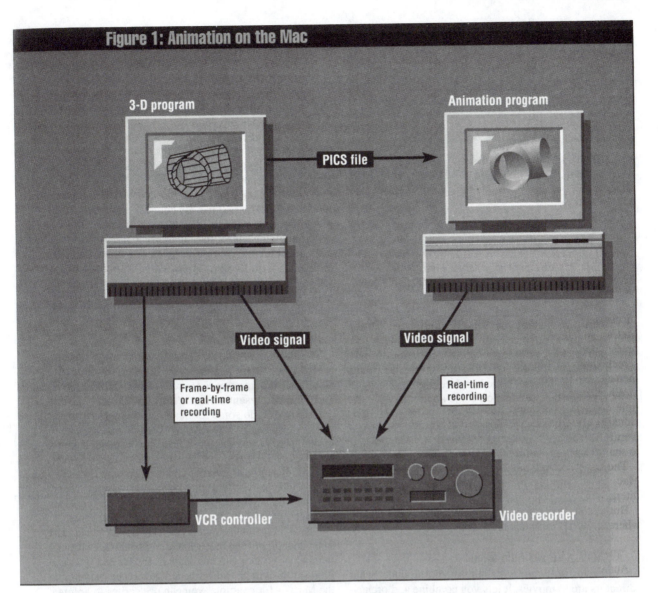

Figure 1: Animation on the Mac

gradually added the ability to import animation sequences. For example, HyperCard 2.0 can call up and run Director routines. Although HyperCard is an authoring package, a lot of animation is now done in what are nominally HyperCard applications, thanks to its XCMD facilities and add-in capabilities.

In short, the line between authoring and animation software started out blurry and has become more so.

Modeling software has tagged along with all this because it makes animation look so real,

even though the massive computation requirements of modeling conflict with the speed requirements of interactivity, another important element of multimedia.

Consequently, today's desktop-media producer faces a multimedia ménage à trois — authoring, animation, and modeling — of intertwined applications. The result is a crop of applications that produce a compelling sensation of reality — the overall result all multimedia applications strive to achieve.

This sense of reality has long been a feature of

Figure 2: A frame from the 1909 animated cartoon "Gertie the Dinosaur," in all probability the first interactive multimedia presentation. At some screenings, animator Winsor McCay would appear live onstage during the film and the animated Gertie would appear to snatch off his hat.

successful presentations. Back in 1909, cartoonist Winsor McCay toured America with a simulated interactive presentation. McCay would stand onstage beside a movie screen and talk about Gertie, the Dinosaur, who would then appear on-screen beside him (see Figure 2).

What made the presentation come alive were its multiple elements: the animation, the well-drawn dimensionality of Gertie, and the illusion that McCay and Gertie were interacting — at one point, for example, she appeared to snatch off his hat and eat it.

The same elements work even better today. With your Mac and multimedia tools, you can now build in real interaction. And you don't have to draw every frame.

But you still have to assemble everything, and that's where authoring software comes in.

Authoring Versus Animation

Authoring software is to multimedia products what directors are to movies. It lets you combine and orchestrate the different media elements you're using into a coherent, sequenced whole.

Authoring software comes with a wide range of prices and capabilities. The most widespread is HyperCard, and the best known of the high-end products is MacroMind Director, but these are far from the only choices. Some

authoring packages are also animation packages, and vice versa. We'll delve into animation first.

Persistence of Vision

Animation is a succession of still images that the human mind interprets as motion because of what is called persistence of vision.

This phenomenon breaks down if the succession of images is slower than 16 to 18 fps (frames per second), because we then start to perceive flicker. Traditional motion-picture and television films stay well above this threshold by running at 24 fps. Professional video normally runs even faster, at 30 fps. By contrast, a standard RGB Mac monitor repaints its display at 66 fps.

Although the Mac's fps capability is more than sufficient for animation purposes, there are some capacity problems: A full-color, 24-bit, 640-x-480-pixel image represents nearly 1 megabyte of data. No Mac is designed to shove so much data swiftly enough through its memory, bus, and storage facilities.

Real-Time Animation

The Mac can handle limited animation in ways that are practical for many applications. The simplest is just to slow down the frame speed to well below the flicker level and treat the presentation as a fast slide show.

But you can do some forms of real-time animation as well — especially when your background is going to remain pretty much the same from one frame to the next (see Figure 3). If you treat the background as one layer, often with a large amount of visual information, you can animate the active foreground elements only and superimpose them on the background, as a separate layer. This approach greatly conserves your computer resources.

These foreground elements are called *sprites* and exist independently of the background. Although the size and number of sprites you can display are limited, you can produce this kind of real-time animation successfully on the Mac — for example, you can use sprites to animate charts and diagrams in a business presentation.

Special Effects

The current crop of software lets you supplant standard slide and overhead-projection shows with livelier fare. You can use special effects, such as wipes and dissolves,

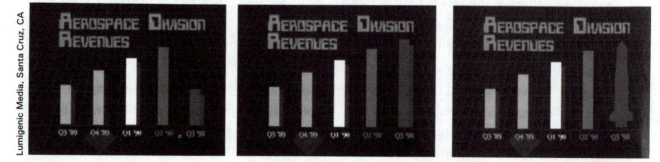

to display textual information and animate charts to convey information more clearly. Animation is also ideal for showing processes that are impossible to photograph or film — the operation of a gas engine's combustion chamber, for example.

Mac Display Versus Videotape

The two most common output formats for Mac-based multimedia displays are the Mac itself and VHS-cassette videotape. Direct animated display on the Mac is constrained by limits of image quality and movement but offers the potential for interactive control.

Output to tape, on the other hand, can provide highly sophisticated image complexity and quality, because each frame is created and recorded individually. You pay for this in the time required to create complete sequences and in the probable loss of interactivity.

With videotape, interactivity is difficult because tape is a sequential-access medium. Designing applications around the need to wait up to two minutes to reach the needed tape segment is troublesome. Also, the VCR must have a serial port, so your computer can control it, and single-frame accuracy, so your computer can find the segments it needs.

Some inexpensive consumer video recorders have limited stop-motion-recording features, which may suffice for some applications.

One last factor: NuBus boards and boxes with NTSC (National Television Standards Committee) video output vary in the quality of image they deliver. The cheapest ones may produce flicker and other problems.

In simple terms, the more animation, the less interactivity, and vice versa. Before you decide which multimedia authoring and animation tools to use, you have to make some decisions about your output format.

If you start with an animation package, you'll want to consider whether the results can be exported into an authoring environment. If you start with an authoring package, you'll need to evaluate its animation capabilities and/or its ability to call animation sequences from other programs. The logical start is HyperCard, the ubiquitous multimedia software.

HyperCard. HyperCard 2.0's design, wherein cards can share common backgrounds, lets you create interactive animated presentations of sometimes surprising complexity, considering HyperCard's past limitations. The release of version 2.0 has opened up HyperCard's scope, and the extensibility of its design lets you add in new capabilities provided by third-party vendors, such as ADDmotion from Motion Works.

HyperCard 2.0 adds many long-awaited features, including variable card sizes; support for 32-bit QuickDraw color (through a new built-in XCMD); multiple windows; styled text; and hypertext, which lets users click on an individual word to get further information.

The underlying HyperTalk programming language and script editor have both been enhanced. HyperCard 2.0 also compiles scripts upon first execution of a handler, making most stacks run faster. The compiled version, however, is not saved on-disk for later runs.

The main limitation of HyperCard for animation is that it replaces entire screens, which slows it down and takes up a lot of memory. But you can use HyperCard's XCMD capability to run animations developed in other products such as MacroMind Director.

ADDmotion. This Motion Works product installs seamlessly into HyperCard 2.0 and adds the ability to do sprite-based animation. ADDmotion adds Film and Paint to the Edit menu and New Film to the Objects menu. The Paint command opens a 32-bit-color Paint window that's accessible from within HyperCard.

You can save paths for animated objects in libraries, and you can save "cast members" — animated figures created in the Paint window or imported from other programs — in libraries as well. ADDmotion lets you play up to four sound resources simultaneously and record directly from Farallon's MacRecorder.

Animations saved in stacks include all the resources necessary to play them through HyperCard 2.0, even when ADDmotion isn't present. This means you can distribute applications to other Mac users who don't have the program.

The user interface is based on visual feedback on the screen and lets you create animations without having to resort to HyperTalk.

Studio/1. At the low-price end is Studio/1 from Electronic Arts, which works in black-and-white only. You can use Studio/1 to create and play back animation or to create sequences you can view from within HyperCard. Its tools are innovative and easy to use.

Figure 3: A standard bar chart gets some visual punch with the addition of animation. MacroMind Director makes it easy to have the bars grow onto the screen and to have the bar on the far right change into a rocket that blasts off and shoots off the top of the screen.

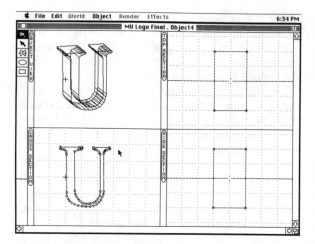

Figure 4: Objects in Swivel 3D Professional's design window are made up of a series of linked lines that describe a polygon. To create the *U* in the *MacUser* logo, the logo was first scanned as a PICT file, which was then pasted into the background of Swivel 3D Professional's object-cross-section window. The letters were then traced over the logo with the polygon tool.

Studio/1 offers two animation techniques. In the first, you create an animated brush that you can drag across the screen. For instance, if you create eight frames of a swimmer, each frame using successive arm positions, and then drag the brush across the screen, your swimmer will swim as you drag the brush along.

The second animation technique does *tweening:* You define the object you want to move and describe its path — including pseudo-3-D moves created by sizing objects — and Studio/1 does the animation work for you.

Animations created in Studio/1 can be exported to HyperCard or to Electronic Arts' own Gallery program, which can play back animations either automatically or under user control.

SuperCard. SuperCard, from Silicon Beach Software, works like an enhanced version of the earlier versions of HyperCard. It has painting and drawing capabilities, color, variable screen sizes, multiple screens, and more. SuperCard can import graphics in several formats, including HyperCard, MacPaint, PICT, and TIFF; can compile animations for increased speed; and can export PICS-format files. The program includes an unlimited run-time license.

Plus 2.0. Spinnaker's Plus 2.0 also works like an enhanced version of HyperCard 1.x. Plus' unique feature is that it works across multiple platforms. Stacks created on the Mac can be opened on PCs, in either the Windows 3.0 or OS/2 Presentation Manager environment. A run-time license costs extra, however.

Like SuperCard, Plus supports color and resizable windows and can open existing HyperCard 1.x stacks. A version that can accommodate HyperCard 2.0 stacks is in development.

Authorware. At the highest end of authoring software is Authorware, from Authorware, the only completely object-oriented multimedia authoring tool for the Mac. Its animation capabilities include the frame-based animation common to most animation packages, plus data-driven animation, in which simple objects can be described quantitatively. These can then be set to behave in different ways, depending on user input, as with the chess pieces in a chess program. Its visual-mapping approach eliminates conventional scripting and enables nonprogrammers to create and deliver interactive multimedia applications by arranging objects along a flow line rather than by using conventional scripting.

The objects represent program events, such as interactions, decisions, and calculations, and multimedia objects, such as displays, sounds, animation, and video. All logic branching is shown graphically, and the result is directly editable. Authorware provides more than 100 built-in system functions and variables, as well as models for both logic and content. Farallon's MacRecorder is included in the Authorware package.

MediaTracks. Farallon's new product MediaTracks promises to be a boon for the concept of individualized training for Mac-related tasks. MediaTracks lets users record screen activity and then later add sound overlays, using MacRecorder, and more PICT or color PICT2 graphics.

For example, you might use MediaTracks to quickly and easily create and polish a training file on how to use Font/DA Mover or how to wade through a complex spreadsheet-based report. The resulting file would be much smaller than a similar MacroMind Director file.

MacroMind MediaMaker. MacroMind MediaMaker is an authoring program that lets you assemble full multimedia productions from a variety of sources — including videotape, videodisc, and CD — as well as Mac sounds, graphics, and animations.

The program is designed for use by people who aren't multimedia specialists. You can do complete productions without recourse to extensive scripting and can use the on-screen collection of tools without referring to the documentation.

MediaMaker has two main parts: Collections and Sequences. The Collections section lets you build databases of elements of any of the listed types. Each element is represented by a "picon" (picture icon).

In the Sequences section, you lay out your picons in the order in which you want them to appear. After you've juggled them around, you can select the Print to Tape feature to record your presentation on videotape, using a display board that provides NTSC video output.

MacroMind Director. Director can create, record, store, and play back complex animation sequences. It can also synchronize playback of sound files, such as those created with MacRecorder, and trigger MIDI devices. A companion product, MacroMind Accelerator, speeds up playback of complex scenes and large animated elements.

The Player utility included with Director compiles animated productions into "projectors" that can be viewed without Director.

Director provides a Paint window you can use to create new sprites, or cast members, and to import a wide range of graphics and sound files. Because it can import PICS files, Director lets you play back animated sequences created in several 3-D-modeling programs. A common use of Director is simply to play back files created in other programs.

Director-based animations are generally of the sprite-over-background type, but more-sophisticated projects are possible. For example, you can create a "movie" of digitized video comprising, say, twenty-seven 8-bit-color 320-x-240-pixel frames. With the images superimposed over a drawing of a movie screen created in Director's Paint window, and playing the movie as a repeating loop, you can achieve a display rate of roughly 22 fps on a Mac IIci. Viewers see this as very fluid motion, although 8-bit color imparts a posterized quality.

A product called MacroMind Windows Player allows you to play Mac-based Director sequences on a PC under Windows.

FilmMaker. Paracomp distributes this French program, which is quite a bit easier to use than Director but which boasts a wide range of abilities nonetheless. It consists of four modules (Animate, Color, Sound, and Present) and five utilities (FilmMaker DA, Mark, Picture Runtime, Sequence Runtime, and Sequence Transfer). FilmMaker provides the ability to render colors in 8-bit palettes, with anti-aliasing, dithering, and remapping options. The Animate module offers tools for performing a wide range of animation effects on graphic objects, such as having the colors change while an object rotates. The Sequence Transfer utility lets you play back animated sequences in HyperCard. The Color module imports PICT, PICS, or EPS files and provides for color manipulations. The Sound module imports sound resources created in other Mac applications. The Present module assembles presentations from sequences and images, offering a wide range of transitions.

Animation Stand. A new $2,000 system, Linker Systems' Animation Stand, can be used for creating real-time animation or working in frame-by-frame mode. It uses the metaphor of the traditional animation device from which it takes its name. It has five modules. The first two comprise a sophisticated painting program that falls between a traditional 2-D program and a 3-D-modeling program and a programming language that falls between HyperTalk and BASIC. The language compiles animation sequences and commands for maximum speed.

The other modules are exposure-sheet and optical-press simulators, along with a transport-controller interface that can work with frame-by-frame controllers.

interFace. This unusual program from Bright Star Technology lets you create "actors," which are animated talking heads. Using up to 120 facial-image positions, you can synchronize the actors' faces either to digitized speech or to text that is "spoken" by MacinTalk. You can draw or import any kind of image — not just faces — but you're limited to 120 animation frames.

interFace primarily allows you to create animations for use in HyperCard, SuperCard, or any other application that supports XCMDs. BrightStar demonstrates the product with a training program on how to use Excel and a Japanese-language teaching program.

MouseRecorder. This unique system, currently under development by Whitney Educational Services, will let you record and narrate live on-screen presentations. It works by eavesdropping on mouse and keyboard signals before they get to the Mac. The signals are recorded on a standard stereo audiocassette tape. One channel of the

Figure 5: The *MacUser* logo in six views as created in Swivel 3D Professional: wire frame, with hidden lines removed, outlined, shaded, smoothed, and fully shaded and smoothed with shadows and anti-aliasing

Figure 6: The *MacUser* logo was scanned into a Mac and turned into a PICT file. The PICT file was then ray-traced and extruded in StrataVision 3d, which was also used to create the crystal ball. The wood paneling was loaded from the attributes file of Studio/8. Using version 1.4.2 of StrataVision 3d on a Mac IIci, the rendering time for this image was approximately 12 hours. Version 2.0, due out by the time you read this, promises to render roughly three times faster.

tape records the ADB (Apple Desktop Bus) signals going to the Mac, and the second channel is available for voice narration. When you play the tape back through MouseRecorder, the Mac sees the mouse and keyboard signals and duplicates whatever you did during the original recording — rather like a player piano for the Mac.

Professional-Quality Output

When it comes to real-time animation, the level of quality required is defined by the purpose for which the Mac is being used. For example, an animation of a piston's movements inside an engine can look pretty good, but nobody would mistake it for traditional cel animation. Producers who create real-time animation on the Mac typically describe their results as being of "high-end industrial" quality, to which they invariably add "but

at a much lower cost and development time than with traditional methods."

There are high-level applications in which the Mac's limited real-time-animation capabilities work well; for example, the standard TV-news weather report that uses a local map with animated temperatures appearing at various locations. This can be done on the Mac and output as an NTSC video signal, with no apologies needed for image quality.

Interactive Presentations

Interactive presentations are an ideal use for real-time animation on the Mac: Applications can allow users to control a presentation, creating a path through and around its various segments.

Architects, for instance, might use a program such as DynaWare USA's DynaPerspective to create a 3-D model of a building and then create a series of views from within it, allowing clients to move from room to room at will. Each view can include animation within a scene. A view out of an office window, for example, might include cars moving on the streets. Virtus' forthcoming Virtus WalkThrough can also be used for similar walk-through (or fly-through) explorations of a so-called virtual reality.

Interactive-animation tools include HyperCard, which can perform simple card-flipping-type animation; SuperCard, which is able to play PICS files created by 3-D drawing programs and move graphic elements around on the screen; and at the high end, MacroMind Director, which offers a scripting language called Lingo that resembles HyperTalk and gives users control of presentations.

Frame Tearing

Designers should keep several factors in mind, including the quality of movement that can be achieved when doing animation on the Mac.

Although you can animate a few sprites fluidly, large numbers of them or larger objects exhibit frame tearing, because the Mac can't redraw an element in its new

Figure 7: Infini-D creates realistic images through surface normal perturbation and constructive solid geometry.

position fast enough for it to be finished before the next frame needs to be displayed.

Frame tearing results in an effect similar to seeing a truck through heat waves on a highway, except that the waviness is sharp-edged.

You might think that putting the speed of the IIfx to work would dramatically enhance the abilities to produce real-time animation, but the difference is only incremental. NuBus is the culprit; it simply can't transfer enough data fast enough to smoothly animate large elements.

Video Anomalies

If a presentation will be videotaped for distribution, you need to be aware of the characteristics of the NTSC video signal: Avoid horizontal lines that are an odd number of pixels wide. Avoid abrupt transitions between primary colors, especially bright reds. And for the best results, use software that performs anti-aliasing (blending of colors to smooth transitions) for both graphics and text.

Unfortunately, although anti-aliasing creates much-better-looking NTSC images, it also often results in perceptible blurring when the same image is viewed directly on the Mac's display. Thus images and animation for video distribution must be designed differently from those that are for direct presentation on the Mac.

Another factor to deal with is color saturation. Fully saturated colors tend to smear on an NTSC monitor. This problem is addressed by one 2-D paint package due out at about the time you read this: Time Arts' Oasis. Among the palettes available in Oasis is one called Video-Legal Colors that limits saturation to 75 percent. It also provides Safe Action and Safe Title, which superimpose overlay grids within which action and titles should be confined to avoid falloff around the edges.

Oasis also supports superimposing two documents simultaneously with selective transparency and also lets you superimpose graphics over full-motion video. This last feature requires a 32-bit-video output board such as the TrueVision NuVista+.

The software and hardware currently available for the Mac let you create high-quality animation with final output to videotape.

3-D Modeling

The new frontiers of computer animation are reached through 3-D-modeling and -rendering software such as Silicon Beach's Super3D, Paracomp's Swivel 3D Professional, DynaWare USA's DynaPerspective, Strata's StrataVision 3d, and MacroMind Three-D.

In developing animation, you first create individual elements in a drawing environment. You keep each element in memory as a mathematical description that the software then draws on the screen, so you can move it, rotate it, and link it with other objects (see Figure 4).

Because 3-D programs keep track of the actual shapes of objects as well as any physical characteristics assigned by the designer, the movements that can then be assigned can be highly realistic. The metal band of a wristwatch, for instance, can be designed to bend inward but not outward.

Once elements have been described — either by entering them with drawing tools, by tracing over bit-mapped images, or by reading in a text file containing a list of coordinates and attributes — the program can display them in a variety of forms. These range from simple wireframe representations to fully modeled and smoothed shapes, with numerous controllable light sources (see Figure 5).

The process of transforming an object from a mathematical description into a realistic-looking image is called rendering and can be approached in different ways. Several techniques, including mapping, shading, and ray tracing, are often used in combination.

Mapping. Mapping involves taking a two-dimensional graphics file that describes a surface texture — for instance, wood grain, marble, or chrome — and projecting it onto an object's surfaces. At more-sophisticated levels, often called bump mapping, mapping includes not only an image but a texture as well. Libraries of texture maps

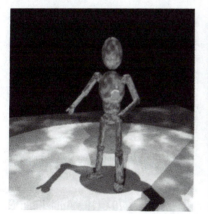

are available from many of the companies that produce modeling and rendering software.

Shading. Shading, which you can perform in several ways, gives objects dimension by controlling the way one or more light sources illuminate objects. More-sophisticated shading techniques provide smooth transitions on curved objects and let objects cast shadows.

Ray Tracing. Ray tracing is a computation-intensive technique that plots a view of every pixel in a scene through a virtual camera's lens. It takes into account the location, strength, and quality (pinpoint or diffuse) of all light sources, along with the surface characteristics of each object in the scene. Shiny objects reflect other objects, and realistic shadows are cast. Ray-tracing one complex image can take hours on a Macintosh (see Figure 6).

Tweening. When it comes time to create motion, some 3-D programs can perform tweening (slang taken from the animator's term *in-betweening*), in which you specify a starting position for an object (and/or the camera), an ending position, a path to follow, and the number of frames over which to distribute the motion. Start the tweening process, and the program automatically creates the movement for you.

Modeling and Rendering Software

Super 3D. Super 3D 2.1 from Silicon Beach Software is a drawing program that gives users extremely precise control over object size and shape and over camera placement. It can control up to four light sources and can save animations as movies that you can play back at speeds approaching or at real time. In addition to using its drawing tools, you can have Super 3D read a text file called a display list and then create objects and images automatically.

Super 3D is designed primarily as a modeling program and doesn't perform high-level rendering, although its files can be imported into several rendering programs.

Swivel 3D Professional. Swivel 3D Professional from Paracomp can map a texture from an external file onto an object, control up to eight light sources, do several levels of shading, and perform anti-aliasing of objects to eliminate the jaggies. It offers powerful tweening capabilities, including accelerating and decelerating moving objects for fluid, natural motion.

Swivel 3D's real-time-animation capabilities are limited to primitive wire-frame representations that let you see how movement will appear, but the program can export PICT, PICS, or Scrapbook files that can be played at full motion in MacroMind Director or, for PICS files only, in MacroMind Accelerator. It can also trigger animation controllers, such as those from Diaquest, for automatic recording of rendered animation sequences.

Part of Swivel 3D's appeal is its intuitive interface; experienced Mac users should be able to start creating substantial 3-D images after an hour's training and another hour's fiddling.

DynaPerspective. This 3-D-modeling and -rendering program really excels in animating architectural models. You can import PICS animations from Director, but the native High Speed Film format is much faster. Files can also be exported to StrataVision 3d and Pixar's MacRenderMan for high-end rendering.

StrataVision 3d. This high-end offering from Strata is notable for both its reasonable price and the fact that it combines modeling and high-quality rendering in a single program. As with DynaPerspective, this saves you from having to switch among modules or programs to perform a complete modeling and rendering session.

Like DynaPerspective, StrataVision 3d provides familiar drawing tools for constructing objects, along with the expected extrude and lathe functions. It can also autotrace imported PICT images and smooth text, if you have Adobe's ATM installed. It offers a wide range of control over rendering options, and when the appropriate ones — such as reflectivity and refraction — are enabled, StrataVision 3d does ray tracing. It's the lowest-priced program that does this. Complex images with clear or reflective objects can take a long time to render, however.

Unlike Swivel 3D, StrataVision 3d doesn't perform

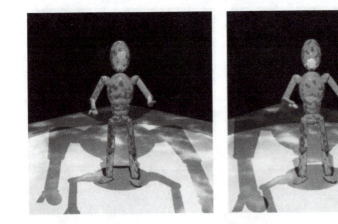

object animation. Strata does, however, offer a companion program, StrataFlight, that can animate the virtual camera's position to create walk-throughs and fly-bys.

Another companion program, StrataVision 3d SRX, adds RenderMan file-output capability to StrataVision (see "MacRenderMan," later in this article).

Infini-D. This new Specular International program, due by the time you read this, promises unique modeling capabilities at a very competitive price (see Figure 7).

It can produce such effects as a truly bumpy surface with appropriate shadows or spreading ripples (called surface normal perturbation) or let you slice into solid objects with various types of "cookie cutters" (called constructive solid geometry). It also supports self-shadowing objects.

Also unique are its mapping capabilities. Specular provides a library of 3-D texture maps. If you map a wood grain onto an object and then cut that object in half, the end cut will have a realistic cross-section of the wood that matches the cut point. The net result is almost eerie in its realism.

Infini-D offers animation functions, including interpolation, not only between object positions but also between colors, surface maps, and even wire-frame shapes. A red ball, for instance, can transmogrify smoothly into a blue cube. The program performs various levels of rendering, including fast ray tracing, and it can also output PICS files for viewing in programs such as MacMind Director.

Ray Dream Designer. A new company named Ray Dream has previewed this powerful 3-D-modeling and -rendering program. Ray Dream Designer offers a wide range of modeling tools and capabilities that are particularly intuitive and easy to use. When you're linking objects, for instance, it presents a Finder-like hierarchical display that makes relationships very clear.

Ray Dream Designer provides a variety of shading functions, including very fast ray tracing. It also performs two-pass anti-aliasing of PICT files, including those created in other applications. The program should be available by the time you read this.

Presenter Professional. Visual Information Development, Inc. (VIDI) offers a series of modeling, rendering, and animation programs. A new bundle plus update, Presenter Professional, combines and supplants the earlier 3-D modeler, presenter, and ray-tracer products. Presenter Professional offers improved ease of use.

The Presenter module performs rendering and animation functions. It has a spreadsheetlike design that permits time-line-based control of individual elements within a scene, and it can import 2-D files created in MacDraw, FreeHand, Illustrator, or other 2-D programs and render and animate them in 3-D. The RayTracer module performs the higher-quality rendering that its name implies. VIDI also offers a collection of surface textures such as marble, wood, and grass in a package called Dimensions Materials.

MacroMind Three-D. Also structured as a series of modules, MacroMind Three-D includes 3DWorks, RenderWorks, and ImageWorks. 3DWorks imports 3-D models created in a variety of CAD and modeling programs and is used to move the objects, camera, and lights to create animated sequences. The Render command then launches RenderWorks, which renders the wire-frame models and in turn launches ImageWorks. ImageWorks can perform anti-aliased compositing of objects and is the module in which the final animation is put together. ImageWorks can create a PICS file for use in MacroMind Director or Accelerator or as a series of PICT images. It can also control video recorders, using Diaquest, Lyon-Lamb, or Videomedia controllers.

Sculpt 3D. Sculpt 3D, from Byte by Byte, is a high-end vector-based object-modeling and -rendering program. It sports extensive editing abilities, allowing objects to be modified at the vertex level. This gives you more freedom than you have when you work with the primitive objects used with most modeling programs. Sculpt 3D is able to perform numerous levels of rendering, including ray tracing.

ElectricImage Animation System. The ElectricImage

ARQ Software Corporation

Figure 8: Pixar's MacRenderMan accepts input files from a wide variety of 3-D graphics programs and performs photo-realistic rendering. Note the motion-blur effects on the ball and flipper.

Animation System, from Electric Image, can't create 3-D models, but it can import files from virtually any modeling program and perform extremely high-quality and high-speed rendering.

The ElectricImage Animation System can control a wide range of video-recording devices and can work with multiple networked Macintoshes to speed the animation process.

MacRenderMan. Pixar has proposed its RenderMan protocol as a graphics-interface standard for a variety of computers, and the protocol has been embraced by many software suppliers.

MacRenderMan accepts files from 3-D-design programs in the RenderMan Interface Bytestream (RIB) format — already widely accepted — and outputs PICT, EPS, and TIFF files. It produces photo-realistic images and offers capabilities never before available on the Mac, including motion blurring, which can greatly enhance the realism of animated sequences.

RenderMan is a rich interface that can be exploited at many levels. The extent to which various developers of 3-D software take advantage of its capabilities will determine the results achieved with different programs.

Levco's RenderMan Accelerator card is built around an Intel i860 processor running at 40 megahertz. It can speed image rendering by a factor of 30.

MacIvory. Symbolics took both the hardware and the software from its popular graphics workstation to create MacIvory, the highest-end rendering and animation system for the Mac. Using a 40-bit CPU on a NuBus card, MacIvory can create broadcast-quality images. If you

want to go all out, $50,000 will buy you the card and the software along with a fully loaded Mac IIfx, a NuVista video card, and a host of peripherals.

The Ideal System

Simple animation and authoring can be done on anything from a humble Mac Plus on up. But without a more sophisticated platform, you'll quickly come up against a host of bottlenecks — including color, speed, screen real estate, and output options.

An ideal system for creating animation and doing authoring would be a Macintosh IIfx with a 600-megabyte hard-disk drive, an 8-millimeter tape or optical-disc drive for backup and archival copies, a multisync monitor, a 24-bit-color display card, a frame-grabber card, and possibly various video sources — VCRs, laserdisc players, and cameras. All this easily adds up to $20,000 or more.

If your needs can be met by the Mac's real-time-animation capabilities and the software mentioned above, you'll probably find that your productivity will increase, because the available tools make the creation of live animation and presentations much simpler and faster than traditional methods do.

If you're tackling high-end, frame-by-frame animation, however, the time factor is worth careful examination. As the quality of images increases, so does the rendering time.

As an experiment, we took a very simple example — a 3-D rendering of the *MacUser* logo created in Swivel 3D — and made it fly around the screen through 30 frames, creating a PICS file for playback in MacroMind Director.

If transferred frame by frame to videotape, the entire sequence would last one second. With anti-aliasing and shadow options turned on, and running on a Mac IIci, rendering the entire sequence took a bit more than an hour.

Complex images can take dozens of hours to render on the Mac. The introduction of MacRenderMan will open the door to the creation of startlingly photo-realistic images on the Mac but may also leave users gasping at the amount of time necessary to render a single frame. A complex image such as "Pinball" (see Figure 8) can take more than 24 hours to render on a Mac II and more than 9 hours on a IIfx. And on videotape, that frame passes by in $1/30$ second.

The solution is to add dedicated hardware. For example, Levco's RenderMan Accelerator reduces the rendering time of "Pinball" to less than an hour. Of course, there's Symbolics' workstation-in-a-Mac for those who *really* want to get serious. The full MacIvory system fills up all six NuBus slots of the IIfx. But nobody said that state of the art was cheap.

Lon McQuillin humbly tries to personify the Renaissance man. Among his pursuits are the design of computer graphics and animations.

Part 6
Video

MANY PEOPLE equate video and television. With the evolution and refinement of equipment and processes and the general reduction of related costs, however, video has become a viable tool for several new applications, including technical presentations. For example, video projectors allow an engineer to show a real or simulated event in a dynamic manner, in either actual, accelerated, or retarded time. By using computer graphics in conjunction with video, dynamic effects can be added to otherwise static visual aids such as flow diagrams, mechanical concepts, charts, graphs, and schedules. In addition, video can be used to add the participation of a colleague or expert who cannot attend a presentation. Teleconferencing, a special application of video, can even bring remotely located audiences to participate in a technical presentation.

Video is generally used to supplement other visual aids rather than replace them. In some applications, such as training, marketing, and public relations, video may be used to replace a speaker. In a technical presentation, however, the speaker is the principal element, and video is simply another visual aid. As with other visual aids, an engineer need not know how to produce a video but should know enough to use the medium to advantage in presenting his or her message to an audience. The articles included in this part identify the unique applications and advantages of video; describe various techniques for preparing, producing, and using it in technical presentations; and summarize the related equipment requirements and costs.

Video has become an accepted part of business and technical activities. Because it can be used to communicate, educate, and motivate all levels of organization, both internal and external, technical presenters are increasingly seeking ways to adapt video to their particular needs. In her paper, "Inside business TV," Donna Rogers describes a typical in-house video production organization and presents examples of video activities at several major companies. She also addresses relevant subjects, such as operating costs, equipment requirements, video procedures, and related communication links. The information included in this article can be readily adapted to technical presentations.

In his paper, "Using the three-column outline to plan your next video," Charles Kemnitz describes an efficient technique for organizing visual and verbal information into an integrated video package, and for testing the related strategy and theme before starting video production activities. His technique starts by defining the requirements, developing a framework, and building a concept. A three-column format is then used to develop the outline. The three columns list the basic verbal information to be presented, the visual to show it, and the desired audience response. Several factors that should be considered in preparing the outline are also described. This tool is particularly useful for planning a video for technical presentations.

Steve Floyd takes the position that video presentations are primarily visual forms of communication, requiring scriptwriters to rely on pictures rather than words to convey information. In his paper, "Visualization: The key to effective scriptwriting," he describes a five-step process for visualizing a presentation before starting to write the script. The result is a better balance between visual and audio elements.

In his paper, "Audiovisual words: The scriptwriter's tools," Saul Carliner stresses the importance of developing a clear understanding of the characteristics of video, such as dynamics, realism, and special effects. He also presents some specific suggestions for writing the associated narration and reviewing the resulting script.

Technical presenters should have some knowledge of the video production process to ensure effective use of the medium. In their paper, "Production: Turning ideas into pictures and sound," R. H. Amend and R. B. Stoner describe the two basic stages of video production. During the first stage, video producers evaluate a script for continuity, visualization, picture requirements, and other related planning. During the second stage, that is, the actual production activities, producers execute the plan consistent with the resources, schedule, and budget. For technical presentations, engineers should coordinate with a producer to assure the resulting video is accurate and reflects the desired message.

Inside Business TV

DONNA ROGERS

"In corporate America, video has become an accepted part of business life. As it continues to communicate, educate and motivate, media managers are seeking even more sophisticated and diversified ways to serve a greater number of internal needs."

Today's corporate executive realizes that video is an effective internal vehicle of information. It communicates, educates and motivates at all levels of an organization. Twenty years ago, the use of the medium for anything other than advertising was unheard of; today, in-house video departments are discovering a need for more sophisticated and diversified ways to serve a greater number of divisions within a firm.

Discovering the advantages and disadvantages to in-house services, management faces tough choices. Is it possible to confront potentially limiting impediments and change these to opportunities? Can they stay current and up-to-speed—firing at full capacity—while operating on a company budget? To what point should in-house talent be utilized and when should professionals be called? What areas should receive special attention: training, sales promotion, satellite transmission?

There are no easy answers. Several video specialists and managers, from a range of corporations throughout the country, offered a sampling of current trends for start-up to established departments, for staffs of two to 20.

According to Dan White, manager of video communications, American Airlines, Fort Worth, Texas, effective staffing is assured by insisting that each member is accomplished at more than one skill. The two basic job titles are producer/writer/editor and photographer/editor. Everyone must be proficient at everything. The latter do lighting and writing—"the whole ball of wax"—while the former write, edit and field produce.

"Our producers don't shoot," White says, "because obviously it is a very technical end of the business. But we don't specialize to the point that the producer is only a producer, a writer only a writer or an editor only an editor. The nine-person staff has come from broadcast television where you had to do those things—you learned to do them early in your career." The company found the abundance of experience to be a big plus in building a smooth operation, which started as a two-man staff in 1984.

Who's the boss? An advantage to producing video in-house is that the corporation has total control from the conceptual stage through the entire production process. "You do not have to relinquish it to an outside production house that might not be as totally up to speed with the subject matter as your staff," White states. Moreover, by designing the training programs, they come in contact with nearly every facet of the company and gain valuable insights.

A bi-monthly video magazine for American Airlines' employees, called *UpdAAte*, and special messages from the chairman of the board and president are the tasks of the multi-pronged department. In addition, all the training programs are produced by the staff; tapes requested by the news media and ad agencies are also provided.

Only ten percent of the work is recorded in the studio. Consequently, the finished product resembles real life because it is shot in the field. Another employee video magazine, *Format*, has "almost a news fla-

ON CAMERA: American Airlines' video communications manager Dan White taping a closing stand-up in San Juan, Puerto Rico.

IN CONTROL: AA producer/editor Susie Robbins applies the finishing touches to a training program in the on-line suite.

Reprinted with permission from *Audio Visual Communications,* pp. 54–61, Feb. 1987.

CLOSE-UP: Chief photographer Max Warren with one of American's three Betacams.

vor to it, a natural sound, similar to when you turn on your television set at home and watch your local newscast," White notes.

"All employees are used for hosts and actors," White explains. "And because the staff has prior commercial TV broadcast experience, it is helpful during production."

According to White, bringing video in-house (in 1984) has saved the company a considerable amount of money while increasing the number of projects produced during the year. For example, last year more than 147 programs were completed, typically ten minutes in length.

Plain economic sense. White comments that easier access is provided to departmental user groups throughout the company. Management of these groups may have contemplated producing a videotape but lacked the expertise to do it themselves, or the money to spend on an outside production house. "We can accomplish the goal in-house with what we feel is equally good quality at a much lower price," he states. "We consider ourselves a production house," he adds, "and they are our clients."

White believes a cost-efficient method of conducting business is the 100 percent chargeback system that went into effect on January 1, 1987. It is zero-based or, in other words, each item is justified on the basis of actual cost and no profit factor is budgeted.

On the other side of the coin, working in-house has its disadvantages. White points out that a full-fledged commercial production facility has digital video effects systems while a corporate studio cannot afford these expensive pieces of equipment. "For our purposes, at this point in time, it is hard to justify the $150,000 cost for an effect or two," he notes.

Because the American Airlines' video department can have three camera crews on the road at a given time, two editing stations are a studio necessity. In the event that one malfunctions, a back-up system is on-line. "We don't have the luxury to sit on our thumbs until one gets fixed because we have that much of a workload," White says.

Quick take off. Three years ago, the personnel division, which had previously contracted video projects to outside vendors, transferred its duties to a separate video department. The primary goal then was to produce *UpdAAte,* the "grand-daddy" video magazine for all employees. As mentioned, it is similar to a news magazine but, White explains, "instead of broadcasting we are narrowcasting to 50,000 employees." At present, a "Flight Attendant Advisory" series is also being produced and another program, titled "Action," is distributed to terminal employees and ticket and gate agents.

Video communication is considered by management to be one of the most effective information tools in the company. According to a recent survey, 50 percent of the staff had seen the last three consecutive *UpdAAte* installments, which are offered on a voluntary viewing basis. White concludes that the program is the video department's major success.

"We're proud of it, and from what we hear, our employees are proud of it too," White reports. "Our chairman of the board and president is dedicated to communications' progress, and he is one of the reasons it exists. We feel it does communicate and that is the name of the game. It lets them know where the company is going, what our goals are and how we can work together to achieve them."

High in fiber. In the late sixties and early seventies, according to Mike Shoup, manager of audio-visual services at Kellogg, Battle Creek, Michigan, the video department originated because of an engineering training need. It has made several evolutions since then but, for the most part, continues to utilize audio-visual presentations for instruction and internal communication. Because it is a division of public affairs, 20 percent of its work is directed at the consumer marketplace.

Not only is upper management frequently taped, but broader aspects such as engineering, R&D and manufacturing considerations between separate sites are covered as well. Subsidiary and international operations are other program subjects.

Budgeting is handled differently at Kellogg because there is no chargeback system in effect, other than outside vendor costs that depend on which client the video group is serving. "Some of the clients utilize what is available while others go outside for talent, scripting or other supportive media," Shoup says.

A variety pack. Generalists are the name given Kellogg's six video staff members. With a full production studio available, each of the individuals must be proficient in more than one medium. Be-

cause there is no resident engineer, equipment repairs and maintenance are contracted on a freelance basis.

The facility is equipped with two complete minicam remote units, a three-camera studio and on-/off-line editing systems. Although a capital plan is designed to upgrade the video hardware inventory on an annual schedule, in-house production can be limited because "we are not always able to take advantage of the latest creative material and technology available," Shoup admits.

The Kellogg video center produces approximately 45 programs each year. Presentations average 15 minutes in length and most frequently are recorded in the field. Twenty percent of the workload is accomplished in the studio, including all post-production projects.

Considering the management point of view, Shoup adds that the greatest success of audio-visual services is "the realigning of priorities in fulfilling the company's needs. In acting as a vehicle between different divisions and operational units, we have been instrumental in improving communication."

From the beginning. "The video department at Colgate-Palmolive evolved; there wasn't a day it all started," notes Bob Murray, director of corporate communications at the firm's New York City headquarters. "It started off simply and went on from there. We've been involved in commercials since the beginning of television. We were one of the first to ever sponsor a TV spot. As far as a means of internal communication among our people, that's been a relatively recent development."

Editing facilities have been on-site since 1950 when commercials on sample reels were centralized and distributed for product managers' reviews. During the past 35 years, thousands of Colgate-Palmolive television advertisements, along with those of competitors, have been stored in the library.

The global challenge. Questions the company strives to answer are: "How do you communicate with 40,000 people in 53 countries around the world? How can you convey a sense of belonging to the same entity and be certain everyone understands the company's goals? As a true multi-national corporation, with sister companies involved in other businesses, the challenge to unite employees is being met with video," Murray states. "Communication is key to us."

Minding the budget is an ongoing process rather than a once yearly review at Colgate-Palmolive. Communication value and what is required to communicate are evaluated regularly. "We are extremely conscious that we don't go overboard in anything we do," Murray emphasizes. "It is important to minimize what is unnecessary."

VIDEO VAULT: Producer/editor Ed Martelle and photographer Bill Elliott in American Airlines' library, which houses more than 30,000 hours of tape shot in three years.

Well-rounded aims. Colgate-Palmolive is diversified in its use of media materials. Audio-visual services manager Jack Hardy employs a wide range of equipment, including a Sony videotape editing system, a small but multi-purpose, two-camera studio that is also utilized for still photography, plus half- and three-quarter-inch videocassette duplication gear.

In-house videotape production basically consists of reports and year-end messages from top management, new product introductions, sales aids and coverage of international meetings that executives cannot attend. Outside consultants are also taped to provide educational and training materials for the sales staff. In-studio and field projects are divided on a nearly equal time basis.

In addition to being a manager, producer and duplicator, Hardy also acts as a sales meeting coordinator, new products photographer and a technician. "I have two or three maintenance contracts with local repair shops," Hardy notes, "but I can do the simple jobs myself. Major work is sent out immediately."

Stars and stripes. Video programming under government contract began in 1983, according to Glenn Wescott, media engineering section head at Sperry Corporation, Clearwater, Florida. While the six-person staff capitalized on training in the beginning, Wescott asserts that it was beneficial to continue growth in other areas.

The Sperry staff has an unusual workload: 40 percent of its programming is developed for in-house applications while 60 percent is completed for external clients, including sponsors' commercial TV spots and projects for other industrial video facilities.

Producing programs for the federal government is a definite plus, according to Wescott. Because of the classified nature of the work, few other in-house or independent production operations are involved in this type of service. For example, Sperry provides the U.S. Army with advanced training material on videodiscs.

"Because military work comes and goes" is the reason Wescott attributes to the evolution of the department from a predominantly internal production center to the considerable amount of outside accounts it has today. At times, the facility is booked solid for six months. The staff, however, often creates time for a variety of projects.

"If they feel like working some extra shifts, it breaks up the monotony and gives them the opportunity to deal with different directors and producers," Wescott comments. "And rather than all technical work, they can see what they produce on TV, which is good for morale."

Staff diversification is another factor that contributes to Sperry's successful communication efforts. "Everyone's an editor, a cameraman, a scriptwriter," Wescott says. The classified government presentations are produced with the department's narrators while company and commercial programs frequently utilize professional voice-over talent.

The equipment arsenal includes complete one-inch videotape production and editing hardware, videodisc systems and a digital effects package.

The budget is based on a 100 percent chargeback system, plus revenue that can be generated independently. The video department also economizes by controlling its own marketing and contracting operations.

A helping hand. Because the department deals with two markets, separate financial approaches are employed. Standard quoting procedures are in effect when working with the government while Wescott establishes the rates for commercial jobs. He reports that assisting other industrial video facilities is the goal.

"While our capital is approximately three-quarters of a million dollars, unfortunately there are many companies that only have a few thousand dollars worth of equipment," Wescott explains. "They can't supply themselves properly yet they put programming on the street. Then everyone pooh-poohs industrial TV.

"We are not looking toward the market to compete with commercial houses," he continues. "Because we do not want poor standards, we try to help the local people produce good TV in any way possible."

EDITING-IN-THE-ROUND: Media engineer Phil Eson working with the Paltex Espirit in the fully-equipped Sperry video post-production complex in Clearwater, Florida.

INSTRUMENTAL TV: Al Bond (above left, standing) in control room, and the shooting stage at Dallas-based Texas Instruments.

In explaining the rate structure, Wescott says, "I took a look at what the commercial houses were charging, at features we had or didn't have in comparison, at how much it costs to pay the staff and at the estimated growth rate." He charges a flat rate of $150 per hour for post-production work, all with one-inch machines. There is no extra charge for equipment, a clock does not start ticking and tickets are not punched once the customer enters the facility. The all-inclusive price is a bargain, according to Wescott.

"We have to limit ourselves though," he continues. "I didn't want to go so low as to allow someone to come in here to do editing with half-inch equipment. We must narrow our market."

Ready for tomorrow. In the last two-and-a-half years, the department has delivered 307 training programs and more than 150 commercials and promotion pieces. Looking to the future, Wescott intends to "re-establish the staff as videodisc producers and also as research and development specialists."

At the forefront as an electronic communications center, the department envisions more media merging due to computers, television and particularly laser optics. "We want to be called formatters," Wescott says. "We want to be a data bank, most specifically for the government. Eventually, we will digitize all types of information. We are not hung up on videotape; we'll record on whatever form the world wants. If they want it on a rock, we'll do it."

Out of orbit. The media center at Texas Instruments' Dallas-based headquarters, which has been in operation for 18 years, is designed to support the corporation throughout the world. Al Bond, manager of the communications complex, originated the department after leaving the Houston Space Center in 1968.

While the group has utilized video in traditional training applications since its inception, it is continually moving into new fields. Satellite videoconferencing, which was adopted in 1980, is used for product announcements, press conferences, employee benefit programs, stockholders' meetings as well as training. The 20-person staff also produces audiotape, slide and vugraph presentations.

"When we began satellite videoconferencing in 1980," Bond recalls, "it was a tough time because of the lack of dishes. We began an ad hoc satellite activity, regardlessly, when we covered the annual stockholders' meeting. We had been transmitting by AT&T lines since 1977, but in that year we carried our meeting live in Europe as we staged it simultaneously in Dallas."

Media center production is accomplished on one-inch videotape recorders while programs are distributed in the half-, three-quarter- and one-inch formats. Studio equipment includes Chyron character generators, CMX editing systems and Ikegami cameras. On an annual basis, 10,000 to 12,000 tape copies are circulated to some 700 monitor/VCR sets around the world.

Auld lang syne. Bond indicates the in-house department is now working with its fifth generation of video cameras. "We initially began in black-and-white and with one-inch Ampex reel-to-reel recorders. At that time," he points out, "it was the only company to guarantee compatibility between its own machines." In addition, TI briefly utilized half-inch, reel-to-reel systems and later moved to three-quarter-inch U-Matic VCRs.

According to Bond, "We were the first industrial corporation in the country to use half-inch VHS. For the most part, we have converted our field network to half-inch playback."

Although the facility attempts to remain reasonably current, Bond stresses that "we don't change with every new model. When you have a seven-hundred-site hardware playback operation, it's not that easy to change."

Distribution of dollars. The budget is based on a chargeback, break-even operation. However, the format varies from the corporate video norm. At TI, management decides which internal groups receive money. "We must carefully look at each department and determine where video is applicable, and only then will the funds be distributed," Bond comments. "Therefore, no department within the organization allocates 'x' dollars to pay us to produce programs for it."

According to Bond, "One of the basic ways we keep costs down is we do it all internally with a well-trained staff. Because there is little turnover (the 'newest employee' is a five-year veteran), they work together to do the job economically. Everyone wears different hats. We sweep floors and we direct shows—we do it all."

ONE-ON-ONE: Sperry senior media engineer James Seals monitors Sony one-inch VTRs.

One of the rare exceptions to this rule is the use of outside talent for on-camera narration. "Many years ago, we learned that because you can design a program does not mean you can speak," Bond says. "Consequently, we allow the producers and writers to provide the script, and we then have the pros deliver the words. In that way, everyone can appreciate the final product."

Generally, the number of video productions exceeds 200 per year. The typical program length is 15 to 20 minutes but could be as brief as 30 seconds or as long as a 40-hour series.

Upping its downlinks. Bond points to a global videoconference last summer as one of the media center's greatest success stories. "If you are looking at size," he notes, "it was our 16-country satellite video symposium last June. Excluding religious groups, it was probably as large as any business has produced. The meeting was transmitted to fifty thousand customers, and we were actually downlinking into countries where it had never been done before. That was a quantum leap."

For the most part, the conference was implemented internally—from conception and organization through production and orchestration—and a minimal number of outside vendors was employed for assistance. Plans are being formulated for the installation of a permanent downlink network at 50 to 60 plants and sales offices, far surpassing the seven reception sites in current operation.

Has Bond attained his goals? "I have known for twenty-five years or better the power of television to communicate and educate," he responds. "When I came to TI, my objective was to make video part of every day life within the corporation. It has taken eighteen years to get where we are. It has been a slow process, but since we started compared to what we're doing now, we have gone much further than I had anticipated."

Bond continues, "In the beginning, however, I had not envisioned satellite global communication—but I'm not afraid to use it. Many people have been hesitant to do that, but Texas Instruments is an electronics company using electronics to that end. And many ideas others are still thinking about, TI has done—years ago." □

Donna Rogers is the assistant editor of Audio-Visual Communications.

As we went to press, Sperry Corporation became known as the Microwave Support Systems Division, a subsidiary of Hercules Aerospace Company.

Using the Three-column Outline to Plan Your Next Video

Charles Kemnitz

SCRIPT WRITERS USE A VARIETY OF TECHNIQUES *to move factual information from the realm of theory to the concrete, visualized format that supports a video production. This article describes an efficient technique for organizing both visual and verbal information and for testing the rhetorical strategy of a video script before the camera rolls.*

The primary challenge in script writing is to get your ideas down on paper in such a way that a director can capture them with the camera. Getting ideas down on paper is important, because as much as 90% of the effort in an audiovisual production occurs before the camera rolls; only 10% actually involves a director and camera crew. And much of what happens before the camera rolls is written into the script.

The script is the blueprint for everything that happens during a video production: Script writers provide the guidelines for every aspect of a production, from pre-production planning to post-production editing. Because we have so much to consider, we naturally look for ways to organize our work efficiently. The three-column outline can help you do that. It organizes both visual and verbal information and tests your rhetorical strategy against expected audience reaction. It is one of the easiest ways to plan a script and to identify problems early, before rewriting is required.

BEFORE WRITING THE OUTLINE

Before you can write a three-column outline, however, you need to do some preliminary work: define the problem, develop a framework, and build a concept.

Define the Problem

To define the problem properly, you must also identify the purpose and the audience for the video. This information guides all future development.

• *The Problem:* People request videos to "solve" problems, such as marketing problems, training problems, and customer relation problems. A problem might be that *equipment operators do not know how to use the auxiliary control valve,* or *customers don't understand the unique advantages of our new loader/backhoe.* Although a video cannot solve a problem by itself, an effective video can be a valuable tool in *helping* to solve the problem.

• *The Purpose:* You should clearly state how the video is going to help solve the problem. Is the video going to educate, affect emotions, or entertain? Videos take one of these three major approaches: they emphasize *instruction, mood,* or *experience.* These three are not mutually exclusive. There is usually considerable overlap between them.

For example, the purpose of a training video might be a performance objective, *train operators in the use of the auxiliary valve.* A customer relations video might have as a goal, *introduce the concept of a totally new loader/backhoe design.* Sometimes a client asks for a video to do more than one thing at a time: *introduce a new loader/backhoe by teaching customers the benefits of our unique design concept.* Notice that *the purpose* in these examples takes the form of a *solution.*

• *The Audience:* You must know who your audience is, what common characteristics they have, and why they are watching your video. You need to understand what they know and do not know about your subject. A good analysis might read like this:

> Construction contractors interested in purchasing loader/backhoes for their current and future equipment needs. Owner-operators are generally in their 40s or 50s. They have only a high school education but years of experience in the construction field.

From the problem, purpose, and audience analysis, you can begin to develop a framework and concept.

Develop a Framework

After defining the problem and researching the content of the video (it is assumed that you have already done this), develop a *framework* for the production. A framework is simply the structure for presentation of the facts. There can be an overall framework such as a drama. Or there can be several frameworks, one for each program segment, such as a montage in the introduction, a demonstration during the body, and a talking head at the conclusion. These separate frameworks can be held together by an overall framework, such as a documentary. For instance, you might open a product introduction with a montage of features and benefits, then cut to a demonstration of the machine at work, followed by an introduction of the design engineer.

Build a Concept

Next develop a *concept.* The concept arises from the *facts* found during

Reprinted with permission from *Technical Communication,* published by the Society for Technical Communication, vol. 35, pp. 198–202, Third Quarter 1988.

research and from the *framework* you just developed:

Facts + Framework = Concept.

Figure 1 shows how facts and framework might interact in the development of a concept. A concept is the narrative storyline that carries your visual and verbal information. The concept should meet or challenge the expectations of your audience. You must choose a concept that best communicates the message within your budget.

When establishing the concept for a video, you become more of a conceptualizer than a wordsmith. You must develop an eye for visual—images that focus the audience's interest and tell stories. In conceptualizing, your tools are, first, an unfettered imagination, and, second, a taste for variety. The more acute the visual concept, the better the ultimate script. To effectively conceptualize, you need to be familiar with the artistic concepts of perspective and composition, realism and surrealism, actuality and abstraction. For instance, a montage can be given a surreal quality through the use of music video techniques. A drama could have the starkness of photographic realism.

The three-column format for developing the concept suits the unique demands of audiovisual media. It concentrates your efforts on developing visual analogues for the message and forces a consideration of imagery rather than rhetoric.

WRITING THE OUTLINE

The preliminary work just discussed gives you a guide to follow when writing the outline. In the outline, you turn your ideas into detailed plans.

Use a three-column format to write the outline: one column for information, one for visuals, and one for the probable audience reaction. Then outline each visual and audio segment. Figure 2 is an example of a three-column outline.

The three-column tabulation of problem, purpose, and audience analysis described earlier is analogous to these three columns.

FACT	FRAMEWORK	CONCEPT
AAA Company has played an important role in the history of this region.	Montage of photographs	A montage of quick camera cuts on sepia-tone photographs used as an opening video segment.
Successful selling is a seven-step process from meeting the client to following up the sale.	Drama	Follow a salesman as he meets with a client and uses the seven-step sales procedure to get an order.

Figure 1—How Facts and Framework Interact in the Development of a Concept

How do you construct the three-column outline? As a general practice, begin with the information column. Decide on the single piece of information you want to pass along to the audience in that scene. Then make a list of things that symbolize or hold some analogous relationship to that "fact" in the visuals column. Draw parallels between the information and the visuals; make actual or implied comparisons. With these, you can translate the abstract/general/complex into visual terms. Use the audience effect column to "check" your creativity. Will your audience react the way you want them to? The goal is never an audience reaction of "What a brilliant video!" The reaction you seek is the one that leads inevitably to your client's "solution."

In more specific terms, consider each column individually:

• *Information Column:* Information in this column corresponds to your *purpose*. If your purpose is to train, then this column contains a statement concerning the training taking place during a particular scene. In other words, state the technical information that is complemented by the visual. In this column, you might also make a note (as a reminder when writing the script) concerning whether this information can be conveyed by the visual alone or needs verbal support. For instance, suppose you want an opening to tell the audience that a particular loader/backhoe is a unique size and suited to specific jobs. You could pose the rhetorical question, Why buy a bigger machine than you need?

• *Visuals Column:* In this part of the outline you answer the question, What does X look like on the screen? X can be anything from state-of-the-art workmanship to a feeling of disappointment at not making a sale. Your problem is to visualize the information and put it down on paper. You can restrict yourself to the flatly representational or you can search for a new, fresh visual that excites your audience and draws them into the

INFORMATION	VISUALS	PROBABLE AUDIENCE REACTION
Establish a common, unsafe practice.	Simulated rollover of construction equipment.	Attention: Ask the question, "What will happen next?"
The danger of failing to wear a seat belt is death.	Close-up of dummy lying crushed under machine.	"That could be me under there!"

Figure 2—Example of a Three-Column Outline

video. To visually support the loader/backhoe question posed earlier, you could show one *inside the bucket* of a gigantic loader used in strip mining operations. The visual is striking, informative, and humorous.

• *Audience Effect Column:* This information is predicated on your earlier audience analysis. How is your *targeted audience* going to react to this visual and verbal information? How do you want the audience to react? If you want them to react differently from your expectations, then you need to rewrite columns one and two. In the loader/backhoe sequence, the absurdity of the comparison comes through. Most construction equipment buyers will laugh when they see it, and you will have made your point.

Considerations for Organizing Information in a Three-column Outline

To organize your visual images, first identify the key point and objective of the video. Choose a mood or a tone to fit the audience and the problem. Pick an approach, an over-all organizing strategy, then proceed to prove each point visually.

Putting the idea into more concrete terms—once you begin building the video sequences in your outline, you may find that you need a visual organizing element so information flows smoothly from one sequence to the next. Video elements are best organized in one of three schemes: *similarity, contrast,* and *contiguity*.

• *Similarity:* put like with like. If you wanted to show that two different makes of loader/backhoes had the same turning radius, you could show an overhead view of a smooth, sandy beach. One of the machines could drive into frame and steer a tight circle, leaving its tracks in the sand, then drive out of frame. The second machine drives into frame and follows the first machine's tracks, matching them exactly.

• *Contrast:* juxtapose images that are not alike to convey information, mood, or experience. Continuing the loader/backhoe example, you want to show that your machine has better visibility than the competition's ma-

chine. So you have someone kneel down in front of the competition's machine and show the view from the operator's seat. The operator cannot see the person in front of the machine. The same view from your loader/backhoe clearly shows the person kneeling in a dangerous location.

• *Contiguity:* assemble visual information that belongs together, as in spatial or chronological order. This organizing technique works well if your video teaches a procedure. Gather the facts you want to put across, using any traditional rhetorical device, then develop a series of contiguous visual images that illustrate each point (and nothing else). For instance, you want to train people how to operate your backhoe; part of that training is to recognize the controls on the machine. You could show an identifying decal, then fade to a picture of the operator control associated with that decal, and last show the operator actually using that control and its effect.

Also consider *mood*. To convey mood, you may want to eliminate facts (the information column) except when a fact can reinforce audience effect. Pinpoint exactly the mood that contributes to the purpose of the video, then gather a group of similar or contrasting images that contribute to that mood.

For instance, to convey a feeling of frustration, contrast the image of a four-wheel-steer loader turning sharply on new sod, leaving it undamaged, with that of a skid-steer loader tearing up a green lawn

Last consider *experience*. An experiential video may use all three organizing techniques (similarity, contrast, and contiguity) during different segments to evoke memories of similar or contrasting experiences. The images should clearly evoke the collective experiences of your audience. One experience that construction equipment owners share is the inability to get large machines under low overhangs. For the loader/backhoe video, find an awning exactly one-half inch taller than the machine. Show the operator driving under the awning at full speed. When the machine does not tear down the awning (hopefully to a

series of exclamations from the audience), you know you have successfully evoked a contrasting experience with the visual image.

STRENGTHS AND WEAKNESSES OF THE THREE-COLUMN OUTLINE

The three-column outline has certain strengths and weaknesses you should learn to exploit and avoid. These are: *unity, progression, proportion,* and *continuity*.

• *Unity:* A good three-column outline is made of whole cloth, addressing itself to a single central idea or theme. For the loader/backhoe product introduction video, you concentrate on presenting five or six major features and benefits that are selling points for the target audience, construction contractors. If another department asks, "Why don't we teach them how to operate the machine at the same time? It'd be easy," you talk the second department into a second video. The second video is instructional. It would be difficult to incorporate instructional material into the sales video. In other words, nail down your key point before you start and stick with it. As your client adds more "cooks to the kitchen," give each cook a separate video; do not try to script a video to satisfy all of the cooks.

• *Progression:* You need to develop an element of forward movement because most videos are linear and require a beginning, middle, and end, just like a good short story. In training videos, adhere to the age-old technical writing advice to "tell them what you're gonna say, say it, then tell them what you said." But a good sales or motivational video can accomplish reinforcement and repetition of visual, intellectual, and emotional elements without redundancy. It introduces and summarizes new material without actually repeating previous scenes. In training videos, repetition is explicit and cognitive; in mood or experiential videos, repetition is implicit and subliminal.

A good analogy for the progression in a three-column outline is that of a nervous system, which requires several stimuli to make a coherent whole.

The goal is to create a particular understanding and recognition that takes place only as the various stimuli are processed.

• *Proportion:* Your outline emphasizes the things that contribute to achieving your purpose and solving the problem. Try not to succumb to the temptation to create such a great visual opening that the rest of the video pales by comparison. Also avoid what short story writers call novel sentences. These are scenes containing information that is good to know but does not actually contribute to your purpose. Novel sentences point outside the story and tell your audience that there is more to the story than you put in it.

• *Continuity:* Your outline should hang together. Look at your concept and decide what you need to show and say. Your job is to arrange the information in a rhythm and flow that will hold your audience's attention. Continuity is essentially the result of staying within your framework and concept. Your script must have both visual and verbal continuity.

But you might want to consider using a *grabber* at the beginning of the show. A grabber is something visually exciting that captures your audience's attention and makes them want to watch the rest of the show.

At points in the show—about every 90 seconds—you might include a *stinger.* A stinger, like a grabber, is something quick, sharp, and memorable that will stick in the viewer's mind. Stingers are especially effective for helping your audience remember difficult and important points.

Finally, you want to avoid *glom* shows. These shows look, feel, and sound the same from beginning to end. You can avoid glom shows by varying the organizing strategies you use to present information or by using different kinds of verbal and visual stingers. Remember, you can change frameworks without changing overall concept or destroying continuity.

AFTER WRITING THE OUTLINE

After finishing the outline, you need to prune it. An ideal corporate video runs approximately 12 to 18 minutes, though you will find that instructional videos can run much longer (hours!), and sales videos will usually run much shorter (6-8 minutes). Generally, you should plan to cut the first draft of your outline by 20%.

The thing to look out for when reviewing your outline is any jerk or jump that causes a momentary sense of bewilderment in your audience. Check your outline for visual organization, unity, progression, proportion, and continuity. Take a careful, step-by-step approach to your outline and *brainstorm, brainstorm, brainstorm.* Your best tool remains your imagination and how it works upon your research. Strive for an artistic vision of the video, and get that vision into the outline.

From the outline, you develop a *treatment.* A treatment is a narrative outline describing in detail what will be seen and heard in the video. Figure 3 shows an example of a treatment. Usually, the treatment is the document you present to clients to give them an

---

VIDEO TREATMENT
We are in the bottom of a strip mine in southeastern Oklahoma. The camera pans from a coal seam in one wall of the mine to an open, barren, dusty area. A huge, 20-cubic-yard loader drives into view with the loader bucket raised about ten feet off the ground so we cannot see what it is carrying. It stops in frame and slowly lowers the loader bucket. When the bucket reaches the ground, our new XXXX loader/backhoe drives out of the bucket. Camera pans to follow loader/backhoe and narrator asks: "Why buy a bigger loader than you need?"

Figure 3—Example of a Treatment

---

initial idea what a video might look like.

The treatment is essentially a short story that follows the outline you have created. Consider the treatment as a condensed version of the script. You need to follow the elements in your outline to write the treatment. While writing the treatment, test ideas for camera angles and moves that will later be described in your script. If you cannot use your outline to effectively describe what the audience will see, hear, and feel, then you have not fully visualized your information and your client will not "see" what is going to be in the video.

Later, use the outline and the treatment to write the script. If you use a two-column script format, writing becomes a relatively simple matter of turning the visual and information columns of your outline into the video and audio columns of the script. The three-column outline is also easily adapted to other script formats, and it, rather than the treatment, serves as an effective blueprint for your final script.

Practice writing a three-column outline for your next audiovisual production. It can help you write a clean, sharp script and get you several steps closer to final approval! Ω

BIBLIOGRAPHY

Amend, R. H., and R. B. Stoner. "Production: Turning Ideas into Pictures and Sound," *Technical Communication* 34, no 1 (First Quarter 1987): pp. 15-18.

Carliner, Saul. "Audiovisual Words: The Scriptwriter's Tools." *Technical Communication,* 34, no. 1 (First Quarter 1987): pp. 11-14.

Floyd, Steve. "Visualization: The Key to Effective Scriptwriting." *Technical Communication,* 34, no. 1 (First Quarter 1987): pp. 8-10.

Matrazzo, Donna. *The Corporate Scriptwriting Book.* Philadelphia: Media Concepts Press, 1981.

Swain, Dwight V. *Scripting for Video and Audiovisual Media.* Boston: Focal Press, 1981.

CHARLES KEMNITZ *is the Communications Designer for The Charles Machine Works, Inc., manufacturers of Ditch Witch® underground construction equipment.*

Visualization: The Key To Effective Scriptwriting

Steve Floyd

AUDIOVISUAL PRESENTATIONS *are primarily visual forms of communication, forcing scriptwriters to rely on pictures, not words, to convey information. The five-step process of visualizing a presentation before writing a script helps scriptwriters develop reliance on pictures.*

The old adage, "A picture is worth a thousand words," is the best advice for technical communicators beginning an audiovisual script. That's because media such as video, slide/tape, and film are primarily visual. Words *support* the pictures.

Unfortunately, the advice is neglected by many technical communicators. All too often, they depend on their most trusted tools, words, rather than the most appropriate tools, pictures. As a result, the overall impact of the message is blunted.

Placing trust in pictures requires that communicators perform the mental process of visualizing their presentations before writing scripts. Visualizing a technical presentation is different from using electronic special effects and dramatic camera angles. Visualizing requires the writer to clearly present complicated and abstract topics through pictures—the writer must find the pictures inherent in the subject matter.

Making a complex technical subject visually interesting—without distracting the viewer from the content or undermining the primary informational purpose of the program—is always a challenge, but it is less difficult if you remember that **form must always follow function.**

In other words, balance the need to maintain the audience's attention with the need for clarity. You can effectively balance these needs by following this five-step process, which emphasizes visualization *before* writing the script:

1. Conduct a front-end analysis
2. Select a creative concept
3. Organize the structure
4. Visualize the content
5. Write a draft script.

STEP 1: CONDUCT A FRONT-END ANALYSIS

The first step is a familiar one—a problem definition that you perform prior to developing any technical document:

- *Analyze the task*, asking what skill or knowledge the presentation should discuss
- *Develop an audience profile*, learning about their jobs, education, experience, interests, and language (so you know the right words and expressions to use)
- *Define objectives*, stating in observable and measurable terms what the audience is expected to do after viewing the presentation
- *Identify critical actions (events)* in the task
- *Design an evaluation plan*, measuring the effectiveness of the presentation.

I call these problem-definition activities *front-end analysis*.

STEP 2: SELECT A CREATIVE CONCEPT

Once the front-end analysis is complete, many people immediately try to draft a script. Trying to draft a script at this point, however, is like writing a novel without knowing the story. You should first select a *creative concept*—an overall direction for a presentation. A concept might be

- A lecture-demonstration
- A series of graphic illustrations
- A dramatization
- A take-off of a broadcast program.

Or you could combine two or more of the concepts listed above to devise a concept for the presentation. Each concept has unique characteristics.

Lecture-demonstrations show an actor or expert delivering an explanation "on-camera." The narrator may also be off-camera and lecture using a *voice-over* technique (voice is heard but the narrator is not seen) while someone else demonstrates the procedure on-camera.

Both lecture and demonstration are direct techniques. A lecture-demonstration is relatively easy to produce and is particularly appropriate for teaching psychomotor skills, such as soldering and driving.

Graphic illustrations use illustrations to show what happens *under* the surface, in much the same way that lecture-demonstrations show what happens *on* the surface. Illustrations show technical processes that would normally be invisible to the eye or too abstract to demonstrate. For example, you could use graphic illustrations to show how computer software processes data.

Recent innovations in computer graphics capabilities make it easier and more affordable to produce attractive and stimulating images than is possible with hand-drawing techniques.

Reprinted with permission from *Technical Communication,* published by the Society for Technical Communication, vol. 34, pp. 8–10, First Quarter 1987.

Dramatizations re-enact scenes from the workplace to demonstrate a specific point. Frequently, one person (an expert) explains a procedure or operation to a second person (a novice). For example, an experienced driver can caution a daredevil novice.

A dramatization should be conversational. Writing conversation requires an ability to write effective dialogue. "Audiovisual Words: The Scriptwriter's Tools," by Saul Carliner, offers some suggestions for writing conversation.

Dramatizations are more difficult to produce than simpler lectures or demonstrations because dramatizations require

- Professional actors, who can convey the feelings of the characters
- Rehearsals, so the actors know their roles
- Sets or location shooting, to create a credible backdrop for the activity
- Additional post-production work (production work that follows the shooting) to handle technical problems that arise during the taping.

Broadcast take-offs resemble broadcast television programs, such as news shows and game shows. Broadcast concepts can be effective if the audience is not motivated or is downright negative towards the subject. For example, you can use a broadcast take-off to explain a new company personnel policy.

Broadcast take-offs require strong writing, directing, and acting skills. Broadcast take-offs also require *high production values*—the time and expense spent on sets, post-production, graphics, and music.

STEP 3: ORGANIZE THE STRUCTURE

The next major step in the visualization process is to structure the content of the presentation around the critical points identified in the front-end analysis.

This step is performed *without* consideration for the concept developed in step 2. The content is considered separately from the concept because the order in which information is presented should be related to the information, not to the concept. Also, organizing the content separately from the concept lets you adapt the content to another medium should you ultimately choose not to produce an audiovisual presentation.

Structuring the content is the same as organizing content for print material: identify the main ideas and the information that supports those ideas and then determine in which order the information will be presented.

Here are some points to consider when organizing the content:

- **Develop a simple and concise structure;** make the hierarchy of information and relationships between different elements as clear as possible. Use a visual scheme, such as an outline or an issue tree (information organized in a tree-like structure, with branches and stems) to show the organization.
- **Clarify information immediately;** viewers cannot scan a table of contents or skip irrelevant sections as they can when reading a manual. For each section, ask, "How can I highlight, reinforce, or clarify this information?"
- **Develop viewer interest throughout the presentation;** viewers may otherwise "tune out" and miss important information. To maintain interest, remind viewers of specific benefits or pay-offs. In other words, ask "Why is this important to the viewer?" as you develop each main point in the presentation.

Figure 1 is an example of a content

```
1. Introduction
   A. Gain audience attention
      (grabber)
   B. Relate presentation to job,
      describe the payoff
   C. Set expectations (objectives)
   D. Give overview of entire
      presentation
   E. Transition to next point
2. First main point
   A. Define
   B. Demonstrate/explain
   C. Provide example
   D. Relate to job
   E. Reinforce (summarize)
      critical points
   F. Transition
3. Another main point
4. Another main point
5. Summary
   A. Review main points
   B. Close presentation
```

Figure 1—Example of a content outline

outline for a presentation. The outline contains the following elements:

- An *introduction,* which orients the viewer to the material and establishes viewer expectations of the presentation. The introduction includes
  —A **grabber** to gain audience attention
  —A list of **objectives** to set audience expectations
  —An **overview** to describe the main points in the presentation
- *Three to five major points* in the body, each with its own
  —**Definition** or explanation of the point
  —**Example,** to highlight the point
  —**Description,** directly relating the point to the viewer
  —**Summary** of the discussion
- A review of the main points in the presentation.

STEP 4: VISUALIZE THE CONTENT

In this step, you develop a visual image of the presentation. Visuals, or pictures, are an essential element of every audiovisual presentation. Visuals have far greater impact on what viewers remember than any words in the script. Think about it. When people talk about scenes from a movie, they discuss the actions, events, scenery, and special effects, but seldom the dialogue.

Not surprisingly, information is presented, clarified, highlighted, and reinforced primarily through visuals. Words support visuals by explaining what visuals cannot show. When a visual fails to achieve its purpose, it is probably distracting the viewers, not adding to their understanding.

Start the visualization step with the outline. Imagine how a series of actions or graphic illustrations could clarify each main point. For example, what visuals would you show during a laboratory experiment?

More specifically, visualize the presentation by considering the following points:

- **Provide special emphasis on each main point** and complex step to reinforce the information.
- **Focus attention on critical steps** by using close-ups, in which the camera zooms in on one subject. Usually, information is presented in a medium shot in which the camera views sub-

---

(SCREECH)

(A TRUCK CLEARS THE COR-
NER. A PEDESTRIAN—OBVIOUSLY
SHAKEN—CROSSES THE STREET
IN FEAR AND WALKS OUT OF VIEW.
THE DRIVER, BOB, PARKS THE
TRUCK IN THE NEAREST SPACE
AND GETS OUT OF THE TRUCK.
BILL, OBVIOUSLY ANGRY, WALKS
OUT TO BOB, WHO'S EMBAR-
RASSED.)

Bill: You're pretty lucky!
       You could've killed that man.
Bob: (Sheepishly) I know.
       That was a pretty close call. But
       I couldn't see him when I was
       going around the corner.

Figure 2—Example of a two-column format script

---

jects the way we normally see them. For example, a medium shot of a person typically shows the head and shoulders, while a close-up shows only critical details, such as the head or a hand.

- **Use color and text** to focus attention. Background colors should highlight the subject; our eyes are naturally attracted to lighter colors. Colors can also highlight graphic illustrations and underline an action.
- **Superimpose text over graphics** or photographs to describe actions or to review information.

If you have difficulty visualizing an action or a procedure, analyze the task again. Return to a location, review a procedure, or interview a subject-matter expert to clarify all of the details.

Visualization before writing offers advantages to you and to your audience. By imagining how the presentation will look before writing the script, you can write narration that effectively enhances the visuals rather than narration that competes with the visuals.

Visual images also have a powerful effect on audience behavior. For example, athletes' ability to visualize their performance *prior* to competition contributes to peak performance. Similarly, providing a visual role model or reference point for someone learning a psychomotor skill, such as driving, can lead to peak learning.

STEP 5: WRITE A DRAFT SCRIPT

After the content has been thoroughly visualized, it is relatively easy to write a straightforward explanation—narration, or script. Use the outline as a road map when writing the script:

1. Sequence by sequence, for each major part of the script, do the following:
   a. Write the visual description
   b. Write the narration that accompanies the visuals.

Saul Carliner's paper, "Audiovisual Words: The Scriptwriter's Tools," later in this issue, offers specific suggestions for writing narration.

2. Carefully review the entire script, checking for continuity and clarity.
3. Rewrite the script, trying to simplify the original draft.

Choose the appropriate presentation or storyboard format when writing the script. Most technical information scripts follow a two-column format, with the visual description on the left side and the narration (audio) on the right, but each organization follows its own format. Figure 2 shows an example of a two-column format script.

CONCLUSION

This five-step process of visualizing information before verbalizing it may seem awkward at first, but stay with it. The process eventually becomes a natural one. It lets a writer use both sides of the brain, going back and forth between the left side (language and logic skills) and the right side (creative and artistic skills).

Visualizing before verbalizing helps you balance information evenly between visual and audio elements. You may even find that a visual sequence really *is* worth a thousand words. Ω

STEVE FLOYD *is president of Floyd Consulting and Design, an Atlanta-based communications consulting firm.*

Audiovisual Words: The Scriptwriter's Tools

Saul Carliner

BEGINNING SCRIPTWRITERS *often rely too heavily on words or overuse audiovisual gimmicks. You can avoid these problems through an understanding of the audiovisual media and by following specific writing suggestions, such as writing in everyday language and using short, simple sentences or phrases.*

Opening One: A narrator begins speaking:

This presentation is an introduction to economics. The objectives of this presentation are to increase viewer understanding of the nature of important market forces, specifically those of supply and demand. This presentation is also intended to increase viewer understanding of the free-market system. Supply and demand are the . . .

Opening Two: Words spin on the screen clockwise, then counterclockwise. The words decompose. Cut to a box. The narrator pops out of the box, saying, *Howdy! Welcome to an introduction to economics!*

Opening One isn't a script; it's a technical paper that someone is reading aloud. How pointless to spend the time and money developing an audiovisual presentation when the job can be done on paper. Opening Two overcorrects the problem of Opening One. The visual introduction puts the graphics equipment on overtime. The script is written so energetically that the audience feels like it's just finished a pot of extra-strong coffee.

Beginning scriptwriters often fall victim to such under- and over-writing. Unfamiliar and insecure with the visual media, writers rely solely on words to convey technical information like a new driver rides the brakes for fear of using the gas pedal. Bedazzled by audiovisual technology and the glamour of the media, other writers try technical and dramatic gimmicks more appropriate for *Miami Vice* than

for *Proper Lifting Techniques.*

This paper offers the following insights and suggestions to help you avoid these writing traps:

1. Characteristics of audiovisual media that you should consider before writing a script
2. Suggestions for writing the narration part of the script
3. Suggestions for reviewing scripts to identify problems, before the problems are preserved permanently on film, tape, or slides.

CHARACTERISTICS OF AUDIOVISUAL MEDIA THAT YOU SHOULD CONSIDER BEFORE WRITING A SCRIPT

In many ways, writing an audiovisual script is no different from writing any other type of technical documentation. Your ultimate goal is the same: *to communicate technical information to a specific audience.*

But certain characteristics of the audiovisual media dramatically affect scriptwriting, differentiating it from other types of technical writing. Understanding these characteristics is fundamental to scriptwriting, and to understanding the specific writing suggestions offered later in this paper.

Special Uses of the Media: Audiovisual media can show certain information more effectively than a manual or online documentation can. Steve Floyd discusses some of the possibilities in "Visualization: The Key to Effective Scriptwriting." If the audiovisual media cannot show information more effectively than other media, don't continue the project.

Sensory Characteristics of the Media: Audiovisual presentations are sensory experiences, appealing directly to sight and hearing, and indirectly to the other senses [1]. This multi-sensory experience can actually enhance learning because viewers receive information from more than one source. Learning the same information through two senses reinforces retention.

Physical Characteristics of the Media: The following physical characteristics of the audiovisual media affect its usability as a source of information:

- An audiovisual presentation is linear, intended to be viewed from beginning to end without interruption.
- Users cannot review information from an audiovisual presentation as easily as they can reread passages in a manual or online information.
- Users view a presentation at the speed determined by the equipment and the producers of the presentation, which may not match their learning speed.

Adapting to these characteristics leads to specific writing approaches. Because users must sit through the full presentation, you must ensure that all of the information is relevant and presented in an interesting manner—otherwise, viewers will tune out. A strong concept for the presentation is the best starting point for ensuring an interesting presentation; interesting narration and visuals naturally flow from the concept. A strong concept is one that is relevant to your audience, not simply one that interests you.

Similarly, the presentation should not outlast the viewers' attention span. Rules of thumb suggest that *an ideal presentation is 12 minutes long* and never exceeds 18 minutes.

Reprinted with permission from *Technical Communication,* published by the Society for Technical Communication, vol. 34, pp. 11–14, First Quarter 1987.

Because viewers cannot easily review sections of the presentation, you must make information clear the first time you present it—otherwise, users won't remember it [2]. Specifically, you should

"Signal," that is, explain the structure of the presentation at its beginning, stating exactly what you plan to do in each section. As you move between sections, first review the material from one section and then let viewers know that a new one has begun and what the new section will discuss.

State main ideas clearly and repeat them regularly throughout the presentation. Make information clear by placing it into a familiar context. Studies in cognitive psychology show that putting information into a familiar context facilitates understanding of it. Repeat ideas using the same phrases and terms that were first used to explain the ideas so viewers can readily identify the repetition. If you use different phrases and terms, viewers may think you mean something different.

Not overload the audience with information. Instead,

- *Stress ideas,* the type of information you can expect viewers to remember. Use facts and examples to support the ideas. Consider, for example, how well you remember stories from *60 Minutes.* You may remember that Ed Bradley investigated welfare hotels in New York City, but probably won't remember the names of these hotels. The problem of welfare hotels is an *idea* that you want the audience to remember. The hotel names are *supporting facts,* used to tell people that welfare hotels really do exist.
- *Limit the number of main ideas to five*—the maximum number of ideas that people can readily remember. Similarly, limit the number of supporting facts for each idea. Lists longer than five items are difficult for people to process, especially if they cannot review the lists to make sure they understand all the items. If you find that your lists of main ideas or supporting facts exceed five, you probably need to reorganize the information.

SUGGESTIONS FOR WRITING NARRATION

A script contains two parts: visuals and narration (called dialogue if it is spoken by more than one person). The two elements work together to make an effective presentation; you cannot rely solely on the words in the narration. To reduce your reliance upon words, you need to adopt new writing practices. This section suggests some of those practices. Figure 1 is an example of completed narration.

Let the narration support the visuals: Narration explains, highlights, reinforces, and clarifies information that is shown in the visuals, explaining what is not immediately apparent in the visuals. In other words, narration plays a supportive role in the script; visuals bear the primary responsibility for conveying information. For example, the narration for a section showing a violently shaking bridge should be

The dynamic forces of physics are causing this violent shaking

and not

The bridge is shaking violently.

The second passage repeats the obvious and insults the viewer.

To ensure that your words support the visuals, take the advice in Steve Floyd's article, "Visualization: The Key to Effective Scriptwriting." Visualize your presentation *before* writing the narration.

Write in everyday language: Narration is intended as a one-way conversation between the on-screen narrator and the audience. Scripts, therefore, should be conversational.

Relax stylistic rules by using spoken grammar. Using spoken grammar specifically involves changes to sentence structure and the use of slang and contractions. Spoken and written grammars differ, and spoken grammar varies widely. The grammar of a Minneapolis computer programmer is different from that of a Pittsburgh steel worker. You need to adjust your grammar to the audience.

Write in short, simple sentences or phrases: Join sentences with connectors such as "and" and "but." Writing this way offers two advantages:

- It's more natural. People speak in phrases or short sentences run together by connectors.
- Simple sentences and phrases are easy to comprehend. The less information in a sentence, the easier it is to process, and therefore the more likely that it will be remembered. Parenthetical clauses and complex sentences, while technically versatile, are difficult for users to process [3].

Use technical terms and acronyms only when necessary: If you must use technical terms and acronyms, repeat them often so that users become accustomed to the terms [3]. All technical writing texts warn against the overuse of jargon. But the problems created by jargon are intensified in scripts. Unfamiliar terms must be defined the first time they are mentioned and should be used often or the audience will forget what they mean. Acronyms pose similar problems because the audience doesn't have an acronym list to decode these strange three-, four-, and five-letter combinations.

(PARKING LOT OF ANY BUSINESS, INC.'S FACTORY. A PEDESTRIAN STARTS CROSSING THE DRIVEWAY AT A CORNER OF THE BUILDING. BILL IS STANDING AT THE FRONT DOOR, WATCHING, APPARENTLY WAITING FOR SOMEONE. QUICKLY, THE PEDESTRIAN JUMPS BACK ONTO THE CURB. IMMEDIATELY AFTER, A TRUCK COMES AROUND THE CORNER. AFTER THE TRUCK CLEARS, THE PEDESTRIAN—OBVIOUSLY SHAKEN—RUNS FOR HIS LIFE, OUT OF OUR VIEW. THE DRIVER, BOB, PARKS THE TRUCK IN THE NEAREST SPACE AND GETS OUT OF THE TRUCK. BILL, OBVIOUSLY ANGRY, WALKS OUT TO EMBARRASSED BOB.)

Bill: You're pretty lucky! You could've killed that man.
Bob: (Sheepishly) I know. That was a pretty close call. But I couldn't see him when I was going around the corner.
Bill: Why didn't you honk your horn to warn him you were coming?
Bob: I didn't think of that. I always figured we weren't allowed to make noise.
Bill: What's more important—the noise or the life?

Figure 1—Example Script.

Avoid lists: The printed list is one of the most efficient devices for presenting information, but it is difficult to use in audiovisual presentations. The structure of the list gets lost in the narration; a lead-in sentence is forgotten long before the narrator reads the second item on the list.

One solution to this problem is to display key words from the list when the narrator reads the list. But displayed words indicate to users that a list is important. In many instances, a list summarizes non-essential information. Displaying list items can give the wrong emphasis to viewers.

Use vocal variety: Vocal variety is providing different sounds. Each change in sound theoretically stimulates the audience, thus increasing attention. You can provide vocal variety by using more than one narrator. In such instances, you use different voices, each easily distinguishable from the others, to handle different types of narration. For example, one voice may read the steps in a procedure while another voice reads the explanation of each step [4]. You can also use sound cues, such as music, sound effects, and pregnant pauses, for emphasis and variety [3].

Make sure that production personnel understand your instructions: The final draft of a script, which contains visual and audio cues, is not intended for ultimate reading by the viewers. Scripts are intended for use by directors, who develop a production, and by narrators and actors, who record the narration. These people often make changes to the script. Some changes are intentional, others are not. To avoid unintentional changes, follow these suggestions:

- Identify stage directions in all capital letters to distinguish them from narration. For example,

(NARRATOR LIFTS THE BOX)

- When writing stage directions, write EXACTLY what you mean. If you want the narrator to lift the box from the left, write:

(NARRATOR LIFTS THE BOX FROM THE LEFT)

If you do not clarify directions, the director will not know that the direction of lifting is significant. Clarifying directions is especially important in technical information scripts, in which the accuracy of all the content, including the visual content, is vital. Scriptwriters are usually familiar with the content and can help avoid technical inaccuracies in the visuals by carefully writing stage directions. Use stage directions sparingly, however. In most instances, stage directions are notes written by the director. You need not use stage directions for every scene; use them only when necessary.

- Change punctuation to help the narrator. Use punctuation that immediately tells people they're reading a script . . . ellipses (. . .) and dashes (— —) [3]. They signal important information:

  Ellipses and dashes indicate emphasis points to the narrator. The ellipses indicate a pause. The dash changes intonation.

  Ellipses and dashes improve content review by telling reviewers that they are reading scripts, so their attention is focused onto the information and off of the grammar.

- Avoid abbreviations to ensure a proper reading. In print, abbreviations reduce the number of words to be printed. That's good because it can reduce production costs. In scripts, however, abbreviations can lead to mistakes. Narrators, who—like typesetters—match input, read exactly what is written in the script. If you want the narrator to read "in the morning," don't write "AM." The narrator will read "AM." Write out the full phrase. Similarly, numbers over one million must be written "one million," or the narrator might get lost in all the trailing zeros.

SUGGESTIONS FOR REVIEWING SCRIPTS

Editing is as valuable a process in writing scripts as it is in writing manuals and online information. Editing can identify writing problems and ensure the clarity of instructions to production personnel.

Different editing techniques are required, however, to review scripts. Unlike a manual, a script cannot be read to be evaluated. One director even suggested, "If a script reads well, it probably stinks."

Instead, scripts editors must *listen* to scripts to evaluate them. One person should read the script aloud while the editor (and scriptwriter, too) listens. Imagine the visuals. While listening, identify concerns such as

- *Difficult and unnatural phrases,* such as tongue twisters and other difficult-to-pronounce expressions. For example, the word "similarly" often causes problems for narrators.
- *Ideas that remain unclear* after the narrator says them.
- *Repetition of information* that should be obvious from the visuals.

The editor can also identify problems with the instructions to production personnel, much as editors identify problems in manuals with the instructions to the typesetters and keyliners. Editors can check the completeness and clarity of instructions. For example, editors might ask whether the box mentioned in the stage directions should be lifted in a certain manner.

REVIEW

Beginning scriptwriters often rely too heavily on words or unnecessarily experiment with audiovisual gimmicks. You can avoid such traps by considering the following when writing scripts:

- **Audiovisual media have unique characteristics.** Audiovisual presentations show certain types of information more clearly than other media, they provide a sensory experience, and their physical characteristics affect the manner in which people learn information from them.
- **Narration supports the visuals,** which bear the primary responsibility for conveying information.

- **Narration should be written**
—In everyday language
—In short, simple sentences
—Without lists
—Without acronyms.

- **Scripts must be easily understood** by production personnel.
- **Scripts should be reviewed by listening** to them, not by reading them. Ω

REFERENCES

1. S. Martin Shelton, "The Influence of the Information Film," in the *Proceedings* of the 32nd International Technical Communication Conference, Society for Technical Communication, 1985.
2. Grant Williams, "Writing Words People Hear," *Video Manager*, May 1985.
3. Joe Farace, "Ten Steps to Better Slide Shows," *Training*, July 1984.
4. Barbara Blazyk and Judi Victor, "Slide Show Guidelines," *Audio Visual Communications*, July 1985.

SAUL CARLINER *is a master's student in the technical communication program at the University of Minnesota. He is an Education Development Administrator for IBM in Atlanta and an Associate Editor of* Technical Communication.

Production: Turning Ideas Into Pictures and Sound

R. H. Amend
R. B. Stoner

SCRIPTWRITERS MUST UNDERSTAND *the video production process to ensure that their scripts are producible. Production has two stages. During production planning, directors evaluate scripts for continuity, visualization, and picturization. During production itself, directors make decisions that could change the script you write.*

Production is the process that transforms the images suggested in your script into real pictures and sound. Just as the publication production processes affect writing for print, video production processes affect scriptwriting. For example, a writer developing a manual that will be printed on a two-color press knows that a second color can be introduced into the publication, with the color being used perhaps to differentiate between different types of information. Video production processes can have a similar effect on your scripts. Therefore, as a print documentation writer must understand publications processes, scriptwriters must understand video production processes.

Directors have ultimate responsibility for designing a video production, responsibilities that include specifying each camera shot, determining the music and other sound effects, and choosing talent to appear in the presentation. However, scriptwriters also play an important role in production design. Scriptwriters must develop producible scripts: scripts that can be realistically produced given the budget, schedule, and equipment available. To write such scripts, you need to understand the processes that follow script writing. This paper discusses those processes:

- **Script evaluation,** during which directors sketch out visual images from the words in your script

- **Production,** during which images become pictures and are united into a coherent whole.

HOW DIRECTORS EVALUATE SCRIPTS

Just as part of your job as a print documentation writer is simplifying the typesetter's and keyliner's jobs, so part of your job as a scriptwriter is simplifying the director's job. What do directors need from you? Consider what directors do between receiving your finished script and starting production: They check the finished script for three visual elements: continuity, visualization, and picturization.

Continuity is "the even, logical succession of events"[1, 508]. Continuity affects production in many ways. For example, continuity affects the appearance of on-camera talent. An on-camera narrator acts as a tour guide. The 10-minute presentation usually takes place during the course of one work day, but the production takes five days to shoot. The narrator needs to wear the same clothing all five production days because a change of clothing between scenes would disrupt the presentation's continuity.

Continuity is important to videography, too. For example, if a person is running towards the camera in a medium shot, the person should continue to be seen running towards the camera when the camera subsequently zooms in.

Visualization is "the framing of a single shot"[1, 254]. The director interprets the scene as written by the scriptwriter and determines where to place people and objects in relation to the camera. Visualization is suggested by the script. Consider the description:

NARRATOR OPENS PROGRAM OUTSIDE ENTRANCE TO THE COMPANY

What type of shot should the director select? A wide shot of the company and the narrator? A close-up of the narrator at the entrance? A close-up, then a zoom out to reveal the rest of the facility? All of these shots would be appropriate for the vague scene description.

Though it is not your job to determine the visualization of each scene, each scene you write must suggest visuals. If you describe scenes in sufficient detail, setting up and shooting is fairly routine. If description is vague, directors must be more imaginative.

Picturization is "the control and structuring of a shot sequence" [1, 518]. Consider how to structure a demonstration of an anti-theft system for cars:

1. The audience sees the narrator a few feet from the car. The narrator walks toward a test car and tells the audience a special key is necessary to open the door and deactivate the system.
2. A close-up of the narrator putting the key into the door lock.
3. A medium shot of the narrator getting in the car.
4. Finally, the narrator sits in the driver's seat.

The picturization of the scene would be violated if the scene begins with

Reprinted with permission from *Technical Communication,* published by the Society for Technical Communication, vol. 34, pp. 15–18, First Quarter 1987.

shot 1 and jumps to shot 4. The audience needs to see shots 2 and 3, when the narrator opens the door and gets into the car.

Consider picturization when writing the script. Avoid "jumps" and other sequences that would be out of continuity, all of which destroy the picturization.

HOW DIRECTORS PLAN PRODUCTIONS

After evaluating the script, directors record the plan for each shot on a *storyboard*. A storyboard is to the video director what a copymarked draft is to a typesetter and keyliner—a plan for production.

A storyboard is a series of pages that show, shot by shot, what the production will look and sound like. A storyboard usually includes

- A drawing or photograph of each picture to be shot
- The narration (dialogue) associated with the shot
- Instructions for transitions between shots and scenes
- Instructions for sound effects and music
- A sequence number indicating the position of the shot in the production (such as 3 in 47).

Storyboard formats vary. "Visualization: The Key to Effective Scriptwriting," by S. Floyd (*TC*, First Quarter, 1987, pp. 8–10), discusses the two-column script format, which some studios also use as a storyboard format. Figure 1 shows another format, the panel format.

DECISIONS

When developing the storyboard, directors make important decisions about the production. These decisions are affected by the two most important production concerns:

- **Budget,** an important concern because video presentations can be expensive, sometimes costing over $2500 a finished minute. Taping is one of the most costly items because it requires an entire video crew—two to ten people and their equipment. Transportation time and costs make location shooting even more costly.
- **Schedules,** an important concern be-

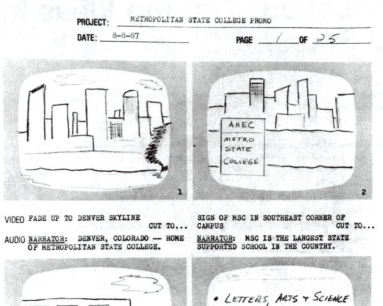

Figure 1: Example of a panel-format storyboard.

cause video production requires many time-consuming processes. Video production schedules, therefore, are a sequence of milestones, much like publications production schedules.

Decisions about the production define the ground rules under which you develop the script. The following sections describe some of these decisions and how they affect scriptwriting.

Decisions about the Talent

Directors decide from where the talent—actors or narrators—is chosen. The decision is not simple because the cost of talent and its quality are often inversely proportional. The cost of talent also limits the number of actors or narrators you can use in your presentation:

- Actors must be paid at rates determined by their union.

- Narrators charge from $40 to $500 an hour. *Voice quality* and *audience familiarity* with the narrator determine the narrator's fee.

Voice quality is the narrator's ability to read smoothly and evenly. The smoother the narrator's reading, the less time-consuming work will be required later during a process called *post-production* to correct technical inadequacies (such as glitches in the voice and differences in sound levels). This, in-turn, reduces production costs.

Familiarity is both an asset and a liability. If an audience is familiar with a certain narrator, the narrator lends credibility to the presentation. If the narrator is too well-known, the narrator may be associated with another organization—and not a believable spokesperson for yours.

This is valuable information for the scriptwriter. If you know how many actors or narrators can be used in the script, you can write parts properly. Furthermore, if you know before writing *which* actors or narrators are used in the presentation, you can "write for the person," that is, write the way that particular person speaks. Everyone has a unique speaking style—intonations, inflections, preferred words. You can incorporate that unique style into the script. For example, if your narrator is a serious person, you would not write cute jokes for that person.

Decisions about Production Equipment

Studio production equipment is an important part of the presentation. As it concerns the writer, the most important equipment is the *video editing system*. Directors use editing equipment to combine the shots taken during production. Shots are usually taped out of sequence for expediency. For example, a script calls for three scenes in a lobby, which are shown at three different times during the presentation. The three scenes are taped at the same time so that the production team does not have to return to the site.

In addition to combining shots, editing systems include special-effects generators that provide the visual transitions between scenes, such as cuts and fades. The transitions that an editor can provide vary among special-effects generators. Basic models perform the simplest transition, a *cut* (in which one scene ends and the next abruptly begins). More sophisticated equipment offers more sophisticated transitions, such as *dissolves* (in which one scene dissolves into another), *superimpositions* (in which one scene is imposed on top of another), *fades* to and from black, and other special effects. The most advanced systems perform a variety of digital effects, such as decomposing a picture. Figures 2 through 4 show equipment found in a video studio.

You should understand the capabilities of the editing equipment. In certain instances, you describe a transition technique in your script. You

can't say DISSOLVE when the equipment can only perform a CUT. (Most transitions, however, are determined by directors when they develop storyboards.)

Although writers need to know about editing equipment, most are more familiar with the problems of videotape format, which is the size of the tape itself. Format has *no impact* on your job as a scriptwriter. But it is an intriguing subject.

Each format has unique features that make it appropriate for specific uses, such as home video, industrial video, and video production. Videotape equipment, including editors and playback equipment, can be used only with one type of format. Formats are not interchangeable, but you can transfer presentations to another format in a process called dubbing up and down.

Formats are indicated by the width of the videotape. Formats are one factor affecting the quality of the videotape. The larger the width, the better the quality of the videotape because the wider tapes record a more complete image. The ¾-inch and 1-inch formats are therefore considered the best for production.

Formats also affect *generational loss*. Generational loss is the loss of quality that occurs when an image is copied from one videotape to another. This is similar to the loss in quality between an original printed page and its copy. The larger the format, the less the generational loss because more information is recorded with the image.

Decisions about Graphics

As mentioned in "Visualization: The Key to Effective Scriptwriting," graphics can show information that is otherwise impossible to photograph, information such as abstract concepts and under-the-covers processes. How much information can be shown through graphics depends on the capability of the graphics equipment. Each type of equipment has its own technical and financial limitations. Graphics can be produced by

Figure 2: A person performing video editing.

Figure 3: A character generator.

Figure 4: The controller of a video editing system.

- **Character generators,** which are typewriter-like machines that produce the letters displayed during a presentation. Character generators can also produce numbers, colored backgrounds, and in some instances, special symbols and generic figures (such as people, animals, and logos). Character generators are common equipment in most video studios.
- **Computer equipment** (designed especially for video), which can produce pictures and animated sequences in addition to the same type of images produced by character generators. Computer equipment is so costly, however, that few studios have it. (The equipment must be specially designed for use on video. Other computer-gener-

293

ated graphics often decompose when used in video presentations.)

- **Graphic artists,** who produce the same work as a computer as well as stylized cartooning, illustrations, and lettering. The work may be produced more slowly than a computer, but labor costs may be more reasonable than equipment costs.

When writing sequences in your script that must be shown through graphics, make sure that the sequence can be produced with the equipment available. If not, make adjustments to the script so that the sequence can be described with the graphics that can be developed.

Decisions about Music and Sound

Music is useful for introducing and closing the presentation and for providing transitions between scenes. Directors usually choose music after shooting is complete. In some instances, however, you may suggest the use of particular music in your script. For example, you may suggest that a certain song be playing in the background for effect, or you may develop an entirely visual presentation around a particular song.

In such instances, you must be aware of the cost of music and the availability of copyrights to determine whether your instructions are realistic. The cost and availability of music depend upon its source. Sources of music include

- **Copyrighted music,** such as a Mozart concerto or a popular song. Copyrighted music can be used only after you purchase rights to use it. Rights can be expensive. In some cases, such as the Beatles' songs, the rights are impossible to purchase. Also, viewers often associate well-known music with images and feelings other than the ones conveyed in your presentation. The "William Tell Overture," for example, connotes images of the Lone Ranger to many people.
- **Music libraries,** which are volumes of original compositions that you purchase for an annual fee and can use without paying additional fees. Music libraries are less expensive than purchasing copyrighted music.
- **Original music,** which you commission. Original music is expensive. In the Denver area, original music composition starts at $2500 for a five- to ten-minute presentation.

Sound effects, too, are an important concern. They support the visuals by adding realism. For example, sounds of machines operating in the background emphasize the activity in a manufacturing environment.

You can record your own sound effects, although recording involves the expense of a production crew. You can also purchase sound effects from a library, just like you purchase music from a music library.

REVIEW

The better you understand how the process of video production trans-

forms your written images into visual reality, the better you can ensure that the final production matches your initial images of the presentation. Specifically, you need to understand the following:

- Before starting production, directors check scripts for continuity, visualization, and picturization.
- Directors mark production plans on storyboards.
- Before developing storyboards, directors make decisions about talent, equipment, graphics, and music and sound. Ω

REFERENCE

1. Zettl, Herbert. *Television Production Handbook,* Wadsworth Publishing Company, Inc. Belmont, CA, 1976.

R. H. AMEND *is an assistant professor of technical communication at Metropolitan State College in Denver, Colorado.*

R. B. STONER, *an Associate Fellow of the STC, is an assistant professor of technical communication at Metropolitan State College in Denver, Colorado.*

Part 7
Multimedia Presentations

ALTHOUGH numerous definitions have been offered for the term "multimedia," it is commonly used to describe "canned" presentations that employ slides interspersed with video on several screens and accompanied by recorded voices and/or music. In reality, however, multimedia presentations can be developed by using the best features of any medium to implement individual segments and integrating them so that they produce the desired message and impact for the target audience. Multimedia presentations are inherently more applicable to marketing or public relations applications than technical applications. Some engineering or scientific presentations, however, could be significantly enhanced by taking advantage of the opportunities offered by multimedia tools. For example, the design review for a new aircraft might benefit from the high quality and resolution provided by slides, the interactive control provided by computer graphics, and the dynamic scenes provided by video. By selecting and integrating the best characteristics offered by each medium, an engineer can optimize the effectiveness of his or her presentation. This part includes articles that outline the basic concept of multimedia presentations and describe various techniques for implementing them.

In their paper, "Perspective on multimedia," Craig Harkins and Emily Schlesinger define the basic concept of multimedia and outline its historical evolution. While the specific applications addressed in the paper are not directly related to technical presentations, many of the ideas and techniques merit consideration, especially as tools for dynamically communicating information and enhancing comprehension. A particularly valuable component of this paper is an extensive list of related information sources.

In their three-part article, "The meaning of multimedia," Larry Tuck and Jennifer Ravi review the definition and benefits of multimedia, the application of computers, and the use of interactive techniques. Although the concept has been used since the 1960s, it has been dramatically expanded and refined with the broad availability of video, computer graphics, and high quality audio. They also describe the use of computers in producing slides, animation, video, audio, and interactive media.

What is multimedia? How can it be used? What will it cost? Jim Heid answers these and other questions in his paper, "Getting started with multimedia." He describes some of the ways it is used and the components that form a complete system. In addition, he defines the various levels of multimedia, including the distinction between passive and active presentations. He also describes the specific processes of authoring, animation, video, and audio, and addresses the issue of cost.

In his paper, "The magic of multimedia," Ken Fermoyle delineates the four major influences in the development of multimedia presentations and their implications for potential users. He also describes several examples of multimedia applications based on using the Macintosh, IBM PC, and Amiga computers as multimedia platforms, and indicates that the hardware and software are evolving along a path similar to desktop publishing. As a result, many engineers already have access to many of the necessary tools.

In their article, "Multimedia: Is it real?," Suzanne Stefanac and Liza Weiman address some of multimedia's promises and problems. They catalog the tools available to multimedia producers, especially those available to Macintosh users. They also consider the factors affecting costs and indicate that multimedia presentations need not require a big-budget investment, making them practical for some technical presentation applications.

The use of multimedia as a presentation tool is considered from a broader perspective by John Murphy in his article, "Multimedia is today's message." He defines the basic applications for multimedia, identifies several related publishers, and describes various platforms for the associated production requirements. He also identifies numerous software packages and their application to multimedia uses. Although not specifically directed toward engineers, many of the techniques and tools described can be readily adapted to technical presentations.

In his article, "The main event," Nick Arnett indicates that multimedia is becoming the hottest tool available for presentations. He addresses what potential users should do to get ready and postulates what can be expected in the near future. To support his position, he cites three real-world examples and identifies the software packages that were used in the production activities.

In the article, "The complete picture: The integration of computer graphics and video," four authors address the subject of multimedia presentations from varying perspectives. John Schwan defines the basic considerations in selecting computer graphics equipment for video applications. Claire Morris and William Jimmerson consider the requirements for computer graphics and video in business presentations for the 1990s. David Hodes presents various user comments relative to the use of videographics and PCs for multimedia applications. All of their ideas can be used by engineers in developing visual aids for technical presentations.

Perspective on Multimedia

CRAIG HARKINS, MEMBER, IEEE, AND EMILY SCHLESINGER, SENIOR MEMBER, IEEE

Abstract—Multimedia is defined as noun and adjective. The development and use of multimedia presentations are surveyed historically, and multimedia is considered a vehicle, technical but at the same time aesthetic, for communicating imaginative as well as factual material, and for entertaining as well as instructing. Aspects of theory, development, research, and practice are discussed, and sources of information about multimedia techniques are listed. Technical and engineering communicators are urged to experiment with multimedia as a dynamic means of imparting facts and establishing comprehension.

INTRODUCTION

Definitions

The word *multimedia* was first used as a simple adjective to mean "involving several media." For example,

A springtime garden, a celebration of the eucharist, and an illustrated travel talk may all be thought of as multimedia presentations.

Usage and activity in the past 25 years, however, have added a technological significance, so that a statement like this is now possible:

The session consisted of three multimedia presentations: a "live" speech with slide projections, a computer-assisted film with music, and a "canned" speech with simultaneous projection of three slide images.

Thus, a two-part definition like the following seems now to be generally acceptable [1]:

multimedia (adjective)—involving several media; referring to any mixture of communication media, including illustrated lectures, film with sound, drama, posters or collages, and slide/tape programs.

But *multimedia* has also become a noun. As such, it means not only a composite, variable, mixed form of communication using slide projection, but also the activities and apparatus involved. Along with the adjectival definitions, therefore, the following nominative operating definitions have also been suggested [2, 3]:

multimedia (noun)—multiple-projection presentations reinforced by sound, and methods or equipment used to make such presentations; a form of communication which uses multi-image techniques combining static and moving projections with live or reproduced sound.

Manuscript received March 31, 1978.
C. Harkins is a Teaching Assistant and doctoral candidate in communication at Rensselaer Polytechnic Institute, Troy, NY 12181, (518) 270-6469; Dr. Schlesinger is a Technical Writer and Editor at the Baltimore Gas & Electric Co., P.O. Box 1475, Baltimore, MD 21203, (301) 234-6137.

Multi-image is obviously a related, but less inclusive term, which refers "generally to two or more images simultaneously projected on a wide screen or on multiple screens. [It] also refers to superimposed images on the same screen; the various images may be juxtaposed or a series of images can be joined to create a panoramic view" [4].

Ancestry

Historically, one may think that Cro-Magnon man combined cave paintings and ritual in early multimedia presentations at religious or other tribal gatherings. Later came dance, drama, opera, cinema, and slide talks—different types of multimedia presentation—for ceremony, education, and entertainment.

In the 1950s, with the improvement of photographic slide equipment, teachers and exhibitors began to project two or more related pictures at one time, superposed on a single screen or separated on adjoining screens. In the 1960s, innovators combined "live" and "canned" sounds with these multi-image presentations. Then, as new audiovisual techniques were developed and experimented with, mixed forms of communication, especially those including slide projection, came to be called *multimedia*.

But such technological multimedia existed in many manifestations for some time before the term was used in print and before it could boast a definition. Presentations were first made and then, often much later, reported on and classified. Theory, as always, both preceded and accompanied experimentation.

THEORY

History

As long ago as 1942, Susanne Langer noted that visual media have great potential for transmitting appropriately selected and organized information [5]. Pointing out that elements of line, color, proportion, etc. can be combined as intricately as words, she remarked on the laws of syntax which operate in the two sorts of construction:

The most radical difference is that visual forms are not discursive. They do not present their constituents successively, but simultaneously, so that relations concerning a visual structure are grasped in one act of vision. Their complexity, consequently, is not limited, as the complexity of discourse is limited, by what the mind can retain from the beginning of an apperceptive act to the end of it.

In 1949, Claude Shannon and Warren Weaver, in *The Mathematical Theory of Communication* [6], considered the trans-

Reprinted from *IEEE Trans. Professional Commun.*, vol. PC-21, no. 3, pp. 118–128, Sept. 1978.

mittal of symbols as presenting three types or levels of problem —those involving technique, semantics (meaning), and effectiveness. According to this analysis, "information" applies not to individual messages (as "meaning" does) but rather to the whole situation of communication, to the statistical character, indeed, of a group of messages. Information thus appears as a manifestation of the entropy, or freedom of choice and adjustment, or degree of predictability—these are synonymous—of the transmittal.

In 1955 George A. Miller presented to the Eastern Psychological Association his now well-known theory on the limitations to an individual's capacity for processing information. From several psychological studies he had determined that there are "seven categories for absolute judgment, seven objects in the span of attention, and seven digits in the span of immediate memory." He argued that the three factors—judgment, attention, and memory—and their built-in limits of seven should be considered when presenting information in any learning situation [7]. For multimedia, the theory has obvious implications in terms of the amount of information to which any given member of an audience can respond.

The next year, N. E. Miller published in *Audiovisual Communication Review* a consensus article dealing with the use of graphics in education [8]. He discussed four characteristics of learning—motivation, stimulus, participation, and reinforcement—and factors affecting the relative effectiveness of films in teaching. Most important was his suggestion that educational films might record, demonstrate, and illustrate— "bring the world into the classroom"—as well as tell a story. This is a clear forerunner of Marshall McLuhan's later claim [9] that

> Multi-screen projection tends to end the story-line, as the symbolist poem ends narrative in verse. That is, multiple-screen, in creating a simultaneous syntax, eliminates the literary medium from film.

McLuhan, indeed, was responsible for introducing much exciting confusion into general discussions of media and related phenomena during the 1960s. His controversial yet thought-provoking ideas that "we now live in a global village" [10], and "The Medium Is the Message" [11], represented a highly original attitude toward communicative processes. In *The Medium Is the Message*, many passages combine rhetoric with insight, as "The medium, or process, of our time— electric technology—is re-shaping and re-structuring patterns of social interdependence and every aspect of our personal life" [12].

Analysis

Other theorists devoted themselves to empirical analysis. Robert Gagné, for example, attempted in 1956 to answer the question, "What is known about the process of learning that can be put to use in designing better education?" His book described eight types of learning and discussed the characteristics of effective instruction for each type [13].

Donald G. Perrin, writing in 1969 [14], used Gagné's ideas as part of the base for important theoretical work which discussed three distinguishing aspects of multiple-image presentations.

First, Perrin considered simultaneous, as opposed to sequential, projection. Films, television, filmstrips, and slides, like writing and speaking, he pointed out, present images in series, so that the meaning of each single image, as it involves the viewer's memory, is determined *in context* by what has gone before. Images simultaneously perceived, however, interact at once in the mind of the viewer, who can process larger amounts of information in shorter periods of time under these conditions than when adjusting to sequential delay, confusions of recall, and so on—comparisons can be made and relationships recognized immediately.

Perrin considered screen size as the second significant factor in multimedia presentation. On screens ordinarily used for film and television, he observed, the general "public" identity of an image is overwhelmingly noticeable and important. When images are shown on larger or adjoining screens, however, each viewer makes his or her own "private" montage of the elements presented. Comparative learning is thus both enriched and made easier.

The third aspect of multimedia presentation analyzed by Perrin was density of information. This characteristic, of course, is a function of the other two. Clearly more material is made available to viewers when two large images are displayed simultaneously than when one small screen appears as a single object of attention; adding speech or other sound further increases the communicative content of a presentation.

Perrin also noted that educators tend to be confused by the divergent purposes, approaches, and products of the two chief groups of people who experiment with multimedia—academic researchers and commercial developers. The latter, he said, attempt "to achieve high visual impact, interest, involvement, motivation, and concentration—in modern terminology, to 'turn on' the audience." Researchers, on the other hand, are more concerned with low levels of motivation. Thus, technical work involves opposite ends of the motivation scale and opposite ends of the real-abstract continuum; and researchers study single variables while industry "appeals to the whole visual sensorium."

Density of information, the third aspect of multimedia presentation discussed by Perrin, has also been considered from other points of view. F. R. Hartman, in 1961 [15], warned against overloading communication channels and viewers' mental (or emotional) capabilities, suggesting that confusion rather than enlightenment can result from a piling-up of information. "Enhanced [i.e., more effective] communication resulting from a summation of [visual] cues occurs only under special conditions," he wrote, and

> Most of the added cues in the mass media possess a large number of extraneous cognitive associations. The possibility that these associations will interfere with one another is probably greater than that they will facilitate learning.

Comparison with Poetry, Painting, and Music

Simultaneity and layering of images, however, are properties which multimedia shares with poetry, as pointed out by C. Low, who worked with Roman Kroiter at Expo '67 [16]:

Poetry uses an amalgam of thoughts, feelings, and word images poured in quick succession as an assault on the unconscious. Some poets seem impatient with the sequential quality of words and phrases and compress language in what seems to be an effort to achieve a kind of simultaneity. Roman Kroiter speaks of multi-screen [presentations] being to single-screen [presentations] what the language of poetry is to the language of prose.

This comparison suggests that visual cues, like spoken and written ones, may sometimes be misunderstood, ignored, or simply not recognized. Consider, for example, the opening of a poem by Dylan Thomas (1914-1953):

> The force that through the green fuse drives the flower
> Drives my green age.

An electric-utility engineer, reading this, might think wrongly at first that in writing "fuse," the poet is describing the stem of a plant in terms of a fault-interrupting device. A road-construction engineer, on the other hand, might not be so mistaken—the metaphor refers to a thread of gunpowder burning to explode dynamite in a burst of flame like a bright blossom. Another error would be to discard the idea that "green age" refers to youth. Or, the sentence as a whole may seem to have no meaning [17].

Because such misinterpretations are possible, critics and scholars offer explanations of meaning so that a poem will be understood by those who fail at first to recognize the poet's cues and symbols.

Paintings, too, often convey a fuller message when explicated, as they consist of visual fragments which may be variously unified, variously perceived, and variously significant. For example, in *The Arnolfini Marriage* by the fifteenth-century Dutch artist Jan van Eyck, familiar household objects appear as symbols of physical, spiritual, and domestic unity. Somewhat differently, composite peasant scenes painted by sixteenth-century Peter Bruegel the Elder consist entirely of separable vignettes: One canvas, for example, shows children playing eighty different games (they roll hoops, ride "horse," gyrate in blindman's bluff, and so on); another canvas illustrates Flemish proverbs, and another, *The Tower of Babel*, pictures several types of construction activity.

Just as explication enriches the meaning of paintings like these, and of poems, so accompanying verbal comment can make multimedia presentations more effective. Similar terms of analysis may be applied in discussing all three vehicles for conveying messages.

Robert Siegler, on the other hand, has described multi-image presentations in terms of music (1968) [18]. The total montage, he wrote, the *masquage*, or selection and arrangement of related images, corresponds to instrumental and melodic articulation. In this view, which emphasizes form rather than substance, the structures of technological multimedia seem to resemble those of musical compositions. Terms related to general and particular design, interplay of form and color, statement and reinforcement of idea, traits of style—such terms apply to complex instructional or entertaining ensembles in multimedia as well as to contrapuntal Bach fugues the conceptual unity of symphonies by Beethoven and Brahms, and the richness of orchestrations by Ravel and Debussy.

These comparisons emphasize the complexity of technological multimedia and reveal the ease with which it can be compared to poetry, painting, and music. They also suggest that there may be a need to develop a critical basis for judging multimedia—a need which Ken Burke [19] has already addressed by discussing methods and standards for analyzing and evaluating multimedia programs and productions.

Techniques

The impact of any presentation will be stronger, also, if writers and producers have learned from tradition, research, and accomplishment how to produce and combine related elements most effectively. Many aspects of multimedia have been examined and reported on.

For example, in 1950 Charles Hoban and Edward Van Ormer [20] conducted a number of experiments with cinema—experiments which showed, among other things, the importance of photographing at viewer-oriented camera angles, of demonstrating common errors, and of repeating key sequences in instructional films. Similarly, eight years later, Mark May and A. A. Lumsdaine [21] called attention to the effectiveness of crude visuals, the advantages of narrated films, and the value of using printed material on film. In the 1960s May [22] did additional work on motivators, reinforcers, cue identifiers, and simplifiers.

Howard Levie and Kenneth Dickie investigated various instructional media in 1963 and concluded that learning is enhanced when presentations are reduced to factors directly contributing to the goal [23]. Two years later James Bornn and William Pelcher introduced their "Media Characteristics Matrix," a highly useful model for determining the potential of various media for accomplishing particular objectives [24].

There is no doubt, furthermore, that the components of any multimedia presentation must be carefully assembled. W. L. Millard pointed out in 1964 that the effectiveness of instructional multimedia depends on the ability of teachers to combine material and techniques into unified, dynamic presentations which serve certain purposes—make comparisons, show relationships, present alternatives, illustrate dichotomies, develop concepts, and so on [25].

Multiple-image techniques must produce specificity, relevance, and simplicity of design, said R. Fleisher in 1969 [26]. William H. Allen, in 1975, demonstrated that teaching media should be designed and used differently for groups with different intellectual ability [27]. And the team of John Gruber, Tony Paris, and Randy Will emphasized in 1976 [28] that

> each step of a multimedia project—from conception to preparation to implementation—requires careful planning, close attention to detail, and collaborative approaches to problem solving.

The work and comments of these investigators suggest that teachers, as well as the researchers and producers noted by Perrin, should investigate techniques and devices of multimedia. Identification of intended uses will reveal areas or lines for research and development, and also indicate whether presentation should be straightforward, impressionistic, or personalized. Is the purpose to thrill, to amuse, to delight, or to mystify? to challenge the reason? to capture attention, establish an attitude, or stimulate memory? to teach facts or explore concepts? Multimedia has many modes, many moods, many messages.

DEVELOPMENT

Types of Display

While theorists consider, educators, demonstrators, and entertainers experiment—and make presentations. By the early 1950s, technology had produced radical changes in the type and number of audiovisual stimuli available for lecture rooms and theaters. For example, T. R. Murroughs discussed two ways of experiencing depth perception [29]:

Normally, as we move forward, we are dependent on what we can see to the right and left. [As] peripheral objects appear to move out ... and curve around us backward to either side, [their displacement] helps us to judge the distance straight ahead and tells us where we are.

Similarly, in the film technique called Cinerama, wide-angle lenses and curved screens were used to impart sensations of tridimensional vision. Murroughs wrote,

In Cinerama we actually sit in the leading car of a roller-coaster as we top the rise of a track. The view is spread out before us. As we "plunge" down the incline, the field quickly enlarges, with more peripheral detail flying outward and around (past us), adding a terrifying realism to the experience [and] causing many in the audience to grip their seats. In the same manner we glide along slowly in a gondola while the panorama on either side slides by slowly, or we fly through valleys between walls of canyons we can almost touch.

Some observers considered developments like Cinerama crucial to the survival of audiovisuals. A. Cornwell–Clyne wrote in 1954 that without technological innovation, film would suffer a slow decline in popularity. Increasing efforts were needed, he thought, to expand the medium's powers of communication and expression [30].

Four years later, evidence of such efforts appeared at the World Fair in Brussels. In the Czechoslovakian "Polyekran," Josef Svoboda produced an eight-screen slide-tape-film presentation. Multi-projection techniques were used also in Walt Disney's "Circarama" presentation and in a U.S. Government-commissioned exhibit of filmed vignettes of American life [31].

Four years after this, for the U.S. Science Pavilion at the 1962 World Fair in Seattle, Charles and Ray Eames produced *The House of Science*, a 13-minute, seven-screen history of science. This production, which reached a wide international audience, undoubtedly was the key inspiration for the subsequent growth of multimedia. It began with an animated cartoon on a single screen, and as new scientific disciplines were introduced, new screens were "opened" to expand the house of science. In a *New York Times* review, Bosley Crowther said that the presentation gave "a sharply stimulat-

ing notion of what this medium [i.e., film] may now be able to do" [32].

In 1964 E. E. Kirkbridge and V. C. Jones provided the first reference to multimedia phenomena in the literature of technical communication. Their paper, "Multivisual Presentations: A Dynamic Communication Technique," was presented at the 11th International Technical Communication Conference in San Diego. Reporting on the use of multimedia at U.S. Naval Ordnance Test Stations in China Lake and Pasadena, California, they claimed increased efficiency in presenting large quantities of technical information quickly and dynamically while retaining maximum audience attention [33].

The 1964–1965 New York World Fair gave further conceptual and practical significance to multimedia. Industrial promoters used special theaters, assemblies of custom-designed equipment, unique instruments and mechanisms, and creative interaction between talent and projected images. Among the major exhibits were the Eames' 13-minute IBM extravaganza in an egg-shaped, screen-paneled loft; Jeremy Lepard's 110-screen "American Journey" for Cinerama; "To Be Alive," Francis Thompson's and Alexander Hamid's moving three-screen presentation for Johnson's Wax; and Saul Bass's 22-minute "Searching Eye" for Kodak. Even the Vatican used multimedia: In addition to its featured attraction, Michaelangelo's *Pieta*, the Papal pavilion had a unique multiple-screen wall on which one-minute films were rear-projected [34].

At an international conference three years later, Louis Perica discussed the "expanding horizons" of technical communication, urging professionals to use not only the printed page but also "media such as photographs, motion pictures, television, microfilm, computers, and communication satellites" to explain and disseminate scientific information. Moreover, he demonstrated the use of animated slides—an audiovisual technique which adds motion to static projections. Animated slides are created by sandwiching polarized material in specific areas of exposed film. During projection, when a polarized wheel is spun in front of the lens, images on the screen appear to move. Such slides are particularly effective when flow directions or other motion patterns are germane to a presentation [35].

"Talking slides" is a more recently developed technique for 35-mm projection. James Bushnell [36] has described his use of a Sound-on-Slide System for industrial indoctrination. In another type of application, Halverson [37] uses sound/slide presentations to keep scientists up to date on developments made in energy research at the Lawrence Berkeley Laboratory of the University of California.

Less than two months after the delivery of Perica's paper to technical communicators, Expo '67 opened in Montreal. Yale Joel, describing this World Fair in *Life*, wrote that it did for audiovisual communication in 1967 what London's Crystal Palace did for architecture with iron and glass in 1851, what the Paris Exposition did for steam engineering in 1889, and what the St. Louis Fair did for the automobile in 1904 [38].

At Expo '67, Joel pointed out, most of the display pavilions showed "technically ambitious" moving pictures, wrapping and immersing viewers in "images that assault the senses and expand the mind," making them understand "more through feeling than through thinking":

> Pictures are thrown at the spectators with or without words, stories are told without logical sequence; viewers are deliberately thrown off-balance mentally and even physically. Film transmits facts, creates moods, tests moral judgments. It speaks at Expo '67 sometimes in poetry, sometimes in gibberish but always in a visual blitz almost blinding in its implications for the future.

At Montreal, three of the more celebrated pavilions were:

Labyrinth (created by Roman Kroiter of Canada's National Film Board), which condensed time by putting more information into each moment (e.g., projecting five images at one time), and expanded space by adding a third dimension to images (e.g., elevating the audience and projecting separate "pictures" onto a floor screen and wall screen simultaneously);

Diapolyecran (an encore by Josef Svoboda), which provided an "optical narration" of the creation of the world, shown on a wall of 112 moving, two-foot-square screen cubes, each of which contained a set of coupled projectors operating in programmed sequence; and

Circle-Vision 360°, which surrounded the viewer with a complete vista of Canadian scenery (filmed from a converted B-52 bomber with nine cameras).

With Joel's article in *Life* was a quasi-theoretical essay by Frank Kappler, who noted that multimedia techniques are "aimed at the generation raised on TV, conditioned by commercials to accept as commonplace the zoom shot, the jump-cut, the freeze-frame" [39].

Kappler predicted a "breakthrough in education." For example, he said, showing a single picture with others continuously juxtaposed will "conjure up a complexity of ideas and relations in which the whole is more than the sum of its parts." Moreover, as much learning is subliminal, such a display of related images, especially if accompanied by sound and changing light, can cram into the mind a network of ideas and impressions for later recall, sorting, and reintegration.

On the other hand, Kappler acknowledged that there may seem to be little variety in audiovisual syntheses. "Amazingly alike," he called them, wondering if mixed-media presentations are "anti-intellectual—they certainly drive hardest at sensations and emotions."

> Could it be that this revolutionary bombarding of the senses can create only attitudes, not philosophies? Can it convey only generalizations (apartheid is hateful, peace would be nice, it's great to be young), not hard facts?

With Joel, Kappler used the word *blitz* to characterize the effect of multimedia presentations. *Blitz* is not only German for *lightning*; it also recalls the idea of *Blitzkrieg* or *lightning-war*, Hitler's devastating deployment of armored tanks—i.e., instruments of technology—to overwhelm the countries and consciousness of Europe. Similarly, Kappler described the "blitzing of the mind" by multimedia, "a softening-up operation which can become a basic part of the educative process. The mind blitzed," he wrote, "is a mind burst open."

Historical Definitions

Technical reporters were slow to agree on definitions for these innovations. In 1969 and 1970 D. G. Perrin described "simultaneous projected images" and also "multiple-image" and "mixed media" (e.g., film plus slides) presentations [14], [40]. Demer and Fentnor, in 1970, spoke of "multiple-image presentations" [41], as did Benedict and Crane, who have already been quoted. The latter also defined *multimedia* —"any variety of two or more projected or non-projected media used together but not necessarily simultaneously" [4].

For Craig Harkins in 1974, *multimedia* meant "presentations which involve a range of audiovisual effects," including demonstration, spoken narrative, film, and simultaneous projection on two or three screens [42]. The next year Pascal Trohanis, writing of "audible multi-imagery," defined multimedia as the "simultaneous projection of multiple images with audio" [43], and in 1977 two two-part "operating definitions"—one adjectival, the other nominative—were proposed, as stated earlier (Introduction).

RESEARCH

Bicultural Relationships

Considering this variety of types and definitions, the similarities with painting, music, and poetry, and the many technical and aesthetic aspects of production and reception, one can conclude that multimedia is a borderline or ambivalent form of communication. A versatile medium, it can be at once, or by turns, factual and imaginative; it states and suggests; it instructs, inspires, and amuses.

Multimedia, indeed, may be considered a bridge across the chasm which often seems to separate science and the humanities. Books concerned with the philosophy of scientific research contrast the disciplinary and emotional characteristics of art and technology [44], and much has been made of the disparity between "the two cultures," science and literature, since physicist-novelist C. P. Snow first spoke on their alienation [45]. Poet Edgar Allan Poe imaged science as a vulture, as a sort of universal Peeping Tom whose "peering eyes" violate natural beauty, and as a destroyer who wantonly preys upon innocence [46].

Similarly, artist Georges Braque wrote, "Art upsets, science reassures," and clergyman Joseph Roux thought, "Science is for those who learn; poetry is for those who know" [47].

Others, seeing only superficial conflict, manage quite easily to bridge the chasm. Physician Frederick Wood-Jones believed that "Whoever wins to a great scientific truth will find a poet ahead of him in the quest [48]; sociologist Theodore Roszak writes, "Nature composes some of her loveliest poems for microscope and telescope" [49]; and teacher-technologist Robert Pirsig [50] finds that

> The Buddha, the Godhead, resides quite as comfortably in the circuits of a digital computer or the gears of a [motor]cycle transmission as he does at the top of a mountain or in the petals of a flower.

Poet e. e. cummings, however, laughs at all of them [51]:

While you and i have lips and voices which
are for kissing and to sing with
who cares if some oneeyed son of a bitch
invents an instrument to measure Spring with?

These authors, of course, are considering other aspects of the problems of development discussed earlier, and of what Perrin referred to as the "confusion between research and commercial production" [14]. Although he noted that the learning of cognitive associations of nonsense words and symbols is certainly not too stimulating by comparison with elaborate World Fair productions, multimedia would appear to be a field in which cummings' monocular, precision-loving scientist and the promoter of supershows can work together for mutual benefit. As costs involved in an elaborate multimedia production are so high, any sponsor would be ill-advised to ignore the possibility that research findings might suggest ways of maximizing the effectiveness of promotion. On the other hand, the inventiveness of today's commercial producer is keeping researchers busy with a constant stream of new variables for study.

Much research has been done on the psychological and physiological aspects of multimedia—that is, about mechanisms and factors involved in visual learning, and the relative effectiveness of single- and multi-image presentations. The major studies are mentioned here.

Processes of Learning

In 1961, at the University of Wisconsin, R. Hubbard demonstrated that a 50-minute taped lecture could be "boiled down" to 20 minutes of Telemation with no loss of learning by students [52].

In 1966 B. Schlanger reported on the relationship between impact and screen size. He found that when screens get bigger, in relation to the size of the audience, the visual impact is increased [53].

In 1967 E. Levonian observed that the arousal of any emotion by a film sequence enhanced retention of material shown, regardless of the nature of the emotion [54]. In 1967 also, the field of advertising provided data on the relationship between the length of an audiovisual stimulus and the retention of information presented: I. Yankelovich reported "no real difference in the communication values of 60-second and 30-second commercials" [55].

During the same year, Werner Severin published two major reports. In "Another Look at Cue Summation" [56] he concluded that additional information made available from *related* (as opposed to simply redundant) audiovisual effects leads to increased cognitive retention. He found, however, that redundancy does give benefits when data are highly complex or transmission rates are extremely rapid. In the other report, Severin showed that careful cross-cueing of related visual material increases learning in a multiple-channel presentation [57]. At the same time, Jerome Conway reported quite similar findings [58].

Potter and Levy, experimenting in 1969 with very short exposure times, found that subjects could remember all images presented for one second or more, but only 15 percent of images shown for 125 milliseconds [59]. Two other investigators, working later (1974) with peripheral vision, concluded that "the ability of the periphery [of the retina] to process information depends upon the region stimulated, the complexity of the task, and the size of the stimulus" [60].

In 1972 Jon Meyers analyzed large-screen multimedia presentations in terms of eye movements and length of eye fixations of viewers. His subjects tended to turn to new images as soon as they appeared and favored viewing large images over smaller ones [61].

Despite these and other studies, however, the mechanism of perception of multiple images is not very well understood, either in general or in particular, chiefly because studies of perception have been made with only simple stimuli. Bruce Goldstein noted that

Multiple-image projection . . . extends the problems of picture perception in space because of the wider visual angles usually employed, and in time because of its property of constant change.

He wondered how persons *observe* a multi-image projection and suggested that the question might be answered by using complex displays to study eye movement [62].

Effectiveness of Learning

The effectiveness of multimedia as a teaching aid has been investigated in a number of ways and without complete agreement.

Several years before Expo '67, A. U. Roshka found simultaneous projection of images a more effective method of communicating with young Russian children than sequential projection [63]. C. Malandin obtained the same result with French children between 9 and 11 years old [64]. In 1963 Allen and Cooney reported that sixth-graders learned more from multi- than from single-image presentations, whereas eighth-graders learned equally well from both types [65].

In 1965 Terrence Snouden compared three types of multimedia presentation: (1) live, (2) completely automated, and (3) automated with a live ten-minute review. Results of his study "seem to indicate that automated presentations are not detrimental to learning" [66].

Until 1969 proponents of multi-image teaching seemed to have an undeniably superior method. Then E. S. Lombard introduced some confusion with a report which compared the use of single- and three-screen presentations in teaching eleventh-grade history. High achievers of both sexes seemed to learn more from single-screen projections. Low-achiever females performed better than did average achievers with three-screen presentations; low-achiever males showed no difference [67].

In the same year M. E. McCombs, after working with first-graders, suggested that "either a verbal presentation or a visual presentation is more likely to be initially effective than is an audiovisual presentation" [68].

R. T. Jones, also in 1969, described the effects of multi-channel audio stimuli on learning efficiency. He found that listeners could recall large portions of four messages transmitted simultaneously—and thus might be thought able to process more information in less time from multimessage than from sequential presentations [69].

In 1970 R. O. Reid found multi-image presentations more successful "agents of attitude change" than linear film formats [70]. An overview article by Williams and Jorgensen [71] concluded that multimedia

is at last coming of age in the educational field. Scores of creative educational productions have been created by commercial and non-commercial producers alike. Manufacturers have begun to solve the knotty problems of hardware, notably in control systems, . . . and they have begun to market hardware which is within reason for educational budgets.

In 1971 B. R. Lawson [72] reported that he had found multi-image techniques of great value at the U.S. Military Academy, but four other workers were less enthusiastic: Charles Bollmann [73] found no significant difference in affective impact between single-image and multiple-image slide shows. Lawrence Atherton [74] found no difference at all in the effects of film and multiple-image presentations on either cognitive or affective learning. Bernard Fradkin [75] wrote,

There is no way to conclude that a multi-image presentation produces greater recall or longer retention than the simple presentation of the same visuals in sequential order.

Tim-Kui Tam and Reeve [76] compared the effectiveness of two- and three-screen sequential and cumulative presentations, keeping total length, narration, slides, and slide order the same; their results were the same in the four cases.

Joy and John Menne [77] found that third-grade students learned more from simultaneous aural-visual presentations than from either aural or visual presentations alone, and D. Ingli [78] reported favorable response (via questionnaire) to his use of multi-image techniques in teaching a basic audio-visual course.

Don Didcoct's conclusions, also published in 1972, were more conservative [79]. Taking "learning" to refer to facts immediately recalled as well as to information and concepts retained, he wrote the following:

1. As compared with a single-image presentation, a multi-image sound-slide presentation does not necessarily result in increased learning . . . for subjects at college level.
2. A multi-image sound-slide presentation is a delicately balanced instrument for instruction and may under certain conditions impair learning.
3. Provided [that the] multi-image sound-slide presentation is designed and developed with the utmost care and presented under ideal environmental conditions, subjects prefer a multi-image presentation to a single-image presentation.
4. Although college-level subjects may prefer a particular multi-image presentation over a single-image presentation, this preference may not result in increased learning.

A study by Yolles which compared material presented in single-screen and three-screen formats confirmed earlier research findings that grade-school children have better cognitive recall with multi-image formats [80].

In 1974 Wilson Brydon [81] published convincing evidence to support his conclusion that

aesthetic considerations being equal, three-screen presentations are a more effective means of communication than single-screen presentations at both the cognitive and affective levels.

In 1975 Meyerowitz and Fradkin [82] agreed. They observed no difference in the effectiveness of single- and three-screen sequential presentations, but found three-screen more effective than single-screen cumulative presentations, and three-screen cumulative significantly better than three-screen sequential presentations. In the same year, Ausburn [83] found that undergraduates learned more from three-screen than from single-screen formats; when he compared visual and haptic (or touch-oriented) learners, he found that the haptics, too, seemed to learn more easily from three-screen presentations.

Early in 1975 the trade journal *Business Screen* published an interview with Carl Beckman, president of the newly formed Association for Multi-Image (AMI) [84]. Speaking as observer and consultant, Beckman said that for five years he had collected empirical evidence on the use of multimedia in classrooms, social gatherings, and professional meetings. He concluded that

use of multimedia improves message retention
image size affects image impact
young persons react more favorably to multimedia than older persons
multimedia is effective in both cognitive and affective learning

These statements are reports of experience—they reflect the past and present of multimedia. Pascal Trohanis, looking in the same year toward the future, suggested subjects for further study in the field of "audible multiple-imagery": such things as the effect on learning and retention of introducing music and "pictorial multiplicity," incidental learning, eye search and scan patterns for various pictorial styles, and the value of following presentations with discussion. His research indicates that audible multiple-imagery is most effective when used for periods of ten minutes or less [43].

PRACTICE

While specialists are reviewing progress and planning new approaches, many beginners are seeking information about multimedia techniques, hoping to use this form of communication for their own purposes. Practical advice is available, and also the help of established practitioners.

Sources of Information

Three books have been written on the subject: *Producing Multi-Image Presentations* by Benedict and Crane emphasizes "how to" information, speaking particularly to do-it-yourselfers [4]. Columbia Scientific Industries' manual, *Synchronized Slide Shows*, is more advanced and contains a great many details [85]. AMI's *Art of Multi-Image* is a collection of articles which variously discuss theory, describe research, and give practical advice [86].

Multimedia techniques are discussed and demonstrated in sessions of the International Technical Communication Conferences held annually by the Society for Technical Communication, and examples of "practical use" abound.

Uses

In productions (1973, 1974) of *Barbara's Polar Bears*, by James Lufkin [87], dissolved slide effects complemented the dialog of live actors whose situations illustrated various attitudes toward the role of research in U.S. business, government, and society.

On a larger scale, many independent producers have joined Walt Disney's pioneering use of multimedia for public entertainment. Theaters at three Busch Gardens show "The Eagle Within" in multimedia [88]. "The Battle of Bunker Hill," a 17-minute, 1.4-million-dollar, Bicentennial, multimedia project, "opened" in Charlestown, Massachusetts in 1975 [89]. "Thrills and Drills," a 12-minute, 12-screen show, ran continuously at a four-day Offshore Technology Conference in Houston, also in 1975 [90]. *Beatlemania*, a 1977 Broadway "hit" show, brought multimedia to the legitimate stage: Memories of the mid-sixties were revived by four Beatle look-alikes, film clips, slides, and an assortment of scrims and screens [91].

John Kucharski specifies "talent, teamwork, and management support" as key ingredients in the successful use of multimedia by the Animal and Plant Health Inspection Service (APHIS) of the U.S. Department of Agriculture [92]. APHIS uses the technique to communicate among 14,000 employees working on 40 separate projects throughout the world.

Several educational efforts are worth noting. Multimedia has been used to give health instruction [93] and to teach English [94]. It has been recommended as a tool for communicating technical information [95–97]. In 1973 G. F. McVey of Boston University specified the components for an effective multimedia system for college instruction [98]. More recently, a survey by C. H. Gardner showed that education administrators consider the medium of great potential "as a possible public relations communication tool for individual schools or school systems" [99]. According to him,

> The basic components of a multimedia presentation system include a rear projection screen, three or four slide projectors, a 16-mm movie projector, a stereo audiotape recorder, an audiovisual programming device, a front projection screen, and an overhead transparency projector which functions as a sort of electric chalkboard. Other components are usually added as the sophistication of the facility grows through use of the system. Such components most likely will include Super 8 and standard 8-mm movie projectors, an opaque projector, a television projector, random-access remote controls, and a wireless microphone system.

This enumeration calls attention to the importance of technological equipment in research and education. When faculty member Peter Mather produced a slide show about the TV Division of Los Angeles City College, a complete multimedia production laboratory was set up [100]. When Ronald

Slawson of Santa Fe Community College (Gainesville, Florida) began to teach "Introduction to Multi-Media Communications," a special room was built for him [101]. The Naval School of Photography recently added a block of instruction in multimedia [102].

Theaters are being built especially for multiscreen shows. Armco has constructed a complete, mobile multimedia center that weighs 7,000 pounds, can be set up in less than eight hours, and fits into a 40-foot van; Dayton Calloway was the designer [103]. Donovan Worland created "Ecosphere," the world's first full-sphere multimedia theater as part of a permanent visitors' center at the Trojan Nuclear Power Plant of the General Electric Company in Prescott, Oregon, about 40 miles from Portland [104]. A recent multimedia presentation was shown simultaneously by A&P Food Stores in many U.S. locations [105]. Other multimedia presentations have been controlled by time-shared minicomputers [106].

AMI, the Association for Multi-Image, has already been mentioned. It was formed late in 1974 to promote the art of multi-image presentation through meetings and conferences, workshops and courses, publication and dissemination of news and information, and provision of consulting services. A committee is working to establish standards for equipment and format that will make possible the interchange and distribution of multi-image programs.

Increasingly sophisticated technology is being used in multimedia. Sales of slide-programming equipment are growing rapidly. Trade journals have introduced regular columnists who specialize in multimedia. Promoters and reporters write glowing descriptions and accounts of multimedia products and productions.

Help for novices in the field of multimedia is available in many of the works mentioned in this survey. In particular, the volume by Benedict and Crane [4] was written for beginners, and *The Art of Multi-Image* [86] is but slightly more sophisticated. The journals *Audio-Visual Communications*, *Audiovisual Instruction*, *AV Communication Review*, and *Multi-Images* contain articles addressed to novices; accounts cited on classroom uses describe very simple presentations; and the two articles by Harkins [42, 95] discuss elementary procedures.

These instructions, however, must be applied creatively and adapted logically in every particular usage. Multimedia has a dual nature—it is both an art and a science, as this survey has tried to show by describing it *vis à vis* the techniques and ideas of various humanistic, sociological, and technical disciplines.

CONCLUSION

Since the 1950s, technological multimedia—or, more simply, multimedia—has become an important form of communication. It is now a subject of research, a tool for education, a means of advertising, and a vehicle for entertainment. Many professionals, practitioners, devotees, and investigators—and a variety of audiences—support it.

Surprisingly, however, multimedia techniques have been very little used to present technical and scientific information, except in classroom teaching. During the early and middle 1960s, technical communicators were among the first to advocate multi-image concepts and productions, but they apparently lost interest because multimedia began to seem too "theatrical."

Undoubtedly the techniques have been put to sensational uses, but the medium is simply a method or expedient. It may obscure or clarify any content, according to the communicator's skill (or his wishes), but *it* is not the content. Only rhetoric, not logic, identifies medium with message or says that the carrier is the carried.

Multimedia, an adjustable composite, is a strong but sensitive instrument which can be used to transmit effectively messages of many different kinds, in a wide range of complexity, with a wide range of intellectual and emotional content, to audiences with wide ranges of capability for feeling and understanding. Perhaps the information and ideas presented in this paper will remind technical and engineering communicators that writing articles and making speeches are not the only ways of disseminating knowledge—will even inspire them to experiment with that almost inexhaustible method and medium, multimedia.

References

[1] See L. K. Burke, "Focus on Multimedia," *Multi-Images*, vol. 2, no. 4, p. 13, Summer 1976.

[2] E. Schlesinger, "What Is Multimedia?" *Journal of Technical Writing and Communication*, vol. 9, no. 1, 1979, to be published.

[3] C. Harkins, *Effective Presentations in Multimedia*, in preparation.

[4] J. A. Benedict and D. A. Crane, *Producing Multi-Image Presentations*. Tempe, AZ: Media Research and Development, Arizona State University, 1973, p. 3.

[5] S. K. Langer, *Philosophy in a New Key*. Cambridge, MA: Harvard University Press, 1951, p. 93.

[6] C. E. Shannon and W. Weaver, *The Mathematical Theory of Communication*. Urbana, IL: University of Illinois Press, 1949.

[7] G. A. Miller, "The Magical Number Seven, Plus or Minus Two," *The Psychological Review*, vol. 63, no. 2, pp. 81-97, Mar. 1956.

[8] N. E. Miller, "Graphic Communication and the Crisis in Education," *AV Communication Review*, vol. 5, no. 3, pp. 1-120, 1957. Special issue published by Department of Audiovisual Instruction, National Education Association.

[9] M. McLuhan, *Counterblast*. New York: Harcourt, Brace and World, 1969, p. 24.

[10] M. McLuhan and Q. Fiore, *The Medium is the Message*. New York: Random House, 1967, p. 63.

[11] M. McLuhan, *Understanding Media*. New York: McGraw-Hill, 1964, ch. 1.

[12] McLuhan and Fiore, p. 8.

[13] R. M. Gagné, *Conditions of Learning*. New York: Holt, Rinehart, and Winston, 1965, 2nd ed. 1970. The eight types of learning are the acquired abilities to respond to 1) signals, 2) discriminated stimuli, 3) chains of stimuli, 4) verbal associations, 5) individual but related stimuli, 6) concepts, 7) rules, 8) problems (pp. 63-64).

[14] D. G. Perrin, "A Theory of Multiple-Image Communication," *AV Communication Review*, vol. 17, no. 4, pp. 368-382, Winter 1969.

[15] F. R. Hartman, "Recognition Learning Under Multiple Channel Presentation and Testing Conditions," *AV Communication Review*, vol. 9, no. 1, pp. 21-43, Spring 1961.

[16] C. Low, "Multi-screen and Expo '67," *Journal of the Society of Motion Picture and Television Engineers*, vol. 77, no. 3, pp. 185-186, Mar. 1968.

[17] *The Poems of Dylan Thomas*, Daniel Jones, ed. New York: New Directions, 1971, p. 77.

[18] R. Siegler, "Masquage: An Extrapolation of Eisenstein's Theory of Montage-as-Conflict to the Multi-Image Film," *Film Quarterly*, vol. 21, no. 1, pp. 15-21, Spring 1968.

[19] L. K. Burke, "A Functional/Experimental Approach to Criticism of Multimedia Programs," dissertation, University of Texas at Austin, 1976.

[20] C. F. Hoban and E. B. Van Ormer, *Instructional Film Research, 1918-1950*. Port Washington, NY: U. S. Naval Special Devices Center, 1950.

[21] M. A. May and A. A. Lumsdaine, *Learning from Films*. New Haven, CT: Yale University Press, 1958.

[22] M. A. May, "Enhancements and Simplifications of Motivational and Stimulus Variables in Audiovisual Instructional Materials," 1965; "The Role of Student Response in Learning from the New Educational Media," 1966; and "Work-Picture Relationships in Audiovisual Presentations," 1965. U.S. Office of Education Contract OE 5-16-006.

[23] H. W. Levie and K. E. Dickie, "The Analysis of Application of Media," *Second Handbook of Research on Teaching*, Chicago: Rand McNally, 1973.

[24] J. A. Bornn and W. Pelcher, "Selecting the Medium for the Message," in *Proceedings of the 22nd International Technical Communication Conference*, Society for Technical Communication, Anaheim, CA, 1975, pp. 254-258.

[25] W. L. Millard, "Visual Teaching Aids: Production and Use," *The Encyclopedia of Photography*. New York: The Greystone Press, 1964.

[26] R. Fleischer, "Multiple-image Technique for 'The Boston Strangler,'" *American Cinematographer*, vol. 3, no. 2, pp. 202-205, Feb. 1969.

[27] W. H. Allen, "Intellectual Abilities and Instructional Media Design," *AV Communication Review*, vol. 23, no. 3, pp. 139-170, Summer 1975.

[28] J. Gruber, T. Paris, and R. Will, "Working Together on Multimedia—Software, Hardware, Client," in *Proceedings of the 23rd ITCC*, STC, Washington, DC, 1976, pp. 332-333.

[29] T. R. Murroughs, "Depth Perception," *Journal of the Society of Motion Picture and Television Engineers*, vol. 60, no. 1, p. 656, June 1953.

[30] A. Cornwell-Clyne, *3-D Kinematography and New Screen Techniques*. London: Hutchinson, 1954, pp. 7-10.

[31] H. Thompson, "Big Country In Loops," *New York Times*, Section II, p. 7, Apr. 20, 1958.

[32] B. Crowther, "Seeing Science on Film," *New York Times*, Section II, p. 1, May 13, 1963.

[33] E. E. Kirkbridge and V. C. Jones, "Multivisual Presentations: A Dynamic Communication Technique," in *Proc. 11th ITCC*, Society of Technical Writers and Publishers (now Society for Technical Communication), San Diego, CA, 1964, 14 unnumbered pp. in Section 16.

[34] O. Collin, "A Pictorial Report on Audio and Visual Exhibition Techniques at the New York World's Fair," *Business Screen*, vol. 25, no. 3, pp. 25-61, Mar. 1964.

[35] L. Perica, "Technical Communications in the Real World," in *Proceedings of the 14th ITCC*, STWP, Chicago, 1967, Paper No. 66, pp. 1-4.

[36] J. Bushnell, "'Talking Slides' to Orient New Employees," *Training and Development Journal*, vol. 28, no. 11, pp. 8-10, Nov. 1974.

[37] ASFA Notes, *Newsletter of the American Science Film Association*, July 1977, p. 2.

[38] Y. Joel, "A Film Revolution to Blitz Man's Mind at Expo '67," *Life*, vol. 63, no. 2, pp. 20-28B, July 14, 1967.

[39] F. Kappler, "The Mixed Media—Communication That Puzzles, Excites and Involves," *Life*, vol. 63, no. 2, p. 28C, July 14, 1967.

[40] D. G. Perrin, "The Use and Development of Simultaneous Projected Images in Educational Communication," Ph.D. dissertation, University of Southern California, 1969.

[41] L. J. Demer and L. H. Fentnor, "Multiple-Image Presentation of Technical Papers—the Concept," *IEEE Transactions on Engineering Writing and Speech*, vol. EWS-13, pp. 2-8, May 1970.

[42] C. Harkins, "'Baffle-Box' Assists in Scaling Down Multimedia Presentation." *Technical Photography*, vol. 6, no. 10, pp. 20-21, Oct. 1974.

[43] P. L. Trohanis, "Information Learning and Retention with Multiple-images and Audio," *AV Communication Review*, vol. 23, no. 4, pp. 395-414, Winter 1975.

[44] For example, L. C. Hawes, *Pragmatics of Analoguing: Theory and Model Construction in Communication*. Reading, MA: Addison-Wesley, 1975. See also W. Weaver, "The Imperfections of Science," *Science: Method and Meaning*, S. Rapport and H. Wright, Eds. New York: Washington Square Press. 1964.

[45] C. P. Snow, *The Two Cultures and the Scientific Revolution.* New York: Cambridge University Press, 1959. Snow said more on the same subject in *Science and Government*, Cambridge, MA: Harvard University Press, 1961.

[46] "Sonnet—To Science":

Science! true daughter of Old Time thou art!
　Who alterest all things with thy peering eyes.
Why preyest thou thus upon the poet's heart,
　Vulture, whose wings are dull realities? . . .
Hast thou not dragged Diana from her car?
　And driven the Hamadryad from the wood—

E. A. Poe, *Complete Stories and Poems.* New York: Doubleday, 1966, p. 771.

[47] Anthologies give no details of sources, only *Pensées sur l'Art* and *Meditations of a Parish Priest*, for these quotations from Bracque and Roux, respectively.

[48] F. Wood-Jones, unidentified article, *Medical Journal of Australia*, Aug. 29, 1931; cited in *Scientific Quotations*, A. L. Mackay, Ed. New York: Crane, Russak, 1977, p. 164.

[49] T. Roszak, *Where the Wasteland Ends.* London: Faber and Faber, 1972, p. 330.

[50] R. Pirsig, *Zen and the Art of Motocycle Maintenance.* New York: William Morrow, 1974 (Bantam Book), p. 18.

[51] e. e. cummings, *Collected Poems.* New York: Harcourt, Brace and World, 1923; ed. 1963, poem 157 ("voices to voices, lip to lip").

[52] R. Hubbard, "Telemation: AV Automatically Controlled," *Audiovisual Instruction*, vol. 6, no. 9, pp. 437–439, Nov. 1961.

[53] B. Schlanger, "Criteria for Motion Picture Viewing and for a New 70-mm System: Its Process and Viewing Arrangements," *Journal of the Society of Motion Picture and Television Engineers*, vol. 75, no. 3, pp. 161–167, Mar. 1966.

[54] E. Levonian, "Retention of Information in Relation to Arousal During Continuously-presented Material," *American Education Research Journal*, vol. 4, no. 2, pp. 103–116, Summer 1967.

[55] I. Yankelovich, "30-Second Spots as Good as Minutes," *Broadcasting*, vol. 73, no. 7, pp. 26–29, Oct. 23, 1967.

[56] W. J. Severin, "Another Look at Cue Summation," *AV Communication Review*, vol. 15, no. 3, pp. 233–245, Fall 1967.

[57] —, "The Effectiveness of Relevant Pictures in Multiple Channel Presentations," *AV Communication Review*, vol. 15, no. 4, pp. 386–401, Winter 1967.

[58] J. K. Conway, "Multiple-Sensory Modality Communication and the Problem of Sign Types," *AV Communication Review*, vol. 15, no. 4, pp. 371–386, Winter 1967.

[59] M. C. Potter and E. J. Levy, "Recognition Memory for a Rapid Sequence of Pictures," *Journal of Experimental Psychology*, vol. 81, no. 1, pp. 10–15, June 1969.

[60] D. D. Edwards and P. A. Goolkasian, "Peripheral Vision Location and Kinds of Complex Processing," *Journal of Experimental Psychology*, vol. 102, no. 2, pp. 244–249, Feb. 1974.

[61] J. R. Meyers, "A Study of Eye Movements and Fixations in a Multi-image Presentation," Ph.D. dissertation, University of Wisconsin at Madison, 1972.

[62] E. B. Goldstein, "The Perception of Multiple Images," *AV Communication Review*, vol. 23, no. 1, pp. 34–68, Spring 1975.

[63] A. U. Roshka, "Conditions Facilitating Abstraction and Generalization," *Voprosy Psikhologii*, vol. 4, 1958, reported by I. D. London, *Psychological Abstracts*, p. 34, 1960.

[64] C. Malandin, "Grouped and Successive Images," Centre d'Etudes et de Recherches pour la Diffusion du Français (CREDIF), Ecole Normale Supérieur de Saint-Cloud, France, undated.

[65] W. H. Allen and S. M. Cooney, *A Study of the Non-linearity Variable in Filmic Presentation.* Los Angeles: University of Southern California, NDEA Title VII, Project No. 422, 1963.

[66] T. Snouden, "Comparison of Three Types of Multimedia Presentations," Ph.D. dissertation, University of Wisconsin at Madison, 1965.

[67] E. S. Lombard, "Multiple-channel, Multi-image Teaching of Synthesis Skills in 11th Grade U.S. History," Ph.D. dissertation, University of Southern California, 1969.

[68] M. E. McCombs, "Verbal and Object Availability in the Acquisition of Language: Implications for Audio-visual Communication," *The Journal of Communication*, vol. 19, no. 1, pp. 54, 63, Mar. 1969.

[69] R. T. Jones, "The Effects of Multi-channel Audio Stimuli on Learning Efficiency," Ph.D. dissertation, Syracuse University, 1969.

[70] R. O. Reid, "A Comparison of Multi-image and Linear Film Format as Agents of Attitude Change," Ph.D. dissertation, Syracuse University, 1970.

[71] D. Williams and E. S. Jorgensen, "The World of Multimage," *Audiovisual Instruction*, vol. 15, no. 6, pp. 50–51, June/July 1970.

[72] B. R. Lawson, "Motivating with Multi-image at the U.S. Military Academy; the Medium for the 70's and Its Public Relations Side Benefits," *Audiovisual Instruction*, vol. 16, no. 5, pp. 54–59, May 1971.

[73] C. G. Bollmann, "The Effect of Large-screen, Multi-image Display on Evaluative Meaning," Ph.D. dissertation, Michigan State University, 1970.

[74] L. L. Atherton, "A Comparison of Movie and Multiple-Image Presentation Techniques on Affective and Cognitive Learning," Ph.D. dissertation, Michigan State University, 1971.

[75] B. Fradkin, "An Investigation of the Effects of Multi-Image Stimuli on Later Recall of Tenth Grade Students," Ph.D. dissertation, University of Pittsburgh, 1971.

[76] P. Tim-Kui Tam and R. H. Reeve, "The Image-Accumulation Techniques as a Variable in Multiple-Image Communication," School of Education, Indiana State University, 1971.

[77] J. M. Menne and J. W. Menne, "The Relative Efficiency of Bimodal Presentation as an Aid to Learning," *AV Communication Review*, vol. 20, no. 2, pp. 170–180.

[78] D. Ingli, "Teaching a Basic AV Course by the Multi-Image Technique," Southern Illinois University, ERIC Document ED 060 634, Apr. 1972.

[79] D. H. Didcoct, "Comparison of Cognitive and Affective Responses of College Students to Single-Image and Multi-Image Audio-Visual Presentations," Ph.D. dissertation, University of Wisconsin, 1972.

[80] R. Yolles, "Multiple Image and Narrative Formats in Teaching Intermediate-Grade Science," Ph.D. dissertation, University of Southern California, 1972.

[81] W. P. Brydon, "Comparing Single-Screen and Three-Screen Presentations," Ph.D. dissertation, University of Southern California, 1974.

[82] P. Meyerowitz and B. Fradkin, "Design of Multi-Image Instructional Presentations," paper delivered at meeting of Association for Educational Communication and Technology, Dallas, TX, Apr. 1975.

[83] F. Ausburn, "Multiple Versus Linear Imagery in the Presentation of a Comparative Visual Location Task to Visual and Haptic College Students," Ph.D. dissertation, University of Oklahoma, 1975.

[84] "Does Multi-Media Really Work?" Unbylined article, *Business Screen*, vol. 35, no. 1, pp. 20, 41, Jan./Feb. 1975.

[85] J. H. McIntyre, *Synchronized Slide Shows.* Austin, TX: Columbia Scientific Industries, 1974.

[86] R. L. Gordon, Ed., *The Art of Multi-Image.* Abington, PA: Association for Multi-Image, 1977.

[87] J. Lufkin, "Barbara's Polar Bears," *IEEE Transactions on Professional Communication*, vol. PC-16, pp. 27–34, June 1973.

[88] M. Glyn, "Multi-Media: Capturing the Eagle Within All of Us," *Audio-Visual Communications*, vol. 9, no. 6, pp. 50–52, June 1975.

[89] "Bunker Hill Multi-media," Unbylined article, *Audio-Visual Communications*, vol. 9, no. 8, p. 13, Aug. 1975.

[90] "Multi-media: Thrills and Drills," Unbylined article, *Audio-Visual Communications*, vol. 9, no. 11, pp. 13, 14, 50, Nov. 1975.

[91] H. A. Grunwald, "'I Wanna Hold Your Hand'—Again," *Time*, vol. 110, no. 6, pp. 54–55, Aug. 8, 1977.

[92] J. Kucharski, "The Industrial Photographer as AV Communicator," *Industrial Photography*, vol. 26, no. 5, pp. 28 and 44–50, May 1977.

[93] W. H. Southworth and G. F. McVey, "The Multimedia Approach to Health Instruction for Prospective Teachers," *The Journal of School Health*, vol. 38, no. 10, pp. 667–672, Oct. 1968.

[94] B. Kaufman, "The Planning of a Multimedia Study: Man's Interest and Fascination with the Sea," *Journal of English Teaching Techniques*, vol. 6, no. 1, pp. 18–24, Spring 1973.

[95] C. Harkins, "Technical Presentations in Multimedia: A Modular Approach," in *Proceedings of the 22nd ITCC*, Society for

Technical Communication, Anaheim, CA, 1975, pp. 269–276.

[96] B. Fradkin, "Effectiveness of Multi-image Presentations," *J. Educational Technology Systems*, vol. 2, no. 4, pp. 201–216, Winter 1974.

[97] R. M. Woelfle, Ed., *A Guide for Better Technical Presentations*. New York: IEEE Press, 1975.

[98] Prof. McVey gave an illustrated lecture at Contech 2: The Second National Conference of Educational Technology Applied to Higher Education, Oct. 14–19, 1973, San Paulo, Brazil. His text was reprinted two years later as G. F. McVey, "Components of an Effective Multimedia System for College and University Instruction," *Audiovisual Instruction*, vol. 20, no. 4, pp. 42–45, Apr. 1975.

[99] C. H. Gardner, "How Educational Administrators View the PR Potential of Multi-Image," *Audiovisual Instruction*, vol. 22, no. 2, pp. 20, 21, 63, Feb. 1977.

[100] "LACC Opens Multi-image Production Lab," Unbylined article, *Multi-Images*, vol. 1, no. 4, p. 4, Winter 1975.

[101] "On Center Screen," Unbylined article, *Multi-Images*, vol. 1, no. 4, p. 6, Winter 1975.

[102] A. Giberson, "Navy Advanced School of Photography Instructs 'Eyes of the Fleet,'" *Technical Photography*, vol. 8, no. 2, pp. 20, 21, 25, Feb. 1976.

[103] J. Briggs, "Multi-image in Middletown, U.S.A.," *Audio-Visual Communications*, vol. 10, no. 3, pp. 10–11, Mar. 1976.

[104] "AV Theater in the Round," Unbylined article, *Audio-Visual Communications*, vol. 10, no. 3, pp. 14–16, Mar. 1976.

[105] J. R. Westberry, "Multi-media: Putting Pride and Price Together Again," *Audio-Visual Communications*, vol. 11, no. 2, pp. 18, 19, 22, 23, Feb. 1976.

[106] "Multimedia Shows Run by Timeshared Mini," Unbylined article, *Computer Decisions*, vol. 7, no. 8, p. 60, Aug. 1975.

The Meaning of Multimedia

There is an old fable about four blind men and an elephant. The four approach the elephant, not knowing what it is, each feeling a different part of the animal's body. The first man feels the elephant's leg and says it is a tree. The second man feels its tail and says it is a rope. The third man feels its ear and says it is a fan. Finally, the fourth man feels its trunk and says it is something without end. None of the four men is able to perceive the elephant as a whole.

Multimedia presentations are a similar creature. The term multimedia implies the use of more than one form of communication, but beyond that there is a divergence of opinion. Does multimedia refer to a show built around multiple projectors and several screens, or is it defined by a computer controlling a compact disc? What is the meaning of multimedia?

By Larry Tuck and Jennifer Ravi

According to Ken Burke, a professor of communications at Mills College in Oakland, OK, multimedia has been around for a long time — from cave paintings to theater, opera and the '60s art gallery environments. But multimedia was a latecomer to business communication, says Burke, coming in through expositions in the mid to late '60s in the form of multi-image shows — multiple slide projectors, multiple screens, and sound; sometimes film was used in the early days. More recently, the definition has broadened to include video, and increasingly other media as well.

Dennis Ducharm of Pratt and Whitney's Media Services department defines multimedia as "the synchronized use of visuals and audio in a coordinated presentation." But while he is a bit loose in his definition of audio and visual, he says, "I always feel there is a slide connection there, that it was built around slides."

Large-scale multimedia presentations like these do more than deliver information in a pretty package; they are also intended to create an experience, leaving the attendee with a positive feeling about a company or its products. They are most commonly used for product reveals, corporate positioning and exhibitions.

Take the example of a product reveal. Many companies — especially in a business like automobiles, where excitement and image play an important part in selling the product — opt for multimedia shows to introduce their new products to public, press and dealers. These companies strive to create an atmosphere of excitement and anticipation through the use of motivational music played through large, high-quality speakers. Special lighting plays up the event. Singers and dancers, lasers, music and video as well as speakers and multi-image all add to the event. Videos are shown and even teleconferencing is used — and often, computer demonstrations are projected onto large screens.

The price range of a ten-minute multimedia presentation of this kind "would start at a quarter of a million dollars," says Bill Mooney, vice president of sales for LSI Communications, a Philadelphia-based producer of multimedia events, and "could go on to a million or more for the same time frame."

Rob Rappaport, president and CEO of the MultiMedia Group, Inc., a Culver City, CA, business communications firm that produces such shows for clients like Toyota, Compaq Computer and Yamaha, estimates that there are about 500 of this type of large-scale multimedia presentations produced annually. But thousands of more modest productions also fall under the heading of multimedia.

Mooney says that slide-based multi-image shows still work in this age of television — and probably will for a long time to come — because TV has set a level of expectation about execution standards, but that the medium is easy to tune out. By contrast, he says, multimedia creates a state of readiness and expectation in the audience, creating an experience in which they are continually surprised.

A New Generation of Multimedia

But a newer definition of the word "multimedia" is also in use. This one is firmly rooted in the computer industry, but the term is fuzzier in definition. There is everything from multimedia computing to multimedia presentations.

One company in particular has driven the use of the term: Apple Computer. Apple has touted the integration computers with video and still images as well as sound and even CD-ROM and laserdisc, any combination of which it refers to as multimedia. And, of course, this is done under the control of the computer. For Apple, the computer is at the center of the multimedia presentation — add to it what you will.

Making Sense of the Elephant

These similar-but-different definitions of multimedia intersect on more than one level. As Rappaport points out, it's the message that's important, and it's dangerous to become too strongly wedded to a specific technology. Above all, multimedia means using whatever tools do the job best.

On another level, the tools themselves are merging, as computers contribute to the production of traditional multi-image/multimedia shows and as media like video and sound are incorporated into computer-based presentations. In another sense, the tools used to create multimedia are merging, as the same computer-based system can be used to make slides, video animation or printed handouts. The lines between the media are blurring.

Starting this month and continuing over the next three issues, Presentation Products Magazine will be exploring the many meanings of multimedia, looking at new, old and future presentation tools, talking to users about applications of those tools, and examining research that's being done on presentation media and techniques. We hope you'll enjoy and learn from this look at what we think is a fascinating and

Reprinted with permission from *Presentation Products Magazine,* pp. 26–34, June 1989; pp. 30–40, July 1989; pp. 44–48, Aug. 1989.

important topic for anyone who makes or uses presentations.

Multimedia User Views

Traditional multimedia shows serve a variety of purposes for major businesses: to introduce new products, help set a corporate image, or keep a sales or dealer organization fired up about the company and its products. Here are a few examples:

Product Reveal

When Toyota Motor Sales, U.S.A. rolled out its 1989 models to 3,500 dealers, guests and Toyota executives, it chose a multimedia presentation.

Toyota has held an annual national dealer meeting, complete with a multimedia show, since the early '70s, according to Michael Bevan, special markets & recognition manager for Toyota Motor Sales, U.S.A. Bevan describes the annual meeting as a bonding event. "This is the one time of the year that we have everybody together," he says. Because of that they try to make it memorable.

And this year certainly was memorable. The product unveiling was held at the Grand Ole Opry in Nashville, TN. The MultiMedia Group, Inc., produced a show which employed the best of many worlds to communicate Toyota's appreciation and anticipation for a great year to its dealers. Bevan summarized the multimedia event as a blend of "humor, irreverence, real good product."

Representing Hollywood was Martin Mull and his family as well as Barry Manilow. Representing celebrity advertising was Lee Trevino and Chi Chi Rodriguez. Representing Toyota were many senior executives as well as corporate marketing manager George Borst.

The event began with a few words from the Toyota Motor Sales, U.S.A. president, Mr. Yuki Togo: "Welcome to a really great American tradition — the family reunion." And the show began.

The extravaganza incorporated an impressive array of media into a presentation about two hours in length. There was an abundance of live performers. Martin Mull hosted the event, playing along in a show-length self-annotated drama about himself, his loves and his cars. His characteristic dry wit sustained interest through the show.

There were 12 dancers who participated in short skits centering around the 1989 models. These new cars were shown on stage — and driven on and off. Behind the stage were projected videos of the products or advertising spots by Saatchi & Saatchi DFS, Pacific, as related to the onstage presentations.

Probably the most spectacular reveal was the 1989 4 × 4 pickup truck, which was lowered from the ceiling onto the stage in a cloud of smoke with simulated lightning and dramatic music. Also impressive was a motivational video from the Georgetown, KY, plant.

Was the show a success? According to Bevan, the dealers and guests "thought it was one of the better shows they had seen."

Corporate Positioning

Two years ago, as one of seven regional telephone companies fresh out of a recent divestiture from AT&T, Bell Atlantic was in need of a new image. It wanted people to know that it "wasn't just a tired old telephone company," says Stan Timson, exhibit manager for Bell Atlantic. Rather, the company wanted to project an image of "a high-tech company with futuristic ideas," he adds.

Bell Atlantic decided to take its new message on the road. Timson worked with Bill Mooney and Mike Hammer of LSI Communications to create a multimedia presentation that bolstered its high-tech image. The presentation, which took three days to set up each time, was shown in exhibit theaters at telecommunications trade shows from New York to California

The eight-minute Future Light Presentation combined video, multi-image and lasers. Product demos were wrapped around the theater. They included information about Public Data Network (PDN), Centrex, ISDN, Consultant Support and Central Office LAN.

The company also had a corporate message piece describing its subsidiaries. Bell Atlantic is involved in many aspects of the communications industry, providing integrated services through the ownership of leasing, hardware and software consulting companies.

When asked why the company chose multimedia, Timson responds, "We wanted live action so there was video. There were certain graphics that we could put up and only work with AV." They also wanted lasers to tie the other media together. There was a central

office projected above with lasers, which connected video terminals together. They also had lasers reaching out into the audience.

The ending was "very functional," says Timson. "We used smoke emanating from a light cone that came down and wrapped the speaker and she vanished," he says.

The exhibit was a popular as well as a financial success, says Timson. "We generated more revenue than we had before with a trade show," says Timson.

The In-House Production Option

Pratt & Whitney started with traditional multimedia extravaganzas and eventually evolved its own style of multimedia.

The Video and Multi-Image Services section of Pratt & Whitney's R&D office in West Palm Beach produced its first multimedia show about six years ago. The event was a celebration of the office's 20th anniversary for the employees and their families. The gathering was a chance to revel in some history and thank the employees for their efforts.

The show was produced using four large screens, 24 slide projectors and three film projectors. "We had the expertise and the manpower internally," says Dennis Ducharm, manager of Video and Multi-Image Services. So it was a natural progression.

A couple more multimedia productions in the next few years included events for training, trade shows, customer/community relations as well as technical reviews for customers. And Ducharm is convinced of their effectiveness. "When you're trying to impact people, multimedia is the way to go," he says.

But a couple of years ago, the section opted out of traditional multimedia largely because of cost, and turned instead to producing mostly video. However, it seems to have found a compromise which works — lower costs, shorter setup times and yet still multimedia. It's a definition based on need and individual circumstances — just as it has "jury-rigged the computer system" for these shows. Now "it's not all the razzle and dazzle and glitz, but it's still multimedia," says Ducharm.

Now it uses video and multi-image as speaker support for technical reviews for the government and NASA. These are usually to a group of 250 people and incorporate, for example, 12 slide projectors, 1 video, and 3 screens. "As far as P&W is concerned, this is the direction we will use multimedia."

Despite the shift from "traditional multimedia," Ducharm remains convinced of its effectiveness: "Multimedia is labor intensive for the setup, but is there anything as powerful?"

Multi-Image Still Going Strong

The term "multimedia" has been widely adopted by proponents of a wide range of presentation hardware and software; at the recent National Computer Graphics Association trade show, it seemed that half the exhibitors on the floor used the word somewhere in their exhibits. But traditionally, "multimedia" has conjured up the image of some combination of slide projectors, 16mm films and audio. On the large end of the scale, it referred to multiple slide projectors, film, theatrical presentation, and events.

In the beginning, there was the slide. By using two or more projectors, usually keyed to a sound track, early presentation producers added drama and variety to their shows at a relatively modest cost — compared to film, which was then the only alternative.

This was multi-image. "Multimedia refers to any combination of media," says Jim Kelner, director of sales and customer support at AVL, a manufacturer of multi-image control systems. "Multi-image refers to programs based on slides. The term multi-image was coined, I think, by one of the production companies, and later adopted by AMI — the Association for Multi Image. It has since been broadened to include other things."

By the 1970s it was very common to include both slides and motion pictures in a single show — say, using motion pictures to show the company's new advertising campaign. With the growing use of video as a business communications tool, there aren't many people using 16mm motion pictures anymore — that's been replaced by video.

But won't low-cost video also replace slides?

"Slide-based multi-image is a dying technology" was one vendor's gloomy prediction recently. His company, formerly active in selling multi-image equipment, has since moved into other areas of the presentation business. "Everything is going to computers and video."

Video probably has taken market away from slides because of its growth, Kelner concedes. But he is quick to point out — as are others involved in multi-image — that slides still offer some advantages over video. "There are still people who insist on the image quality provided by slides," says Kelner. "Today people usually use a mix, slides where they're appropriate and video where it's appropriate. If you need full motion, slides don't really make it . . . but for image quality, video can't compete with slides."

Growing Market

The market for both slides and slide projectors has been increasing at a slight but measurable rate, says Don ver Ploeg, a spokesman for Eastman Kodak. The reason, he says, is that compared to other media technologies, color slides offer superior clarity and concentration of information.

And, in the context of event-type multimedia shows, high impact. Notes Kelner, "We're finding a lot of users who have used video for their presentations, and when they do their audience surveys afterward, people will say, 'It was nice, but I still remember that big slide show we had last year.' I think it's going to be that way for a while."

One drawback of multi-image is the complexity of the equipment needed for the presentation. "Multi-image is a labor-intensive medium," says Theo Mayer of Metavision, a Los Angeles-based production firm. "It's as complicated to set up the equipment rig as it is to produce the program. And when you get done, the big cookie is 'I can't believe you did that with slides.' There's something wrong with that."

Cost is clearly another factor. While the largest multimedia shows can cost hundreds of thousands of dollars, slide-based shows are generally less expensive to produce than video. While staging might be slightly more expensive with multi-image, Kelner says ("I'm not sure it is, but it might be"), video production carries a much higher price tag, both in terms of man-hours and the equipment needed for production.

Video projection and creation are going to be very expensive for a long time to come, agrees ver Ploeg of Kodak. "Look at film. People for years have been predicting that video would replace film, but it hasn't happened. Video projection of NTSC signals still hasn't gotten to the point where people are comfortable with the quality. We expect that will continue to be true for the foreseeable future for both slides and motion pictures."

Even Mayer admits that there's at least one thing that slides do better — "rock-solid, crisp, 35mm projection."

"My favorites are simple two-projector shows," he says. "Using the medium for what it is." For example, he points to multi-image projects he did for a camera company, where the whole point was to showcase the quality of pictures produced by the company's 35mm cameras. "For a car company, it doesn't make much sense to use a static medium to show cars zooming down the road. For a show of pictures from the Louvre, it might make sense."

There are more choices now, says ver Ploeg. "It is possible to integrate electronic information with still photography. You can create text slides from information stored on a floppy disc. Obviously, we at Kodak are not looking at this with blinders on . . . [but] we think electronic and photographic media will coexist for a long time to come."

Next month, we'll look at how new technologies are contributing to the evolution of multimedia.

The Computer Revolution Brings A Merger of Presentation Tools

The repertoire of media available to today's presenter is extensive and varied, from the proverbial flip chart and felt pen to computer animation and live video. It's no wonder that, as we pointed out last month, examining multimedia sometimes seems like the old fable of the blind men and the elephant.

There was a time when a presenter might create his own materials for the simplest of presentations—drawing a graph on a flip chart, or having his secretary type an outline on transparency film. If he wanted something more sophisticated, he turned to an expert. Depending on what form of media he required, he might have to talk to any one of several different artists or technicians, each with his own unique set of tools and each speaking his own jargon.

Within the company there was a graphics or audiovisual department, making slides from hand-drawn or typeset art with a camera and copy stand; down the hall was the corporate film-making (later video) department, with its own set of cameras and editing equipment; in yet another place was the printing plant, with typesetting machines and printing presses. While all might fall under a single corporate department, each function was really quite distinct—and each required a substantial investment in specialized equipment and employee training.

The computer is changing all that. As spreadsheets gave birth to the first, simple charting programs, and word processors evolved into desktop publishing systems, personal computers have helped give new life to presentations. Their real significance, however, is that they are the key to merging all presentation media for greater impact and lower cost.

The same personal computer that creates desktop publishing documents can also be used to make overheads and even slides, and applications are emerging for animation and even editing and control of full-motion video. What's more, the same images can be repeated consistently through all aspects of a communications campaign, saving time, effort and money. Over time, the distinctions between video, graphics and print become increasingly blurred—perhaps meaningless.

"These days, people want to do slides, print and animation," says Sharon Adcock, sales manager, Pansophic. "They can't afford to have separate systems to do all those things…. The days of the proprietary, single-application system are pretty much over."

"People are making more efficient use of their graphic arts," agrees Laura Malone, marketing manager at Time Arts, a publisher of high-end graphics software. "They can take one good image created on the computer workstation and use it for video as well as picking it up for slides and even for print."

Slidemaking

In just over a decade, computer graphics has taken a significant role in the audiovisual industry. For example, The Hope Reports of Rochester, NY, says that "computers have been the driving force behind the continued growth of the slide medium, keeping the medium in the leader position among all [AV] media."

"We're beginning to see a real increase in how much they're accepted," says Laura VandenDries, communications manager, AT&T Graphics Software Labs. "Just a few years ago, people were still much more comfortable with the old way of doing things—art boards and copy stands. In the last couple of years there has been a lot of attention paid to and a lot of articles in the general business press on the importance of quality presentation visuals."

Hope Reports estimates that close to half the nation's 2,000 contract slide producers now have computer graphic

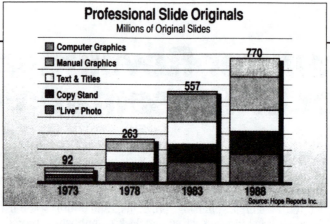

Professional Slide Originals
Millions of Original Slides

- ■ Computer Graphics
- ■ Manual Graphics
- □ Text & Titles
- ■ Copy Stand
- ■ "Live" Photo

92 — 1973
263 — 1978
557 — 1983
770 — 1988

Source: Hope Reports Inc.

capability. Many of those service bureaus offer users the option of sending data created on their own computers to the bureau for imaging on a high-quality film recorder, as well as providing turnkey slide creation for those who prefer it. But as Hope Reports points out, in-house production units have taken to computer graphics even more wholeheartedly than contract producers.

"Going to a service bureau costs $10 to $50 per slide to get them made," says VandenDries. "A company that has a number of slides to make can very quickly rack up pretty substantial charges. Today, you can get a film recorder for $6,000 that will produce good quality slides—quality as good as most people need, better than most people need."

Use of computer graphics for slidemaking was virtually unknown ten years ago, and as recently as 1983 represented only 3 percent of the slides produced, according to Hope Reports. In 1988, however, 16 percent of all original slides were produced using computer techniques, Hope says, and the percentage is expected to continue to grow.

Overheads

Another area where computers have played an important role is in the revival of the overhead projector. Computer graphics software first gave businesspeople the means to create charts and graphs quickly and easily, but until recently film recorders were a high-priced luxury, and most users were not aware of the possibility of outputting slides through a service bureau.

Nearly everyone had access to a printer or plotter, though, and as laser printers came into common use, it became possible for just about anyone with a computer to create high-quality overhead transparencies.

And because of their convenience and familiarity, many users still prefer overheads to other presentation alternatives. "We still get more requests for overheads than we do for slides," notes John Serrone, data manager at AT&T Microelectronics. His department can offer users at the AT&T division anything from slides to on-screen shows; still, for day-to-day presentations overheads are the favorite.

Desktop Presentations, Inc., of Mountain View, CA, predicts even more dramatic growth for overheads than for slides, largely due to computer-created transparencies. The research firm says that use of electronically created overheads will skyrocket from 138 million transparencies produced in 1987 to more than 800 million by 1992.

"Desktop Media"

Apple Computer, which almost single-handedly created a new industry with its desktop publishing marketing campaign a couple of years ago, last month launched a new campaign introducing the newly-coined term: "desktop media."

"It's an umbrella term for a set of solutions that incorporates desktop publishing, presentations and interactive multimedia," explains Ames Cornish, marketing manager for desktop presentations at Apple. "It includes all the communications tools that someone would need, whether for print, overheads

or delivery direct from the computer.

"One of the amazing things is how powerful it is to have all these things integrated," says Cornish. "Adding those things together, the whole is greater than the sum of the parts."

But the computer is only a tool, and despite the best efforts of companies like Apple, it remains a lot more difficult to use than, say, a piece of chalk or a felt pen. As nearly anyone who has used a computer will tell you, uninformed use of even the best computer software won't guarantee good or even acceptable results.

The often-quoted example is of desktop publishing, where many early users, not having a background in graphic arts, used the computer to create lots of incredibly unattractive desktop-published documents. Will the same happen with presentations?

One who is not entirely comfortable with the proliferation of computer-based tools—and particularly with the hype and hoopla surrounding them — is Rob Rappaport, president of The MultiMedia Group of Culver City, CA, a major producer of large-scale multimedia productions. "Sometimes people mistake the technology for a solution in

**Apple
MacIntosh SE/30**

Polaroid PalettePlus SlideMaker

itself," he says.

As a tool, computers have helped multi-image producers to do their jobs better, Rappaport says. Aside from actually creating graphics, he says, "computerization has helped us to make control of the show, things like lighting and sound, much more reliable." But Rappaport warns against putting too much reliance on technology alone. "The tools can't do the job," he says. "The most important component of the computer is the person using it."

Ames Cornish of Apple thinks that with time, experience and education, users will make more effective use of tools for presentations, just as they have with desktop publishing. "If you talk to average desktop publishers today, they understand a lot more than they did a couple of years ago," he says. "They know what a font is, points and picas—those used to be terms only printers used. The same thing will happen with presentations."

New Media

About four years ago, a new form of presentation began to emerge, one that took advantage of the capabilities of the computer to both create and display presentations. Known by a variety of names—desktop presentations, electronic presentations, interactive multimedia, to name a few—the technology remains in flux, but promises an exciting future.

"Slidemaking is still more popular [than electronic presentations] because it has a 50-year history behind it. It has a strong mainstream feel," says Herb Baskin, president of General Parametrics. "[But] now that we have all these wonderful computer graphics programs, people would like to be able to use those graphics without the cumbersome process of going to film."

General Parametrics, with its Videoshow product line, was one of the innovators in this product category. Videoshow, and similar products from Autographix, Computer Support Corp., Kodak, Agfa-Matrix and others, allow sequencing of computer-generated graphics or captured video stills into a "slide show" for presentation on a computer monitor or video projector. Since then, many presentation software packages have added "slide show" functions.

As these computer graphics presentations have merged with still video, interactive videodisc and other, even more esoteric technologies, a new form of multimedia has emerged. Next month, we'll look at the impact of these new media on presentations today and in the future.

Making the Most of Multiple Media

The staff of the Graphics and Media department at CitiCorp.'s Consumer Bank Western Division are maximizing the time and costs invested in presentation materials by exchanging images between desktop publishing, slides/ transparencies and video. Also, for speakers who want handouts of a slide presentation, the graphics department can laser print the exact presentation.

"Mostly it is the transfer of information from the high end to the desktop world—color to black and white," says Don Hulslander, manager of Graphics and Media.

He is one of a staff of two full-timers in his department who produce presentation materials. "We're generalists and we have our own specialties," as he puts it. He specializes in the Time Arts graphics and his associate, Vicki Maurer, specializes in the Zenographics Mirage software. They use Lumina for pizzazzy openers and Mirage as the workhorse for slides, charts and text.

The majority of the materials they prepare are for senior management to use as lecture support to the public. These are generally flashier than internal presentations, says Hulslander. Internal presentations by management to their superiors are simpler, usually overhead transparencies—and a lot of black and white. As Hulslander notes, "You don't go into a boardroom asking for money looking too glitzy."

CitiCorp.'s Graphics and Media department in San Francisco has two desktop publishing workstations, two graphics workstations and an animation workstation in their IBM-based shop.

They also make extensive use of a color scanner with the graphics workstations. This enables them to scan in information on the desktop systems and print out on a laser printer or film recorder, for example. Hulslander says that scanners are "very important because we don't have a lot of time."

Hulslander says they start with a disparate array of information for presentations—sometimes it's a headline from the Wall Street Journal, sometimes it's cartoons, children's books or CitiCorp. brochures or other materials. They can scan in an image and modify it using Lumina. "One of the big advantages of having something like this in-house is the ability to make changes up until the last minute," says Hulslander. Scanned images touched up with Time Arts Lumina software "become part of the show in a half hour," he adds. — *J.R.*

Graphics Boards Bring Video To The Desktop

The first generation of personal computer displays could offer only limited, low-resolution graphics. Graphics have improved rapidly, however, as manufacturers adapted graphics technologies developed for high-end workstations and provided them to personal computer users at reasonable cost.

The heart of a computer graphics system is the graphics adapter or controller, the graphics board. Traditionally, personal computer makers have designed this as a separate, modular board so that the user can pick the level of graphics quality he needs for his application, or substitute a higher-quality graphics board if his needs change. Naturally, a whole industry has emerged to provide these boards, with manufacturers vying to produce the highest-quality display at the lowest price. The buyer benefits from this competition.

One of the earliest standards for personal computer graphics was called CGA, for Color Graphics Adapter—the color display option offered by IBM for its original IBM-PC computer. CGA offered a resolution of only 320 x 200—that is, the picture, made up of small dots of light called "pixels," was 320 dots wide and 200 dots high. CGA could display only four colors on the screen at one time.

Still, it was color, and software vendors began to develop the precursors of today's presentation graphics software, programs like Ashton-Tate's original Chartmaster and Lotus' 1-2-3, the innovative spreadsheet that first combined graphics with number-crunching.

A few years later, IBM upped the ante with EGA, the Enhanced Graphics Adapter, which provides 640 x 350 resolution in 16 colors—and personal computer graphics really began to take off. EGA remains the most commonly used graphics standard.

The latest standard from IBM, introduced with its current PS/2 model line, is VGA, the Video Graphics Array, which offers up to 640 x 480 resolution in 16 colors, or up to 256 colors in 320 x 200 mode. VGA is also designed to be compatible with older standards—VGA hardware can "downshift," so to speak, and operate in CGA or EGA mode so that older software can still operate with the new display system.

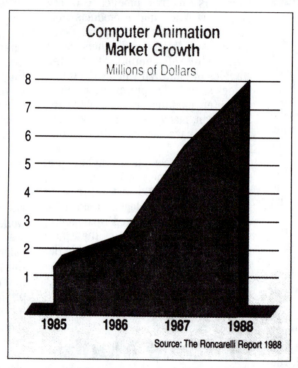

Computer Animation Market Growth

Millions of Dollars

Source: The Roncarelli Report 1988

Because of the modular design of IBM and compatible personal computers, third-party manufacturers have been able to provide IBM-compatible graphics products that offer even better resolution than the existing standards. Currently, a generation of Super-VGA products offer resolutions up to 1,024 x 768 pixels, enabling them to show extremely lifelike, near-photographic-quality images.

Apple's Macintosh computers were designed from the beginning with high-resolution graphics in mind, but until the Mac II was introduced, their value for video was limited because the original Macintosh had only a black-and-white screen. The Mac II accepts color graphics boards from Apple as well as from other vendors, and in the last two years Apple color graphics products have proliferated, including the NuVista from Truevision as well as competing products

from Data Translation, Matrox, Raster-Ops, SuperMac, and others.

The biggest problem has always been how to share these images with other people. Color printers have become more widely available and lower-priced, and film recorders and slidemaking services provide one way to share images with other people, but many users have looked for a way to share images live, create computer-generated slideshows without the slides, or do animation. That involves mixing computer images with video. Unfortunately, that's not an easy task.

Although a computer display may look a lot like a television set, and personal computer graphics today are clearer and more highly detailed than TV pictures, compatibility between the two remains a problem. The standards and technologies used for computers and video are almost completely different.

Actually, most CGA cards can be connected to a regular TV monitor—when IBM introduced the CGA standard in the early '80s, many home computer users were still hooking up their computers to their television sets. But the resulting graphics are blocky and the signal often jittery. "CGA graphics are very unsatisfactory" for video recording, says Michael Laurie of Matrox, a manufacturer of video boards for personal computers. Standard EGA and later cards are not designed to be connected to standard video at all. But here again, independent developers have stepped forward to provide solutions.

A key step forward in the merger of personal computers and video was the introduction in 1985 of the TARGA board by the AT&T EPICenter—now Truevision Inc. The TARGA board allowed graphic artists to capture full-color video images, enhance, retouch, or modify the image, and view it in full color. TARGA has become the standard of comparison for PC video graphics.

Software vendors like AT&T Graphics Software Labs and Time Arts have developed sophisticated software to take advantage of TARGA and other high-end graphics boards, allowing personal

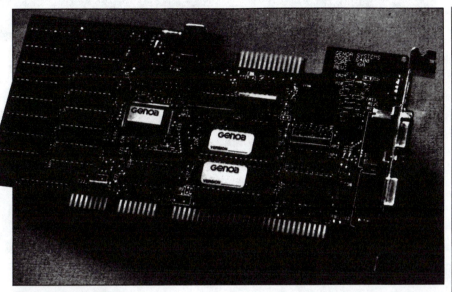

Genoa's SuperVGA

computer users to do work that once required expensive, high-powered graphics workstations. Products like AT&T's TOPAS and Time Arts' Lumena can retouch and modify still images imported from video, create near-photographic-quality pictures or make three-dimensional titles; AT&T's Animator can turn them into animation. Other makers of TARGA software include Digital Arts, Zenographics and Software Clearing House.

Says Laura VandenDries, communication manager at AT&T Graphics Soft-

ware Labs, "If you go to a service, it costs $800 to $1,200 a second for finished animation. For a company that does a large amount of training tapes, that can mount up very quickly.

"They're not doing Walt Disney," she explains. "It's more like having the company logo flying across the screen, spinning as it goes. That kind of thing is relatively easy to do with a PC. You still have to go into an edit suite with it to cut it together with your live

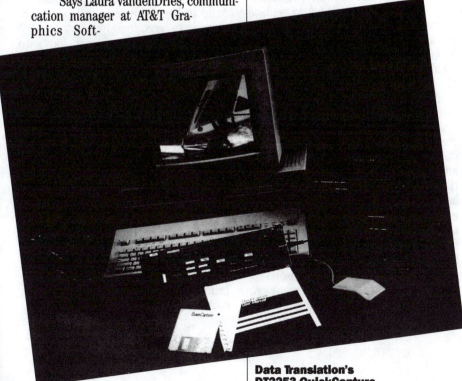

**Data Translation's
DT2253 QuickCapture**

footage, but that's relatively simple."

Input, Output

There are actually several elements that may be involved in a computer/ video graphics system. One is the graphics board itself—VGA or other— which allows you to display computer images on a computer monitor. This is basic to any computer system, whether used primarily for graphics or for any other purpose.

Another piece of hardware that's becoming popular is the frame grabber, sometimes called a video digitizer. These cards convert video signals to digital computer format for display on a computer screen, or manipulation with graphics software. Not all frame grabbers work in "real time": they take a relatively long time to translate the image and so can capture only single frames or still images. The trade-off, as you might expect, is cost: real-time frame grabbers are more expensive. Lower-priced frame grabbers are useful to capture images to be included in computer-generated slides—the story on page 32 tells how one company takes advantage of this feature. Real-time digitizers can be used for interactive applications and video titling, among other applications.

Finally, there are output devices that allow export of the computer image to standard video for display on a video monitor or recording with a VCR. This may be incorporated into the graphics board, as it is in VGA/video cards available from Willow Peripherals and US Video, or may be in the form of a separate device, a scan converter. Often, various features are combined in a single product; some cards combine input and output capabilities, for instance.

Valarie Gardner, director of communications for Willow Peripherals of Bronx, NY, says many of its video output boards are sold to people who simply want to use a large television monitor or video projector to show computer images for software training.

Another, growing application is as an alternative distribution medium for computer graphics. Many presentation programs now on the market can produce on-screen slideshows with various transition effects, and even animation is becoming more accessible. But it's not always convenient to show a

TinToy
Photographed on Matrix MultiColor

The Commodore Amiga 500

computer-based presentation.

With graphics cards like Willow's VGA-TV, or interface devices from Covid, Extron and RGB Technologies, it's fairly easy to create a slideshow on the computer and make a training or sales tape and send it out to potential customers or salespeople in the field. "You can send a videotape in the mail, which you can't do with a computer presentation," says Gardner.

"The TV and VCR will always be the delivery system for these things," says Rick McEwan of RasterOps, which makes video capture and output cards for the Macintosh.

Desktop Video

Another player today in computer/video is Commodore's Amiga. Designed with a set of custom graphics chips, and priced affordably, the Amiga has found its way into cable television facilities, corporate video departments and independent production houses for uses like animation, special effects and titling. The company has found a niche by stressing low-priced, easy-to-use solutions to video problems. "The only way these things are going to be popularized is if they're easy for anyone to use," says Keith Masavage, Commodore's director of marketing.

Commodore, in fact, claims to have invented the term "desktop video" to refer to the use of low-cost PCs and easy-to-use software to create video. "Just as desktop publishing allows your computer to function as a typesetting machine, desktop video allows it to function as a

character generator, animation workstation, etc.," explains John Sievers, Aegis Development, which publishes animation software for Amiga and Macintosh II computers.

"All of the things that you can do with a $15 million system for broadcast television can also be done, at least in a rudimentary fashion, on a personal computer," Sievers says.

Promoters of desktop video, including both Commodore and Apple Computer as well as third-party peripheral manufacturers and software publishers, don't expect desktop video to replace conventional video; rather, it will open up a new form of communication to people who aren't now using video at all. "What video lets you do is put together all the pieces," says Ames Cornish, marketing manager for desktop presentations at Apple. Still images, computer graphics, motion, sound, can all be combined in one medium. But there are still some problems to be overcome, he adds. "How do you present video to more than one person? People are using Sony and Barco video projectors, but those are expensive." When low-cost, high-quality, large-screen video displays are available, Cornish predicts, desktop video will become far more attractive for mass use.

"There are many good reasons to use video as either speaker-support material, or as the entire presentation in

itself," notes Bill Coggshall of Desktop Presentations Inc., a research firm in Mountain View, CA. Because video can include movement, color, sound, and live action and graphics, he says, it is a very versatile medium.

Coggshall predicts that in-house (desktop) video will be used for training, presentations, education, sales, employee motivation and other "information distribution applications" like documentation. "In addition, we speculate that videotapes may become as widely and commonly distributed as [and in lieu of] reports and correspondence," he says.

Desktop video is not likely to replace professional video, however. For many in-house uses, says Coggshall, "programs created with a desktop video system will be appropriate, but for high-profile or broadcast material professional studios will be used."

Says Carol McGarry, product marketing manager, Data Translation, "The applications people are looking for are fairly simple at this point. It isn't spinning logos—it's simple titling."

The merger of computers and video represents a key area for the development of presentations. Just as desktop publishing is changing the way words are put on paper, desktop video may bring a whole new dimension to business communications.

Interactive Multimedia: Presentations for Tomorrow

Ever since the first computer operator turned from his terminal and said to a colleague, "Come look at this," the computer display has been a presentation medium.

But it hasn't always been a very good one. The typical computer monitor is far too small for viewing by more than a few people at a time. Besides, even in today's increasingly computer-literate world, most presenters are at least a little uncomfortable with trying to talk and run a program like Lotus 1-2-3 at the same time.

Over the years, developments in hardware and software have made computers more useful in creating presentations. Just five years ago, most users had to be content with fairly crude graphics programs, perhaps printing the resulting charts on overhead transparency film using a pen plotter. Today, graphics software is much more sophisticated and (at least sometimes) easy to use. Widely available output options include laser printers, high-resolution color printers, and moderately priced desktop film recorders for slidemaking. Even videotaped presentations can be created using hardware and software products that are now becoming available.

Using the computer to *create* presentations, which can then be *displayed* with conventional equipment, like overhead or slide projectors or distributed on videotape, is often an ideal solution. Use of personal computers can result in faster production time, lower cost and greater creative control and flexibility.

But there are other times when the presenter would like to have the immediacy of telling his audience, "Come look at this," and let them figuratively look over his shoulder as he puts the computer through its paces.

"Slidemaking is still more popular [than electronic presentations] because it has a 50-year history behind it. It has a strong mainstream feel," says Herb Baskin, president of General Parametrics. "[But] now that we have all these wonderful computer graphics programs, people would like to be able to use those graphics without the cumbersome process of going to film."

A variety of products are helping to make such direct-from-the-computer

In Focus Systems 480GS/2 PC Viewer

presentations possible. They include both display hardware, such as LCD projection panels and CRT-based video projection equipment, data storage devices (computer graphics require lots of digital storage space), and a whole array of hardware to control and interconnect these devices—not to mention software.

Add to these the ability to connect and control noncomputer audio and visual information sources—like videodisc or tape players—and you have what has become known as interactive multimedia.

Interactivity

The "interactive" part of that phrase refers to the fact that making the computer part of the presentation can also make the presentation more dynamic, more flexible. The presentation is no longer "carved in stone," but can be changed at will to match the needs of the moment. Not only sequencing, but also content can be changed. And at that point, it becomes nearly impossible to distinguish between creation and display of presentations.

In the past when an art department created original drawings or typeset charts and photographed them to make slides, which were later shown by a presenter with the aid of an audio-visual technician, the gap between creation and display was pretty wide. As personal computers and desktop film recorders made it possible for the presenter or his secretary to create slides relatively easily, the gap began to narrow.

More recently, LCD display devices (used in conjunction with overhead projectors), larger video monitors and video/data projectors have made it possible to share computer screen images with larger audiences. With these computer-based presentations, it is no longer possible to separate the creation and display functions. Both are part of the same process.

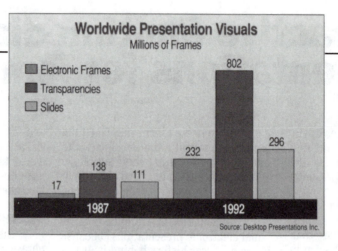

Worldwide Presentation Visuals
Millions of Frames

Electronic Frames
Transparencies
Slides

802
232
296
138
111
17
1987 1992

Source: Desktop Presentations Inc.

Media, more media

The other half of the phrase "interactive multimedia" refers to combining a variety of formerly independent media, with the computer in control.

Some of this integration is happening naturally as audio and video move from analog to digital formats, where sound or visual information is stored as what amounts to computer code. In the audio world, for example, analog phonograph records are quickly being replaced by digital compact discs. Current broadcast television standards are analog, but production is moving to digital formats. And new advanced television standards, like HDTV, may eventually make TV distribution digital as well. Being digital, these media are easily controlled, edited and modified by computers or computer-like equipment.

New media, many of them based on compact-disc-like technology, permit combining audio, visual and computer data on a single medium. Compact Disk Interactive (CD-I), Digital Video Interactive (DVI) and the even newer CD-ROM/ XA are a group of related technologies that use laser compact discs to store all types of information. All are still under development.

With the appropriate interface hardware, even analog media, like conventional videotape or disc players, can be controlled by computers and the information stored in them can be merged with computer-generated text, graphics and sound. There's a flourishing industry involved in the production of interactive videodisc, and several companies have explored the benefits of using videotape and other media in combination with computers.

The result: A wealth of possibilities for creating interactive multimedia—and a whole new realm of possibilities.

Real stuff

So much for definitions. But how do these catchy phrases and flashy products apply in the real world?

To take one example, there's interactive training. Studies show that students learn more when they are actively involved rather than just passively observing. Personal computers have made it possible to create interactive training systems at relatively low cost, and users find the results are well worth the investment.

"We have seen a decrease in training time of as much as 50 percent while increasing retention by 80 percent," says Jane Curtis, director of

Hewlett Packard

field development, Massachusetts Mutual Life Insurance Co. "One trainer told me the program saves him 10 hours a week. Also, we have seen an increase in first-year commissions. At one field office, the system contributed to a 40 percent increase in agent productivity. That we can measure, and that makes our case for interactive multimedia."

One of the benefits of multimedia training is the ability to simulate real-life situations and events more realistically than otherwise possible.

"Interactive multimedia simulation not only imparts knowledge, but instills confidence. A trainee can make mistakes and not ruin an expensive piece of equipment. It's good experience, and it builds confidence that he takes with him to the floor. Although the development cost was high, the program is now paying for itself through decreased downtime and training costs," says William Ziemba, manufacturing training specialist at Ford Motor Co.

Electronic brainstorming

Use of interactive presentations can also stimulate the creative process.

"People have to go through this in stages," says Ames Cornish of Apple Computer. "The first is understanding the benefits of doing a presentation from the computer.... Just showing slides can bore everybody to death." With Hypercard—Apple's interactive authoring software—you can present just the cards appropriate to your audience.

"A lot of people are going a step further," Cornish adds, "doing meetings and using the computer to organize and analyze." These "elec-

Apple Macintosh II/Laserdisk Monitor

tronic meetings" employ the computer, usually with some kind of outlining software, to help groups of people organize their thoughts and develop ideas on-the-fly. While the concept is still in its infancy, it represents a significant area for growth of computer-based presentations. (See the accompanying story for more about interactive meetings.)

Interactive selling

Any good salesman knows that one useful sales technique is to draw the customer into the sales process—asking questions, getting the customer to focus on the presentation by becoming involved. Computer-based interactivity is finding its way into this process as well.

One example is in video kiosks that provide product information in stores and shopping malls. Touche-Ross estimates that 50,000 such systems will be installed by 1990. User studies have shown that shoppers who use the information kiosks are likely to spend more money and buy more items than those who don't. Such kiosks have been used to sell blue jeans, audio recordings and home improvement products.

Competitive Solutions, Inc., of McLean, VA, has recently announced an interactive system called Expert ReaListor that would enable a real estate agent to sit clients down in front of a computer screen, evaluate their needs and qualifications with computer assistance and then "walk" them room-by-room through any of thousands of homes using words, color photographs, maps, floor plans and audio descriptions, all captured, stored and accessed through a desktop computer.

Computer-based presentations lend themselves particularly well to product information handouts. Software vendors in particular have been quick to take advantage of demo disks—no surprise. An interesting sideline—while software publishers usually discourage users from distributing copies of their disks to friends, with demo disks copying is usually encouraged and actually helps to spread advertising at no cost to the advertiser.

Software companies aren't the only ones who are finding interactive computer disks a worthwhile medium for publicizing their products. For example, automobile companies are using them as a way to distribute information on their new cars to upscale, computer-using consumers.

IBM's Storyboard software has long been popular with developers of interactive disks and other training programs. Originally created for in-house use at IBM, it has gone through a series of revisions and has been available to outside developers for several years, but it has been overshadowed recently by other presentation software, especially on the Macintosh.

But in IBM's latest version, Storyboard Plus Version 2.0, the program offers many of the same features included in the more highly publicized presentation programs, including the ability to combine computer graphics and text with animation, still video, music and voice (on properly-equipped computers). The program runs on all IBM and compatible personal computers.

"IBM is the world's largest user of interactive video," notes Peter B. Blakney, manager of sales support for IBM's Advanced Education Systems unit. "We've developed over 200 courses for internal use." IBM systems are also used by major companies from Ford to Quaker Oats to GTE, as well as by the government, he adds.

"Up to now, though, we've hid our light under a bushel, so to speak," he says. But in recent months IBM has begun a new push to play a larger role in the software business—with multimedia as a major target. Storyboard Plus is a cornerstone of the strategy, though other efforts will also be important in the long run, like IBM's cooperative venture with Intel Corp. to develop

Digital Video Interactive (DVI), a multimedia storage system based on compact disc technology.

Problems

John Serrone, manager of data management at AT&T Microelectronics, has worked with General Parametrics' VideoShow and points out a number of benefits. "It's better, faster than slides—and we can change almost instantly—you can't do that with slides, especially a few minutes before the presentation."

But he also points out a problem that remains to be addressed—one that's not limited to this particular product, but applies to all interactive multimedia and for that matter to computers in general. The problem is what Serrone calls "technophobia."

"When people see the VideoShow and see a viewgraph, they figure it's easier to use the viewgraph than to figure out what to do with this remote control," he says. This, he says, is the greatest barrier to broader acceptance of computer-based presentations.

Another possible problem, a tough one that will no doubt resolve itself in time, is the lack of standardization. To take just one example, three different standards currently exist for interactive computer storage based on the compact disc format.

We've seen how the definition of "multimedia" has changed through the years, from the graphics and sound of slide-based multi-image to computer-produced video and now interactive multimedia. As we said at the beginning of this series, the common thread remains the intelligent use of available technologies, alone or in combination, to produce the best possible presentations.

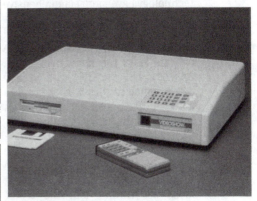

General Parametrics VideoShow

Computers the Key to More Productive Meetings

A new kind of meeting is taking hold in many companies. Using personal computers, presentation peripherals and appropriate software, these companies are taking advantage of the latest technology to facilitate the exchange of ideas.

The purpose of these meetings may be to produce a final document or come to some consensus about a problem; to create a strategic proposal, presentation or annual report, working out the details of a program, system or design.

While products such as overhead projectors and even low-technology flip charts and blackboards have been fixtures at planning sessions for years, they had some limitations: It wasn't always easy to change or reorganize information, someone needed to keep detailed notes, and there was no quick way to distribute copies of the results. Electronic copyboards, which make convenient-sized copies of anything written on their surfaces, helped solve some of those problems, but only with computer technology have more comprehensive solutions become possible.

The technology involved in electronic meetings is a mix of the old and new: the familiar overhead projector, but now used with an LCD panel, which in turn is hooked up to a PC or Macintosh computer, often running outlining or graphics software.

The computer is much more powerful, say many, than more standard meeting formats, because ideas are thrown into the arena, reworked and restructured and then formalized.

Bernie DeKoven has been in-

volved with computer-enhanced meetings for several years. He works for clients as far-ranging as Apple Computer and the American Red Cross. He coined the phrase technography and created the job title of technographer.

According to DeKoven, a technographer is a meeting facilitator

The Macintosh computer, overhead projector and LCD panel are the basic components of an electronic meeting system.

who keeps track of ideas generated during the meeting, makes annotations by participants and works with them during the organizing stages and finally prepares a final document which represents the consensus of the group.

He believes that a technographer's role is to "transform the meeting process, making sessions not only shorter and more productive, but qualitatively better." One of the ways in which this is done is "the ability to edit on the spot, which makes it easier for a group to reach a consensus quickly," says DeKoven.

Brainstorming is made easier through the use of outlining software because it lets you expand, contract,

create headings and subheadings and organize the information fairly easily. Programs include: More II by Symantec for the Macintosh, and GrandView by Symantec, MaxThink by MaxThink and PC Outline from Brown Bag Software, all running on the IBM PC and compatibles.

Another key element of the process, according to DeKoven, is "at the end of a meeting, a technographer can produce a written report for immediate distribution." Again, the computer helps make that faster and easier.

Many who run and participate in computer-aided meetings agree that the meetings typically consist of three stages. These break down into scoping out the conceptual space of a problem, analyzing the problem including generating alternatives and converging upon a course of action, and identifying conclusions and clarifying next steps, according to Marc Gerstein, managing director for Delta Consulting Group in New York.

Gerstein's group is unique in a method they pioneered which enables participants—usually between six and 12—to utilize his or her own input device.

Gerstein's company consults to top managers in making strategic decisions. And often the basis for decision making, or actual decisions, occurs during interactive meetings.

This enables participants to have a lot more input than they might in other meetings; everyone can see ideas that have been offered and keep track of them and later move them around. — J.R.

Getting Started with Multimedia

WHAT IS IT? CAN YOU USE IT? WHAT WILL IT COST?

BY JIM HEID

Multimedia is many things—literally and figuratively. Literally, multimedia is the integration of two or more communications media. It's the use of text and sounds, plus still and moving pictures, to convey ideas, sell products, educate, and/or entertain. It's built around the premise that anything words can do—words, sounds, and pictures can do better. The more, the media.

Some see multimedia as the harbinger of an era when computers will routinely convey information with sound and animation, as well as text images, and when television will become more interactive. But others see it as the victory of sound bites and flashy visuals over the printed word.

Multimedia brims with potential—and the potential for misuse. It can also be technically complex and, in its most advanced forms, quite expensive. This month, I explore the world of multimedia, spotlighting some ways in which it is used, as well as the components that form a complete system.

The Multi Levels of Multimedia

Not long ago, combining sound and visuals—say, jazzing up a slide show with music or narration—was about all there was to multimedia. It didn't require much equipment or technical expertise. You could create title slides

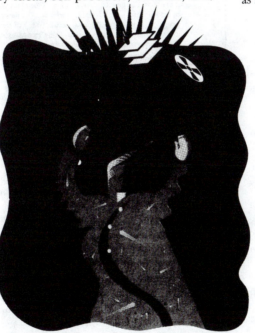

and other visuals using a presentation program such as Aldus Persuasion or Microsoft PowerPoint, and then use a stereo cassette deck to record and play the sound track.

Today, multimedia generally means using an *authoring* program such as HyperCard or MacroMind Director to create and playback a production. You can have animation,

such as bullet chart items that slide on screen, and *transitions,* such as dissolves and fades between visuals.

A more advanced level of multimedia might use animation sequences that illustrate complex concepts, such as how a steam engine works or how heart valves operate. At this level, your role as producer becomes more demanding. Designing and executing a complex animated sequence requires artistic skills and some knowledge of animation techniques. It also requires software with more advanced animation features than Hyper-Card's—more about that later.

A sophisticated multimedia production might also incorporate still images captured from a video camera or VCR. A corporate presentation might include shots of a new factory; a medical tutorial might show images captured from a videotape of a surgical operation. For this, you need additional hardware as well as some knowledge of video and lighting techniques.

Then there's sound, which authoring programs support in a few ways. They can play back short sound passages you record directly into the Mac using an add-on such as Farallon Computing's MacRecorder or using the recording circuitry built into the Mac LC and IIsi. So, an animated heart can beat to the sound of a recorded heart, and an animated steam engine can chug as a locomotive

STEVE LYONS

sound plays. Authoring programs can also control electronic synthesizers and CD ROM players, which can store hundreds of megabytes of data as well as CD-quality audio. Thus, a corporate presentation can play to the sound of CD-quality background music, or a music tutorial can display text on the screen while a piece plays.

Passive versus Active

Multimedia productions can be linear affairs—watched from start to finish, like a slide show or a TV program. But the most significant—and exciting— aspect of Mac multimedia is *interactivity*. The most advanced multimedia productions are nonlinear and interactive. Instead of passively watching from beginning to end, you use the Mac to interact with the production, setting your own pace and branching to different topics and areas as they interest you. With interactive multimedia, the Mac and programs like HyperCard become more than devices for presenting various media—they become tools for navigating the media themselves.

The primary tools of multimedia navigation are on-screen buttons and other *hot* areas that, when clicked, take you to other screens, display windows containing additional information, or play sounds—or even video sequences. Interactive multimedia often makes use of yet another piece of hardware, a *videodisc player,* whose discs look like a cross between an LP record and a compact disc.

Videodiscs and interactive multimedia complement each other beautifully, primarily because a videodisc is a *random-access* medium. One side of a videodisc contains 54,000 numbered *frames,* and a player under the Mac's control can skip to any one of them almost instantly. A videodisc player can display frames continuously to show up to 30 minutes of moving footage, or it can freeze on any one frame to show a still. Thus, a multimedia production can include moving and still pictures stored on the same videodisc—along with up to 30 minutes of two-channel audio per side. Those two audio channels can be used for stereo, or they can be used separately to hold, for example, narration in two languages.

INTERACTIVE HISTORY Two cards from the HyperCard stack that controls ABC News Interactive's Martin Luther King Jr. program. Clicking on a topic displays a card containing a timeline (top). Clicking on a subject on the timeline displays a card (bottom) with descriptive text and buttons that display accompanying visuals from the videodisc. Another click displays definitions of any terms, or biographies of any names, that appear in boldface.

Multimedia in Education

Clearly, interactive multimedia has tremendous potential as an educational tool. Nowhere is this better illustrated than in a series of videodisc packages produced by ABC News Interactive and distributed by Optical Data Corporation. In the Martin Luther King Jr. package, you use HyperCard to explore a videodisc brimming with footage of civil rights protests, vintage news reports, and King's speeches, as well as still photographs, maps, and charts (see "Interactive History"). You can watch the entire "I Have a Dream" speech while reading King's prepared text on the Mac's screen— and you can see where, halfway through, he diverged from the prepared text to capture the attention of millions. Other ABC News Interactive presentations include The '88 Vote; In the Holy Land; The Great Quake of '89 (distributed by The Voyager Company); and AIDS.

Each package in the ABC News Interactive series also includes a documentary maker with which students can assemble their own documentaries based on the videodisc's images and news footage. It's this feature that raises a red flag among interactive multimedia critics, who wonder whether students are learning about Martin Luther King Jr. or learning how to produce TV documentaries and splice sound bites.

Another example of interactive instructional media is Warner New Media's Audio Notes series, which uses CD ROM discs and HyperCard to let you listen to and learn about music. One three-disc package presents Mozart's opera *The Magic Flute;* a one-disc package called The String Quartet: The Essence of Music presents a Beethoven string quartet. As the latter plays, you can use an accompanying HyperCard stack to view any of several measure-by-measure commentaries on the music, each assuming different levels of musical knowledge (see "Listen and Learn"). Other parts of the stack contain graphical data on Beethoven and other tutorials on music theory.

Optical Data Corporation produces complete packages that teachers can use as is or as tool kits for creating their own instructional mate-

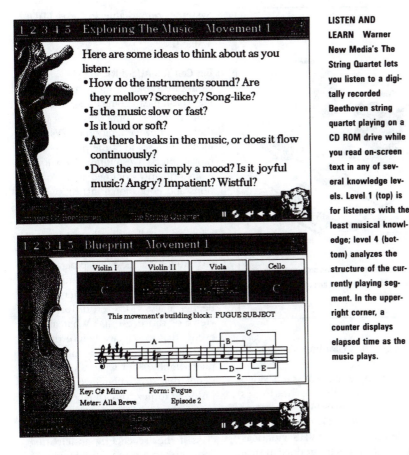

LISTEN AND LEARN Warner New Media's The String Quartet lets you listen to a digitally recorded Beethoven string quartet playing on a CD ROM drive while you read on-screen text in any of several knowledge levels. Level 1 (top) is for listeners with the least musical knowledge; level 4 (bottom) analyzes the structure of the currently playing segment. In the upper-right corner, a counter displays elapsed time as the music plays.

rials. The Planetscapes package, for example, contains a videodisc laden with images of planets taken by the Voyager spacecraft, of North America taken by Landsat satellites, and of space shuttle missions and components. Accompanying stacks let you browse the Voyager images, view Landsat images by clicking on a map, and learn about the space shuttle. You can also create your own stacks that use the videodisc images.

To give educators a hands-on overview of interactive multimedia's potential—and to allow them to create their own multimedia materials— researchers at Apple's Multimedia Lab spent two years developing a package called The Visual Almanac. It comprises a CD ROM containing 25MB of HyperCard stacks and digital audio; a workbook; and a videodisc with 7500 sounds and moving and still images. The HyperCard stacks include 14 student activities in science, the arts, social studies, and mathematics. There's also a composition work space that lets you search for images and sounds based on keywords you type, and then assemble them into stacks. The Visual Almanac is available for the re-

markably low price of $100 from Optical Data Corporation.

Of course, interactive multimedia has applications beyond the classroom. Interactive information kiosks in airports, at shopping malls, or on trade show floors can provide directories and profiles of cities, stores, or exhibitors. Interactive presentations can spotlight a concept, company, or product in an engaging way. Interactive museum exhibits can enliven any subject. In San Francisco's Exploratorium, you can "fly" over the Bay Area, viewing aerial footage and setting your own course using a trackball.

Examining the Pieces

Here's a closer look at the software and hardware components involved in multimedia ("Multimedia to the Max" illustrates how they interrelate) and a partial list of products from each category.

• *Authoring* This key software lets you direct a production's cast of audio and visual characters. HyperCard is the most popular authoring program; its relatively-easy-to-learn programming language, HyperTalk, lets you create simple animations and

establish links between on-screen hot spots and other cards. HyperCard users might want to try Motion Works' Addmotion, which adds animation features to HyperCard 2.0, enabling you to author and animate without switching between programs. Silicon Beach Software's SuperCard provides better color support than HyperCard and gives you the ability to create stand-alone applications, people who don't have SuperCard. Spinnaker Software's Plus offers similar features and is also available for IBM PCs. You can move Plus productions between Macs and PCs.

More-specialized authoring programs include Farallon Computing's MediaTracks, MacroMind Director, and Authorware's Authorware Professional. MediaTracks lets you produce Macintosh training materials by recording screen activity and then adding graphics, digitized sounds, and on-screen navigation buttons. Director 2.0 has a HyperTalk-like language, Lingo, for creating interactive animations containing navigation buttons; Director also includes a player application that lets others use your productions without their having to own Director. Authorware Professional is a high-end ($8050) package that provides sophisticated animation features and lets you create interactive productions, without programming, by drawing links between the production's components. Authorware also provides training, customizing, and production services for its clients.

• *Animation programs* In addition to Director and Authorware, there's Bright Star Technology's interFACE, which lets you create *agents*—talking heads whose mouths move and facial expressions change as digitized speech comes from the Mac's speaker. Agents can guide users through a production or act as on-screen teachers. They can be used with HyperCard, Director, and other programs that support HyperCard external commands (XCMDs).

Electronic Arts' Studio/1 is a monochrome paint program that lets you create simple black-and-white animations that you can play back within HyperCard and other authoring programs. Three-dimensional animations made with 3-D drawing packages such as Silicon Beach's Su-

Multimedia to the Max

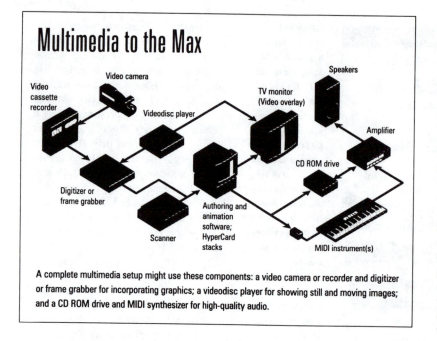

Video cassette recorder

Video camera

Videodisc player

Digitizer or frame grabber

Scanner

Authoring and animation software; HyperCard stacks

TV monitor (Video overlay)

Speakers

Amplifier

CD ROM drive

MIDI instrument(s)

A complete multimedia setup might use these components: a video camera or recorder and digitizer or frame grabber for incorporating graphics; a videodisc player for showing still and moving images; and a CD ROM drive and MIDI synthesizer for high-quality audio.

per 3D and Paracomp's Swivel 3D Professional can also be played back within authoring programs. For more advanced animations with special effects, you can use Linker Systems' The Animation Stand. It can also produce broadcast-quality video animations in conjunction with high-end animation hardware such as Diaquest's DQ-Animaq, a Mac II board that controls professional video recorders on a frame-by-frame basis.

• *Video hardware* For recording a presentation on videotape, you might use a board—such as one from RasterOps or Radius—that outputs *NTSC-compatible* video. For capturing video images from camcorders, videodiscs, or VCRs, you might use a *digitizer* such as Koala Technologies' MacVision. For faster digitizing, use a *frame-grabber* board such as RasterOps' FrameGrabber 324NC, Orange Micro's Personal Vision, or Workstation Technologies' WTI-Moonraker.

With boards such as Aaps' Digi-Video, Computer Friends' ColorSnap 32+, and Radius's RadiusTV, you can display video from VCRs, videodiscs, or TV stations in a Mac window. To combine Mac graphics and text with a video source, you need *video-overlay* hardware such as Mass Microsystems' ColorSpace IIi, Truevision's

NuVista+, or Computer Friends' TV Producer Pro.

Then there are VCRs and videodisc players. NEC Technologies' PC-VCR is an S-VHS VCR you can attach to the Mac and control much like a videodisc player. VHS videocassettes cost far less to produce in small quantities than videodiscs, so a PC-VCR could be an economical alternative if you produce your own video materials. But a videocassette isn't a random-access medium; it can take as long as 2½ minutes to locate a given point on the tape. As for videodisc players, Pioneer and Sony make the largest assortment with RS-232C serial interfaces that connect to the Mac.

• *Audio* To record and play back CD-quality audio, you need Digidesign's Audiomedia board. Throw in a big, fast hard drive, too, because CD-quality stereo requires 20MB of disk space per minute. As an alternative, use a MIDI interface to connect a Mac to one or more MIDI synthesizers and play back MIDI *sequence* files, which use a small fraction of the space. Passport Designs' Music Data company sells hundreds of pre-recorded MIDI sequences (MIDI Records and MIDI Hits). You might also consider Digidesign's Mac-Proteus board, a version of E-mu

Systems' remarkable Proteus MIDI instrument that you can plug in to a Mac II.

But Can You Afford It?

Going for broke in interactive multimedia can mean going broke. Indeed, until the prices come down—or until there is a new generation of TVs that incorporate some of the pieces—full-blown interactive multimedia will remain primarily confined to corporations and educational institutions.

But the larger issue is the impact interactive multimedia will have on education. Will the ability to create, in essence, their own documentaries give students a better grasp of current events? Or will their technological wizardry merely distract them from deeper analysis or even distort their views of those events? What about events that took place before the inventions of sound recording and photography? And what about disciplines that don't benefit from computer graphics and old newsclips, such as literature? Can students really concentrate in an expository writing class if they're eagerly anticipating zooming around the Solar System in their next period? After a few years of nonlinear, TV-based education, will they be able to concentrate at all?

Educators and curriculum planners who seek the answers to those questions—and to brace themselves for the consequences—should definitely be in on the ground floor of interactive multimedia. Not only will they need to prepare for what appear to be tomorrow's teaching techniques, but also to shape those techniques into being more than technological diversions. Instead of asking, as some media critics have, How can we use television and the computer to shape the form and content of education? they should be asking, How can we use education to shape the form and content of television and the computer? ▣

Jim Heid is a contributing editor of Macworld. *His latest book is* Macworld's Complete Mac Handbook *(IDG Books Worldwide, 1991).*

The Magic of Multimedia

BY KEN FERMOYLE

In little more than a decade, microcomputers have transformed the communications process, particularly in the areas of publishing, processing and data transmission. In the last few years, computer technology has begun to work its magic on the presentation process as well. Powerful graphics tools and sophisticated display products have combined with the personal computer to enhance the way presentations are produced and delivered.

Most recently, audio, still imaging, animation and full-motion video have been added to the formula, creating what is commonly called multimedia. The definition of multimedia is a moving target. Multimedia denotes different things to different people, depending on their backgrounds, viewpoints, and sometimes their own self-interests. Just a few years ago, a multimedia presentation consisted of multiple slide projectors, screens and a soundtrack with maybe some film, video or fireworks thrown in.

One of the more workable definitions comes from INTECO Corporation, an international primary research and strategic consultancy firm based in Connecticut. INTECO says multimedia is, "The addition of one or more media, such as images (moving or still), animation or sound (input commands or output prompts) to existing text and computer graphics."

This broad definition covers systems ranging from a basic microcomputer configuration with an analog-based laserdisc attached, to complex interactive systems, and on to the latest digital compression technology. The common denominator in each case is a CPU in command of the authoring and playback of multisensory material. In other words, multimedia, by any other name, would smell as sweet.

In what appears to be the broadest multimedia survey done to date, INTECO interviewed 75 system developers and 220 end users in the United States and did comparable primary research in Europe. Interestingly, researchers did not use the word "multimedia" at any time during the interviews to define the applications which were studied.

The survey delineated four major themes in the development of multimedia.

■ **1.** Applications for multimedia systems and courseware are already growing to cover five major segments:

• Desktop Applications (e.g., enhancing presentations)

• Industrial Training (e.g., retraining workers or providing remedial skills)

• Education (e.g., promoting learning through student interaction with course material)

• Sales and Information Kiosks (e.g., providing information and increasing sales)

• Industrial/Scientific Operations (e.g., presenting complex information graphically or helping workers to perform tasks)

■ **2.** Multimedia growth is being driven by measurable benefits to developers and users. (e.g., Factory workers are trained faster and doctors are better able to diagnose with the aid of multimedia.)

■ **3.** Developing multimedia materials requires new combinations of microcomputer and creative skills. Interactivity is essential to effective multimedia, for example, and multimedia creators must learn to use it to its maximum advantage.

■ **4.** Multimedia is stimulating new hardware markets (e.g., CD-ROMs, sound processors and video cards).

These points have been made as "educated observations" by many in the industry over the past two years, but INTECO's research validates them. And other evidence corroborates the study. On the hardware side, for example, stimulus to the market is obvious. Increasing numbers of new, lower-cost and higher-performance multimedia-related product introductions are reported regularly in the new product pages of this and other industry publications. The recently announced collaboration between Tandy and Microsoft on a $3,000 Multimedia 386 PC with built-in CD-ROM player, digital sound support, eight-bit VGA display, color monitor, 30-megabyte hard drive and two megabytes of RAM, is one good example of the surge of new products predicted by INTECO.

Thanks to the architecture of its Macintosh computer products, Apple Computer has enjoyed a well-documented lead in graphics processing. In multimedia as well, the innate talents of the Mac have helped make the machine a favorite with many producers. But the Mac is being challenged with increasing vigor by IBM and the entire DOS-based vendor world. The affordability and upgradability, combined with rapid strides in hardware and software development, have brought IBM and compatible products into the limelight as multimedia authoring and application platforms. In addition, Commodore's Amiga line, which was designed with extensive video- and audio-processing capability and boasts the most affordable entry-level price, has come on strong, particularly with video-oriented multimedia creators.

A Mac-Powered Showcase

Some of the most striking and inventive uses of multimedia to date have Macs for brains. Although not originally designed as multimedia machines, even low-priced Macintosh models have built-in support for coordinated images and sound. That, and Apple's policy of working with independent multimedia product developers, have attracted a growing number of third-party authoring tools (e.g., Director from MacroMind, San Francisco), music creation and editing programs, plus other hardware and software multimedia products.

"We encourage third parties to work on a particular area and do that very well," says Mary Bushnell, Apple's product specialist for HyperCard.

Reprinted with permission from *Presentation Products Magazine,* pp. 32–41, Nov. 1990.

''What you get then is a modular environment so you can put together the pieces the way you want.''

In New York's Jacob K. Javits Convention Center, NYNEX Information Resources Company (NIRC) chose Macs to author and run one of the most extensive and diverse groupings of multimedia applications in the country.

A 14x18-foot Sony JumboTron monitor is the focal point of the main thoroughfare at Javits Center. Fed via Macs with continuous editorial and advertising video, it gives visitors information such as show schedules, and provides advertisers with an attention-getting, 252-square-foot display for their products.

Supplementing the JumboTron is a VideoWall consisting of four Pioneer projection cubes. The wall is used as a multiple-screen, portable video and sound system to display full-motion video and computer graphics at various locations in the facility. ''Content is managed by a Macintosh IIci, and programming appears on individual screens or across the entire multiscreen wall, to produce a dramatic, large-scale image,'' reports Harold Dakin, NIRC vice president, New Business Operations.

In addition to the large displays, six VideoGuide interactive, touchscreen kiosks with Apple's MicroTouch monitors provide Javits Center visitors with information on shopping, dining and entertainment. VideoGuide combines stereo-quality audio, full-motion video,

and high-resolution text, graphics and animation. Each kiosk includes a printer, so users can obtain incentive coupons or printouts of directions to the various establishments.

''The unique and potent advertising package available at the Javits Center today is the result of a state-of-the-art production platform conceived and designed by NYNEX and made possible by the power and flexibility of the Macintosh,'' says Dakin.

''We at Apple view the NYNEX installation at the Javits Center as a significant showcase for the Macintosh in a public-access environment,'' reports Jack Murphy, Apple Computer's vice president of eastern operations. ''This groundbreaking application, which illustrates the broad capabilities of Macintosh in both the development and display of multimedia presentations in a commercial setting,'' could establish a new standard for public-access systems ''and open potential markets for Apple worldwide.''

MacroMind's Director Software, Version 2.0, was the authoring system used to create the presentations by combining output from a variety of software programs with digitized sound and video. NYNEX utilizes MASS Microsystems' ColorSpace II and ColorSpace FX video graphics display cards to combine computer graphics and video on one screen.

Another company, Steelcase Stowe & Davis, a Michigan-based office furni-

ture manufacturer with outlets across the country, uses multimedia to train its retail staff and as a sales tool for its 13,000 products. Running on a Macintosh IIcx with SuperCard (from Silicon Beach Software, San Diego, Calif.) and a 600-mb hard disk, the application provides a visual presentation of the manufacturer's extensive product line. Serving as an "interactive catalog," it allows mix-and-match manipulation of the products and option selections to help customers make choices. Santa Fe Interactive, Santa Fe, N.M., developed the application.

Atop the IBM Multimedia Platform

IBM has been actively developing products and applications for multimedia almost since the capability of the hardware and software was just a gleam in theorsts' eyes. As a result, some of the most elaborate and impressive productions to date have been done on the IBM platform.

''IBM's commitment to multimedia is evidenced by the fact that our multimedia organization was formed this summer to guide IBM's worldwide strategy in developing multimedia products,'' says Peter Blakeney, manager of market support for the company's Multimedia Solutions division. ''We see the market growing at a compounded rate of 40 to 45 percent annually, reaching $35 to $40 billion by 1994.''

Multimedia Glossary

Antialiasing—Technique for smoothing audio or visual data; removes the jagged look of graphics images, for example.

Authoring System—See Authorware.

Authorware—Development environment that provides software tools used to create multimedia presentations, specifying elements (e.g., still images, video, animated sequences, sound) you want to include and combining them. Most authoring systems provide interaction capability.

AVC (Audio Visual Connection)—Original IBM authoring package for the PC and PS/s.

Bit—A single binary digit.

Byte—A group of eight bits.

CD-I (Compact Disc Interactive)—Multimedia form of CD that allows integration of data, still graphics, motion video and audio.

DSP (Digital Signal Processor)—Special computer chip designed to perform such complex operations as processing sound and video.

DVI (Digital Video Interaction)—Intel technology allows storage

of compressed full-motion video on CDs and magnetic media.

Full-Motion Video—Video sequences that emulate those normally seen on TV because they have enough images (30 frames per second) to impart smooth motion.

Genlock—Abbreviation for "generator lock." Aligns data transfer rates to combine standard broadcast (TV) video signal with computer-generated RGB images.

HDTV (High-Definition Television)—Possible next standard for U.S. TV; raises screen resolution standard from current 525 lines per frame of NTSC to 1,100 or more.

NTSC (National Television Systems Committee)—Group that formulated standard for U.S. color TV: 525 horizontal lines per frame at 30 frames per second with interlaced scans.

Pixel—Basic unit of an electronic image on a monitor or TV screen: the more pixels available, the higher the resolution.

RGB (Red-Green-Blue)—Standard for projection of color images on a computer monitor that uses three separate signal lines for red, green and blue picture data.

One of the most ambitious and far-reaching presentations to date is a multimedia event developed for IBM by AND Communications, multimedia specialists based in Los Angeles. Based on the poem *Ulysses* by Alfred, Lord Tennyson, the project has has been shown to audiences at trade events and seminars throughout the world.

IBM commissioned the project to show the potential of multimedia technology, according to Blakeney.

Ulysses combines a dazzling array of text, graphics, video, still images and sound (including renditions of portions of the poem by different people) that can be accessed interactively by various tools built into program menus.

The program was put together in just 27 days, reports Morgan Newman, a cofounder of AND. It was created on a PS/2 Model 70 using LinkWay software and the new M-Motion Video Adapter/A to provide motion, still video and audio capabilities. "Of course, there was considerable preproduction involved to prepare the pieces that make up the presentation," says Newman.

IBM's goal has been to explore a variety of applications and resources for the technology. "We plan not only to offer our own hardware and software products but will encourage third-party developers," Blakeney states. As an example of multimedia's role in training, he points to an IBM-based project prepared for Bethlehem Steel to provide steelworkers with a new kind of on-the-job learning environment.

The interactive application makes training available to them in the mill for the first time. "Employees interact with realistic simulations of job situations so they are prepared for similar situations when they arise," reports a Bethlehem spokesman. "Both younger and older, more experienced employees are eagerly participating in the voluntary program, which translates into a bottom-line payoff: better teamwork, greater accuracy and higher productivity."

A collaboration between Pacific Bell Telephone and Jasmine Productions of Santa Monica, Calif., is another type of hands-on training application for an IBM platform. Titled "Refresher Defensive Driving," the interactive application runs on both touchscreen and non-touchscreen systems. "Refresher Defensive Driving" covers the basics of defensive driving in a hands-on approach that includes simulated city and highway driving, plus sections on seatbelts and safety inspections. "We saved millions [of dollars] in vehicle damage, worker's compensation and third-party lawsuits so we decided to share our course with business and government," says Harlan Chiu, who heads Pacific Bell's training group. "We recommend it for any company or department with a fleet of 25 vehicles or more." Chiu added that Pac Bell's 13,000 drivers traveled more than 114 million miles in 1989 with only 557 accidents—down 242 percent from 1986, prior to the making of "Refresher Defensive Driving." The application runs on any Info-Window, Sony View or IBM PC with laserdisc player.

Other examples of multimedia technology in action include: Chase Manhattan, which uses touchscreen multimedia technology and simulations to train bank tellers; Peters, an East Coast sporting goods retailer, which is employing a kiosk-type presentation as a merchandising tool; an online interactive system used to help Sears catalog merchandise buyers check catalog sales and inventories; and an interactive program called "Target," which is working in high schools across the country to help prevent substance abuse among high-school students.

Amiga's Loyal Followers

Although some might consider Commodore's Amiga an also-ran in the PC race, the machine's talents give it leading-edge status among dedicated users who work the multimedia vineyards. "There are no better machines for multimedia than Commodore's Amiga 2500 and 3000 models!" insists Keith Nealy, head of The Nealy Group. "They come with a custom chip set that makes them faster than IBM or Macintosh machines." He points out that the chips remove the burden of graphics, sound and other functions from the CPU. The chip set includes a dedicated graphics coprocessor with "blitter" circuitry that quickly alters images on the display.

A second chip has built-in animation "sprites," graphical shapes that can be defined and moved across the display background with simple commands. A third chip helps provide stereo sound and shuttles data through the Amiga's bus network.

Nealy's New York-based firm "creates sales and marketing communications solutions, including large multimedia shows, for Fortune 500 and other companies." Nealy explains, "We pioneered the use of computer-generated imagery in large multimedia shows, and Amiga gave us the capability of animating TV-style graphics. We began using Amigas professionally five years ago when the 1000 model first came out, and we now have about eight of them—at least one of every Amiga model ever made.''

About three years ago, Nealy originated a product/service called 'CompuVision,' which creates animated slide shows for the firm's clients by combining a mixture of off-the-shelf Amiga hardware and other add-on components.

Nealy is currently creating an animated production for Lancome Cosmetics that combines live-action, steady-cam photography of the company's robotic manufacturing line with extensive use of 3-D animation. Among the products Nealy uses are Commodore's AmigaVision authoring software; Deluxe Paint III from Electric Arts in San Mateo, Calif.; AmiLink video editor from AmiTech Computers, Boystown, Nev.; Sculpt 4D, Byte by Byte, Austin, Texas; Turbo Silver from Impulse, Minneapolis, Minn.; and Pro Video Post from Shereff Systems, Beaverton, Ore.

Jim Carey, founder of Desert Sky, a film and video production company based in Agoura, Calif., is another dedicated Amiga user. He originally purchased an Amiga 2000 for creating images. After working with it on several projects, he learned to take advantage of its animation capabilities. Carey first used his Amiga professionally to bring storyboards to life for a 30-minute General Motors documentary featuring the company's solar-powered race car.

"While working on the GM video, I realized I needed to illustrate how the components of the car came together," says Carey. "The best way for GM executives to visualize my concept was to show them an animated storyboard." Quality of the storyboard was such that, with a little cleaning up, it became the final animation actually used in the documentary. General Motors distributed 80,000 copies to schools as part of an education package, and segments of the video have been featured on broadcast television programs about alternative energy sources.

Carey creates animated computer storyboards from scratch using Deluxe Paint III, sometimes importing logos and product shots using NewTek's (Topeka, Kan.) Digi-View video digitizer or images from his large collection of clip art. He has used this technique for other clients, including Hughes Aircraft and Lockheed Corporation.

Carey's experience of creating storyboards and producing videos with the Amiga led him into multimedia presentations. When Miller Brewing, for example, planned a presentation for the annual distributors meeting, the firm was going to once again use slides to illustrate points made by the keynote speaker. "I talked them into letting me create a live multimedia presentation with animated bar charts showing sales growth," Carey reports. "Using the Amiga, I could make changes in the presentation at the last minute and it was no problem." Which was fortunate, because the speaker did decide to alter the presentation and Carey was "literally making changes as he was speaking."

Using Elan Performer, an interactive presentation package by Elan Design of San Francisco, Calif., Carey was able to store all types of pictures and animation. The Amiga multitasking feature allowed him to call up Deluxe Paint III and Digi-View and animation or graphics files at the click of a mouse.

Desert Sky recently added Edit Decision List Processor from MicroIllusions in Granada Hills, Calif., to its library of video software tools. Linked to an offline editing system, this package enables Carey to create and alter a decision list as he edits. He can then clean it up, take the file on floppy disk to a video facility and use the list to control the online editing system. "It saved twice its cost on just one project," he reports.

The DTP Analogy

Industry analysts and presentation specialists agree that right now multimedia is at the same point desktop publishing was five years ago: still in its relative infancy but poised to explode into a major market segment.

Not all of the hardware peripherals and software tools are in place yet, and much of what is available is still too pricey for most end users. Think back to 1985, however, and recall that a PostScript laser printer cost about $6,500; there are dozens of models available today for less than half that price.

What will gradually emerge for multimedia is the same kind of multilevel range of tools now found in desktop publishing, graphics and other special market areas. There will be everything from affordable systems for individuals who want to be desktop DeMilles and add multimedia pizazz to their presentations, to high-end hardware and software professionals who want to create big-budget extravaganzas for Fortune 500 companies. ❑

Multimedia: Is It Real?

Suzanne Stefanac and Liza Weiman

Buzzwords intrigue but rarely inform. The term *multimedia*, for instance, is bandied about as though we all agree on its definition and implications. The truth is, consumers, developers, and computer pundits have stretched its meaning until there is little left beyond hyperbole and generalizations.

To find out what lies behind the hype, we interviewed pioneering multimedia developers. We looked at the available production tools and publishing options. Finally, we drew up a definition of multimedia as it applies to the Macintosh.

In the course of our research, we listened to pie-in-the-sky visionaries, sympathized with frustrated developers, and talked to everyone from computer retailers to high-school teachers. We catalogued the tools available to multimedia producers (see ''Multimedia Production Tools''). Finally, after juggling the pieces of the puzzle, it became clear that asking whether or not multimedia computing lies in our future is a bit like our grandparents' wondering whether or not movies should have sound. It's not a question of if, but when.

Using his solar-powered multimedia lab, ecologist Mike Hamilton catalogs the wildlife of the James San Jacinto Mountains Reserve.

Multimedia Defined

- Multimedia incorporates several media types—text, graphics, audio, animation, video—in a single computer document; sound and/or motion are integral to a true multimedia project.
- What distinguishes multimedia from film or video is interactivity. Watching a detective movie is one thing, solving the mystery is another. Computer-based multimedia allows the user to determine the pace and path through ever-branching options.
- Multimedia developers must be much more than programmers, video producers, and audio wizards. They must also be interface designers able to knit several media into easy-to-navigate documents.

Where Is the Knowledge Navigator?

Apple touts the Macintosh as a multimedia machine, primed for easy integration of multiple media. As a teaser, John Sculley often shows a video about the Knowledge Navigator—a simulated notebook-size computer that talks and listens, fetches clips from video databases, and even schedules appointments for its lucky owner. But as much as we'd like to believe that voice-activated electronic butlers are in our near future, it just isn't true.

The problem, as usual, lies not in the vision, but in the technology. Before full-motion video can be delivered over networks and real-time animation is available in 32 bits, we'll need CPUs and graphics processing capabilities many times faster than those currently available. Storing these memory-hungry multimedia files is another problem in search of a solution. And although Articulate Systems' Voice Navigator looks promising, voice-recognition technology for personal computers is still in its infancy—a fact even Apple tacitly admits.

In a recent Desktop Media campaign aimed at the business community, Apple featured a multimedia presentation of a cartoon car with propellers—the HeloCar—that included page layout, 3-D graphics, 35mm slides, and overhead transparencies. Screens with HyperCard-like buttons let you choose between an animation of the car flying, schematic drawings of the interior of the engine, or dynamic cost breakdowns. What this slick presentation lacked was 32-bit color animation, full-motion video, and voice annotation.

Reprinted with permission from *Macworld,* pp. 116–123, April 1990.

Other Platforms

While Apple is promoting the Macintosh as a platform for multimedia, other computer companies are pitching their own development schemes. Next, for instance, is attracting some first-rate developers (see "Multimedia on the Next"), and IBM's InfoWindows with TouchScreen Display and Sony's View System have been popular in the corporate community for years. Intel's continued promise of affordable DVI (digital video interactive) standards might make the compression and playback of full-motion (30-frames-per-second) video a reality on the IBM PC—someday.

The problem with Intel's compression scheme is that although you might be able to play back compressed video on a personal computer, it takes a large, expensive computer to compress the data in the first place. Personal-computer owners can inexpensively play multimedia releases that were very expensive to produce. Just as with records or CDs, consumers can play them, but not make them.

Apple, in contrast, has stated publicly that its vision requires that the Macintosh be able to both make and play multimedia on the desktop. Apple believes that users should have the option to be both consumers and producers.

In keeping with this vision, HyperCard was Apple's early entry into the multimedia sweepstakes. Because it enables users to write their own simple interactive applications and because it can be used to access and control other media—CD ROM and videodisc players, for instance—HyperCard struck a chord. As HyperCard creator

Bill Atkinson notes, "We were surprised at the way people jumped right in and started using HyperTalk. This was probably due to a pent-up demand for user programmability."

The Multimedia Pyramid

Of course, even if Apple's vision becomes a reality, most people will probably encounter multimedia presentations in boardrooms and salesrooms, schools and training centers, museums and shopping-mall kiosks. Some people will learn to add a few sounds or short animations to presentations. Very few, however, will be in a position to invest the time and money necessary for creating truly professional multimedia projects. Multimedia production, particularly if video is involved, will still be in the hands of the few who have access to advanced technology. (For an overview of the costs involved in optical publishing, see "Be Your Own Producer.")

On the other hand, homemade videos are quite popular, and the fact that the quality isn't professional doesn't seem to matter to most people. Multimedia presentations are similar.For a small investment in time and money, you can incorporate text, graphics, and sound into your documents. For a little more money, you can add color, animation, and CD-quality sound. For still more money and time you can use digitized video still-frame pictures. For a lot more money you can include full-motion video. And at the highest end of the spectrum, you can incorporate real-time video complete with computer-generated text and graphic overlays.

The following profiles illustrate

some of the ways in which people have found multimedia to be a cost-effective solution to problems in their workplaces. (For a listing of multimedia production tools arranged loosely according to the time and money investments involved, see "Multimedia Tools," the table running along the bottom of these pages.)

In-House Expertise

In a fast-changing world, keeping employees and customers up to date on technical information is a challenge many companies face. Multimedia-based training, because it can incorporate voice, music, animation, and interactive course work, has proven popular in many corporations.

Codex Corporation, a subsidiary of Motorola, is a data-communications and networking-services company that needed to find a way to effectively train its sales representatives, applications engineers, systems engineers, customer-service representatives, and customers in the basics of digital telecommunications technology.

Working with interactive multimedia developers Madeline Butler and Paul Raila, Codex produced Basics of Digital Voice Technology, an 8MB interactive HyperCard stack that replaces an eight-hour lecture-type course with a three- to four-hour interactive one.

Incorporating VideoWorks II animation, sounds digitized with Mac-Recorder, and animated characters produced with HyperAnimator, this interactive training course simulates a year in the life of a communications manager in a medium-size company.

"We chose to make it a simulation

MULTIMEDIA PRODUCTION TOOLS

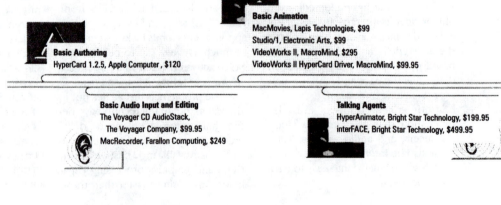

Products on this line are arranged according to the amount of time and money you must invest if you want to include sound, animation, or video in your multimedia documents. – Suzanne Stefanac

Basic Authoring
HyperCard 1.2.5, Apple Computer , $120

Basic Animation
MacMovies, Lapis Technologies, $99
Studio/1, Electronic Arts, $99
VideoWorks II, MacroMind, $295
VideoWorks II HyperCard Driver, MacroMind, $99.95

Basic Audio Input and Editing
The Voyager CD AudioStack,
    The Voyager Company, $99.95
MacRecorder, Farallon Computing, $249

Talking Agents
HyperAnimator, Bright Star Technology, $199.95
interFACE, Bright Star Technology, $499.95

with specific challenges," says lead designer Madeline Butler, "so that people would have an inherent motivation to finish the course without feeling pressure related directly to their own jobs."

Basics of Digital Voice Technology is currently running in over 30 sales and service centers nationwide, and Codex is developing an interactive stack-based course on ISDN (Integrated Services Digital Network). Having developed expertise while working on Basics of Digital Voice Technology, Codex is creating the ISDN project entirely in-house, using outside contractors only for the design of some of the graphics. This time the developers are adding a Computer-Eyes digitizer for capturing the images of real people, and they've added MacroMind Director to their toolbox for creating animations.

Getting the Big Picture

Ecologist Mike Hamilton, resident director of a University of California ecological reserve for teaching and research, wanted to create an interactive ecological database both as a teaching aid and as a model for cataloging the natural world. In order to do so, he built a solar-powered multimedia lab.

Using a Mac II with an Apple 8-bit video board, a 40MB hard disk, a 20MB hard disk for backup, a Digital Vision ComputerEyes Pro Color video digitizer, a Pioneer 4200 videodisc player, and a 19-inch Sony multisync monitor, Hamilton has created the Macroscope Ecology Laserdisc, a videodisc with over 25,000 still images, full-mo-

tion video sequences, and a 30MB HyperCard stack that contains detailed ecological information. His goal is to create a scientifically accurate and comprehensive database appropriate to the needs and levels of interest of groups from kindergartners to Forest Service professionals.

Using still-frame video cameras from Canon—the Xap Shot and RC701—and an 8mm Sony HandyCam camcorder, Hamilton and his crew of researchers take video images in the field of all relevant biological phenomena, including close-ups, full-motion sequences of animal behavior, and a highly structured 150-frame view of an area that becomes a *videomap*, a series of overlapping images that record everything a person can see while standing in a particular spot. Information about the location, format, date, time, and photographer of each image, as well as species information and ecological classification, is added back at camp on HyperCard cards using a Macintosh Plus. Parabolic microphones and cassette tape recorders are used to capture animal sounds, and MacRecorder is used to digitize the sounds and add them to the stack back at the office.

Hamilton makes three to four videodiscs per year for a total of about $3000. Working with Laser Edit, a post-production video service bureau in Los Angeles, he transfers his video in animation mode, one frame at a time, onto professional 1-inch-format videotape. This is then recorded on an Opticle Disc Corporation Mark II WORM drive to make a one-shot check disc that is LaserVision 2–compatible (play-

able on standard industrial players). This plastic disc cannot be replicated, but at $300 per check disc it is an economical way to prototype the version of the Macroscope that Hamilton will eventually produce in large quantities (see "Be Your Own Producer").

"The important thing," Hamilton says, "is that this doesn't have to be a big-budget enterprise. You can make multimedia happen on a Mac Plus with less than $1000, once you own the basic equipment. This stuff is like Hollywood, it will absorb as much money and time as you're willing to put into it—your budget may go way beyond $1000 if you don't plan carefully."

Selling in the Twenty-First Century

The people at Live Marketing have been creating presentations for large corporations for the past 17 years. They've used slides, multiple slide projectors, video walls, lasers, and even puppets and holography to sell, persuade, and communicate with people in the chaotic and competitive trade-show environment.

Now Bill Stimwich, vice president of marketing, has added Mac-based multimedia presentations to the company's portfolio. Using MacroMind Director as the basic design tool, he adds images created with Swivel 3D, Aldus Persuasion, MacDraw, and Canvas 2.0. He scans in 35mm slides with a Barneyscan color slide scanner and uses a flatbed Howtek scanner for color photographs. For certain digital-audio effects Stimwich uses MacRecorder, but the bulk of the sound track usually runs on a tape deck hidden within the booth, synchronized to the

Midrange Authoring
ArchiText, BrainPower, $395
Course Builder, TeleRobotics International, $395
Guide, OWL International, $295
Interactive Teacher, American Intelliware Corporation, $295
SuperCard, Silicon Beach Software, $199
*Plus, Spinnaker Software Corporation, TBA

Basic Video Digitizers
Xap Shot, Canon USA , $799
ComputerEyes, Digital Vision, $249.95
ComputerEyes Pro Color Digitizer, Digital Vision, $450
HyperVision, Pixelogic, $195
MacVision, Koala, $400

Midrange Audio Input and Editing
Audiomedia, Digidesign, $995
Sound Designer, Digidesign, $999

Midrange Animation
MacroMind Director, MacroMind, $695
MacroMind Director Accelerator, MacroMind, $195

Midrange Editing
*Media Tracks, Farallon Computing, TBA

TV Window on Mac
MicroTV, Aaps Corporation, $395
*BigTime TV, Hyperpress, TBA

*Not shipping at press time.

presentation running on a 37-inch monitor and a Mac II.

Stimwich's productions also include a live actor working in sync with the multimedia presentation, which runs repeatedly from five to six minutes throughout a trade show. Calling his work twenty-first-century slide shows, he uses animated effects including montages of photos, color animations, and visual magic to draw attention to his presentation subject.

For presentations projected onto screens, he feels that 8-bit color achieves the nearly photographic realism he requires. "There's nothing limiting about the Mac environment today. Right now, I can do everything I can ever imagine wanting to do, with readily available tools."

Access to Video Archives

What happens to that video clip after you've seen it on the evening news? Until very recently, at ABC News it went into storage; to see it again, you had to pay thousands of dollars for each minute of video footage. Then in late 1988 David Bohrman and his team at ABC News Interactive began publishing videodiscs for the education market.

Bohrman had been executive producer for ABC's coverage of the 1988 presidential election, so it made sense that ABC News Interactive's first release would be The '88 Vote, Campaign for the White House. Subsequently, in tandem with distributor Optical Data Corporation, ABC News Interactive released the videodiscs AIDS, The Great Quake of '89, and Martin Luther King Jr. (including the

MULTIMEDIA ON THE NEXT

The Next machine incorporates several hardware and software features that are generating interest among multimedia developers. With its 25MHz 68030 processing unit, 8MB of RAM (expandable to 16MB), direct memory access (12 I/O processors), and digital signal processor (DSP), the Next is designed to capitalize on multimedia's need for enhanced speed, memory, data-storage, and sound capabilities.

The Next comes bundled with Mail, an electronic-mail application that lets you send voice messages over built-in Ethernet circuitry. FrameMaker, a desktop publishing program from Frame Technology Corporation, lets you add voice annotation to documents.

Developers like Metaresearch have created software and external peripherals that let users input and manipulate CD-quality sounds and

video images, taking advantage of the Next's DSP chip. Digital Ears consists of a sound utility plus hardware; it opens, plays, records, and saves sound onto SCSI disks or optical discs using a simple graphical interface. Digital Eye grabs still-video frames or sequences from laser disc players, still and full-motion video cameras, and other standard NTSC video sources. Files created with either product can then be included in documents and in other applications.

MediaStation, an application from Imagine, enables users to capture, store, and process images, animations, and sounds in an environment that allows several media to be accessed and retrieved at once. Storing data on optical discs allows visual archives, digital audio libraries, electronic storyboards, and desktop presentations to take advantage of high resolution visual and sound files.

entire "I Have a Dream" speech). In the works are Mission: The Moon, an overview of the Mercury, Gemini, and Apollo space flights; and The Powers That Be, a three-part project that will look at the Supreme Court, Congress, and the presidency.

Using HyperCard as a navigational front end, the ABC Interactive videodiscs enable teachers to create customized lesson plans that draw upon

the recorded video information. In a supportive move, Apple approved the use of its educational marketing funds for the purchase of videodisc players and monitors in schools.

Development, Distribution, and Tools
The Voyager Company is striving to create a consumer market for interactive multimedia. Its products range from educational to entertainment,

MULTIMEDIA PRODUCTION TOOLS (continued)

Video In
ColorSnap-32, Computer Friends, $1595
FrameGrabber 324NC, RasterOps, $1995
Neotech Image Grabber 1.8,
    Advent Computer Products, $1499
Personal Vision, Orange Micro, $2899
PixelGrabber, Perceptics, Corporation, $1720
ProViz Video Scanner, Pixelogic,
    $1100 black and white, $1700 color
ProViz Video Color Converter, Pixelogic, $595
Quick Capture, Data Translation, $1245
Color Capture, Data Translation, $2995

RGB and Composite Convertors
NTSC Converter for Macintosh II, Julian Systems, $699
NTSC Converter, Orange Micro, $799
VIDI/O Box (NTSC and PAL versions), Truevision, $995, $1995
D5-SCX Video Scan Converter, Dimension 5, $1695

Video In and Out
NuVista (1MB, 2MB, 4MB RAM), Truevision,
    $2995, $4495, $6495
ColorSpace II, Mass Microsystems, $1995
ColorSpace FX, Mass Microsystems, $2995
Genlock Converter, Julian Systems, $999
*Moonraker, Workstation Technologies, TBA

Video Out (RGB and NTSC)
ColorBoard 224, 232, 264,
    RasterOps Corporation, $3995, $2995, $995
SFX, RasterOps Corporation, $1495

from videodisc based to CD-audio based. Categorizing itself as analogous to a book publisher, Voyager adds supplementary material to videos and other work brought in from around the world, generates its own video productions, and distributes its own and others' works.

Voyager also sells videodisc and CD ROM drivers that can be converted to SuperCard, and the company supports videodisc models that the Apple-supplied drivers don't. Voyager sells The Box as well, an external device that provides computer control over some models of consumer videodisc players, enabling consumers to play interactive multimedia on the same player they bought to play movies.

Voyager's in-house staff of artists, writers, video producers, graphic artists, and HyperCard scripters collaborate with recognized authorities in specialized fields such as music or art history to produce the informative, visually appealing, and comprehensive products for which the company is known.

But making the products is one task, distributing them is another. Not yet comfortably positioned in computer software stores, video rental outlets, bookstores, or record stores, Voyager sells some of its titles in videodisc stores and most of them through direct sales. Voyager would like to see a new kind of store emerge, where you could buy multimedia hardware and software in one convenient place.

Another pioneer in the multimedia market is Warner New Media, which is producing innovative CD-audio and CD ROM products, including a three-CD set on Mozart's *The Magic Flute*.

Is It Realistic Yet?

Multimedia is not just a buzzword or a set of products. It isn't even really an emerging market niche. Rather, it is the convergence of many new, and some not so new, technologies that promise to work together in new and meaningful ways. Multimedia promises to enrich our computing environment and bring multisensory, nonlinear approaches to our work and play. It is expensive, demanding, and potentially rewarding. The appropriate question may not be "Is multimedia real?" but rather, "Is it realistic?"

Until optical read-write technology is easily affordable and technically reliable and real-time schemes for compression and decompression of video become feasible, the storage and transportation of multimedia documents will continue to present problems. One frame of digitized full-motion video, after all, currently takes up 700K of space. Because full-motion video requires 30 frames per second, digitized video on the desktop remains unrealistic.

The amount of disk space and RAM a sound file takes up depends upon the sampling rate that it was recorded at. The sampling rate reflects the number of digital "snapshots" taken of a sound in one second. The higher the sampling rate used, the more room the recording will take up, and the higher the quality of the recording.

Eight-bit sound, adequate for speech and some sound effects, re-

Publisher Jane Wheeler of The Voyager Company recognizes that distribution is a crucial link in the development of a multimedia market.

quires up to 22K of RAM per second to play. CD-quality sound, in which 16 bits are assigned to each sample, is ideal for music and high-quality effects, but needs up to 85K of RAM per second. Until recently, these storage requirements have made it difficult to add more than brief sound bites to presentations. New software compression schemes now make it possible to compress sound files for storage, but they do degrade the quality of the recording on playback. The Audiomedia board from Digidesign uses a DSP chip and specialized driver software that allow real-time playback of CD-quality sound from a hard disk, making the Mac competitive with the Next ma-

Voice Annotation
*Voice Navigator, Articulate Systems, TBA
*Voice Navigator XA, Articulate Systems, TBA

High-End Animation
DQ-Animax, Diaquest, $2995
*Electric Image Animation System, Electric Image, TBA
The Animation Stand, Linker Systems, $7495

High-End Authoring
Authorware Professional, Authorware, $8050

Videodisc Authoring
Mentor/MacVideo, Edudisc, $695
Video Builder, Telerobotics International, $695
Video Builder Color, Telerobotics International, $995
VideoDisc Writer, Whitney Educational Services, $589

Videodisc Production Tools
HyperCard Videodisc Toolkit, APDA-Apple Computer, $40
Interactive Video Design Toolkit, Electronic Vision, $395, enhanced version $495
Storyboarder, American Intelliware Corporation, $495
The Voyager Video Stack 2.1, The Voyager Company, $99.95

High-End Video Editing
AVID/1 Media Composer, AVID Technology, $50,000–$85,000
Midas 1, Seehorn Technology, $10,000
The Worx, Julian Systems, $9995
Quicksilver Videodisc Editing System (software), Sundance Technology Group, $3950

*Not shipping at press time.

333

chine in audio processing, and making it feasible for desktop developers to include long sound sequences in multimedia documents.

But even if you've digitized still frames of your videos and only used snippets of sound to save disk space, a five-minute multimedia presentation incorporating 3-D animation and 8-bit color could easily take up several megabytes. Traveling with your presentation means lugging an expensive and fragile hard disk around or investing in a removable system.

Worse still, if you want to show your presentation to more than a few people, you're going to have to spend the money to transfer it to videotape and rent either an expensive video-projection unit or a 37-inch monitor. Either way, you have to decide whether the increased interactivity, animation, and sound really make the added cost and setup time worth it.

Multimedia, however, *is* the wave of the computing future. The speed, storage, and price constraints that currently limit multimedia productions will largely evaporate as the technology evolves. While the task of producing full-fledged multimedia productions will continue to fall to dedicated developers, more and more individuals within the education, training, and business fields will undoubtedly invest the time and money to create lower-end multimedia presentations that will assist them in communicating with and persuading their target audiences. **M**

BE YOUR OWN PRODUCER

CD ROMs hold over 600MB per disc and are perfect for the distribution of multimedia applications that incorporate animations, graphics, audio, and text. For those who want to use full-motion video in their multimedia presentations, videodiscs are the preferred medium.

CD ROM pressing plants accept data on tape or hard disk. Mastering prices range from $1500 to $3000; replication and packaging prices depend on the total number of discs pressed. To produce from 1 to 500 discs, the price per disc starts at $2.35. As the volume goes up, the price per disc goes down. A one-time-only disc for last-minute testing and fine-tuning, costs from $500 to $1000.

Videodiscs haven't had the impact on the consumer market that their cousins, VCRs, have, but they do offer high-resolution pictures and CD-quality sound. In addition, you can access the information on a videodisc in a random manner, and the disc doesn't wear out. Authoring systems such as HyperCard and SuperCard control the access to specific frames or sequences.

Consumer versus Industrial

There are two kinds of videodisc players on the market. Consumer-level players are generally priced from $400 to $2000 and are designed primarily for viewing movies and listening to audio CDs. Industrial-level players, starting at $1000, have access times three to four times faster than consumer-level players. They also have an RS-232C port for interfacing with a computer, and an internal microprocessor that allows programmed control of the videodisc.

CLV versus CAV

Videodiscs can be recorded in two different ways. The CLV (constant linear velocity) format provides up to 60 minutes per side of straight play on a 12-inch videodisc but has limited interactive capabilities. You don't have random access to the material (except on the most expensive industrial-level players), and you can't freeze frames or view material in slow motion, although you can still scan through the material sequentially and access it by chapter.

In contrast, CAV (constant angular velocity) format provides only half the storage capacity—30 minutes of full-motion video per side of a 12-inch disc—but it does offer interactivity. Depending on the player you're using, you can randomly access any of the 54,000 frames per side, view material in fast or slow motion, step through frames backwards or forwards one frame at a time, or freeze single frames. Because of this interactive capability, most computer-based multimedia applications use the CAV format.

Both CLV and CAV offer dual audio channels of 30 minutes per side, allowing you to record two audio tracks for the same video sequence. You can have narration in two languages or at two levels of difficulty, for instance, without having to rerecord identical video tracks.

Interactive Programming

There are three levels of interactivity available to videodisc producers.

Level I lets you place automatic stops and chapters in your material during mastering. User interaction is limited to playing and searching for material using these markers, and input is usually limited to a remote-control device. Level I disks are meant to be played on standard consumer-level players.

Level II contains programming, located in one or more segments of the second audio channel, that controls the videodisc. The data is preloaded into the videodisc player's memory. Normally Level II programming requires the use of an industrial player.

Level III videodiscs are completely controlled by an external computer. All the data necessary to do this is loaded into the computer from another source, not encoded on the videodisc itself. Since a computer is controlling the play, more-sophisticated levels of programming control video playback and user interaction. Level III allows more flexibility in programming, since the controlling information is not pressed onto the videodisc.

Production Costs

While the main cost of any videodisc production is in scripting, shooting, and editing, the following is a list of the major postproduction costs.

Premastering You must deliver your material to a videodisc production facility on professional-quality videotape. This necessitates the use of a postproduction facility where, at the very least, you transfer your videotape to a 1-inch or ³/₄-inch master at a cost ranging from $150 to $400 per hour.

Producing a check disc A one-time-only copy of your material for use in development of the final version, or for limited distribution, costs from $300 to $750 dollars, depending on your choice of a glass or plastic disc and of the manufacturing plant.

Mastering the disc Transferring your data to a glass master costs from $1800 to $7000 depending on turn-around time (faster costs more) and interactivity level (Level II costs more).

Replicating the disc Stamping out multiple copies of a mastered disc costs from $9 to $21 per side, depending on the total number of copies.

MULTIMEDIA

is today's message

by John A. Murphy, Wohl Associates

Multimedia is the hottest new concept in the personal computer (PC) world. With multimedia, you use a PC to construct and convey messages—such as presentations and programs—that mix a number of media, including audio, video, text, numerical data and graphics.

These days, almost every computer trade show boasts multimedia exhibits. And PC vendors are starting to get into multimedia in a big way.

For example, Apple Computer has been pushing multimedia as a market for its next-generation Macintosh PCs. And IBM has embraced multimedia as a future direction for its PS/2 family of PCs. In fact, IBM's PC chief, George Conrades, put on a multimedia show to dazzle attendees at his keynote address during the Comdex trade show last November.

Unfortunately, many proponents of multimedia have been heavy on hype and light on details. They've spent time and money building impressive demos and a few specialized applications. But they've done little to explain the advantages of multimedia to the average user.

What's it all about?

Multimedia promises to become the medium of the future. Its impact is equal to that of film or television. And, like most print and film media, it may be applied to a variety of markets—such as entertainment and education—and applications—such as training, presentations and promotions.

Yet, PC-based multimedia is richer and more versatile than conventional media. Information conveyed by print, film or video is basically fixed and linear in its physical and logical structure.

Text, narrative and picture images are cast in a straight, one-dimensional, start-to-finish format. You cannot change or abridge the information, and your ability to search its content or interact with it is limited.

In addition, you must work with information in a passive manner. Unaided by any intelligent store and search capability, you are restricted to browsing or scanning on a simple linear forward-reverse basis.

Multimedia, on the other hand, exploits the PC's ability to capture, manipulate and display information. When outfitted with the appropriate accessories, the PC offers a fully interactive, intelligent interface to store, script and search on a user-customized basis.

Multimedia can be structured with multiple-index, cross-reference or other automatic-access links. These links provide tremendous flexibility in finding specific information.

The automatic-access abilities of a PC are really put to good use in a

Reprinted with permission from *Office Technology Management* (formerly *Today's Office*), pp. 7–14, Feb. 1990.

form of multimedia called hypermedia. Here, the message is multidimensional and associative in its scope and content. It is built with a variety of independent but interrelated sequences or scenarios, plus a myriad of logical cross-reference points to link them together.

A PC with multimedia capabilities lets you interact with the web of related information sequences that make up a message. It also allows you to connect with other external or remote messages or information sources.

And you can branch out from link points in one information sequence to explore another sequence—one that might elaborate on, or update, a related subject.

Making multimedia

If a picture is worth a thousand words, what is a multimedia presentation worth? That really depends on how well it is produced.

Multimedia has to conform to the ways the audience may think, and to any knowledge they have about the topic.

In the wrong hands, multimedia can easily become muddled media. Garbage in, garbage out is an apt description of the misuse of multimedia.

Most of us have suffered through homemade movies that are cute to their creators but deadly dull to everyone else. And most of us have been forced to witness disasters produced on desktop publishing systems.

To be effective, multimedia must have something to say. But it must also say it in a way that is entertaining, informative and easy to access. And it must be structured

with the appropriate associations so that related topics can be easily accessed and explored.

In addition, multimedia has to conform to the many ways the intended audience may think, and to any knowledge they may have about the topic. Only then can the PC become what science historian James Burke calls "the connection machine, a symbiotic, knowledge filtering and buffering engine that interacts with and augments the user's own memory and thought processes."

Such freedom of expression and choice is not easy to achieve. A multimedia publisher must combine the creativity of an artist and author; the talents and knowledge of a scholar; the skills of a film director, editor and technician; and, since the making of most multimedia doesn't come cheap, the bankroll of a Hollywood mogul.

Few individuals or small organizations have all these abilities and resources. So, in this the formative stage of multimedia, most of us will have our first exposure to multimedia as viewers rather than as producers.

Multimedia publishers

A few mass-media publishers and communicators have already jumped on the multimedia bandwagon. Here's a sampling of these pioneer multimedia producers and their "edutainment" products.

● McGraw-Hill/Datapro has repackaged and published, in multimedia fashion, some of the information contained in its computer hardware and software, print media and product directories.

● ABC News InterActive has published interactive video disk documentaries for high school and college students.

● Time Warner New Media has introduced a multimedia version of Mozart's opera *The Magic Flute*. The music, contained on a compact disk (CD), is synchronized

Major markets and applications for multimedia include the following:

● *Business and manufacturing:* Applications include customer presentations, product demonstrations and training; point-of-sale order-entry systems; worker, operator and production-line training and simulations; parts,

---

Multimarket multimedia

---

repair and installation manuals; and client, management and stockholder reporting.

● *Education:* In this area, multimedia is used for self-instructional and learner-customized coursewares; news and current events tutorials; and training of teachers.

● *Technical:* These applications include medical training, diagnostics and record keeping; weather mapping and simulations; geophysical surveying, mapping and simulations; and dynamic computer-aided design and computer-aided manufacturing drawings and simulations.

● *Consumer:* Multimedia can be used for home tutorials and training; armchair travelogues; current events, fine arts and other "edutainment" presentations and programs; interactive, multidimensional dictionaries, thesauruses and encyclopedias; and multiscenario novels and game playing. —*J. A. M.*

with the opera's libretto, which is displayed on an Apple Macintosh.

• Apple, Lucasfilm and *National Geographic* magazine are planning a joint multimedia venture, Geography TV (GTV), to be published this spring. GTV is a geohistory of the United States that combines stills from *National Geographic's* photo library with narrative, text and illustrations.

• Grolier Electronic Publishing has published a multimedia CD of its encyclopedia. It couples word-subject nodes with pictures, slide shows and audio tracks.

• *Encyclopedia Britannica* has a similar version of its *Compton's Encyclopedia.* The text is augmented, multimedia-style, with illustrations and audiovisual animation sequences.

In addition, a few independent producers and some educational and corporate organizations are busy creating multimedia.

• The Harvard Business School plans to issue a series of multimedia programs on management issues. And it has just released a program dealing with crisis management during the Three Mile Island mishap.

• Robert Abel, an independent producer of television commercials, has created *Guernica,* a prototype multimedia documentary that covers the life, works and times of Picasso.

• *Palenque* is a pilot program developed by the Bank Street College of Education. It takes the user through a travelogue of ancient Mayan sites, using text, still pictures, audio and video images.

• The Children's Television Workshop, better known as the Sesame Street folks, has developed a pilot program called *Words in the Neighborhood.* Designed for preschoolers, this multimedia venture links a video "wordbook" with footage taken from the Sesame Street TV show.

And companies as diverse as Bethlehem Steel, DuPont, General Motors, Domino's Pizza, General Telephone & Electric, Arthur Anderson, General Electric, Ogilvy & Mather and Steelcase have begun to use multimedia in corporate training, reporting, promotion and analysis programming.

Presenting platforms

A variety of technologies constitute multimedia systems. Here's a glimpse of what they now include and what they may include in the future.

Currently, developers of multimedia hardware and software are focusing on the installed base. Sony, Philips, NEC and other manufacturers of consumer CDs and video game players already have mass-market multimedia products for home users on their drawing boards. However, most multimedia today is intended primarily for PC users.

> **The preferred multimedia platforms are the Apple Macintosh and an 80286 or higher-level PC from IBM or a vendor of compatible PCs.**

Consequently, the preferred multimedia platforms are the Apple Macintosh and an 80286 or higher-level personal computer from IBM or a manufacturer of compatible machines. The Commodore Amiga and computer workstations such as NeXT's have good to excellent graphics and audiovisual capabilities. But they are not as widely used as Apple, IBM and IBM-compatible PCs.

Almost all platforms lack full multimedia navigation and presentation skills in their off-the-shelf configurations. Most require pe-

---

Any multimedia message is ultimately a personal learning adventure. Consider a multimedia encyclopedia, in which the entries are show-and-tell scenarios composed of text, figures, still pictures, narration, music, animation and video.

Gothic interludes

Each scenario has a number of subject nodes at various word or presentation points. These subject nodes are links to other related scenarios.

How would you use this encyclopedia? Let's say that while reviewing a general sequence dealing with Gothic architecture, you encounter a brief blurb on "flying buttresses." Mystified, you interrupt the presentation at this node point and branch off to review an allied sequence that elaborates on this subject.

You then return to the point of interruption and continue with the general presentation. Soon you realize that the scenario doesn't delve deeply enough into why gargoyles were used as decorative art on Gothic cathedrals and town halls.

Branching off once again, you bring up another sequence devoted to these stone demons. Navigating further, you decide to call up visuals that illustrate the use of gargoyles in drain spouts, friezes, rooftop sculptures and other utilitarian capacities.

You can continue this self-paced walk through the nooks and crannies of Gothic architecture, stopping at link points to seek out additional text, narrations and audiovisuals. —*J. A. M.*

ripheral equipment to perform multimedia tasks.

One factor to consider is that a mixture of text and still-image graphics is the least expensive multimedia package to implement. Multimedia consisting of audio and full-motion video is the most expensive.

Your hardware requirements will vary, depending on whether you simply want to view someone else's multimedia message or create your own. To *view* multimedia presentations, you'll need more PC memory and greater storage capacities, plus a variety of peripheral equipment.

These peripherals might include a CD drive, video disk or tape player; an audio output card and speakers; a high-resolution color computer display or television monitor; and a video output or audio-video decompression card. One or more of these hardware op-

Your hardware requirements will vary, depending on whether you simply want to view someone else's multimedia message or create your own.

tions might add from $1,000 to $5,000 to the cost of a platform.

Hardware requirements for *creating* your own multimedia presentations are even greater. They depend on the complexity of the application and the media mix it employs.

Typical add-ons include a document image scanner; a still-video frame grabber or full-motion video capture card plus camera or camcorder; an audio digitizer; an audio-video compression card; and other audio-video, input and editing hardware. These devices can add from $2,000 to $30,000 or more to the cost of a platform.

Selecting software

Software that can be used to drive these platforms includes Apple's HyperCard and Silicon Beach Software's SuperCard for the Macintosh. Software for IBM and compatible computers includes Brightbill-Roberts' HyperPad, IBM Educational Systems' LinkWay and Cognetics' Hyperties. Owl International's Guide is available for both IBM and Macintosh PCs.

These and similar programs are used to build and link multimedia sequences with input from a variety of internal and external sources. For example, text, data and images can be obtained from software packages, such as word processing, desktop publishing, graphics, computer-aided design and map-making programs.

Input can also come from more specialized software programs, such as the following:

● *Presentation or slide show software:* Packages available for the Macintosh include Microsoft's PowerPoint, Cricket's Presents and Aldus's Persuasion. Similar software for IBM and compatible machines includes IBM's StoryBoard, Lotus Development's Freelance Plus, Xerox's Presents and Software Publishing's Harvard Presentation Graphics.

● *Animation software for building live action, text, title and image presentations:* Packages for the Mac include MacroMind's VideoWorks and Electronic Arts' Studio/1 or Studio/8. AutoDesk's Animator runs on IBM and compatible PCs.

● *Three-dimensional (3D) modeling software that can create, rotate, zoom and animate complex object or figure simulations:* Paracomp's Swivel 3D is designed to be used with the Macintosh, while Pansophic Systems' 3D Graphics Generator runs on IBM and compatible machines.

● *Audio or voice processing and audiovisual animator software that mixes sound tracks or combines audio with animation:* Packages for the Mac include Farallon Computing's SoundEdit or Screen-Recorder, Digidesign's Sound Designer and MacroMind's Director. Paul Mace Software's Grasp runs on IBM and compatible PCs.

Maximizing multimedia

The potential of multimedia is vast. It may one day revolutionize the way information is assembled and absorbed.

But multimedia is still in its infancy. A number of design and applications problems have yet to be resolved. For example, developers have to figure out how to cost-effectively digitize, store and display full-motion video.

Multimedia is also beset by a lack of widely accepted standards. And with multimedia developers and producers going their own ways for their own markets, it will take time to resolve these issues.

So, for the present, taste but don't drink too deeply from the multimedia well.　ō

Murphy is a senior staff consultant and director of special projects for Wohl Associates, an office and computer systems consulting firm in Bala Cynwyd, Pennsylvania.

THE MAIN EVENT

BY NICK ARNETT

Multimedia

is becoming

the hottest

ticket

in town.

Here's what

you can do

now, and

what's ahead

You're an architect, and your firm's been hired to design an office building with a sophisticated air-conditioning and ventilation system. So what's the best way to present your design ideas to the client for approval?

You could build a scale model and create a series of sketches or slides that show how the ventilation system will work. Or, you could wow the client, instead, with a three-dimensional, sight-and-sound, interactive, animated multimedia presentation.

Multimedia—that's the term given to a presentation combining sound, graphics, and text generated on a desktop computer for display on a computer or video screen.

Multimedia scintillates, educating and informing in a way that directly involves the viewer. For example, The Sigel Group, an architectural firm in a Philadelphia suburb, uses multimedia to show off its building designs for its high-technology and pharmaceutical clients. The buildings typically feature complex ventilation systems, and what better way to demonstrate those systems than with the movement of a multimedia presentation? (For more on The Sigel Group's multimedia work, see the sidebar "The Amiga: Designed for the Job.")

If you've ever played a computer game, you know what multimedia can do. Microsoft's Flight Simulator is one of the most popular mass-market multimedia presentations; aside from being a game, Flight Simulator teaches you how to navigate an airplane and read an instrument panel.

For the viewer, multimedia combines the audiovisual impact of television with the power of a computer. For those whose business it is to make sales presentations, create advertising storyboards, or provide any kind of information, multimedia is another means of publishing, another method of persuasion.

Multimedia's emergence is the result of several technologies maturing and mingling: the visual desktop computer (the Macintosh, the PC under Windows or OS/2 Presentation Manager, the Amiga), CD ROM disks, audio compact discs, scanners, laser videodiscs, and large storage systems such as Write-Once, Read-Many (WORM) optical disks.

With this technology and with software such as Autodesk Animator and MacroMind Director, you can create animated presentations for a fraction of what it used to cost to hire a video firm or advertising agency to do the job. And as a desktop publisher, your investments in computer equipment, along with your experience in combining text and graphics, give you an advantage over those people moving to interactive multimedia publishing from paper, video, film, or other media. After all, desktop publishing, with its merging of words and images, was one of the first applications of multiple media on the personal computer.

The Amiga: Designed for the Job

The Sigel Group, an architectural firm near Philadelphia, designs buildings for high-technology and pharmaceutical companies that require buildings with complex air-conditioning and ventilations systems.

To produce color prints and video animations for proposals and works in progress, the Sigel architects use the Commodore Amiga. "Our color output has never failed to astound people," says Paul Miller, the Sigel architect who brought Amigas into the company.

Miller typically spends two to three hours producing each minute of finished animated video. To create the animations, he downloads files to the Amiga from the firm's mainframe computer.

Sigel's systems each cost about $6,000, and consist of an Amiga 2500 with 3 megabytes of memory and a 45-megabyte hard disk, a video card to remove interlace flicker when producing videotapes, a Xerox ink-jet printer for color stills, and a "genlock" video card for animation overlays.

For producing videos, Sigel uses industrial-grade VHS and Super-VHS videocassette recorders and an editing controller, which together cost about $2,700, Miller says. The company also has Hewlett-Packard DeskJet printers for 300-dots-per-inch printing. Sigel's primary software tool is Electronic Arts' DeluxePaint III, which can produce simple animations with color cycling or more complex "real" animations, Miller says. The architects create 3-D modeling effects with Modeler 3D and Videoscape, both from Aegis Corp.—*Nick Arnett*

For Brooklyn Union, Multimedia's a Gas

At three Brooklyn Union Gas offices in Greater New York, you can walk into the lobby and find an IBM PC AT with an IBM InfoWindow interactive video touch screen and a Pioneer 6000 laserdisc player. By touching the InfoWindow screen, you can select information on gas safety, energy conservation, and financial assistance for bill payment.

Instead of staring at static computer menus, though, your eyes are treated to a kinetic, interactive presentation of full-motion video, photographs, computer graphics, text, and sound. And you can choose to hear the presentation in English or Spanish; at some kiosk locations, the choice is between English and Yiddish.

At the same time, the computer gathers public-opinion data about the quality of Brooklyn Union Gas's services, information which is then compiled and forwarded to the company's marketing department.

The multimedia system was developed by Fusion Media, an interactive videodisc design and development firm in New York, using the IBM InfoWindow touchscreen. The system is based on IBM PC ATs with 20-megabyte hard disk drives and Pioneer 6000 laserdisc players. To develop the presentation, Fusion Media used IBM's InfoWindow authoring software, the InfoWindow Presentation System, and PC Paintbrush; the firm also developed its own software to handle the gathering of public opinion from the touchscreen responses.

"We like the fact that interactive video is easy to use," says Dan Cavanaugh, Brooklyn Union Gas' director of district offices, customer inquiry department. "It's informative, and graphics can communicate certain information much better than text alone. Full-motion video, with the user's choice of language, lets us communicate with our customers in the most familiar way."—*Nick Arnett*

WHAT YOU CAN DO NOW

The natural next step from desktop publishing is to desktop presentations—electronic slide shows created with such software programs as Software Publishing Corp.'s Harvard Presentation Graphics for IBM PCs and compatibles, and Microsoft's PowerPoint and Aldus Persuasion for the Mac.

If you can desktop publish good-looking, easy-to-read documents, chances are you have the skills to create quality desktop presentations. Most presentation programs already allow you to cycle through slides rapidly, creating a sense of animation; from there, you're on your way to a multimedia presentation.

Beyond simple desktop presentations, there's interactive video and animated audiovisual presentations. Until just recently, you usually had to use a two-screen system for interactive video—one for computer text and graphics, the other for motion video, which would play back video from a laser disk. Today, however, new frame-grabber video cards, such as VideoLogic's DVA-4000/MCA card for IBM PCs, and Orange Micro's Personal Vision card for the Macintosh, enable you to show both computer animation and motion video on a single screen.

For the most part, interactive video has also required two kinds of disk drives: magnetic for digital data, and optical for analog video. New video technologies such as Intel's Digital Video Interactive (DVI) for IBM PCs and PS/2's and compatibles, and Philips and Sony's stand-alone Compact Disc-Interactive system are eliminating the need for a videodisc player. Each can produce full-motion video on a standard CD ROM disk by means of powerful, but time-consuming, data compression.

With all this potential, what kinds of multimedia presentations can desktop publishers expect to create today? And which of the three main multimedia desktop systems—PC, Macintosh, and Amiga—is best for the presentation you have in mind?

Desktop publishers with Macs or Amigas should find it reasonably easy to move into low-end multimedia presentations, as these two systems supported graphics and audio since their introductions. To generate a multimedia presentation on a PC, however, you'll need more hardware and greater expertise, as the PC isn't as graphically equipped as the other two systems are.

Creating multimedia presentations on the desktop, in fact, may require some steep investments in hardware and software. The necessary audio and video peripherals and storage subsystems can add $10,000-$20,000 or more to the cost of a desktop system. For an extra $10,000 worth of equipment, you can have animation and sound—but the cost of adding video to your system jumps your overall additional expenses up toward $20,000 and beyond.

On the Macintosh, for example, you can create a simple, 5-minute narrated overview of your company to show to new employees by combining black-and-white slides, animated titles, and text in MacroMind Director, and adding sound with Farallon Computing's MacRecorder. With those two programs, you could fit that presentation onto one 800K floppy disk and run it under Apple's Hyper-Card. If you want something more advanced, though—an animated color presentation, say— you'll need a little more expertise and a lot more storage (about 2 megabytes).

On a PC system, Autodesk Animator combines power with an ease of use greater than that of its software predecessors for PC animation. Animator's ability to play an animation directly from a PC's hard disk drive, rather than by first loading the animation into memory, means you can make longer-running animations on PCs with less memory than you'd need on a Mac or an Amiga.

Some say the Amiga is better suited to multimedia than are the Mac or the PC. The Amiga computers include three coprocessors: one for sound input and output, and two for graphics. With these coprocessors, real-time animation with sound doesn't slow the computer down. In addition, the Amiga uses the standard NTSC video as its display-monitor output, so moving a computer-generated presentation onto videotape is easy.

Two of the Amiga's best-known multimedia programs are the Viva, from MichTron, and Intuitive Technologies' UltraCard.

Say you want to create an interactive product catalog for a furniture company. You can shoot a video of each product in use (such as a man reclining in a chair), store the audio and video on a laser disk, create a touchscreen menu in Electronic Art's DeluxePaint III software, and pull it all together as an Amiga presentation using the icon-driven Viva. (According to MichTron, a Windows 3.0 version of Viva is in the works for the PC, and a high-end Viva Professional program is in development for the Amiga.)

At $49.95, UltraCard is Amiga's alternative to HyperCard; both programs are similar, but Ultra-Card includes more sound capabilities.

The Amiga's major benefit, aside from video capabilities, is the low cost of its hardware and software. If you're planning to add computer graphics to a video presentation, for example,

the Amiga offers the best performance at the lowest price. The Amiga 500 costs about $1,000 in a retail store, while the Amiga 2000, with nine expansion slots, costs about $2,160 (both prices include color monitors).

Also, the Amiga is fairly easy to use. Its graphical interface lacks the refinements and consistency of the Mac's interface, but it's less cumbersome than the PC's text-oriented software.

With some practice and expertise you can create more advanced animated audiovisual presentations with such software as the OWL Int'l Guide for PC and Mac systems, Paul Mace Software's Grasp for the PC, and MacroMind Director and HyperCard for the Mac. For the Amiga, Commodore is co-developing an authoring program that it hopes will have many of the same capabilities as HyperCard.

Of course, software alone won't turn you into a super audiovisual communicator, just as Aldus PageMaker won't automatically transform you into a successful page designer. If designing multimedia presentations were a breeze, in fact, the cost of turning to an independent producer for your presentation would be a lot lower.

Computer animators charge anywhere from $75–$150 an hour, or about $2,000–$4,000 per finished minute of animation, depending on the complexity of the piece. Prices for a complete multimedia presentation can range from $3,500 for a 30-second, noninteractive presentation with animation and sound, to $5 million for an advanced interactive training video.

WHAT YOU CAN'T DO NOW

Despite the promise of creating dazzling presentations, multimedia, like any new technology, has some kinks to be worked out. The Mac and PC operating systems aren't designed to handle the mixture of audio, video, and life-like animation. It's extremely difficult to make sure animation and audio happen at the right speed; there just aren't standard software tools that synchronize the presentation or animation. (The Amiga, though, is better equipped to handle multimedia.) If your computer slows down as you're reading text, that's a drawback; but if a presentation's sound, animation, or images begin to drag during playback, that's a catastrophe—especially if they lose synchronization in the process.

The lack of sophisticated operating-system support, among other things, has made the current crop of audiovisual authoring software difficult, time-consuming, and inefficient. Programs such as MacroMind Director still must undergo a significant revision before the number of desktop multimedia producers ap-

Multimedia Fits at Levi's

Levi Strauss & Co. turned to multimedia when it needed a centerpiece for a "consumer day" exhibit in the lobby of its San Francisco headquarters.

"It was an event that was primarily to educate employees on what we do for consumers," says Brett LaDove, a research assistant for the blue-jeans manufacturer. "I thought multimedia would be a good way to get our point across in a very cutting-edge way," he adds. The presentation did generate a great deal of excitement. "Because multimedia is new and fresh, it was effective. People really wanted to see it."

To create the multimedia presentation, Levi Strauss hired Printz, a San Francisco desktop publishing and multimedia service bureau. Printz created a 4½-minute, 3-D animation on a Macintosh II. "Levi's gave us a bunch of clothing and color swatches that we scanned in black and white, colorized, and texture-mapped to 3D objects," explains Charles Wyke-Smith, one of the principals at Printz.

Two of Printz's animators, each working for about a week, created a 20-megabyte animation, with a musical sound track, that ran on a Mac II and a 19-inch color monitor placed in the Levi Strauss lobby.

Wyke-Smith says typical costs for desktop animation from a service bureau are $500–$1,000 per finished minute, which includes 3-D elements, complex backgrounds, textures, and other effects. It's rare for an animation to run more than six or seven minutes, he says.

Printz's multimedia production system is a Mac II with 8 megabytes, two Apple 13-inch color monitors, a 45-megabyte removable hard disk drive, and an Apple scanner. For video, the agency uses a Computer Friends TV Producer video card and a MacVision digitizer. Their multimedia software includes MacroMind Director, Paracomp's Swivel 3D, Avalon Development Group's PhotoMac, and AppleScan. —*Nick Arnett*

proaches that of desktop publishers. Current software lacks sophisticated object-oriented controls. For example, to make animation easier, you need to be able to make a circle and define how it moves, so that you can create a ball that already "knows" how to bounce.

Memory is another big hurdle for multimedia to overcome. The huge amounts of data necessary for animation, audio, and video rapidly overwhelm the memory and storage capacities of current desktop computers. Storage costs can be hefty, too; to compress DVI video animation on CD ROM costs $250 per minute.

On the Macintosh and Amiga, memory isn't usually a problem—as long as you can afford to buy it. Typically, 4 or 5 megabytes on your system will suffice, but it's difficult to get by with less.

In any case, multimedia usually means large files. The most popular solution to the storage problem is the 45-megabyte removable hard disk, which makes it easy to archive and transport multimedia presentations. When files become too big, it's time to start thinking about storing them on a writable-optical or a CD ROM disk.

It's unlikely that multimedia's memory and storage constraints will soon vanish. As multimedia authoring software moves into its second generation, storage demands will skyrocket. After all, a multimedia document can't be archived on paper; the computer disk is its only link to life.

WHAT THE FUTURE HOLDS

In the short term, easier-to-use multimedia software will create applications such as multimedia electronic mail—a feature that the NeXT computer system already boasts. You'll see an increasing array of multimedia marketing presentations such as Buick's "soft ad," in which the car maker distributed Mac disks that showed off, with animation, its new models. Other possibilities include employee training materials and a host of other reference and educational presentations.

One day, maybe ten years from now, multimedia will bring us to "virtual realities." Imagine that your desktop computer could instantly transport you anywhere from the North Pole to the lunar surface, or that you could meet and talk, simulated face to simulated face, with anyone using a similar computer anywhere.

Virtual reality is the most mind-boggling example of the power of adding audiovisual capabilities, or multimedia, to personal computers. For virtual realities, the computer creates a simulated three-dimensional world on your screen. Autodesk is already developing a software system that lets you "explore" a building that you've designed with a CAD system. Other virtual-reality projects are in development at VPL Systems and NASA, among others.

One of the most important multimedia projects on the horizon is the integration of television and computer technologies to create a new communications medium unlike either one. Full integration of the television and the computer, though, depends heavily on the development and widespread acceptance of a digital network for communication, a technology that's unlikely to mature for a few years.

When that day arrives, though, the personal computers we now use may be as unrecognizable to us as an audio compact disc would've been to Thomas Edison, who thought his gramophone would be used to transmit messages, or the telephone to Alexander Graham Bell, who figured the phone would be a dandy way to listen to opera at home. □

NICK ARNETT is president of Multimedia Computing Corp. in Santa Clara, California, publishers of the monthly newsletter Multimedia Computing & Presentations.

The Complete Picture: The Integration of Computer Graphics and Video

COMPUTER GRAPHICS: KNOW THE BASICS
BEFORE YOU BUY
by John Schwan

Pixels. Rasters. Bytes. These are strange words spoken in the brave new world of computer graphics, where humans and high technology combine to create everything from simple titles to 3-D video animation—not to mention printed handouts, overhead transparencies and slides.

Simply put, computer graphics means using a computer to create graphics that can be output to a variety of media. From that point, however, you step off the beaten path into a veritable jungle of software, hardware and accessories with a range of prices and capabilities. With proper preparation, though, you can map out a trail that will lead you safely to the computer graphics system right for your needs.

To determine your corporation's presentation-graphic requirements, ask current and potential internal users questions such as:

—What kinds of graphic media do you intend to use in upcoming presentations (i.e., slides, video or overheads), and what will you need in the near future?

—How often do you have requests for presentation graphics?

—Do you anticipate these requests increasing or remaining constant in the future?

—Are quality and turnaround time of equal importance or does one factor outweigh the other?

Once the responses are in, you should have a clear idea of your internal market for graphics, as well as the volume and frequency of the graphics requests you'll have to fill. This brings you to the first fork in the road: whether to handle computer graphics requests in-house or farm them out.

In general, volume is the deciding factor. If current demand for presentation graphics is minimal and seems destined to remain so, it probably makes sense to send the work out. But because more business executives are realizing the impact that presentation graphics can have on customers, stockholders and other audiences, chances are you'll see an increasing demand for slides, color hardcopy, animation and other sophisticated graphics. Then it quickly may become more economical to handle this work yourself rather than paying a premium for outside service.

John Schwan is director of product marketing and development at Pansophic Systems, Inc.

If the decision is made to work in-house, the next question is: Who will do this work and how? At this point, it's time to review your survey results. If you find that most users are looking for simple text and chart slides with accompanying hardcopy, it will pay to look into computer graphics packages.

Type, bar, line and pie charts as well as free-form art are all standard features in these easily learned, easily used packages. In addition, they can accept data from popular business software packages such as Lotus 1-2-3 and from word-processing files such as WordPerfect. With minimal training, a business user can create professional-looking slide presentations.

These software packages are relatively inexpensive, retailing for $400–$500, and run on computers found in most businesses—IBMs, IBM clones and Apple's Macintosh line. In addition, these packages work with low-end, affordable film recorders (devices used to develop slides). So you can place packages throughout your company, providing users with quick, easy access to computer graphics.

If users' demands are for more-varied, more-sophisticated graphics, a number of options are available, depending on your implementation time frame, budget and anticipated future needs. The common denominators here are that you're looking for a computer graphics system capable of more than one function, and that you'll be buying hardware as well as software.

For the short term, a good solution might be to piece together graphics software from several different vendors. Such a system might have AT&T's TOPAS package, which can create and animate 3-D objects; a slide/chart package from Zenographics; Truevision's TIPS, a ''paint'' package for drawing; and a TARGA image-capture board, which is something like an artist's palette.

The advantage to such a bit-and-piece solution is economy: It's inexpensive compared to the cost of an integrated workstation, in which a single software and hardware system is used to perform a range of functions. It may cause more than a few headaches down the road, however.

For example, suppose you're dealing with several vendors who aren't necessarily working in concert with each other. One may upgrade a piece of software you're using to the point that other pieces in your system can't keep up with it or don't maximize its benefits. Or you may need to add another function to your system, then learn that the available software isn't compatible with your components. In computerese, this means that your system doesn't have much of a migration

Reprinted with permission from *Video Manager*, pp. 15–35, July 1989.

path; that is, it's difficult to move to a higher level of capability.

Migration may not be a worry if you don't expect to outgrow your system's functions for a few years, but if you expect to take on increasing volume and different functions in the near future, an integrated computer graphics workstation may be your solution.

Autographix, Genigraphic's Infinity and Pansophic's StudioWorks systems fall into this category. Each offers a variety of functions so that you can meet multiple needs. For example, StudioWorks allows a user to create a drawing, such as a corporate logo, turn it into a 3-D object, add a "texture" to it that's been scanned into the system from a photograph, then either output it to video, output it as a slide or drop it into a page layout.

Systems like these benefit your productivity in several ways:

First, the user learns only one "interface," or computer menu, of options and instructions, whereas in a multivendor solution, there may be several to master.

Second, the integration of systems means that a drawing created with one function may be used in all the other function modules, eliminating redrawing.

Third, the system may be used with a variety of accessories, so you can recoup some of your investment in an earlier, less-functional graphics system.

Also, the hardware and software in workstations are sold together, so most vendors work with you to make certain you have the right equipment for your needs.

Finally, you're working with one vendor, ensuring almost painless additions and enhancements to your system.

Whether you opt for the bit-and-piece solution or an integrated graphics workstation, bear in mind that these aren't quite as simple to use as the inexpensive slide packages. Yet they are designed to emulate the tools that graphic artists and designers are familiar with, so the learning curve to begin producing graphics can be as low as two weeks. Depending on their size, shops often cross-train personnel or hire a computer graphics artist.

When you've decided whether a low-end or high-end computer graphics solution is best for your company, your next step is to do your homework. Attend seminars held by organizations such as the National Computer Graphics Association (NCGA) and SIGGRAPH. Go to trade shows and talk to users about what worked for them. Read the trade journals. When you have a better idea of what's available and how it's being used, then it's time to start talking to vendors and to step into the world of computer graphics.

BUSINESS PRESENTATIONS FOR THE 1990s
by Claire Morris and William Jimmerson

Thousands of new business and marketing plans are being developed each year, designed to capture a bigger piece of

Claire Morris is market communications manager at General Parametrics Corp.
William Jimmerson is director of marketing at General Parametrics Corp.

the competitive pie, and a majority of them will be prepared on personal computers.

As users discover and depend on the efficiency and cost-effectiveness of computers in planning and analysis, there is a natural tendency to explore applications of the computer to business presentations. The desktop computer is a powerful, computational/graphical system that can dramatically improve the preparation of visual material and increase the overall effectiveness of business presentations.

While the computer is an excellent tool for authoring a presentation, it falls short of acceptable performance in the conference room. It is far too cumbersome and complicated to use in front of an audience and to move from the office to the conference room and back again. It lacks the necessary speed for smooth transition of images. And most computer monitors' limited color range and low-level resolution make it less than desirable as a presentation vehicle.

For these reasons, the electronic presentation system has evolved to work in tandem with the computer and capitalize on its power and capability. Today's electronic presentation systems eliminate the expense of producing high-quality slides and overheads via outside service bureaus. The $15 to $60 price tag traditionally associated with preparing a single image is now reduced to pennies, and increasing numbers of presentations pundits are beginning to see the purchase of electronic presentation systems and accompanying software as a smart investment.

The investment factor can be largely justified by sheer numbers. The Department of Commerce estimates that more than 16 million people in business and industry give presentations on a regular basis. Herbert Baskin, president of General Parametrics Corp., notes that there are more people in corporate America working on presentations than on spreadsheets. A study conducted by 3M in the mid-1980s indicates that an astonishing 15 million meetings, 8 million sales presentations and 10 million classroom presentations are conducted every working day.

When these numbers are combined with the fact that a 30-minute slide presentation by an outside service can cost several thousand dollars, there is an obvious case for investing in a system that will allow the work to be done professionally for far less. In addition, image updates bring significant additional costs that can quickly eat up audiovisual budgets.

But thanks to modern technology, presenters do not have to be computerphiles to turn out a professional presentation. Some of the new electronic presentation systems allow even computer beginners to successfully teach themselves how to create professional-looking presentations.

The ultimate success of any presentation, of course, is greatly dependent on the complete integration of every element of the presentation, from concept to final creative flair. The content, quality and form of the electronic presentation are all key elements in its impact.

A study conducted by the University of Minnesota shows that when visuals are used, people remember 43 percent more of the content than they do with presentations lacking visuals. When color is used, audience retention and comprehension increase by more than 50 percent. The same study also concluded that without high-quality visuals, even an excellent presenter may be less effective than a poor presenter who uses good visual support.

The future of electronic presentations is promising. John Sculley, president of Apple Computer, said he expects the desktop-presentation market to far outpace the desktop-publishing market in size and importance. Desktop Presentations, Inc. (DPI), a market research firm, projects annual growth of slide production and transparencies at 22 percent and 42 percent, resepectively. These numbers are overshadowed by DPI's forecasts that electronic presentations will grow by almost 1,400 percent over a five-year period.

The increasing scope and importance of information as an emerging industry in its own right has greatly influenced the presentation industry. Some studies indicate that the invention and marketing of information services and systems are becoming more important than traditional manufacturing industries. As an example, it is estimated that more information was printed during the 1970s than in all of recorded history up to that decade.

Where there is information, there is a need to communicate it. We believe electronic presentations will emerge as the most effective means of doing so.

VIDEOGRAPHICS AND PCs: USER COMMENTS
by David Hodes

Personal computers keep getting less expensive but more sophisticated, and their range of application in the field of video graphics continues to expand. PC-created video graphics displaying 3-D animation is available for less than $5,000, and ordinary users are engaging in the kind of inspired, artistic experimenting that used to be the province solely of graphic artists.

Fred Hurteau of Micro Digital Graphics started out in computer programming in 1969 and then, after switching to commercial art, became interested in electronics and photography. "The photography, design and electronics just kind of came together in video," he said.

The Amiga 2000 he has been working with has been a real hit with his clients, Hurteau said. With one of the programs he uses, he can input a two-dimensional object into the computer, and it will extrude the object into a three-dimensional object. He can then make the object tumble and spin, or zoom in and out of it. A 3-D ray-tracing program lets the operator define a 3-D object in space by defining points connected with lines, similar to a CAD/CAM approach. The operator defines colors, surface textures, ambient light and other various light sources.

"Then, you can tell the computer to take that information and render a picture of the object from that position, with that light," Hurteau said. "It will calculate all those rays of light, where they should fall, how they would bounce, where

David Hodes is a video producer and scriptwriter from Overland Park, Kansas. He has written articles for Billboard, Millimeter, Television Broadcast and other video-related trade publications.

you would get multiple reflections and shadows—the whole thing."

The Amiga will automatically anti-alias or, with a paint program, smooth over the edges. Hurteau said he recently purchased a Deluxe Paint III Program ($150), which more than paid for itself after just two uses. "The first animation I did with Deluxe Paint took me just 20 hours, when I would have had to spend several days otherwise," he said. "I'm not even sure I could have gotten it done. I would have had to use two or three different programs to do various parts of the animation."

Deluxe Paint III also allows the operator to create and play full-color animation for several seconds at 30 frames per second (or any other frame rate specified). Any frame can be cut and pasted anywhere else in the animation, and it can be played forward, backward or in a "yo-yo" mode.

Hurteau said that the Amiga software market is expanding rapidly. "The Amiga was designed by an ingenious person who put a lot of capabilities into it, not knowing whether anyone would ever use them," he said. "When the software people got hold of it, they started realizing its potential."

Jim Sweeney of Design Mirage has two Amiga 2000s, a 2000HD and two 2500s. He recalled how he saved $3,000 on a rush print order for a customer. Using Professional Page, he scanned images into a computer, imported them into the professional page documents and then modemed them to a Linotron to produce four-color separations. "In two days, we produced four-four color separations and chromalins," Sweeney said. "And with rush charges and Federal Express charges plus our design time, total start-to-finish expenses came to $1,100."

It's a fun business, Sweeney said, and his company is starting to get more productions and opportunities to demonstrate its capabilities every day. "What we get in the way of local commercials is people who say: 'Look, I had that bad, low-budget commercial five years ago. It was awful and I knew it was. Now I have a few bucks, and I'd like a commercial closer to what I'm seeing on TV.'"

Computer graphics can have many different uses, a point to which Raymond Johnson and Mac Bright will testify. Johnson is director of video for Atlanta Gas and Light Co., and Bright is the owner of Tech Three, a graphics-design firm. Both are users of StudioWorks, Pansophic Systems' integrated multimedia graphics workstation.

"Most of the time, we use the system for producing high-quality 35mm slides for business and sales meetings," Johnson said. "In fact, that's how I justified its purchase to management. Originally, we weren't doing any slide work in-house, but sent it outside and paid high prices for it. Once we began doing slides ourselves—and we're doing three times as many as we did before—the cost savings paid for the workstation in six months.

"The other major function for the system is video animation," Johnson said. "We're responsible for creating a quarterly employee newsletter, The Flame Report. Our video artist, Beth Vickers, completes animation sequences for the newsletter in StudioWorks, then we output them to $3/4$-inch tape."

Johnson said that his department is filling more requests for flat-art illustrations with StudioWorks and sometimes transfers those to the workstation's video module for use in video presentations. "My advice to a shop considering buying a workstation would be to dedicate someone on your staff to learning how to use it," he said. "Without that commitment, it'll just sit and gather dust."

Bright isn't worried about his artists learning the capabilities of StudioWorks; instead, his concern is educating customers to look beyond slides to support their presentations. "We do a lot of slide work on the system but also many other interesting things that most people aren't aware a graphics workstation can do," he said. "For example, we do design visualization. We recently had a major client that needed packaging for a new product. Within four hours, we had created 12 different full-color packaging visualizations.

"Some clients need canned VHS presentations for their sales force to take on the road, so we're transferring more slides to video on the workstation. We're not animating them but using still images that remain on the screen for however many seconds the presenter requires.

"Most people don't understand what computer graphics are good for," Bright said. "You need to educate them about what a system can do, how it helps them accomplish their goals. Sometimes that means you must become more aware of how you can utilize the system. Then you can become a better teacher to your clients."

About half the clients of Producers Color Service, a post-production house in Southfield, Michigan, are in the corporate market, said Scott Gray, supervisor of the facility's Electronic Graphics Department. Producers Color uses Cubicomp's PictureMaker and Vertigo.

"The availability of personal computers has heightened the mainstream knowledge of computer technology," Gray said. "A lot of industrial and agency people are familiar with the style of electronic photographs now. The trend will continue, as both client and consumer discover the best way to reflect their message through animation."

The Sketch Department of Ogilvie & Mather in New York has also acquired Cubicomp equipment. This division uses PictureMaker to create artwork for external presentations, such as sample commercials for prospective clients. "A fresh wind is blowing through the corporate market," said Dick Chandler, O&M vice president and manager of the Sketch Department. "We are able to attain a high-end look at the cost of a moderately priced system."

Chandler has produced innovative animation pieces: For a potential ad campaign for recyclable products, a soft-drink bottle "metamorphosed" into a mayonnaise bottle and then into six other shapes. "With the PictureMaker," he said, "I can animate anything—light source, backgrounds—or even place objects within texture-mapped cylinders. As an artist and animator, I want a computer that doesn't confine or limit my imagination."

Independent producer David Henry used an AT&T PC-based TARGA 16 for title graphics and simple animation that he said "blew away the clients." Henry works mostly with industrial clients but recently completed work on a tape about resume writing. "Since the tape described writing a resume," he said, "I wanted to animate a hand 'writing' the text over a paper-textured background."

Henry used the camera-capture feature of the TARGA to capture his hand and two-thirds of his forearm, holding a red pen against a white background. He painted in all the white around the hand and forearm with TARGA's black, which the system recognizes as invisible when it is saved and loaded as a window and then a certain icon is activated by a mouse.

Then he made 20 TARGA files of a porous paper background, each one revealing more of the title in red Helvetica font with a touch of an airbrush effect to broaden the characters. "I then made 20 TARGA window files of less and less of my forearm, to adjust for the movement of the hand from the lower right-hand corner of the screen to the middle and then left to right, line by line, as it wrote the title over the paper background," Henry said.

With the animation feature on his Paltex Abner, Henry adjusted the preroll time of his VTR so that, after loading each window file separately over each new background and recording it, he would have enough time to load the next window. Then he set the editor in motion and loaded a background file with slowly revealing title, followed by a new hand/pen window. Henry carried the paper texture and animated-hand idea throughout the 20-minute resume-writing program, using it at three different places.

"I used to go to a local production company for bids on their $150,000 Wavefront system," Henry said, "but my clients were shocked at the $1,000–$3,000 price tags on those five- to seven-second effects." Henry was able to duplicate some of the simple effects and graphics needs of his clients, using the camera-capture and paint effects on the TARGA. "We would charge it off as part of the edit-session —around $100 an hour," he said. "Once you build a background or do one effect routine that you save to disk, you can turn out some pretty amazing stuff in four to five hours."

Users of low-cost computer graphics systems caution that low resolution on the video encoding is still a concern, and that most of the available animation looks a little stilted. But as higher-powered microprocessors for PCs, presentation-graphics software and high-quality genlock devices are further developed, users will benefit even more from investigating the world of low-cost, high-impact, video graphics.

Glossary of Presentation Terms

THIS GLOSSARY presents terms related to technical presentations as well as supporting services, such as text processing, typography, illustrations, photography, video, visual aids, computer graphics, and desktop publishing. These terms include the special terminology, abbreviations, and acronyms used in the papers in this book as well as terminology encountered in other related sources. It is not, however, intended to be a comprehensive dictionary of all possible terms associated with technical presentations.

ACETATE. Transparent plastic sheet used over artwork for protection or as an art separation overlay for reproduction of lettering or color.

ACHROMATIC. Without hue or color. Term used to refer to material, especially printed material, that is black, gray, and white.

ADDRESSABLE RESOLUTION. Describes the number of pixels that are placed on each line of film by the film recorder's light source. The greater the number of pixels, the clearer the image on the finished film medium.

ALIASING. The jagged, stair-stepped effect along the edges of a computer graphic image.

ALPHANUMERIC. A character set containing letters, digits, and other symbols such as punctuation marks.

ANTI-ALIASING. An image-processing technique that reduces stair-stepped effects and makes the edges of a graphic image appear smoother and less sharp.

APERTURE. Size of an opening; the diameter of a lens in relation to its focal length.

AMBIENT. (1) Overall light directed around rather than at a subject. (2) Noise in the background or existing sounds at a location.

ANALOG FILM RECORDER. A film recorder that registers an image as it comes from the computer screen and at the same resolution as the screen.

ANALOG MONITOR. A unit that uses an analog signal and displays an infinite number of shades of the primary colors.

ANIMATION. A process of using a series of still pictures on film or video to simulate the effect of motion when projected.

ANSI. American National Standards Institute. The organization responsible for most standards used for U.S. audiovisual and computer equipment.

APPLICATION PACKAGE. A commercially available computer program designed for a specific task or function.

ARROW. Leader line with arrowhead extending from nomenclature (callout) to an identified part on a drawing or photograph.

AREA CHART. A type of graph that uses the space below the line in a line chart to represent the desired values.

ARTIST'S CONCEPT. A pictorial representation of an equipment item, system, or scenario as it will look when it is designed and manufactured.

ARTWORK. An original illustration or design, or a group of drawings, photographs, hand lettering, or type, prepared for reproduction by any of the standard reproduction processes.

ASCII. American Standard Code for Information Interchange. A seven-bit plus parity code established by the American National Standards Institute to achieve compatibility between data services.

ASPECT RATIO. The ratio of the horizontal to vertical dimensions of a frame or image.

AUDIO. The audible part of a presentation, including background sounds, music, dialog, and narration.

AUDIO TRACK. The portion of a videotape that carries the audio signal.

AUDIOVISUAL. All forms of prepared or recorded material projected, displayed, or replayed during a presentation.

AUTHORWARE. Development software programs used to produce interactive multimedia presentations, specifying elements that need to be included.

AXIS. A line that is fixed, along which distances are measured or to which positions are referenced.

BALANCE. In page composition, the overall visual relationships and weights of elements on a page.

BANNER. A large headline or title. Originally, banner referred to headlines that extended across the full page, but it is now used to indicate any large head.

BAR CHART. A type of graph that employs horizontal or vertical bars of various lengths to indicate comparative quantities or values.

BETA. A videocassette format developed by Sony; not compatible with the VHS format.

BIBLIOGRAPHY. A listing of books or articles on a particular subject.

BIT. A single binary digit.

BIT MAP. A method of creating images by assigning an individual memory location for each picture element (pixel) on the screen. Image resolution is expressed in pixels (picture elements per inch).

BLEED. (1) Any image that extends to the edges of a page or beyond the final trimmed edge of the printed sheet. (2) Any area of continuous-tone artwork that extends outside the crop marks or beyond the trimmed edges of a printed page.

BLOCK DIAGRAM. Diagram showing the relationship and information flow of one part or function to others in a complete unit.

BLOWUP. Mechanical or photographic enlargement of an illustration or photograph. In electronic imaging, the process is often called ''scaling up'' the image.

BOARD ART. Camera-ready illustrations, photographs, graphic designs or type, or a group of these, mounted on a poster board.

BOLD. A way of emphasizing or highlighting text characters or words on paper or on a video display screen.

BOLDFACE. A thicker, more prominent version of a typeface; used for emphasis, headings, subheadings, key numbers, and other special elements.

BRILLIANCE. The quality of an image that has the right amount of tone in all areas.

BUILD-UP. A method of presenting topics by adding one line at a time as the topic is presented.

BULLET. A dot or filled-in circle, in varying sizes, used to emphasize, code, or separate data in text, tables, and illustrations.

BYTES. A number of bits, usually eight, that represents the number of pixels displayed on a screen, both horizontally and vertically.

CAD. Computer-aided design. The production of drawings and plans for architecture and engineering systems using computer graphics processors.

CADD. Computer-aided drafting and design.

CAE. Computer-aided engineering.

CALLOUT (illustration). Numbers, letters, symbols, or words placed within an illustration to identify parts, assemblies, or locations, or to relate illustrated items to text or legend.

CALLOUT (text). References in text to identify related illustrations appearing in the same document. Callouts are also used in composing a page to determine the proper placement of illustrations.

CAM. Computer-aided manufacturing.

CAPTION. The title, explanation, and/or identification accompanying an illustration.

CCD. Charge-coupled device. A light-sensitive electronic plate replacing video tubes in most camcorders.

CCTV. Closed-circuit television.

CD-ROM. Compact disc–read only memory. A format standard for placing digital data on a compact disc.

CEL. A single drawing or frame in an animation.

CELL. The storage position of one unit of information; such as a character, bit, or word. Also used to describe a precise location in an electronic spreadsheet where specific row and column coordinates intersect.

CHARACTER. An individual letter, number, symbol, or punctuation mark in a type font.

CHARACTER GENERATOR. A device used to electronically produce titles and other graphic displays directly on a video monitor for use in production.

CHARACTER RECOGNITION. The identification and conversion of printed characters to digital electronic signals by automatic means.

CHART. An illustration containing key words, graphs, and/or illustrations to emphasize pertinent information used for presentation.

CHART ELEMENT. An essential component of a chart or graph, such as a bar or pie segment, line grid element, legend, or coordinate axes.

CHROMA. Chrominance. The amount of hue, and its relative brightness, as measured in an video signal.

CHROMAKEY. Inserting video imagery over a chosen color in a second video source using a special-effects generator or computer.

CLUSTERED BAR GRAPH. Type of bar chart that compares two or three sets of data by grouping the elements together.

COLLATE. (1) To organize and assemble printed sheets in proper order. (2) To verify the order or arrangement of printed sheets within a publication.

COLOR BARS. A standard color chart for adjusting the color balance in a video image.

COLOR CORRECTION. Steps taken in photographic or printing processes to correct the balance among colors to produce copy as nearly like the original as possible, or with special effects.

COLOR CYCLING ANIMATION. Changing the colors of individual pixels to give the effect of movement.

COLOR SEPARATION. The method of separating color art for printing by process color using four continuous-tone negatives (one of each of the process colors used in printing: cyan, yellow, magenta, and black).

COLOR TRANSPARENCY. A transparent photographic color positive used for projection, color separation, or display.

COLORIZATION. Adding color to a black-and-white image, or changing the colors in an existing image.

COLUMN CHART. A chart consisting of vertical bars or columns.

COMPOSITE. The combination of three separate color signals (red, green, and blue), plus timing or sync (synchronization) signals, for delivery over the airwaves or through a single cable.

COMPOSITE ART. The integration or superimposition of line and continuous-tone art within one illustration.

COMPOSITION. The arrangement of all graphic elements in a specific space. In printing, the typesetting of copy into a specified form for publication.

COMPUTER GRAPHICS. The family of computer applications that can produce plots, diagrams, pictures, and type composition either on paper or film.

COMPUTER SIMULATION. The representation of physical processes using a computer program.

CONTINUOUS TONE. An image, such as a photograph or artist's concept, having tones of various gray values or shades of color.

CONTRAST. The relative brightness of the white areas of an image compared to the black areas.

CONTRAST RATIO. The ratio of the brightest possible area to the darkest possible area of an image.

CONTROL TRACK. The portion of a videotape on which all the sync information is recorded.

CONVERGENCE. The proper alignment of the vertical and horizontal lines in a video projection.

COPYFITTING. Estimating the length or number of pages that a manuscript will occupy in the corresponding typeset and printed form.

COPY STAND. Device for holding and adjusting a camera while copying flat art or still items.

CPU. Central processing unit. The part of a computer that controls the interpretation and execution of instructions, including arithmetic and logic capabilities.

CROP. To cut off or mask undesired portions of art work, photographs, or computer displays to fit in the allotted space, eliminate unwanted detail, focus attention on the remaining image, or reduce an illustration to the desired size without losing key details.

CROSSHATCH. A type of shading using parallel lines drawn at right angles to each other.

CRT. Cathode ray tube. A type of display screen, video screen, or monitor used to display text and graphic information.

CUE CONTROL. A device for advancing slides or controlling a tape or film to sample the contents or move to a desired section.

CUT. An abrupt, instantaneous transition between two scenes in a video production.

CUTAWAY. A view with an opening cut in the illustrated item revealing the internal structure.

CYMK. Cyan, yellow, magenta, black. The primary colors of ink that are mixed in four-color printing to produce printed images.

DAT. Digital audio tape.

DECIBEL. Abbreviated dB. A measurement of sound.

DEFINITION. The sharpness or resolution of a picture.

DESKTOP VIDEO (DTV). Video production using low-cost video equipment and desktop computers.

DIGITAL AUDIO. Recording technique that stores sound as a train of numbers for improved fidelity.

DIGITAL DATA. Numerically storable data that can be computer-maintained and processed.

DIGITAL FILM RECORDER. A film recorder that reads files from graphic software and converts the digital data into an analog signal.

DIGITAL SCANNER. An optical reader that scans and converts images into digital form.

DIGITIZE. To convert an audio or video signal from its "analog" form into digital code numbers.

DISPLAY SIZE. The actual screen area of a monitor on which a user can work.

DISPLAY TYPE. Type used for headings, headlines, titles, and signs.

DISSOLVE. An optical effect in which one scene appears to blend into the next.

DISTORTION. A modification of the original signal appearing in the output of audio equipment that had not been present in the input.

DITHERING. Using a paint or image-processing program to blur the transition from one color to another in a computer picture.

DOS. Disk operating system. A set of software instructions used to manage the hardware and logic of a computer.

DOUBLE BURN. The method used to burn both halftones and line copy onto one plate.

DOWNLOAD. To transfer programs and/or data files from a computer to another device or computer.

DRIVER. A computer program used to control external devices or to run other programs.

DUPE. Duplicate slides, videotapes, or audio tapes produced from the original.

DYNAMIC RANGE. The highest and lowest signal levels in a device.

EDIT. Manipulating a manuscript, illustration, videotape, or slides to alter the content (added to, deleted, replaced, extended, shortened, or otherwise changed from its original form), to attain the final form of the production.

EDITING. The process of preparing a manuscript for publication; includes making any necessary editorial changes and marking type style and size on the manuscript.

EIA. Electronic Industries Association. The association that determines recommended audio and video standards in the United States.

ELECTRONIC EDITING. Inserting or assembling program elements on videotape without physically cutting the tape.

EMULATION. To duplicate the behavior of a product or standard.

ENHANCE. The act of altering a basic image to conform to better design standards and visual understanding.

ENLARGEMENT. Copy made larger than the original.

EXPLODE. Style feature to emphasize one element of an illustration by moving it out of the main image.

EXPLODED VIEW. View showing parts and subassemblies physically separated, with each part projected along the center line of its assembled position.

FADE. A gradual increase or decrease of a video or audio signal.

FADE-IN, FADE-OUT PHOTOGRAPHIC. Optical effect in which the picture gradually appears or disappears on a black screen.

FAT BIT EDITING. The ability to magnify a portion of a computer display and edit it pixel by pixel.

FCC. Federal Communications Commission. The government agency that controls the use of broadcasting.

FEEDBACK. The regeneration of sound caused by a microphone's pickup of output from its own speakers, causing a ringing sound or high-pitched the squeal.

FILE. An organized collection of information directed toward some purpose.

FILL PATTERN. Solid color or design provided in software and used to fill in, or give texture or pattern to, elements of a chart or drawing.

FILM NEGATIVE. A transparent material, usually acetate coated with a photosensitive emulsion, in which the light and dark portions of the original are reversed, used to produce a positive print.

FILM POSITIVE. A positive image on photographic film obtained by contacting a photo-negative onto another piece of film, usually emulsion to emulsion.

FILM RECORDER. Device for producing slides or prints of a video or computer graphic image.

FINAL ART. Art that is ready for reproduction.

FINAL PROOF. The last proof before sending material to a printer, showing all corrections.

FLIP CHART. A visual aid, usually prepared on inexpensive light paper or cardboard, most often with the aid of crayons, felt pens, or markers and placed on an easel for viewing by an audience.

FLOW DIAGRAM. Sequential diagram showing direction of flow or sequence of functions.

FLUSH LEFT. Text justified along the left margin, or an instruction to align art or text on the left margin.

FLUSH RIGHT. Text justified along the right margin, or an instruction to align text or art on the right margin.

FOCUS. When the image of an object appears crisp and sharp, it is "in focus." When it appears fuzzy and dull, it is "out of focus."

FOCAL LENGTH (FL). The distance between the focal point of a lens or mirror of projection equipment and the corresponding principal plane.

FOIL. A positive print on transparent acetate, used as master for producing duplicate copies or for illuminated display or projection. Same as overhead or viewgraph.

FONT. The style and size characteristics of type used in both hot-metal and cold-print composition.

FORMAT. The physical makeup of a document in terms of page size and layout, typeface, placement of tables and illustrations, margins, headings, and other similar physical factors.

FRAME. A complete video or computer graphic image.

FRAME ANIMATION. Animation procedure in which a screen image is drawn by a computer, then recorded on a frame of film or videotape.

FRAME GRABBER. Device used to digitize, capture, and store an image from an external video source for display in a computer.

FREEHAND. Drawn by hand without the aid of mechanical or optical aids.

FREQUENCY RESPONSE. The frequency range in audio and video systems over which signals are reproduced within a standard amplitude range.

FRONT-SCREEN PROJECTION. An image projected on the audience side of a light-reflecting screen.

FULL BLEED. An illustration that runs off all four sides of a page after trimming.

FULL-MOTION VIDEO. Video sequences that emulate those normally seen on television.

GAIN. The ability of a device to amplify an input signal.

GANTT. A chart used for project management that shows the scheduling of steps required to complete a project during a specific period of time.

GENLOCK. A service that synchronizes one video source with another for mixing and recording.

GRAPH. Chart or diagram showing values in graphical rather than tabular relationship.

GRAPHICS. Artwork, illustrations, or tables as differentiated from text. Visual communication based on pictorial, symbolic, or abstract forms.

GRAPHIC ARTS. The fine and applied arts of representation, decoration, and writing or printing on flat surfaces, together with the techniques and crafts associated with each.

GRAPHICS DISPLAY. A high-performance display terminal designed for specialized applications.

GRAYSCALE. An even range of gray tones between black and white.

GRID. Horizontal and/or vertical guide or background lines on a graph or chart.

HALFTONE. A reproduction of a continuous-tone image produced by photographing the image through a screen that breaks the gray tones into dots of varying size that represent gray values.

HANDOUT. Copy that is provided during a presentation to supplement or clarify major points.

HARD COPY. Printed output in a permanent readable form on paper or other medium.

HEAD. A headline, as distinguished from the body text.

HIGHLIGHT. The lightest or whitest parts in a photograph.

HORIZONTAL RESOLUTION. Defines the number of pixels that are available horizontally across a video screen.

HUE. A single identifiable color, such as green, red, or blue. The perceived color of an object.

HYPHENATION. Techniques employed by a word processing or composition system to perform line-ending decisions.

ICON. A symbol or pictorial representation of an object or idea. A symbolic image in software that a user selects for executing a desired function.

ILLUSTRATION. Any device by which information is presented to the reader in pictorial (graphic) form rather than in text form.

IMAGE. A likeness or symbol; a visual conception of an idea. In photography, the object counterpart produced by an optical system.

IMAGE PROCESSING. The use of software to manipulate a scanned computer graphic image to add color, increase the contrast of shape boundaries, or eliminate unwanted patterns.

IMAGE RECORDER. A peripheral device that captures and stores digital signals representing graphic images.

IMAGING AREA. The area on which an output device can place

information, most often expressed as the bounds of a rectangle.

IMPROMPTU PRESENTATION. The delivery of a presentation on the spur of the moment without previous preparation, relying solely on skills and information available at the moment.

INDEX CHART. Variation of a line chart used to show relative change; compares two or more series of data measured in different units.

INITIAL CAP. A capital letter used to start paragraphs, chapters, or sections.

INPUT. Information or data transferred from an external source into a computer; a device used to transfer data into a computer.

INPUT/OUTPUT (I/O). The processes involved in transferring information into or out of a computer-based system.

INTERFACE. The connection between two devices, such as a computer and a peripheral.

IRIS. The aperture of a video camera. An iris setting on a video camera is the same as an *f*-stop setting on a 35mm camera.

ITALICS. The slanted variation of a typeface.

ITVA. International Television Association.

JAGGIES. The ragged edges of shapes in computer graphics. Same as aliasing.

JITTER. Instability of an image due to sync or tracking problems.

JUSTIFICATION. The horizontal spacing of lines of copy within an area so that it begins flush left or ends flush right.

JUSTIFY. In composition, to space out lines of type to a selected uniform length, having straight left and/or right margins.

KERNED LETTERS. Letters having a part of their face projecting over the body of adjacent letters.

LAN. Local area network. A communication system that allows computers, terminals and peripheral devices to interconnect and exchange files.

LANDSCAPE FORMAT. A page that is laid out to be turned and read horizontally or sideways.

LANTERN SLIDE. A variation of slide-frame projection employing an optical lantern and lenses.

LAVALIER. A small microphone meant to be worn on the lapel; also called a lapel mike.

LAYOUT. A drawing, sketch, or on-screen representation of a proposed printed piece or visual showing the arrangement of text and pictorial elements on a page.

LEADING. The space between lines of type added to the type depth itself. Pronounced *ledding*.

LEGEND. Explanatory information or definitions attached to a chart.

LINE (video screen). A horizontal row of characters on a display or terminal screen.

LINE ART. Illustrations containing only black and white, with no shades of gray.

LINE CHART. A graphic interpretation composed of lines connecting data points.

LINE DRAWING. Representation of an object's image by entering a solid-line outline of surfaces.

LIQUID CRYSTAL DISPLAY (LCD). A type of display that forms characters by subjecting a liquid crystal solution to an electrical charge.

LCD PANEL. A panel that allows text and graphics information from a personal computer to be displayed onto a large screen or wall using a standard transmissive-type overhead projector as the light source.

LUMAN. A measurement of light emitted by a light source at the point of emission.

LUMINANCE. The range from black through gray to white in a video picture.

MAP CHART. A pictorial graph showing shapes and locations of a designated geographic area.

MASK. An opaque device used to prevent an image area from being exposed.

MASKING. Special-effects procedure involving high-contrast film and multiexposure techniques.

MASS STORAGE. Large-capacity secondary storage systems, such as recording tape and magnetic and optical disks.

MASTER. The original copy of a publication or presentation. In offset printing, the paper plate made from an original.

MATTE. Dull, flat surface without gloss or sheen.

MATTE WHITE SCREEN. Smooth-surfaced screen with high sharpness and relatively low reflectance for wide-angle viewing.

MANUSCRIPT. Original hand- or typewritten text with or without illustrations, prior to final art, composition, and layout.

MEDIA. The vehicles that store or transmit information, classified by source, input, and output.

MEDIUM. (1) The means, technique, or agent employed to produce artwork. (2) A communication channel such as TV, radio, and magazines.

MEMORY. Area of a computer system that accepts, holds, and provides access to information.

MENU. A program-generated list of options presented on the display screen, from which a user can select desired procedures for the computer to execute.

MOIRÉ. In printing, an undesirable wavey pattern produced as a result of improper screen angles during the preparation of a halftone.

MONOCHROME. Images that are only black and white.

MONTAGE. The arrangement of scenes with no continuity of action but a continuity of meaning. Two or more illustrations combined, often with overlapping or blending, to produce the effect of a single illustration.

MORGUE. File of material held for reference or reuse.

MORTICE. Technique of squeezing a video picture and surrounding it with a black or other border.

MULTI-IMAGE. Process of presenting viewers with several images simultaneously.

MULTIMEDIA. The integration of video, graphics, and audio through a computer.

MULTISCREEN PROJECTION. The use of more than one screen or viewing surface for the display of more than one image at a time or in sequence.

NARRATION. Oral description accompanying a slide show, video, or film and usually a part of its sound track.

NEGATIVE. Photographic film or paper print in which the white and black tones are the reverse of the original.

NOISE. In audio systems, noise is electrical interference or any unwanted sound. In video, it refers to random spurts of electrical energy or interference.

NOMENCLATURE. The names, abbreviations, symbols, and/or index numbers assigned to an item or its components.

NTSC. National Television Standard Committee. A broadcast engineering advisory group, which is part of the Federal Communications Commission, that sets standards for video hardware and broadcasting.

ON-SCREEN SLIDE. A graphic image projected directly from a computer terminal.

OPAQUE. Not transparent or translucent.

OPAQUE PROJECTOR. A device that projects a light downward on to an opaque subject image and projects the resulting reflected image on to a screen.

OPERATING SYSTEM. The software that manages a computer's capabilities, performing such functions as process scheduling, file and memory management, and command interpretation.

OPTICAL CHARACTER RECOGNITION (OCR). The technology that enables a scanner to ''read'' printed or typed characters and convert them into digital editable form as input into a composition system or word processor.

OPTICAL DISC. A type of videodisc storage device that records and reproduces digital information using a laser beam.

ORIENTATION. The arrangement of a plotted or printed image horizontally or vertically on a page.

OVERHEAD PROJECTOR. A device that projects an image on a screen by projecting light upward through a transparency into a reflective device and then on to a screen.

OVERLAY. (1) In the preparation of artwork, a transparent or translucent sheet laid over base art to add nomenclature or to provide art for color separation. (2) In offset lithography, a transparent or translucent covering on the copy on which directions or work to be overprinted are placed.

PAL. Phase alternating line. A video standard used in Western Europe, Latin America, Great Britain, South Africa, Australia; it is not compatible with NTSC.

PALETTE. The selection of colors available on an artist's tablet or in a computer graphics system. The collection of colors supported by a particular graphics standard.

PAN. A camera move that swings along the horizontal x-axis.

PASTEL. A soft or light color used by artists.

PASTEUP. The process of manually composing various textual and visual elements, such as type and artwork, on a surface in master form for subsequent reproduction.

PERIPHERAL. A device connected to and operating under the control of a computer.

PHOTOCOPY. A copy produced either by exposure of photosensitive paper through a negative or by direct image projection.

PIE CHART. A graphic interpretation of data in which parts are compared in size or quantity to a whole.

PICA. A unit of typographic measurement equal to 12 points, or one-sixth of an inch.

PIXEL. The basic picture element of a computer screen. The smallest addressable point of a bit-mapped screen that can be independently assigned color and intensity.

PIXELIZATION. Using image-processing software to break up a continuous image into rectangular blocks to give it a ''digitized'' look.

PODIUM. The place from which a speaker delivers his or her presentation to an audience.

POINT. The smallest unit of typographic measurement, equaling 0.01384 inch.

POSITIVE. A photographic image that corresponds to the original, such as black letters on a white background; the reverse of negative.

PORTRAIT FORMAT. A page that is laid out to be read vertically or upright.

POSTERIZATION. Limiting the number of different shades in an image to produce a high-contrast, poster-like effect.

POST-PRODUCTION. Any video process that takes place after shooting—usually editing, audio enhancement, or special effects and graphics integration.

POV. Point of view. The viewing angle and apparent position of a camera.

PRESENTATION. A talk, briefing, or lecture to an audience, usually involving the use of visual aids.

PROCESS CAMERA. Graphic arts camera used to photograph line or halftone copy or to produce color-separated negatives for printing or other reproduction processes.

PROCESS COLOR. Any of the four primary colors used to produce all of the colors in a publication; cyan, yellow, magenta, and black.

PRODUCTION. The complete process of converting copy and art into finished, printed pieces.

PROGRAM. A sequence of instructions supplied to the computer to perform specific functions or tasks.

PROPORTIONAL SPACING. Typed, printed or displayed text in which each alphanumeric character is assigned a weighted amount of space.

RAGGED LEFT/RIGHT. Refers to unjustified (ragged) right or left margins.

RANDOM ACCESS. The ability to retrieve, in any sequence, slides, audio, or videotape segments, regardless of the original chronology.

RASTER. The scanned area of the monitor screen in which an image is displayed.

RASTER-FILL. The process used by a graphics device to fill in the spaces between the raster lines of a video image to provide an image that appears to have a finer, or higher, resolution.

RASTER GRAPHICS. A graphics system in which the computer image is treated as a collection of dots.

RAY TRACING. A highly realistic method of 3-D computer-graphics rendering that traces the path of a ray of light from the camera POV to every element in the scene.

READ-ONLY MEMORY (ROM). A storage location where information is permanently stored and cannot be altered.

REAL TIME. The actual time it takes events to occur; action on a computer screen that corresponds in a one-to-one ratio to real clock time.

REAL-TIME ANIMATION. The ability to produce and display animation at a selected speed.

REAR-SCREEN PROJECTION. The display of an image on a viewing surface by placing the projection equipment behind, rather than in front of, the screen.

RECORDER. A device for storing a signal, program, or other information.

REDUCTION. Copy made smaller than original.

REGISTER. Positioning printing images in exact alignment with each other upon the same sheet of paper.

REGISTRATION MARKS. Lines on artwork used to facilitate the accurate alignment of superimposed images.

REMOTE CONTROL. A device for controlling a machine or some function of a machine at a distance through wired or wireless means.

RENDER. Production of a detailed drawing as compared to a rough sketch. Production of a realistic 2-D scene from 3-D object information.

RENDERING. The addition of shading or gradations of tone or color to a basic line drawing to add dimension and realism.

RESOLUTION. The smallest resolvable detail in the image. The degree of graininess of a slide, video, or computer image as measured by lines or pixels.

REVERSE. Light characters on a dark background rather than dark characters on a light background.

REVERSE VIDEO. A mode of displaying selected characters on a CRT screen in a manner that is exactly the opposite of that screen's normal display color.

RGB. Red, green, blue. The primary colors of light that are mixed to produce a screen display or video image.

RGB MONITOR. A color display screen for computers using the primary RGB colors.

ROTATION. The turning of a computer-modeled object or image relative to an original point on a coordinate system.

ROUGH ART. Preliminary art used for visualization purposes and generally not published.

ROUGH COPY. Unedited, uncorrected preliminary text.

SANS SERIF. A typestyle that lacks serifs, such as Helvetica or Universe.

SATURATION. The intensity of a particular hue.

SCALE. To reduce or enlarge images or type to the desired size.

SCAN CONVERTER. A device that changes the scan rate of a video signal.

SCAN RATE. The speed at which an electron beam scans a picture tube.

SCANNER. An electronic device that converts images and type into digitized signals that can be read by a computer. Also an electronic device used in the production of color separations.

SCATTER CHART. A variation of a line chart that shows the correlation of two or more sets of data.

SCENE LIST. Description, sequence, and length of scenes in a slide, video, or film presentation.

SCHEMATIC. A symbolic diagram showing the functions and interrelationships of a unit or system.

SCREEN. (1) A process in which a graphic is converted into a collection of tiny dots or lines. (2) A flat surface upon which a picture or series of pictures is projected or reflected.

SCREEN DUMP. To send the image displayed on a screen to a printer.

SCREEN SHOT. A photograph of a computer screen or video monitor.

SCSI. Small computer systems interface. A standard peripheral interface used in Apple's Macintosh line that can connect up to eight devices in a daisychain and can prioritize the demands of the peripherals for the host's attention.

SEMILOG CHART. A variation on a line chart used to show a relative change between data expressed in different units.

SEPARATION NEGATIVE. One of a set of color negatives used to publish full-color images.

SERIF. The finishing strokes and cross strokes at the ends of the main strokes of letters in some type faces.

SINGLE SCREEN. A presentation in which all the images are displayed on the same screen area.

SKETCH. A freehand drawing containing essential detail.

SLIDE. A 35-mm transparency or the corresponding on-screen projected image.

SOFTWARE. All items in a system that are not hardware.

SOUND TRACK. The portion of a slide, video, or film presentation that reproduces the associated sounds.

SPECIAL-EFFECTS GENERATOR (SEG). An electronic device used to produce wipes, split screens, inserts, image keys, and mattes.

SPLIT SCREEN. A video effect that shows images from two different video sources, each on one half of the screen.

SPOT. Common art term for a small sketch or subordinate illustration.

SPOT COLOR. One or more colors used occasionally for emphasis or for accenting design.

STACKED BAR CHART. Bars that are placed on top of one another instead of adjacent to one another.

STAGING. The process of setting up and executing a presentation.

STAIR-STEPPING. Refers to the discontinuous nature of a line drawn by a raster display at any angle other than vertical, horizontal, or 45°.

STILL VIDEO. A special camera that allows individual video frames to be stored on magnetic discs similar to computer floppy discs for viewing one frame at a time.

STORAGE DEVICE. An erasable storage medium in a computer system for recording and retrieving information.

STORYBOARD. A series of preliminary sketches to show all of the text, graphic, and sensory elements of a presentation in a continuous flow.

STYLE. A mode of expressing thought through language or design; a manner of writing.

SUBHEAD. A secondary heading, usually set in a smaller type size than a main heading.

SUPERIMPOSITION. Laying titles or graphics over another image.

SURFACE CHART. Similar to a line chart, in which data points are connected, but the grid area enclosed is filled or shaded.

SWITCHER. A device for instantly controlling two or more video signals to produce special patterns, such as dissolves, wipes, and image keys.

SYNC GENERATOR. Device that generates stable sync pulses for use by all the components in a video system.

SYNTHESIZER. An electronic device for producing artificial sounds and audio effects.

TABLE. Presentation of numerical and/or descriptive data, physically arranged and organized to show the interrelationships of data.

TALKING HEAD. Video slang for the typical head-and-shoulders shot of a person in a video presentation.

TAPE FORMAT. Any of several tape widths and recording methods.

TELEPROMPTER. A device used off-camera to provide cues to a person being televised.

TEXT. The body content of a page or visual, as distinguished from titles, headings, captions, notes, indexes, and other auxiliary content.

TEXT CHART. A presentation using words, symbols, and/or tables.

THEME. The communicative vehicle by which a particular primary idea or message is transmitted to a reader or audience.

THROUGHPUT. The average amount of time for a film recorder to produce an image on a slide.

TIFF. Tag image file format. The format used for transporting computerized versions of scanned images.

TILE. A rectangular computer image repeated in a pattern over the screen image.

TILT. A camera move that swings along the vertical y axis.

TIME BASE. The timing of a portion of a video signal, particularly the horizontal and vertical sync pulses.

TIME CODE. An electronic counter or index of videotape duration.

TINT. The various strengths of a solid color.

TINT BLOCK. A solid or screened plate to provide a background or spot color to a printed page.

TRACK. The path of a recorded signal on film, tape, or disc.

TRANSFER. Reproduction of a slide presentation on film or videotape.

TRANSITION. A gradual change from one scene to the next.

TRAP. The overlap allowed for different color images to print on the same sheet; used to compensate for misregistration or printing plate movements and avoid white space between colors.

TURNKEY. A system or installation that is complete and ready to run without further additions or modifications.

TYPEFACE. The style (appearance, characteristics, and size) of type.

TYPE SIZE. The distance from the top of the highest projecting character of a font to the bottom of the lowest projecting character.

UPLOAD. To receive data from another computer by direct interfacing or through a modem.

USER INTERFACE. An on-screen menu that is used to execute a program.

UNJUSTIFIED. Copy that has full lines set so all are not the same length, that is, having a ragged, uneven right-hand edge.

VCR. Video cassette recorder. A device for recording a video signal.

VECTOR GRAPHICS. Graphic images that are a collection of simple geometric shapes.

VERTICAL RESOLUTION. The ability of a monitor to display horizontal lines from top to bottom.

VIDEO. Information displayed on the screen of a CRT or television monitor.

VIDEO DISPLAY. Television-like devices displaying words, letters, or graphs as the result of computer information.

VIDEO OUTPUT. Any visual representation produced and displayed or projected by mechanical, electrical, or electronic means.

VIDEO PLAYER. A device used to reproduce sound and pictures from a videotape on a video monitor or receiver; it cannot record images or sound.

VIDEO SIGNAL. A waveform carrying video information.

VIDEO TAPE RECORDER (VTR). A device that accepts signals from a video camera and a microphone, and records images and sound on videotape, usually in the form of open reels.

VIDEO TRACK. The portion of a videotape that reproduces the television image or picture.

VISUAL. The photographic and art portions of a document, slide, video, or film.

VISUAL AID. Video, slides, charts, photographs, models, or posters used to supplement and illustrate written or spoken information.

VISUAL PANEL. A single projected or displayed visual image or frame.

VOICE-OVER. Narration or dialog that is heard without the speaker being seen.

WIDOW. A partial line of type at the top of a page or column; to be avoided in good typesetting. Also, a single word or syllable standing alone as the last line of a paragraph.

WIPE. A film or video transition in which one image replaces another along a border that moves across the screen. An optical effect in which the upcoming scene appears to push the preceding one off the screen.

WYSIWYG. What you see is what you get. A computer monitor that displays information exactly as it appears on a hard copy or film.

X-AXIS. The horizontal line of a chart, usually listing the values for data points.

Y-AXIS. The vertical line of a chart, usually listing the values for data points.

Z-AXIS. The front-to-back line on a 3-D chart.

ZOOM. Use of a special lens that permits a camera to appear to move toward or away from the subject, thus enlarging or reducing an image's apparent size.

Author Index

Editor's Biography

Robert M. Woelfle (M'60–SM'75) received the B.S. degree in electrical engineering from the Fournier Institute of Technology in 1955, and the M.S. degree in electrical engineering from the University of Notre Dame in 1963.

He is currently Manager of Proposal/Presentation Services for the Greenville Division of E-Systems, Inc., Greenville, TX. In this position, he is responsible for managing the coordination, editing, and publication of proposals, brochures, reports, and other marketing-oriented documents, as well as the planning, development, and implementation of management and technical presentations. Prior to joining E-Systems, he served in several communications-related positions for the Bendix Missile Systems Division, Mishawaka, IN. His most recent position was Long-Range Planning Coordinator, responsible for defining, documenting, and coordinating divisional planning activities and related publications. In a prior position with Bendix, he served as Senior Proposal Coordinator. He has also worked for 16 years as a part-time instructor in the Business Department of Eastfield College in the Dallas County Community College District.

Mr. Woelfle is a Registered Professional Engineer in the State of Indiana. He is a member of the Society for Technical Communication and the Association of Old Crows. He served for 15 years as a member of the Administrative Committee of the IEEE Group on Professional Communication, and three years on the IEEE Membership Development Committee. In 1964 and 1965, he was a nominee for the "Young Engineer of the Year" sponsored by Eta Kappa Nu, and in 1966, he was named "Engineer of the Year" by the South Bend–Mishawaka Section of the IEEE. He served as editor of that section's publication, *The Nucleus*, for five years and authored several papers presented at various IEEE conventions. In addition, he is the author of one chapter in the *Handbook of Technical Writing Practices* (Wiley-Interscience, 1971).

ORDER TODAY!

TO ORDER: **Check the items you wish to order, fill in data below and mail to:**
IEEE, C. S. Department, 445 Hoes Lane, PO Box 1331, Piscataway, NJ 08855-1331 USA
Prices and availability subject to change without notice.

Please send me:

		Mem	**Non Mem**
☐ **Real-World Engineering** (PP02733-PVH) Lawrence J. Kamm		$15.00	$ 19.95
☐ **Writing and Speaking in the Technology Professions** (PP02782-PVH) edited by David F. Beer		24.00	29.95
☐ **Writing Reports to Get Results** (PC02154-PVH) Ron S. Blicq		22.50	27.95
☐ **A Guide for Writing Better Technical Papers** (PP01537-PVH) edited by Craig Harkins and Daniel L. Plung		22.50	27.95
☐ **Writing for Career Growth** (HL04523-PVH) David McKown		14.95	19.95
☐ **Presentations That Work** (HL04531-PVH) Carole M. MableKos		9.95	14.95
☐ **High-Tech Creativity** (HL04564-PVH) Michael Dick		14.95	19.95
☐ **Teaching On Television** (HV02493-PVH) Ralph Ayers		39.95	129.95

PAYMENT:

☐ **Check enclosed:** payable in U.S. dollars drawn on a U.S. bank - made payable to **IEEE.**
You must include handling charges (and State sales tax in CA, DC, NJ and NY) in your check.
Canadian residents, please add 7% GST (Reg #125634188).

☐ **Charge My:**   ☐ Visa   ☐ MasterCard   ☐ American Express   ☐ Diners Club

Account Number: _____ **Exp. Date:** _____

Sign here to validate your order: _____

FOR FASTEST SERVICE... 1-800-678-IEEE FAX (USA) 908-981-9667 PHONE (USA) 908-981-0060	**Handling Charges**

Handling Charges

For orders totaling:		Add:
Up to	$ 50.00	$ 4.00
$ 50.01 to	75.00	5.00
75.01 to	100.00	6.00
100.01 to	200.00	8.00
200.01 and	over	15.00

If you pay by check: You must add handling charges to your payment. **Otherwise,** we will automatically add correct handling charge and sales tax to your total order. OVERSEAS MUST SHIP VIDEOS VIA AIRFREIGHT. CALL FOR CHARGES.

☐ If you charge your order, you can have it expressed. Check here. We'll add charges to your account.

Name _____

IEEE Member No. _____

Company _____

Address _____

_____

_____

You must include your membership number to receive member price!